ATLAS
OF
THE GREAT CAVES
OF THE WORLD

ATLAS
OF
THE GREAT CAVES
OF THE WORLD

Paul Courbon — Claude Chabert

Peter Bosted — Karen Lindsley

CAVE BOOKS — ST. LOUIS — 1989

Published by CAVE BOOKS, 756 Harvard Avenue, St. Louis, MO 63130
        Roger E. McClure, Publisher
        Richard A. Watson, Editor

Cave Books is the publications affiliate of
the Cave Research Foundation.

Cover Design by Roger E. McClure

Library of Congress Cataloging-in-Publication Data

Atlas des grands gouffres du monde. English.
    Atlas of the great caves of the world / Paul Courbon ... [et al.].
      p.   cm.
    Enl. and translated from: Atlas des grands gouffres du monde.
    Bibliography: p.
    Includes index.
    ISBN 0-939748-21-5
    1. Caves--Maps.   I. Courbon, Paul.   II. Title.
GB601.A8813   1989
551.4'47'0223--dc20   89-15722
                                                              CIP

The original French book was authored by Paul Courbon and Claude Chabert, who
released several editions, the most current being in 1986. This English translation is by
Peter Bosted, who also updated the book and added new material. Karen Lindsley is
responsible for the compositing and design and indexing of this book.

This book was composed on an Apple® Macintosh Plus™ computer in ITC Bookman®
font utilizing Microsoft® Word for word processing and indexing, and Aldus PageMaker®
for page design. The font editing program Fontographer™ by Altsys Corporation was
used to modify the Bookman font in order to create special accent characters used in
foreign words. The final printout of this book was done on an Apple® LaserWriter Plus™
PostScript® printer.

ISBN 0-939748-29-0 (Hb)

ISBN 0-939748-21-5 (Pb)

Printed in the United States of America

PREFACE
TO THE ENGLISH EDITION

The exploration and mapping of caves, as with most pursuits, is subject to a certain amount of record-keeping. Although depth and length records are hardly ever the ultimate goal of an explorer or mapper, we all have a natural curiosity about the longest, the deepest, the greatest.

The ranking of caves in this Atlas obviously depends upon two factors: not only the existence of the caves, but also the perseverance of the cavers who have explored and surveyed them. Many regions, such as southeast Asia, have much more potential than has yet been realized. On the other hand, there are some countries, such as Britain and Czechoslovakia, in which a rather modest potential has been exploited to an extent far greater than might be expected, thanks to the diligence of many cavers.

Depth and length are only two of many possible criteria for judging the merits of a cave. These are simply the most easily quantified ones. The relative position on this list gives only a hint of the true attraction of a cave. Some of the deepest are nasty unpleasant holes. The longest inevitably contain great stretches of monotonous passages. Yet there is a mystique about sheer size that overrides any consideration of quality.

What determines the depth or length of a cave? Overall depth is limited more or less by the vertical extent of soluble rock above base level, with a small amount of leeway provided by diving in water-filled passages. It is therefore possible to eliminate most karst areas as potential sites for deep caves. But there are many mountain regions where limestone reaches thousands of meters in vertical extent with no known caves of significant depth. There must also be sufficient recharge to form caves. Even the deepest caves sacrifice some of their potential depth, because the tops of ridges and peaks rarely have sufficiently concentrated groundwater recharge. In addition, some potential caves have not contained water for a long enough time to enlarge them to traversable size. Perching of water on insoluble beds helps to maintain the flow through the same passages long enough for them to enlarge sufficiently, without being abandoned prematurely in favor of lower routes. On the subjective side, perseverance and mastery of technique are important factors, as shown by the tremendous increase in the number of entries in the deep-cave list in recent years. Finally, there is a large element of luck: entrances choked with debris from frost-shattering, breakdown along fault zones, and sumps perched far above base level are common obstacles in alpine caves that interfere with even the best-laid plans.

The length of a cave depends on entirely different factors. The most important are: (1) large areal exposure of soluble rock; (2) favorable relationship between recharge areas and potential discharge points; (3) depth and duration of river entrenchment within the soluble rock; (4) presence of a protective insoluble cap-rock to limit the erosional destruction of passages, but with enough breaching of the cap-rock to provide many points of groundwater recharge; and (5) prominent bedding, which allows diversion routes to form independently of the ancestral ones, rather than simply deepening the existing passages, as is the tendency in prominently fractured rocks.

A comparison of depth vs. length records is interesting. The deepest cave has many contenders not far behind, but the longest, Mammoth Cave, is more than three times longer than the runner-up! Such an overwhelming disparity is unheard of among any other of the world's natural features. Part of the reason is that we cannot see the entire sample of existing caves. Surely there are some as yet unknown that could fill the gap to some extent. However, Mammoth Cave is truly unusual in that it combines ALL of the above-mentioned factors with the further advantages of accessibility, nearness to population centers, lengthy period of exploration, and pleasant cave conditions.

One of the fascinating aspects of caves is their diversity. Despite all that we know about the geologic variables that control the pattern of caves, it is (fortunately!) still impossible to predict exactly where a cave will lead or what features it will contain. The maps in this Atlas by themselves provide an education in the many ways that solvent groundwater interacts with its geologic setting. At the most basic level, caves can be grouped into two major categories: stream caves with branches joining together as tributaries, and maze caves with passages interconnecting in labyrinths of closed loops. Many cave mappers consider maze caves to have an unfair advantage in achieving length. It is true that a maze will make the most efficient use of a given amount of groundwater recharge, but this recharge (usually floodwater or diffuse flow from above or below) tends to be of limited extent. According to the list of long caves in this Atlas, maze caves do not have a clear-cut advantage - thanks mainly to Mammoth Cave, which was formed by bona fide underground streams. Although the Russian gypsum networks and the maze caves of the Black Hills lurk near the top of the list of the 25 longest caves, at least 65% of both the total length and number are represented by stream caves, while the figures for diffuse-flow mazes and floodwater mazes are only 26% and 9% respectively.

If depth and length are considered such important aspects of a cave's merit, there must be standards for how these are to be measured. There is general agreement, in principle at least, that total depth should be measured from the highest accessible point to the lowest, and length should not include resurveys or superfluous shots. But attempts to impose further criteria have run afoul of the fact that the specific configuration of every cave demands a unique survey layout. The only valid guideline is that each one be surveyed thoroughly and accurately. If branch shots are taken in a large passage or room, they are valid measures of length if an explorer must follow a similar trail to view the cave. It is doubtful that the relative ranking of caves could be substantially disrupted by honest variations in survey technique. Any blatant mis-representations would not stand the test of time.

The example of Carlsbad Cavern comes to mind. Many people who are familiar with the cave wonder where its 30 km of passages are concealed. Obviously much of its length comes from ramifying surveys within huge rooms. Without these the cave could never be portrayed accurately, and they give a valid indication of the distance a person needs to travel in order to see it. Besides, no one will deny that Carlsbad is finer than any cave having a 'true' length of 30 km.

This book is far more than a mere list of records. It is a catalogue of the world's greatest caves and the dedication that went into their exploration and mapping. Readers will inevitably feel a twinge of regional or national pride - but beneath it all there will be a deeper feeling of unity with speleologists throughout the world. Their hopes, frustrations, and achievements are also our own.

Dr. Arthur N. Palmer, Professor of Geology,
State University of New York, College at Oneonta.

ABOUT THE AUTHORS

Paul Courbon

Born in 1936, he is a surveying engineer.

He was one of the driving forces in developing modern sport caving. He particpated in the through trip in the Pierre Saint-Martin in 1969 using a rope and "décrocheur" (a device to retrieve ropes) in 1969. A through trip in the Coumo d'Hyouernedo in 1970 used the same technique. Also in 1970, along with the S. C. de La Tronche, he was the first to use jumars to ascend in a deep European cave, the Lonné Peyret (-717 m).

In 1971 he made one of the great solo explorations of modern caving, entering the Pierre Saint-Martin by the Lépineux pit with 90 kg of ropes and ladders (similar to those used by the first explorers) and rigged and derigged the drops to arrive at the Verna room (727 m lower) twelve hours later.

In 1972 he was the first to explore a deep pit in Europe using single rope techniques. This pit (puits d'Aphanize, 328 m deep) became the deepest in-cave pit in the world.

In 1975 he participated in the first international expedition of the F.F.S. to Guatemala.

In 1978, with F. Poggia, M. Chiron and R. Astier, he made the first top-to-bottom trip in the P.S.M. (-1321), then the deepest in the world.

In 1979, with C. Chabert and U. S. cavers from the A.M.C.S., he dug open the upper entrance of the Sistema Purificación (Mexico), then the deepest through trip in the world.

From 1977 to 1979 he was director of the large expedition commission of the F.F.S. and participated in the first two French expeditions to Papua New Guinea.

His publications include:
- *Atlas des Grands Gouffres du Monde* (1973), published by the author (stock exhausted).
- *Atlas des Gouffres de Provence et des Alpes de Lumiére* (1975), published by the author (stock exhausted).
- *Atlas des Grands Gouffres du Monde* (1979), Ed. J. Laffitte, Marseille.
- *Atlas Souterrain de la Provence* (1981) with the S.C. Sanary.
- About 60 articles and reviews in various proceedings and periodicals (*Spelunca, Karstologia, Caving International, NSS News, Grottes et Gouffres*, etc.).

Claude Chabert

Professor of philosophy born in 1939.

Has been caving since 1961 with the Club Alping Français. Former president of the Spéléo-Club de Paris, and current president of the commission on long and deep caves of the International Union of Speleology. Also president of the Documentaion Department of the I.U.S.

Has concentrated on the study of caves in Asia: exploration expeditions to Turkey (1966-1971, 1974, 1976-1980, 1982), to Lebanon (1974), in Afghanistan (1975), to India (1983), and to Indonesia (Borneo, 1983, Sumba, 1985). He has also made trips to Mexico (1972, 1976, 1977, 1979, 1983) and Brasil (1984-1985).

In France, he has published two department-wide cave inventories, that of the Yonne with Georges Maingonat (1973-1977) and that of Nièvre with Alain Couturaud (1980-1985).

He has made interesting through trips with Paul Courbon in France, Great Britain, and Mexico. He has also collected through trips in other countries including the United States.

He has collaborated on several caving journals, as well as working of the Documentation Commission of the F.F.S. Between 1970 and 1971 searched the major French caving libraries to establish, with Michel de Courval, the definitive bibliography of the works of E.-A. Martel.

Peter Bosted

Experimental high-energy physicist born in 1954. Caving since 1979. Mapped several of the larger caves in California and participated in expeditions to Mexico (1980, 1981, 1982, 1987), Papua New Guinea (1982), and Belize (1986). Interest in photography has brought him to cave across the U.S.A. and in Canada, Mexico, Belize, P.N.G., France, Spain, Switzerland, and Great Britain. Published numerous articles and cave maps.

Karen Lindsley

Born in 1948, she has been caving since 1968 when she first rappelled into the Devil's Sinkhole in central Texas. She participated in surveys in Carlsbad Cavern in the early 1970's. Since then she has participated in Cave Research Foundation expeditions at Mammoth Cave in central Kentucky and in other areas. Karen majored in geology at college, but has since turned her interest to running a computer typesetting business which she operates from her home. She has used her computer skills in the publishing of the Cave Research Foundation's annual report for the past six years.

FORWARD TO THE 1986 FRENCH EDITION

The present French edition of 1986 builds upon the previous 1972 and 1979 editions by Paul Courbon. The objectives of this edition have been broadened to include maps of long caves as well as deep ones and to give descriptions of the karst and caves of most of the countries of the world, whether or not they contain long or deep caves on the international scale. Due the great increase in the number of deep caves, maps are generally included for only those over 700 m deep, rather than 500 m deep as in the previous editions.

The 1986 edition has tried to correct some of the imperfections (typographical mistakes, errors, etc.) of the previous editions, at the same time undoubtedly adding several more of its own. More attention has been given to proper naming of caves and localities, with preference in general given to the local language or dialect. In general we have not capitalized the local name for cave or spring in a cave name (sima, gouffre, sistema, etc.) so that it can be distinguished from the identifying part of the name (Berger, Pierre Saint-Martin, etc.).

The quality of maps of the long and deep caves is unfortunately quite variable, and the survey methods used and accuracy achieved can vary significantly. Even for relatively accurate surveys (angles both vertical and horizontal to 1 degree, measurements to the nearest 0.1 m, special care in measuring deep pits) the expected errors are large enough that one should not be too definite in pronouncing that a cave surveyed to -1008 m is really deeper than one surveyed to -992 m. Caves with entrances near their upper and lower reaches, or caves with internal loops, provide opportunities to check depth figures, while the increasing use of modern instruments like laser rangefinders for measuring deep pits can provide much higher accuracy.

For lack of time, we have only been able to update all of the national lists up to the end of 1985, and have not been able to verify all of the uncertain numbers. The rapidly growing number of long and deep caves makes it increasingly difficult to keep track of everything that has been discovered around the world.

We have had to limit the number of maps and the length of the long and deep cave lists more than we would have liked for reasons of publication costs. A truly comprehensive book would be many times longer than the present volume!

We apologize in advance that inevitably this lack of time and space will lead to some faults in the present edition.

Paul Courbon and Claude Chabert.

FORWARD TO THE
1989 ENGLISH EDITION

When we were approached about publishing an English translation of this book at the 9th International Congress of Speleology in Spain in the summer of 1986, we perhaps would not have accepted had we known how much work was involved! The exigencies of full time jobs and other ongoing projects meant the book could not be completed in a few months as we had originally hoped. The translation was completed in late 1987, but it served mainly as base for the many updates and additions that have since been incorporated right up until the publishing date. A lengthy index has been added to facilitate finding the information on your favorite caves.

The book is organized into five continents (Africa, America, Asia, Europe and Australasia) with the countries in each continent organized in alphabetic order. Descriptions are first given of the deepest caves, then the longest. Maps are generally included (if available) for all caves over 700 m deep or 30 km long. Maps are also included for some caves of special interest. Caves that have maps are indicated by having their names set in italics and the length or depth set in bold-face type.

The chapter on the American continent (especially Canada, the United States, Mexico, and Central America), which will most likely be the chapter of greatest interest to readers of the English edition, has been completely updated. In addition to updating the information, more maps and descriptions of caves on the American continent have been included. Thanks to all the people who answered letters and phone calls to accomplish this goal (see acknowledgements).

While it was not possible to completely update the lists for the rest of the globe, some changes were made to include major new discoveries, such as the new deep caves in the U.S.S.R., France, Italy, Switzerland, etc. In many cases the new depths are given in the text, while the maps still show lesser development since new maps were not available at publication.

Claude Chabert sent a considerable amount of new information on some of the smaller countries of Africa, Asia, and America which has been included. He also sent several corrections and some information on new deep caves.

Although we have made every attempt to weed out errors, inevitably some misspellings of names and places will occur, and on occasion inaccurate information will get passed along. In some cases there were ambiguities in the translations of expressions, geologic terms, and the description of localities. Please send any corrections to the publishers. We appreciate the assistance of the numerous cavers that have contributed time in proofreading the text.

Peter Bosted and Karen Lindsley

ACKNOWLEDGEMENTS

We extend our heartiest thanks to all of the collaborators who made this book possible, not only for the maps and descriptions which they sent, but also for their strong support and commitment.

We would like to especially thank Günter Stummer (Austria) who helped on many many occasions; Carlos Puch (Spain); Luigi Ramella (Italy) and Alexander Klimchouk (U.S.S.R.) for their remarkable cooperation; Tony Waltham (Great Britain) who worked with us from the very beginning; Richard Watson and Arthur Palmer (U.S.A.) equally for their great efficiency; Steve Worthington (Canada) who even anticipated our requests and spent many hours proofreading the entire book; Pete Lindsley (U.S.A.) who acted as a proofreader and our technical consultant; Bob Gulden (U.S.A.); David St Pierre (Great Britain); Ic· Giurgiu (Romania); Franco Urbani (Venezuela); and Andrej Kranjc (Yugoslavia) who were always available.

Special thanks in the preparation of the English edition go to Pierre Strinati for information on some of the lesser-known countries of the world (Botswana, Namibia, Nigeria, Niue, Palau, Vanuatu, Iraq, Laos, Barbados, Anguilla, the Antilles, Paraguay, El Salvador, Cambodia, Vietnam, etc.); to Arthur Palmer for supplying a new preface and much information on caves of the U.S.A.; and to Ian McKenzie for descriptions of Canadian caves. The English edition was also proofread by A. Bosted, M. Bowers, E. Coffman, D. Desmarais, C. Doland, M. DiSoto, R. Miller, V. Rennewitz, J. Sowers, B. Ruble, W. Rausher and J. Tinsley.

We would also like to thank the following for supplying information on specific caves or areas:

Algeria: Bernard Collignon and Paul Benoit.
Argentina: Victor Hugo, Demaria Pesce.
Australia: Ross Ellis and Francis Le Guen.
Austria: G. Stummer and Hubert Trimmel.
Belgium: Guy De Block.
Belize: Tom Miller and Steve Worthington.
Bermudas: Tom Ilifre.
Brazil: Pierre Martin.
Canada: Paul Griffiths, Steve Grundy, Ian McKenzie and Steve Worthington,
China: Tony Waltham.
Columbia: Martinho Rodriguez.
Czeckoslovakia: Anton Droppa and Otakar Stelcl.
Dominican Republic: Steve Worthington.
Equator: Jean-Pierre Besson and Alain Gilbert.
France: Patrick Cabrol, Maurice Chiron, D. Colliard, Alain Couturaud, J. -J. Delannoy, Patrick Degouve, Jean-Claude Dobrilla, Michel Douat, Marice Duchêne, L. -H. Fage, Jean-Louis Fantoli, Bernard Faure, R. Fradin, G. Gros, Thierry Krattinger, P. Lesauinier, Baudouin Lismonde, Richard Maire, Daniel Martinez, Serge Puisals, Pierre Rias, Jacques Rieu, Christian Rigaldie and Pascal Vauchier.
Fiji: Michael Bouke and Pierre Strinati.
Gabon: Gérard Delorme.
Great Britain: Tony Waltham.
Greece: Joël Rodet.
Guatemala: Steve Knutson.
Papua New Guinea: Jean-Louis Fantoli.
Honduras: Steve Knutson.
India: Daniel Gebauer.
Indonesia: Louis Deharveng.
Ireland: John Gunn,
Israel: Amos Frumkin.
Italy: Furio Bagliani, Gilberto Calandri, Suzanna Martinuzzi, Francesco Salvatori, André Depallens, and especially Luigi Ramella.
Jamaica: Alan Fincham and Tony Waltham.
Japan: Natsumi Kamiya.
Lebanon: Hani Abdul-Nour.
Libya: Attila Kósa.
Madagascar: Jean Radofilao and Eric Gilli.
Malaysia: Tony Waltham.

Morocco: Michel Chassier and Camille Lamouroux.
Mexico: Carlos Lazcano, Americans Bill Farr, Mark Minton, Terry Raines and Peter Sprouse, and Belgians Richard Grebeude and Jean-Claude Hans.
Nepal: Daniel Gebauer.
Norway: D. St Pierre.
New Zealand: Kevin Pearce and Cathy Worthy.
Oman: Tony Waltham.
Papua New Guinea: Mike Bourke and Pat Génuite.
Philippines: Claude Mouret.
Poland: Jerzy Mikuszewski.
Portugal: Frenchman Ch. Thomas.
Puerto Rico: Kevin Downey, D. St Pierre and Steve Worthington.
Romania: Ică Giurgiu.
Solomon and Samoa i Sisifo: R. M. Bourke.
South Africa: Steve Craven and Jaques Martini.
Spain: Carlos Puch and Pat Génuite.
Sweden: Rabbe Sjöberg.
Switzerland: Alfred Bögli, Philippe Rouiller and Albin Vetterli.
Syria: Jean-Claude Dobrilla.
Thailand: Luis Deharveng.
Turkey: Oral Ülkümen.
U.S.A.: Cal Alexander, Bill Balfour, Jed Blakeley, Rick Bridges, Joel Buckner, George Dasher, Mike Dyas, Sheck Exley, Scott Fee, Angelo George, Lester Good, Bob Gulden, Paul Hauck, Scott House, Ted Lappin, Pete Lindsley, Philip Lucas, Doug Medville, Mark Minton, Gerald Moni, Keith Ortiz, Arthur Palmer, James Quinlan, John Rosenfeld, Joe Saunders, Richard Schreiber, Jeff Sims, Paul Stevens, Bill Torode, Steve Worthington, Jerry Vineyard, Richard Watson, Michael Wiles, Alan Williams, and Richard Zopf.
U.S.S.R.: A. Klimchouk and his collaborators.
Venezuela: Carlos Galan and F. Urbani.
West Germany: Thomas Rathgeber.
Yugoslavia: Andrej and Maja Kranjc, Franz Maleckar, Tonci Rada and Georges Robert,
Zaire: Yves Quinif,
as well as: John Ganter (large rooms), Robert Carroll (granite caves), André Caillaud (topographic maps).

To all, a big thanks!

CONTENTS

I
INTRODUCTION

NOTES ON THE HISTORY
OF SPELEOLOGY

It would be difficult to write even a succinct history of speleological exploration without making major omissions. A strict definition of what we are calling speleology would begin in the 19th century, with a prehistory dating to the 17th century. If one talks of underground exploration in a larger context, one should also go back to prehistoric time: the Indians in Salts Cave (Kentucky) 2400 years ago, the Mayans in Mexico and Guatemala, the bison sculptors of the Tuc d'Audoubert in the Ariège, etc. deserve the title of cavers as well as we do.

Whatever the goals were, (search for water, exploration, search for minerals, religious activities...), underground incursions have a long history. If we confine ourselves to written tradition, it becomes thinkable to put together a history of speleology. If one includes oral tradition and folklore, the task becomes almost impossible.

Therefore, let's simplify! The development of modern speleology parallels that of the other sciences. We can fix its emergence around the 17th to 18th centuries. For example, Jacques de Clugny scientifically described in 1666 the grotte d'Arcy-sur-Cure in the Yonne, France. One often cites the date of 1748, when the mathematician Nagel descended to the bottom of Macocha propast, Czechoslovakia, on the order of the emperor of Austria.

The choice of dates is somewhat arbitrary, the decision being made at the time when the motivations for exploration became primarily scientific. The scientific motivation has been dominant up to the present day, although after WWII the sport element has become significant as well. The romantic return-to-nature movement at the end of the 18th century formed the nucleus of a phenomenon amplified by the industrial and technological revolution of the 19th century. Increasingly, caves became the object of multidisciplinary studies; geology, mineralogy, and hydrology were combined with paleontology, archeology, biology...

The first systematic explorations of cave systems began in the 19th century (Mammoth Cave, U.S., beginning in 1802, or Baradla barlang, Hungary, beginning in 1794, for example). Karst regions were also explored: the Carniole (corresponding today to the karst of Yugoslavia and Trieste) was the birthplace of European Speleology. In 1839, Lindner bottomed the grotta di Padriciano and in 1841 the abisso di Trebiciano. Between 1850 and 1857, Adolf Schmidl explored the great caves of the Yugoslavian karst, followed several years later by A. Hanke, J. Marin-

itsch, and F. Müller. The through-trip in the Bramabiau by Martel in 1888 also marks an important event in our history. With his annual exploration campaigns, he left a legacy of new concepts and a field of investigation pushed to the far reaches of the planet: in effect Martel's career is a snapshot of speleology in the 20th century! Aside from cave diving, there is nothing which we discover today that was not lived, known, or predicted by him.

Thanks to Martel, the thing that makes us strong and distinguishes us from the Indians at Salts Cave, for example, is our capacity for archiving. We accumulate manuscripts and documents, we gain a knowledge which we are charged with preserving against the vagueries of oral tradition. This is the goal to which this Atlas is dedicated.

During the 20th century, there has been both an expansion and an explosion of underground exploration conquests. The evolution of methods of exploration is not sufficient to explain this phenomenon. Speleology is under the influence and domination of the industrialized nations in which the number of participants is rapidly expanding. This has been strongly influenced by the desire for a return to nature and the increase of leisure time in these nations. Paradoxically, the taste for outdoor activities (mountaineering, speleology) is coincident with a rapid colonization of the last frontiers; roads now penetrate deep into the mountain ranges, cable cars carry people to the summits, and runways are springing up in the most remote tropical forests,.... Return to nature and mechanization are both sides of a coin that can be found in speleology. This situation precedes by only a small amount the day when visiting a cave will become quite ordinary, as has happened in mountaineering. Twenty years ago, we were astounded by a new cave over 500 m deep. Today, a new 1000 m deep cave is greeted with relative indifference. Is this the sign that we are approaching the end of our history?

More prosaically, the graph below illustrates the pace of discoveries[1]. In 1947, only two caves were over known to be over 500 m deep (the Dent de Crolles, France, and the anou Boussouil, Algeria). Six years later, two more caves joined them; the Geldloch and the Piedra de San Martín. By 1960, five more were added. After that comes a strong

---

[1] Deep caves permit an accurate comparison; oral and written traditions are well distinguished. For horizontal (long) caves, a similar graph would only be accurate beginning in the 1960's.

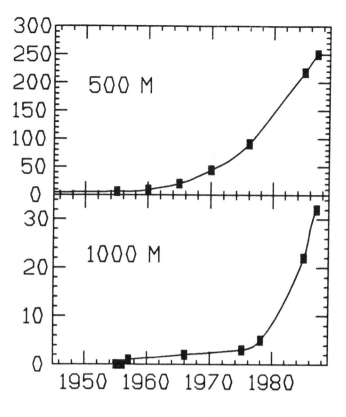

The X axis shows the years. The Y axis shows the number of deep caves over 500 m and 1000 m deep.

acceleration. In 1972, the first edition of the Atlas described 56 caves over 500 m deep. In 1976, their number reached 92. In 1988, it has passed the 250 mark (taking into account that some of the original entries have disappeared due to their integration into another system).

This curve is not yet asymptotic and should soon hit some kind of first plateau. It will then be more appropriate to plot the total length of all surveyed caves as a measure of the progress of cave exploration.

Finally, we list below not only the currently deepest and longest caves in the world, but also a chronology of the record for the deepest cave, using published data rather than announced exploration results whenever possible, and a list of the caves presently over 1000 m deep.

A tentative list of the largest underground volumes, a list of the great through trips (of special interest to sport cavers), and a partial list of some of the great hydrological systems of the world round out this historical section.

Bibliography: Shaw (Trevor) - *History of Cave Science*, A. Oldham. ed., Wales, 1979, 2 vol., XVI-500-XXV p., 88 fig.

Courbon (Paul) - L'aventure spéléologique moderne, 1982, 4 p. (Convegno Internazionale sul Carso di Alta Montagna).

Chabert (Claude) - Le karst de haute montagne dans la spéléologie moderne, Atti Convegno Internazionale sul Carso di Alta Montagna, *Imperia* 1982, vol. 1, 1983, pp. 173-179.

Chabert (C.) and Courbon (P.) - Les progrès de l'exploration spéléologique, *Karstologia*, 1983 (1):5-8.

The lists below showing the deepest and longest caves in the world were updated at publishing time to provide the most current information available.

THE DEEPEST CAVES IN THE WORLD
(as of 30 March 1989)

1. **réseau Jean-Bernard** ....(–1494, + 41) 1535 m (Haute-Savoie, France)
2. **Vjačeslav Pantjukhina** .... –1508 m (Bzybskij, U.S.S.R.)
3. **Laminako ateak** (Navarra, Spain) .... –1408 m
4. **sima del Trave** (Asturias, Spain) .... –1380 m
5. **Snežnaja-Mežonnogo** ....–1370 m (Abkhazian S.S.R., U.S.S.R.)
6. **sistema Huautla** (Oaxaca, Mexico) .... –1353 m
7. **réseau de la Pierre Saint-Martin** .... –1342 m (France/Spain)
8. **réseau Rhododendrons-Berger** .... –1241 m (Isère, France)
9. **V. V. Iljukhina** (Arabika, U.S.S.R.) .... –1240 m
10. **sistema Cuicateca** .... approx. –1230 m (Oaxaca, Mexico)
11. **Schwersystem** (Salzburg, Austria) .... –1219 m
12. **complesso Corchia-Fighiera** .... –1215 m (Toscana, Italy)
13. **gouffre Mirolda** ....(–936, +275) –1211 m (Haute-Savoie, France)
14. **sistema Arañonera** (Huesca, Spain) .... –1185 m
15. **Dachstein-Mammuthöhle** .... (–757, + 423) 1180 m (Oberösterreich, Austria)
16. **sistema Soaso** (Huesca, Spain) ....–1180 m
17. **Jubiläumsschacht** (Salzburg, Austria) ...... –1173 m
18. **sima 56 de Ándara** (Cantabria, Spain) ..... –1169 m
19. **gouffre de Bracas de Thurugne 6** .... –1166 m (or rivière Soudet)(Pyrénées-Atlantiques, France)
20. **anou Ifflis** (Djurdjura, Algeria) .... –1159 m
21. **sistema Badalona** (Huesca, Spain) .... –1149 m
22. **sistema del Xitu** (Asturias, Spain) .... –1148 m
23. **Akemati** (Puebla, Mexico) ....–1130 m
24. **Kujbyševskaja** (Arabika, U.S.S.R.) .... –1110 m
25. **Schneeloch** ....(–969, + 132) 1101 m (Salzburg, Austria)
26. **sima G.E.S.M.** (Málaga, Spain) .... –1098 m
27. **Jägerbrunntrogsystem** .... –1078 m (Salzburg, Austria)
28. **sistema Ocotempa** ....–1063 m (Puebla, Mexico)
29. **pozo della Neve** (Molise, Italy) .... –1050 m
30. **torca de Urriello** .... (–1017, + 5) 1022 m (Cantabria, Spain)
31. **Herbsthöhle** (Austria) ....–1020 m
32. **Siebenhengste-Hohgant-Höhlensystem** .. –1020 m (Bern, Switzerland)
33. **abisso Ulivifer** (Toscana, Italy) .... –1007 m
34. **Lamprechtsofen** .... (+ 995, –10) 1005 m (Salzburg, Austria)
35. **système de la Coumo d'Hyouernèdo** ...... –1004 m (Haute-Garonne, France)

THE LONGEST CAVES IN THE WORLD
(as of 30 March 1989)

1. **Mammoth Cave System** .... approx. 530,000 m
   (Kentucky, U.S.A.)
2. **Optimističeskaja** (Ukraine, U.S.S.R.) ..... 165,000 m
3. **Hölloch** (Schwyz, Switzerland) .... 133,050 m
4. **Jewel Cave** .... 123,771 m
   (South Dakota, U.S.A.)
5. **Siebenhengste-Hohgant-Höhlensystem** .... approx. 110,000 m
   (Bern, Switzerland)
6. **Ozernaja** (Ukraine, U.S.S.R.) .... 107,000 m
7. **système de la Coumo d'Hyouernèdo** ..... 90,496 m
   (Haute-Garonne, France)
8. **sistema de Ojo Guareña** .... 89,071 m
   (Burgos, Spain)
9. **Wind Cave** .... 82,074 m
   (South Dakota, U.S.A.)
10. **Zoluška** (Ukraine, U.S.S.R.) .... 82,000 m
11. **sistema Purificación** .... 71,583 m
    (Tamaulipas, Mexico)
12. **Fisher Ridge Cave System** .... 71,500 m
    (Kentucky, U.S.A.)
13. **Friars Hole Cave System** .... 68,824 m
    (West Virginia, U.S.A.)
14. **Organ Cave System** .... approx. 60,510 m
    (West Virginia, U.S.A.)
15. **Hirlatzhöhle** .... 57,000 m
    (Oberösterreich, Austria)
16. **Mamo kananda** .... 54,800 m
    (S.H.P., Papua New Guinea)
17. **système de la Dent de Crolles** .... 54,094 m
    (Isère, France)
18. **red del Silencio** (Cantabria, Spain) .... 53,000 m
19. **Lechuguilla Cave** .... 53,000 m
    (New Mexico, U.S.A.)
20. **Easegill Cave System** .... 52,500 m
    (Cumbria/Lancs., Great Britain)
21. **sistema Huautla** (Oaxaca, Mexico) .... 52,111 m
22. **gua Air Jernih** (Sarawak, Malasia) .... 51,660 m
23. **réseau de la Pierre Saint-Martin** .... 51,200 m
    (France/Spain)
24. **Kap-Kutan/Promezutocnaja** .... 50,300 m
    (Uzbekskaja S.S.R., U.S.S.R.)
25. **Raucherkarhöhle** .... 48,033 m
    (Oberösterreich, Austria)
26. **réseau de l'Alpe** (Isère/Savoie, France) ... 46.173 m
27. **Crevice Cave** (Missouri, U.S.A.) .. approx. 45,385 m
28. **complesso Corchia-Fighiera** .... 45,000 m
    (Toscana, Italy)
29. **Cumberland Caverns** .... approx. 44,444 m
    (Tennessee, U.S.A.)
30. **Ogof Ffynnon Ddu** .... 43,000 m
    (South Wales, Great Britain)
31. **Eisriesenwelt** (Salzburg, Austria) .... 42,000 m
32. **Bol'šaja Orešnaza** (RSFSR, U.S.S.R.) .... 41,000 m

CHRONOLOGY OF WORLD DEPTH RECORD

Mo/Day/Yr

Date	Cave	Country	Depth
1723	**Macocha propast** [1]	(Czechoslovakia)	−138 m
1839	**grotta di Padriciano**	(Italy)	−226 m
4/6/1841	**abisso di Trebiciano**	(Italy)	−329 m
1909	**Nidlenloch**	(Switzerland)	−376 m
8/10/1923	**Geldloch**	(Austria)	(−368, + 96) 464 m
1934	**antro di Corchia**	(Italy)	−480 m
8/5/1944	**système de la Dent de Crolles**	(France)	(−365, +147) 512 m
8/7/1945	**système de la Dent de Crolles**	(France)	(−365, +184) 549 m
5/4/1947	**système de la Dent de Crolles** [2]	(France)	−603 m
8/14/1953	**sima de la Piedra de San Martín**	(Spain)	−689 m
9/25/1954	**gouffre Berger**	(France)	−903 m
7/29/1955	**gouffre Berger**	(France)	−985 m
7/11/1956	**gouffre Berger**	(France)	−1122 m
8/13/1963	**gouffre Berger**	(France)	−1135 m
8/24/1966	**réseau de la Pierre Saint-Martin** [3]	(France/Spain)	−1171 m
8/10/1975	**réseau de la Pierre Saint-Martin** [4]	(France/Spain)	−1273 m
8/23/1975	**réseau de la Pierre Saint-Martin** [5]	(France/Spain)	−1321 m
7/5/1979	**réseau Jean-Bernard** [6]	(France)	−1358 m
3/2/1980	**réseau Jean-Bernard**	(France)	−1402 m
2/28/1981	**réseau Jean-Bernard**	(France)	−1455 m
2/15/1982	**réseau Jean-Bernard**	(France)	−1494 m
11/11/1983	**réseau Jean-Bernard**	(France)	(−1494, + 41) 1535 m

[1] today connected into the Amatérskej a Punkevní jeskyně system (-192 m).
[2] connection with upper entrance, P 40.
[3] connection with upper entrance, Basaburuko lezia.
[4] connection with upper entrances, gouffres M3/M13.
[5] connection with upper entrance, gouffre de Beffroi.
[6] connection with upper entrance, B21.

Note: We have not established a chronology for the longest cave. A careful study of published maps shows that the record has always been held by Mammoth Cave. Hölloch "officially" took first place for a while in 1955, but a recalculation of the lengths using the "discontinuity" principle restored first place to Mammoth Cave.

CHRONOLOGY OF 1000 M DEEP CAVES

(The numbers in parentheses are the current depths.)

gouffre Berger	1956	−1122 m	(1241)
réseau de la Pierre Saint-Martin	1965	−1000 m	(1342)
réseau Jean-Bernard	1976	−1208	(1535)
Schneeloch	1978	−1086 (−954, +132)	(1101)
sima G.E.S.M.	1978	−1074	(1098)
Lamprechtsofen	1979	−1005 (−10, + 995)	(1005)
système de la Coumo d'Hyouernèdo	8/28/79	−1004	(1004)
sistema Huautla	5/9/80	−1220	(1353)
Laminako ateak	8/80	−1192	(1408)
sistema Badalona	1980	−1105	(1149)
Snežnaja	1980	−1200 approx.	(1370)
sistema del Xitu	1981	−1139	(1148)
Schwersystem	1981	−1105	(1219)
Jubiläumsschacht	1981	−1050	(1173)
Dachstein-Mammuthöhle	9/3/81	−1180 (−757, + 423)	(1180)
torca Urriello	1982	−1022 (−1017, + 5)	(1022)
Jägerbrunntrogsystem	1982	−1061	(1078)
complesso di Corchia	1983	−1210	(1215)
nita Nanta	1983	−1026	(1080)
sima 56	8/83	−1169	(1169)
sima del Trave	1984	−1172	(1380)
anou Ifflis	1985	−1007	(1159)
gouffre Mirolda	?	−1046 (−936, + 110)	(1211)
V. V. Iljukhina	1986	−1220	(1240)
Kujbyševskaja	1986	−1110	(1110)
Vjačeslav Pantjukhina	1986	−1025	(1508)
Bracas de Thurugne 6 (or rivière de Soudet)	1986	−1137	(1157)
sistema Arañonera	1987	−1185	(1185)
Mäanderhöhle	1987	−1028	(1028)
Siebenhengste-Hohgant-Höhlensystem	1987	−1020	(1020)
sistema de Ocotempa	1987	−1063	(1063)
sistema Cuicateca	1988	−1038	(1230)
Akemati	1988	−1130	(1130)
sistema Soaso	1988	−1180	(1180)
pozo delle Neve	1988	−1050	(1050)
Herbsthöhle	1988	−1020	(1020)
abisso Ulivifer	1988	−1007	(1007)
Lamprechtsofen	1988	−1005 (+945, −10)	(1005)
système de la Coumo D'Hyouernèdo	1988	−1004	(1004)

RECORD CAVE DIVES

Sumps are no longer a major obstacle to exploration, and many significant dives have been done in the 1980's, predominantly by American, Australian, and French divers. Presently, the longest explored underwater caves are in the area of Florida and the Bahamas.

1. **Cathedral Falmouth Cave System** .... 10,229 m (Florida, U.S.A.)

2. **Lucayan Caverns** (Bahamas) .... 9,184 m

3. **Peacock Springs Cave System** .... 6,507 m (Florida, U.S.A.)

The above list is for caves that contain air pockets. It is also of (primarily sporting) interest to list some of the longest continuous dive distances without air pockets:

1. **doux de Coly** (Dordogne, France) .... 3,125 m

2. **Cocklebiddy Cave, sump N° 2** .... 2,550 m (Nullarbor Plain, Australia)

3. **the ocean resurgence of Port-Miou** .... 2,210 m (Bouches-du-Rhône, France)

As far as depth, irrespective of the special techniques used (special gas mixtures, etc), the following depths have been reached:

1. **nacimiento del Río Mante** ....−238 m (Tamaulipas, Mexico)

2. **fontaine de Vaucluse** ....−205 m (Vaucluse, France).

2. **fontaine supérieure de Tourne** ....−140 m (Ardèche, France).

THE TALLEST CAVES

We here list some of the great underground ascents. The figures given do not correspond to the total elevation that must be actually climbed up and down, but to the difference in elevation between the extreme points of the possible climb, whether the point of departure is the entrance of the cave or a low point inside the cave:

1. **Lamprechtsofen** (Austria) .... + 995 m
2. **Hölloch** (Switzerland) .... + 828 m
3. **grotte de Gournier** (France) .... + 680 m
4. **Pološka jama** (Yugoslavia) .... + 519 m

THE GREAT HYDROLOGIC SYSTEMS

Not being able to make a precise list, we content ourselves with indicating the three longest dye traces made to date:

1. **Homat Bürnü düdenieri-Yedi Miyarlar** ...... 75 km (Turkey)
   (Homat Bürnü düdenieri-Oluk köprü, 56 km, not confirmed)
2. **gouffre de la Belette-fontaine de Vaucluse** (France) .... 46 km
3. **Skocjanske jame-Il Timavo** .... 40 km (Yugoslavia/Italy)

For the deepest dye traces, we have found the following systems (certain values were obtained by deduction):

1. **V. V. Iljukhina-Reproa** (U.S.S.R.) .... 2,308 m
2. **Ural'skaja-Mačaj** (U.S.S.R.) .... 1,800 m
3. **gouffre du Pourtet-Bentia** (France) .... 1,622 m
4. **Lamprechtsofen** (Austria) .... 1,600 m
5. **gouffre Touya de Liet-fontaine des Fées** (France) .... 1,598 m
6. **houet Faouar Dara-Antelias** (Lebanon) ..... 1,573 m
7. **Laminako ateak-Illamina**(Spain/France) . 1,538 m (if the pozo Estella and the sima An 6 belong to this same system, one would get 1621 m and 1728 m respectively)
8. **gouffre des Trois Dents-Iscoo** (France) .... 1,520 m
The potential of ghar Parau (Iran) is thought to be about 1650 m.

THE GREAT THROUGH TRIPS

Entering by one of the upper openings (insurgence or pit), leaving by one of the lower entrances (resurgence or dry cave), following the course of the water, a through trip must be one of the great joys of caving. It is a quintessential activity, combining at once sporting and aesthetic values, one in which the caver becomes a real traveler. In our own experience, we found the through trip in the sistema Purification to be the perfect example of an underground journey.

We list here the nine deepest through trips, a few of which have been done as pull-down trips taking all necessary material from the upper entrance: some of them need special techniques (the sumps of the Coumo d'Hyouernèdo,

the great pits of Cueto), others require a very good knowledge of the route (the numerous blind pits in the Dent de Crolles, or the mazes in the sistema Purification).

1. **sistema Badalona** (Spain) .... –1149 m
2. **système de la Coumo d'Hyouernèdo** .... –966 m (France)
3. **Nettlebed Cave** (New Zealand) .... –867 m
4. **sistema Purificación** (Mexico) .... –852 m
5. **sistema Cueto-Coventosa** (Spain) .... –805 m
6. **complesso Fighiera-Corchia** (Italy) .... –710 m
7. **système de la Diau** (France) .... –701 m
8. **Råggejavre-Raige** (Norway) .... –617 m
9. **système de la Dent de Crolles** (France) .... –603 m

THE LARGEST UNDERGROUND CHAMBERS

The definition of an underground chamber is not easy, as was made clear in the recent Seminaire sur les Grandes Volumes Souterrains (Seminar on Large Underground Cavities), held in Paris in 1984 (cf. *Mem. S.C. Paris*, 1985, N° 12). The questions come from both problems of morphology and topography. For example, should a very large passage be counted as a room? Should large pits open to the sky be counted? If so, when does a pit become a doline? If we are to count dolines, then what about volcanic craters?

Underground rooms, pits, and dolines can either be compared by their surface area or their volume. Either way, an accurate topography of lengths, widths, and heights are needed, enough to fully define the shape of the volume. Taking all these problems into account, the following lists of great rooms by surface area (as established by John Ganter in November, 1985) and by volume are to be regarded as being indicative, but certainly not definitive!

A – by surface area

1. **Sarawak** (Chamber) (Malaysia) .... 162,700 m²
2. **torca del Carlista** (Spain) .... 76,620 m²
3. **Majlis al Jinn** (Oman) .... 58,000 m²
4. **Belize chamber** (Belize) .... 50,050 m²
5. **salle de la Verna** (France) .... 45,270 m²
6. **gruta de Villa Garcia** (Mexico) .... 40,820 m²
7. **gruta de Palmito** (Mexico) .... 39,730 m²
8. **Kocain** (Turkey) .... 37,200 m²
9. **Carlsbad Cavern Big Room** (U.S.A.) .... 33,210 m²
10. **sótano de las Golondrinas** (Mexico) ...... 33,110 m²
11. **Chiquibul** (Chamber) (Belize) .... 32,090 m²

B – by volume

1. **Luse** (Papua New Guinea) .... 50 Mm³ (doline)
2. **Ora** (Papua New Guinea) .... 29 Mm³ (uvala)
3. **Minye** (Papua New Guinea) .... 26 Mm³ (pit-doline)
4. **Da Xiao Cho Koo** (China) .... 25 Mm³

5. **sima mayor de Sarisariñama** (Venezuela) .. 18 Mm³
   (pit)
6. **sótano del Barro** (Mexico) .... 15 Mm³
   (pit)
7. **Kavakuna** (Papua New Guinea) .... 15 Mm³
   (doline)
8. **Korikobi** (Papua New Guinea) .... 14.6 Mm³
   (doline)
9. **Sarawak** (Malaysia) .... 12 Mm³
   (room)
10. **sótano de las Golondrinas** (Mexico) .... 5 Mm³
    (pit)
11. **Korikobi** (Papua New Guinea) .... 5 Mm³
    (room)
12. **Benua** (Papua New Guinea) .... 5 Mm³
    (room)

THE GREAT NON-LIMESTONE CAVES

It is well established that great caves do not necessarily occur in limestone. They also form in non-limestone soluble rocks such as conglomerate, gypsum, and chalk, as well in non-soluble rocks such as granite, gneiss, sandstone, and lava.

We here present only some brief summaries, ordered by the type of rock in which they occur. For more details, refer to our article "Les grandes cavités en roches pseudokarstiques et non-karstiques", in *Spelunca* 1980 (3) ; 109-115.

Conglomerate:

The deepest cave in conglomerate is the bofia de Torremàs (Lleida, Spain, -198 m), and the longest is Orešnaja (U.S.S.R.) at 28 km. See listings under Spain, Nepal, Turkey, and the U.S.S.R.

Chalk, tufa (after Joël Rodet):

The distinction between chalk and limestone is sometimes difficult to make. The deepest cave is **Weebubbie Cave** (Nullarbor Plain, Western Australia, Australia), at –134 m. The longest are:
1. **Mullamullang Cave** ....10,800 m
   (Nullarbor Plain, W.A., Australia)
2. **grotte de Rouffignac** .... approx. 7000 m
   (Dordogne, France)
3. **Cocklebiddy Cave** ....6,500 m
   (Nullarbor Plain, W.A., Australia)

Granite, gneiss (after Robert W. Carroll, Jr.):

Granite caves are generally tectonically formed along fissures or in boulder piles. The deepest, Greenhorn Cave (California, U.S.A.) reaches -152 m. The longest are:
1. **T.S.O.D. Cave** (New York, U.S.A.) .... 3,977 m
   anorthosite
2. **Bodagrottorna** (Gävleborg, Sweden) .... 2,606 m
3. **Bat Cave** (North Carolina, U.S.A.).............. 1,694 m
   gneiss

4. **M.B.D.A.T.H.S. Cave** .... 1,615 m
   (New Hampshire, U.S.A.) granite
5. **Greenhorn Cave** (California, U.S.A.).......... 1,557 m

Sandstone, quartzite:

For a reference to the eight deepest caves, see the section on Venezuela, where a depth of -362 m has been obtained. To these, we could add the cova del Serrat del Vent (Barcelona, Spain), 215 m deep. Brazil and Columbia also have significant sandstone caves. The longest are:
1. **cova del Serrat del Vent**........................ 4,273 m
   (Barcelona, Spain)
2. **fontaine du Vignal** (Ardèche, France) .... 1,900 m
3. **Mogoto Cave** (Transvaal, South Africa) ..... 1,615 m
4. **sima de la Lluvia** (Bolivar, Venezuela)........ 1,352 m

Gypsum (after Stephen Kempe):

The five longest are all in the U.S.S.R., the longest reaching an astounding 157 km. One also finds significant gypsum caves in the U.S.A., Italy, and Libya. The deepest are:
1. **tunel dels Sumidors** (Valencia, Spain) .... –205 m
2. **Shakta A** (Piemonte, Italy) .... –200 m
3. **cueva de los Cuñados N° 2**....................... –129 m
   (Almeria, Spain)
4. **cueva de los Cuñados N° 1**....................... –122 m
   (Almeria, Spain)

Lava, basalt:

When classifying lava tubes, even if one does not count trenches and craters, one still runs into the problem of segmentation by collapse pits; when is a pit a skylight and when is it a section of trench? The following lists are for the longest segments in a system, taking into account that sometimes the definition of a segment is somewhat arbitrary. The deepest are:
1. **Leviathani Cave** (Kibwezi, Kenya)............... –408 m
   Total depth of the system is 465 m.
2. **Ainahou Ranch Cave** (Hawaii, U.S.A.) .... –352 m
3. **cueva del Viento, Breveritas segment**...... –262 m
   (Canarias, Spain)
   Total depth of the system is 478 m.

The longest are:
1. **Manjung-gul** (Je Ju Do, South Korea)....... 13,268 m
   This represents the entire length of the system.
2. **Bilremos-gul** (Je Ju Do, South Korea) ...... 11,749 m
3. **Kazamura Cave** (Hawaii, U.S.A.) .... 11,713 m
4. **Leviathani Cave** (Kibwezi, Kenya) .... 11,152 m
   The longest segment is 9152 m.

Salt (after Ica Giurgiu and Amos Frumkin):

The largest salt caves are found in Israel (see section on this country). To these, we could add the **Cirgue de Bol'šoj cave** (Palmir, U.S.S.R., –120 m and 1029 m) and the **peştera 6S de la Mînzăleşti** (Vrancei, Romania, 3160 m). Algeria and Spain also have significant salt caves.

THE GREAT VERTICAL PITS

The image of a well (well and pit are the same word in French!) corresponds truly to the image that one forms in thinking of the great vertical cave. Two obstacles are brought together for the caver to face, darkness and emptiness, obstacles which have historically evoked legends and strong emotions, inciting fear and caution. The conquest of deep pits has long been held to be one of the more prestigious activities in underground exploration. One only has to go back forty years or less to remember what a challenging adventure it was to descend the clot de la Henne Morte (Haute-Garonne, France) or the big pit in the gouffre de la Pierre Saint-Martin (Navarra, Spain), where some of the most ingenious cavers of the era were brought together to perfect new techniques. The atmosphere of danger was reinforced by the tragic accidental death of Marcel Loubens in 1952 while testing one of these devices.

Until the era of modern speleology, pits often stopped incursions into caves. Occasional spectacular exceptions were made by locals or scientific investigators (descending into such caves as Padirac, Macocha propast, Le Buu tem, hoyo del Aire,...). Even recently, much questioning, fear and agonizing has preceded the exploration of great pits such as Aphanize (1971-1972), Mont Caup (1969-1973), and Hochlecken-Großhöhle (1972-1975).

It was the single rope technique which did more than anything to demystify the descent of deep pits, much to the chagrin of some. In this regard, the ascent of the Sótano de las Golondrinas using prussic knots created a sensation in Europe, but it was not until 1970 that a great European pit, the Lonné-Peyret (see réseau des Arres Planères), was explored using jumars. The great pit of Aphanize (328 m) was the first to be conquered using this technique. Relatively speaking, this was only yesterday! How short are our memories!

Around 1929 Henri Brenot, solicited by the alpinists who later formed the Spéléo-Club de Paris, created the "monkey", which was used in the 1930's by Félix Trombe in the caves of the massif de Paloumère. This system, too revolutionary for its time, was not widely adopted. The hemp ropes used were subject to rotting and did not inspire confidence. In addition, cavers were under the influence of their predecessors, notably E. A. Martel, who didn't hesitate to attack great pits, such as the abîme de Rabanel in Hérault (125 m) in 1889, the aven Jean Nouveau in the Vaucluse (163 m!) in 1892 (using the escarpolette technique), or the Gaping Gill in the Yorkshire Dales (100 m under a waterfall) in 1895 (using both the escarpolette and the ladder techniques combined). The escarpolette technique involved a wooden winch which rolled out thick hemp rope. On the end was the escarpolette on which the caver sat while his numerous assistants raised or lowered him on the surface. Other assistants unrolled a telephone line to maintain communication with the explorer. The mechanical winches used in the 50's and 60's did little more than perfect these techniques.

The "motorcorde", a motorized device for ascending ropes, invented by B. Dressler in 1973 and perfected by others, has not become generally accepted.

We know today of more than sixty pits over 200 m deep and around twenty that pass the 300 m mark. Their depths are often controversial, not so much for the accuracy of the measurements, but for the question of the pit's morphology. Should sloping drops be mixed with free fall lengths? Most pits are interrupted by ledges or have sections against the wall. What is the distinction between a ledge merely interrupting a vertical drop and one which separates two pits or adjacent pits? There is such a problem with the sima de la Zapatilla (Huesca, 358 m), where one descends numerous slopes and small cliffs before arriving at the 176 m free fall. Yet a rock thrown from the entrance goes directly to the bottom at 358 m without hitting any walls. One could also cite the case of Altes Murmeltier (Salzburg, 476 m), interrupted by numerous ledges, but with one vertical section of 307 m. Because of their configuration, we have decided to use the shorter of the two lengths in each case in our list.

Actually, the number of deep vertical drops that present no ledges and are not subject to dispute is quite limited. The queen, of course, is Las Golondrinas (333 m), which is totally free. The Aphanize (328 m) is free except for one small niche with a 1 m pendulum for a re-belay. El Sótano has a 310 m free section after the last section against the wall. One can descend 306 m in the Tras la Jayada.

Höllenhöhle	(Austria)	450 m
Minye	(Papua New Guinea)	417 m
Provatina	(Greece)	389 m
El Sótano	(Mexico)	364 m
Stierwascherschacht	(Austria)	350 m
sima Aonda	(Venezuela)	350 m
Mavro Skiadi	(Greece)	342 m
Las Golondrinas	(Mexico)	[1] 333/376 m
Tomasa Kiahua	(Mexico)	330 m
Aphanize	(France)	[2] 328 m
Lépineaux	(Spain)	320 m
Nare	(Papua New Guinea)	310 m
pozzo Mandini	(Italy)	310 m
Xonga	(Mexico)	310 m
Vicente Alegre	(Spain)	309 m
Altes Murmeltier	(Austria)	307 m
pozo Trasla Jayada	(Spain)	306 m
Pot II	(France)	302 m
Touya de Liet	(France)	302 m
Juhué	(Spain)	302 m
Enrico Revel	(Italy)	[1] 299/316 m

[1] Depending on the reference point chosen.
[2] Ascending section not explored

Note: The profiles of pits shown on this and the following pages are all drawn at the same scale.

Bibliography: Trompe (F.) Gouffres et cavernes du Haut-Comminges, *Travaux Scientifiques du Club Alpin Français*, Paris, 1943, t. 2.
Raines (T.) (ed.) Sótano de las Golodrinas, *Bull. Assoc. Mexican Caves Studies*, 1968, 2.
Génuite (P.) in *L'Aven*, bull S.C. Seine, 1984 (44).

AUSTRIA

Höllenhöhle (Tennengebirge, Salzburg) .... 450 m

Formerly called Hades Schacht (1511/274), this pit was discovered and explored during expeditions organized by the Speleoklub "Bobry" of Zagan (Poland) in the Tennengebirge. The average cross section of the pit is 6 by 10 m, narrowing to 2 m between -180 and -250 m. One finds snowbanks on the ledges, outcrops, and at the bottom of the pit. It is interrupted by ledges at -68, -158, -250, and -390 m.

Map: received from H. Zyzanski, from the drawing by his club.

Bibliography: *Salzburger Höhlenbuch*, 1985, Vol. 4, pp. 454-455 (profile).

Stierwascherschacht .... 350 m
(Höllengebirge, Oberösterreich)

This is the great pit of Hochlecken-Großhöhle discovered by the Austrians in 1972, but first descended in 1975 by several French teams. This is the deepest known pit in the world that is not open to the surface. A spacious ledge (Cap Kennedy) interrupts the pit after 50 m, but the last 300 m is mostly free, with only a few rounded steps beginning 120 m from the bottom, the last 100 m of which are very wet.

Bibliography: see Hochlecken Großhöhle (Austria). *Spéléo Darboun*, N° 3, 1978, MJC Cavaillon.

Altes Murmeltier (Tennengebirge) .... 307 m

Formerly called the Stary Swistak (1511/302), and prematurely announced as being 480 m deep, this pit was explored in 1981 by the Polish cavers of the S. K. Bobry (Zagan). As can be seen on the map, this cave is quite vertical, both for the first 171 m of depth, and also for the pit which has many ledges about half way down.

Bibliography: *Salzburger Höhlenbuch*, Band 4, pp. 477-478 (profile).

FRANCE

puits des Pirates .... 328 m
(Mendive, Pyrénées-Atlantiques)

This is the large pit in the gouffre d'Aphanize in the Arbailles range, located near the saddle of the same name, a dozen meters south of a road serving these mountains.

After a doline was opened up by heavy rain, the Spéléo-Club de Pau reached to -155 m in 1971, stopping at the top of the immense pit, which was descended in September 1972 by P. Courbon, J. -P. Combredet, and R. Gomez. This was the first great European pit done with jumars, and has a free fall depth of 328 m. A recess 1 m from the rope allows the drop to be split into lengths of 90 m and 238 m if desired.

Bibliography: see Aphanize (France).

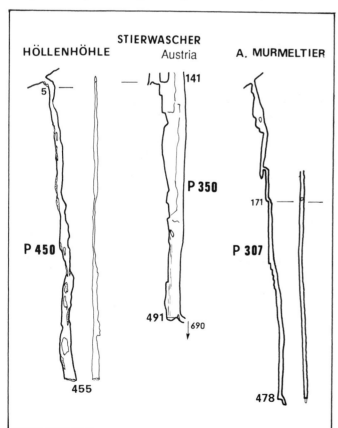

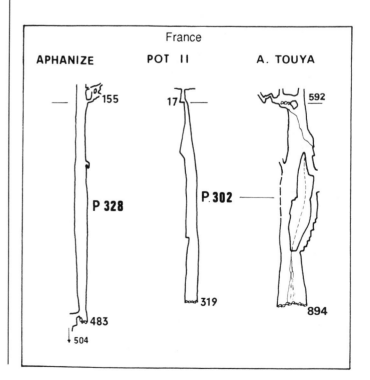

Pot II (Saint-Andéol, Isère) ....302 m

This abyss is found in the community of St-Andéol in the immense lapiaz des Erges, on the Vercors plateau. Discovered the 10th of July, 1969 by the Association Spéléologique du Vercors, it was descended only three days later. Rock debris at the bottom blocks all hope of further progress. The initial depth of 337 m was adjusted to 319 m. The cave is composed of two pits, a 17 m one and a 302 m one interrupted by ledges at -80 and -214 m. It was first descended by winch.

Bibliography: *Spelunca*, 1969, N° 1, pp. 52-53. Lismonde, (B.), J. -M. Frachet, *Grottes et Scialets du Vercors*, t. 1, 1978, pp. 92-94.

Touya de Liet (Accous, Pyrénées-Atlantiques) .....302 m

The terminal pit in this cave is 302 m deep and is broken by numerous ledges. The longest free fall is 129 m. It was first descended in 1974 using jumars.

Bibliography: see Touya de Liet (France).

GREECE

Provatina (Pindhos Oros, Ioaninna) ....392 m

In 1976, a French team confirmed the survey of the first explorers (392 m). In 1977, an American team measured 376 m and cast doubt on the depth of the pit. In 1979, the Swiss R. Wenger and Ph. Rouiller made an accurate survey of the pit and confirmed the English figure, finding 389 m versus their 392 m for this pit formed by two orthogonal conduits which join at the -170 m level.

Bibliography: Rickstone (C.) 1968, Provatina Abyss, *Wessex Cave Club Journal*, August 1968, 10 (118) p. 109. Sombardier (P.) and Poggia (F.) Le grand puits de la Provatina, *Spelunca*, 1977, N° 4, pp. 159-160.

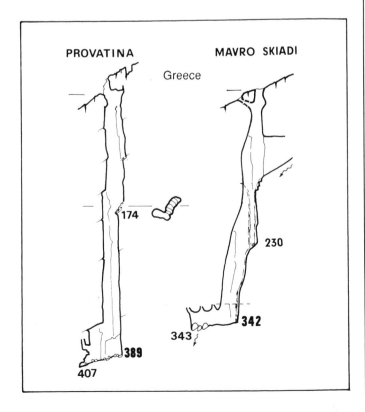

Mavro Skiadi (Lefka Ori, Kriti) ....342 m

The gaping opening to this cave is at 2000 m altitude, on the slopes of Mont Aghios Pneumani (Lefka Ori) in Crete. Access is by a 10 km mule trail leaving from the village of Melidoni.

First noted by the Cretian E. Platakis, a reconnaissance was made in 1974 by the Italians, with exploration in 1975 by the GRES Paris VI. In 1976, this same team, reinforced by the G.S. le Havre, completed the exploration and made an accurate survey. The pit is 342 m deep, with a horizontal displacement of 50 m. Ledges are found at –131 m and especially at -234 m.

Map: from J. Rodet, J. Quinqueton and R. Marion

Bibliography: Michaud (B.) Minotaure 75, *Spelunca*, 1976, N° 4, pp. 155-158. *Speleo Drack*, N° 11, 1977.

ITALY

abisso Enrico Revel ....299 m
(Alpi Apuano, Lucco, Toscana)

This cave is located in the Alpes Apuanes, near Carrare, in the same area as the antro del Corchia, one kilometer north of the ridge joining Pania Secca to Pania della Croce. The opening is sloping. The drop is 316 m from the high side of the opening, and 299 m from the low side.

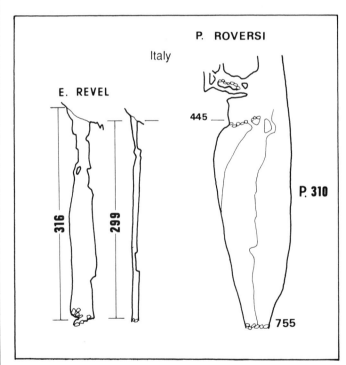

Explored the 20 and 21 of July, 1931, by the Groupe Spéléologique Florentin using rope ladders. It was the first 300 m pit to be descended, well before the descent of the Lépineux pit.

Bibliography: Noir (J.) *Mémoires du S. C. Paris*, 1976 (2).

pozzo Mandini ....310 m
(Monte Tambura, Alpi Apuane, Toscana)

This is the terminal pit in the abisso Paolo Roversi, discovered in 1977. The pozzo Mandini begins at the -445 m level and is 310 m deep. It is interrupted by numerous ledges. It was first descended in 1979.

Bibliography: see abisso Paolo Roversi (Italy).

MEXICO

El Sótano (Ayutla, Querétaro) ....364 m

El Sótano is located not far from the tiny village of Ayutla in the state of Querétaro. A small trail climbs towards the Cerro de la Tinaja from the village, descends to the Rio de la Atreja, then climbs past the cave, passing by the ranch of el Barro. It is an 8 hour hike from Ayutla to el Barro and another three hours to the cave.

The cavity was first explored using jumars on January 28 and 29, 1972, shortly after its discovery by a Texas group from the A.M.C.S.

The immense sloping entrance (150 of depth between the high point and the jumping-off point) has a cross section of 420 by 210 m, and is located in thick jungle. The descent is made from the most easily accessible side. The Americans originally announced a total depth of 410 m. The French expedition of the Spéléo-Club des Causses (1980) lowered the surveyed depth to 364 m, of which 310 m are free-fall, using the same tie-off point as the Americans.

It should be noted that the descent is not made from the highest possible point, and that the rope touches bottom at a point above the bottom of the talus slope, which is 455 m below the high point of the entrance (according to original American survey).

The bottom of the pit is 100 by 200 m. From here a 30 m gallery is found, where the Mexicans have placed a statue of the Virgin Mary and a register. With a volume of 16 million cubic meters, this cavity ranks third among the most voluminous in the world.

Bibliography: *A.M.C.S. Activities Newsletter*, N°. 9 (1979) p.80, profile, plan.
Bulletin N°. 3 (1980) of the Spéléo-Club des Causses.

Las Golondrinas (Aquismón, S.L.P.) ....333 m

The opening to this pit is entirely surrounded by jungle, and is an impressive 50 by 65 m, with the west side being 35 m higher than the east one. The pit is in the shape of a bottle, gradually belling out to a phenomenal floor area of 135 by 305 m. Adding to the beauty of the cave, hundreds of green parakeets and swallows (golondrinas) make their nests inside. To reenter the cave at night, the swallows fold their wings and plunge downwards at more than 100 km/hr before reopening and reaching their nests. The parakeets try to imitate them, but lack their grace.

The normal tie-off point offers a free-fall of 330 m, perhaps not the longest, but certainly the most impressive one known. By tying off on the high side of the pit, a drop of 376 m can be experienced. The latter is not popular due to the dense vegetation.

Located in 1957 by a group of French and Mexican mountaineers, the cave was explored in 1967 by eight Americans from the A.M.C.S, led by T.R. Evans. Unlike the Europeans, who had grown accustomed to using winches, the Americans revolutionized the exploration of pits: they climbed a 13 mm nylon rope using prussic knots, taking an average of two and a half hours to climb the 333 m.

With a volume of five million cubic meters, this cavity is among the largest in the world.

Bibliography: see Sótano de Las Golondrinas (Mexico).

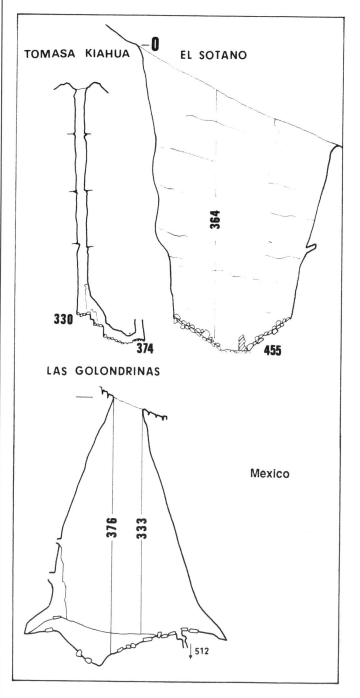

sótano Tomasa Kiahua ....330 m
(Sierra de Zongolica, Veracruz)

The opening of the cave is at 1380 m altitude, in the Sierra Madre Orientale, not far from the village of San Jose

Independencia, at the bottom of a doline, between two trails.

Long known to the local villagers, it was first descended in 1981 by P. Ackerman and G. Rouillon. According to the furnished survey, the pit does not have any ledges.

Bibliography: Ackerman (P.), Rouillon (G.) Sótano Tomasa Kiahua: le voyage vertical, *Grottes et Gouffres*, 1983 (**88**):20-28.

nita Xonga (Chilchotla, Oaxaca) .... 310 m
Cave explored in 1985 by the Sydney S. S. ending at a 310 m pit.

PAPUA NEW GUINEA

Minye (Nakanaï Mountains, New Britain) .... 417 m
Explored by the French expeditions of 1978 and 1985, this pit has numerous ledges and its descent is made difficult by the abundant vegetation. The entrance slopes downward for a depth of 100 m. The depths are with respect to the bottom of the entrance slope.
Map: from the survey notes furnished by Pat Génuite.
Bibliography: see Papua New Guinea.

Nare (Nakanaï Mountains, New Britain) .... 310 m
Descended in 1980 by the French national expedition. Interrupted by numerous ledges. To make a film more interesting, a 230 m free-fall was descended.
Map: from *Spelunca*, 1981, supp. to N° 3.
Bibliography: see Papua New Guinea.

SPAIN

pozo Lépineux (Larra, Navarra) .... 320 m
Presently capped by a locked monument to keep out visitors, this pit is located near and slightly below a hairpin bend in the Isaba-Arette road. Discovered in 1950 by Lépineux, Cosyns, and Occhialini, the pit was bottomed using a winch the following year. It was used for the first explorations of the Pierre Saint-Martin and made famous by the accident in which the unfortunate Marcel Loubens was the victim. Interrupted by several ledges at -86, -169, and -213 m.

Bibliography: Tazieff, (H.) 1952, *Le Gouffre de la Pierre Saint-Martin*, Arthaud.
Casteret (N.) 1961, *Aventures Souterraines*, tome II, Perrin.
Queffélec (C.) 1968, *Jusqu'au fond du gouffre*, Stock.

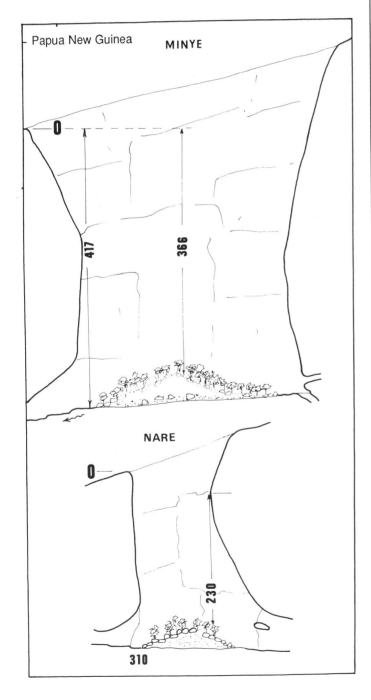

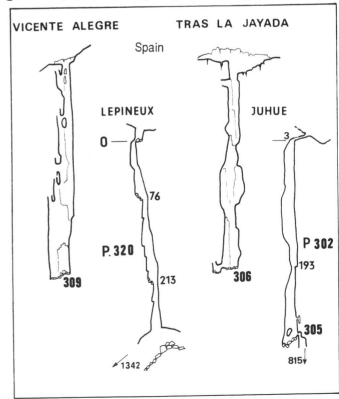

pozo Vicente Alegre..................................309 m
(Picos de Europa, Asturias)

This is the first pit in the sima del Trave. Discovered at the end of an expedition in 1982 and descended to -160 m by the Speleo-Club de la Seine, the pit, as well as the system it leads to, was fully explored in 1983. It carries the name of a caver who died while diving. Ledges near the top and bottom.

Bibliography: see sima del Trave (Spain).

pozo Tras la Jayada..306 m
(Picos de Europa, Asturias)

Located in the Massif de Cornión, one reaches the pit taking the Ario path from the Freira lake. Discovered in 1980 by the Sectión de Investigaciones Espeleologicas del Centro Excursionista "Aguila" (Barcelona). The pit has insignificant continuations that bring the depth of the cave to -313 m. One can avoid the landing at -125 m, making a 306 m free drop if desired.

Bibliography: Puch (C.), 1981, *El Topo Loco*, N°. 3/5, p. 40.

pozo Juhué (Arredondo, Santander) ....302 m

The tight opening to the cave (0.40 by 0.80 m) was located in April 1966 by G. Juhué. In August of the same year, the S. C. Dijon used a winch to descend the 302 m pit that followed, discovering a vast system that connects with the cueva Coventosa 815 m deeper. One ledge at -193 m.

Bibliography: see sistema Cueto - Coventosa (Spain).

VENEZUELA

sima Aonda (Auyantepuy, Bolivar)......................350 m

Explored in January 1983 by the Sociedad Venezolana de Espeleológia, who were transported to the entrance by helicopter.

The 320 m by 80 m opening borders the Meseta d'Auyantepuy in the "Lost World", not far from the Brazilian frontier. The walls are vertical, the east one being

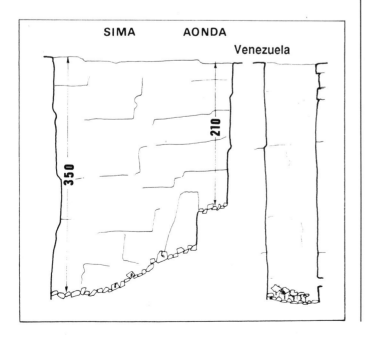

205 m deep, the central one 350 m deep (see map). With a volume of seven million cubic meters, it is one of the most voluminous in the world.

Bibliography: see sima Aonda (Venezuela).

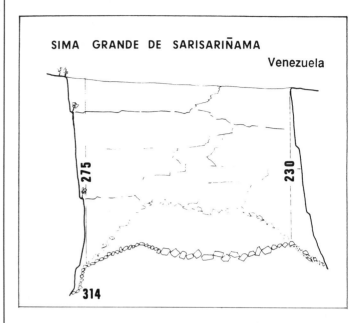

sima mayor de Sarisariñama ....275 m
(Sarisariñama, Bolivar)

This enormous orifice (300 m by 340 m) splits the Meseta de Sarisariñama much as the proceeding one does in the "Lost World". The enthusiastic exploration was done in 1974 to 1975.

This pit is the second most voluminous cavity in the world with a volume of 18 million cubic meters. It is 275 m deep, with a free-fall depth of 230 m, but the rigging point used in the exploration has numerous ledges and a longest free distance of 70 m. It was ascended using rock climbing techniques.

Bibliography: Bellard-Pietri (E. de) 1974, Exploration préliminaire du plateau de Sarisariñama, *Spelunca*, N°. 4, pp. 99-101.
Urbani, (F.) 1978, Les karsts gréseux du Venezuela, *Spelunca*, N°. 1, pp. 24-28.
See also sima mayor de Sarisariñama (Venezuela).

II
AFRICA

ALGERIA
DJEZAIR

There are many small limestone ranges scattered throughout Northern Algeria, spread roughly equally from East to West. The most important and spectacular range is the Djurdjura, with alpine peaks towering to 2321 m in the heart of the Grande Kabylie.

The Romans seem to have left some signs of having visited a few of the caves. In 1867 Captain Rivière descended to − 100 m in **rhar el Djemaa**, near Guelma (Constantine). Further explorations in this country continued in 1910 when a team of biologists (Peyerimhoff, Jeannel, and Racovitza) came to study various caves. Finally, the Spéléo-Club d'Alger was formed in 1935 and Mr. Dollfus began organizing expeditions in the Tlemcen region.

With the organization of a research arm of the hydrology department, a systematic series of explorations began in 1937 under the direction of Jean Birebent (1903-1970). The most impressive of these was the exploration of l'anou Boussouil (1937-1947), where the depth of − 495 m was reached.

The War of Independence (1954-1962) put a stop to all caving activities. Since 1963 P. Courbon, followed by French and Belgian cavers including Y. Quinif and B. Collignon, have been continuing explorations, with an increasing number of French caving clubs since 1975. The most remarkable finds have been the extensions to l'anou Boussouil, in underground Tafna (**rhar Bou Ma'za**) and, above all, the descent of **anou Ifflis**, the deepest cave discovered on the continent to this date.

Beginning in 1980, local Algerian cavers have become more active in the exploration of their caves, helped by the creation of a club in Bejaia.

DEEP CAVES:

1. *anou Ifflis* .... **−1159 m**
(Ras Timedouine, Djurdjura Range, Bouira)

The anou Ifflis is currently the deepest cave in Africa. The entrance is found at 2160 m altitude, north of the Bouira-Tizi Ouzou highway, in the municipality of Tikjda (wilaya de Bouira). The narrow entrance is located at the foot of the uppermost cliff of the crest of Ras Timedouine, about 3 km NE of Tikjda. (coordinates: x 630.3, y 353.1; Tazmalt 1/50,000 map).

The exploration of the cave began in 1980, when the Association Spéléologique de Montreuil reached − 87 m

after digging away an obstruction. In 1981, exploration was halted at − 300 m when a tight section was reached. This obstacle was passed in 1983, when two expeditions with the clubs of Li Darboun and Ragale and the Comité Spéléologique de l'Ile-de-France pushed the cave to a sump at − 975 m. In 1985, the E. C. Grácia (Barcelona) was able to pass the sump, (which had been diminished by a long dry period), and were stopped at a pit at − 1007 m. In 1986 they continued to a depth of − 1159 m.

The cave gets its name (Leopard cave) from the numerous golden spots that adorn the passages in the − 200 m to −530 m region. Below − 200 m, the cave is fed by many water sources that combine to form a stream at − 920 m with a dry season flow of about 10 l/s. The presumed resurgence is **ansor Arbaïloune**, located to the NW of the cave at an altitude of 950 m.

Map: taken from *Spelunca*, 1984(15).

Bibliography: Collignon (B.) and R. Maire - Le massif du Djurdjura. Elements de synthèse sur l'hydrogéologie et la spéléologie. *Spelunca*, 1984 (15): 25-38.
Rodriguez (M.) - Explorations Djurdjura 2000, August, 1983. Expédition du COSIF, *Spelunca*, 1984 (15): 29-34.
Fage (L. -H.) - Explorations de Couscous 83.. Anou Ifflis, − 975m, *Spelunca*, 1984 (15): 35-38.

2. *anou Boussouil* .... **−805 m**
(Terga m'ta Roumi, Djurdjura Range, Bouira)

The entrance to anou Boussouil is at 1700 m altitude, located at the bottom of a vast basin, on the north flank of the Terga m'ta Roumi, 200 m south of the Bouira-Tizi Ouzou highway, between Tikjda and the saddle of Tizi n'Kouilal, wilaya de Bouira (coordinates x 633.85, y 353.13, Tazmalt 1/50,000 map).

Although the cave has been known a long time, it was first explored in 1933, when mountaineers M. Fourastier and A. Belin descended to − 68 m, stopping at the top of the second pit. Belin and the Société Spéléologique de France then reached − 153 m in 1938 and − 194 m in 1939. Together with the Club Alpin d'Alger, Belin reached − 330 m in 1941, − 420 m in 1946, and − 495 m in 1947. The terminal sump was subsequently resurveyed to be at − 539 m, − 520 m, − 515 m, and − 470 m! B. Collignon, Delail, and Menault discovered a passage to bypass the sump and were able to push the system down to − 780 m in 1980. With the help of the Association Sportive et Culturelle de Bejaia and the Groupe Spéléologique de l'INSA de Toulouse they reached a new sump at − 805 m. Fossil passages explored in 1981 and 1982 brought the length of the cave to 3200 m.

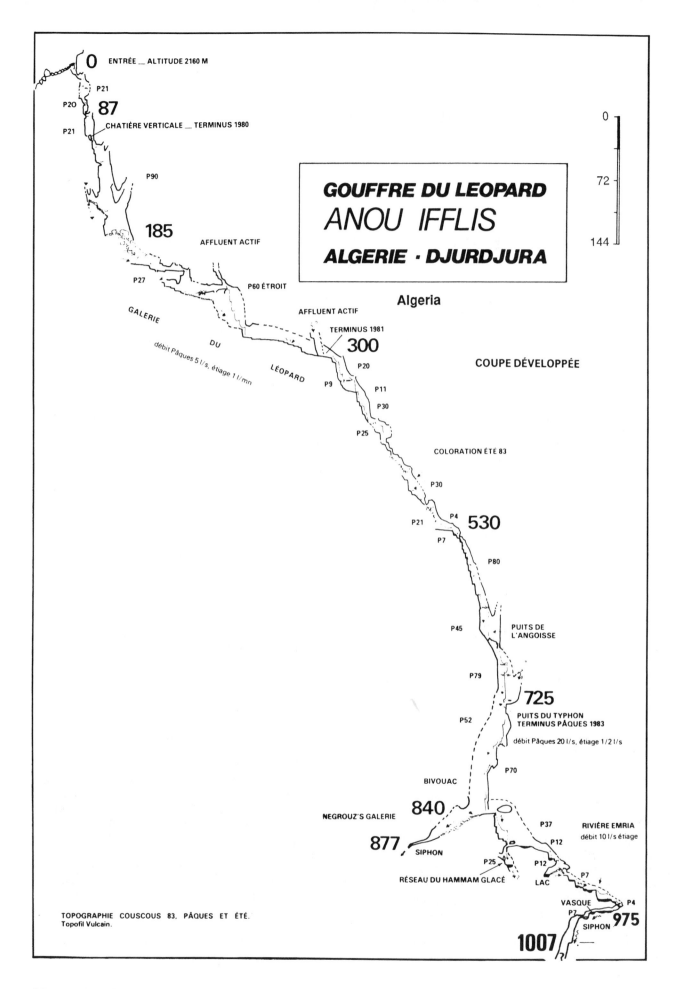

0 ENTRÉE — ALTITUDE 2160 M

P21
P20 **87**
P21
CHATIÈRE VERTICALE — TERMINUS 1980

P90

185
AFFLUENT ACTIF

P27
P60 ÉTROIT

GALERIE
AFFLUENT ACTIF
DU
TERMINUS 1981
300
P20
LÉOPARD
P9
P11
débit Pâques 5 l/s, étiage 1 l/mn
P30
P25

COLORATION ÉTÉ 83

P30

P21 P4 **530**
P7
P80

P45 PUITS DE
L'ANGOISSE

P79
725
PUITS DU TYPHON
P52 TERMINUS PÂQUES 1983

débit Pâques 20 l/s, étiage 1/2 l/s

P70
BIVOUAC
840
NEGROUZ'S GALERIE
RIVIÈRE EMRIA
877
P37 débit 10 l/s étiage
SIPHON
P12
P25
P12
RÉSEAU DU HAMMAM GLACÉ
P7
LAC
VASQUE P4
P7
1007 SIPHON **975**

GOUFFRE DU LEOPARD
ANOU IFFLIS
ALGERIE · DJURDJURA

Algeria

COUPE DÉVELOPPÉE

0

72

144

TOPOGRAPHIE COUSCOUS 83, PÂQUES ET ÉTÉ.
Topofil Vulcain.

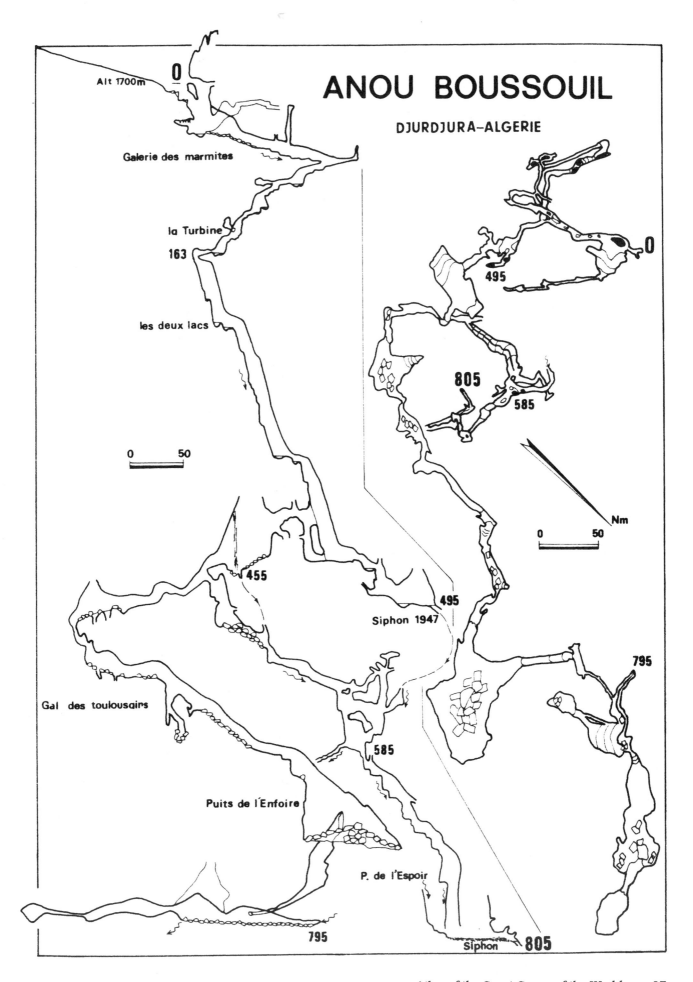

ANOU BOUSSOUIL

DJURDJURA-ALGERIE

Alt 1700m

0

Galerie des marmites

la Turbine

163

les deux lacs

0

495

805

585

0   50

455

495

Siphon 1947

Nm

0   50

795

Gal des toulousairs

585

Puits de l'Enfoire

P. de l'Espoir

795

Siphon **805**

The resurgence for the cave is at Tala el Hammam Boudrar (alt 740 m, 8 km as the crow flies, dye trace of April 8, 1968 by Birebent). The cave is always active: it receives rain water and snow melt which are collected by the entrance doline.

Map: after GS INSA (1980-1982), sent by B. Collignon.

Bibliography: Birebent (J.) - Explorations souterraines en Algérie. Campagne 1946-1947, *Annales de Spél.*, 1948, III (2-3): 49-144.
Collignon (B.) - Explorations spéléologiques dans le Djurdjura, *Spelunca*, 1981 (3): 35-40.

3. **anou Achra Lemoun** .... –323 m
(Ras Timedouine, Djurdjura Range, Bouira)
Alt. 2160 m. Explored in 1980 (A. S. Montreuil) and 1983 (COSIF) (*Spelunca*, 1984 (15) profile).

4. **anou Bou Hadjar** .... –273 m
(Ras Timedouine, Djurdjura Range, Bouira)
Alt. 2160 m. Three entrances, explored from 1977 to 1983 (AS Montreuil, COSIF, SC Rosny) (*Spéléologie algérienne*, 1982-1983, profile).

5. **anou Inker Temdat** .... –255 m
(Azerou Thaltatt, Djurdjura Range, Bouira)
Alt. 1710 m. Sump reached in 1942 (Belin, Marichal). Survey extended to –255 m in 1974 by Quinif.

6. **rhar Dar el Beida** (djebel taya, Guelma) ..... –215 m
Alt. 1020 m. Explored in 1953 by Birebent (–195 m) and in 1971 by P. Courbon (*Subterra*, 1976 (67) profile).

7. **anou Timedouine** .... –205 m
(Ras Timedouine, Djurdjura Range, Bouira)
Alt. 2150 m. Explored in 1973 by a Belgian expedition. Contains a 190 m pit.

8. **takouatz Guerrissene** ....(+65, –123) 188 m
(Ras Timedouine, Djurdjura Range, Bouira)
Alt. 2000 m. Explored in 1933 (Belin *et al*), dives in 1977 (SC Rosny, –123) and 1982 (Goergler, Shebbab, +65).

9. **anou Akouker** .... –173 m
(Ras Timedouine, Djurdjura Range, Bouira)
Alt. 2210 m. In 1977 by the SC Rosny (*Spéléologie algérienne*, 1982-1983, profile).

10. **sistema Sotterranéo di Dahrak** ....–173 m
(Nader, Guelma)
In gypsum. Explored by G. S. Imperiese CAI in 1987.

11. **anou Machouftch-Manguebetch** .... –162 m
(Terga m'ta Roumi, Djurdjura Range, Bouira)
Alt. 1780 m. In 1974, by a Franco-Belgian expedition (*Spelunca*, 1976 (1), profile).

12. **anou Heizer** (djebel Heizer, Djurdjura) ...... –156 m
In 1985, by the S. C. Lodève.12... (Alt. 940 m. In 1938 by the C.A.F.)

13. **ifri Smedane** ....+142 m
(Azerou Thaltatt, Djurdjura Range, Tizi Ouzou)

14. **bir Sidi Safiane** .... –138 m
(djebel Sidi Safiane, m. Traras, Beni-Saf)
In 1982 by Courbon and Petitbon.

15. **bir Tichtiouine 1** .... –135 m
(djebel Tichtiouine, m. de Tlemcen, Tlemcen)
Alt. 1150 m. In 1953 by Larat and in 1980 by the SC Tulle (*S. C. Infos*, 1981 (7) profile).

16. **rhar Es Skhoun** ....(+ 10, –122) 132 m
(Azerou El Kebir, Biban Range, Bordj Bou Arreridj)
Alt. 630 m. Difficult thermal cave explored from 1977 to 1982.

17. **rhar el Djemaa** (djebel Taya, Guelma) .... –129 m
Alt. 800 m. Explored in 1941 by Barone and the Eclaireuers de France (*Subterra*, 1976 (67) profile).

18. **rhar Tintoun** .... –120 m
(djebel Brek, Babors Range, Setif)
Alt. 130 m. Explored around 1947 by J. Birebent.

19. **rhar Sidi Amar** .... –120 m
(grand pic de l'Ouarsenis, El Asnam)
Alt. 1700 m. In 1964 by P. Courbon (*Spelunca* 1965 (3) profile).

LONG CAVES:

1. *rhar Bouma'za* ....18,400 m
(Terni, Tlemcen Range, Tlemcen)
Better known as the underground river of the Tafna, the cave is found in the dolomites and limestones of the Terni Plateau (upper Kimmeridgian) (commune de Sebdou; x 132.9; y 163.42; Alt. 1110 m; Terni 1/50000 map). The cave is 500 m from the aïn Taga, at a place called El rhar, 800 m SW of the forestry cabin of Merchiche. The entrance is visible from a long distance and essentially has always been known. The first explorations were led by Henry in 1933, then by Dollfus, Dupuy, Henry, and Soupault until 1936. The cave was mapped almost to the terminal sump for a length of 3887 m in August and September 1947 by Jean Birebent with the help of M. Philbert, Camous, and Trabut. The sump was passed in 1982 by Collignon, Petitbon, Pablo, and Benoit. They brought the length to 14,600 m in 1984, then to 18,400 m in 1985 making use of a seven day underground camp involving cavers from the S. C. Orsay Faculté and the C.A.F. Roanne.

Map: after P. Benoit, B. Collignon, and B. Pablo.
Blibliography: Birebent (J.) - Explorations en Algérie. Campagne 1946-1947, Annales de Spél., 1948 III (2-3): 49-144.
Spéléologie algérienne, 1982-1983, partial plan.

2. **kef el Kaous** ....4070 m
(Honaine, Traras Range, Tlemcen)
Explored in 1953 (Larat) and 1984 (Collignon)(*Spelunca Mémoires*, 1983 (13) plan).

3. **anou Boussouil** ....3200 m
(Terga m'ta Roumi, Djurdjura Range, Bouira)
See above.

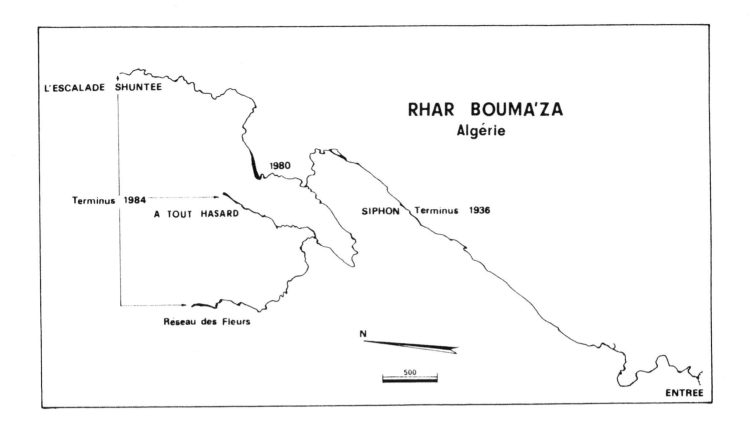

RHAR BOUMA'ZA
Algérie

L'ESCALADE SHUNTEE

1980

Terminus 1984

A TOUT HASARD

SIPHON Terminus 1936

Reseau des Fleurs

N

500

ENTREE

4. **rhar el Kahal** ....2190 m
(djebel Sidi Blal, Tlemcen Range, Tlemcen)
Explored by Birebent (1948), extended by Collignon *et al.*, in 1982.

5. **rhar Amalou** ....2000 m
(Azerou es Srhir, Bibans, Bordj Bou Arreridj)
In 1980 and 1981 by Collignon and Goergler (*Karstologia*, 1983 (2) plan).

6. **sistema Sotterranéo di Dahrak** ....2000 m
(Nader, Guelma)
See above.

7. **aïn Berd Zaa El Kbira** ....1805 m
(djebel Bouchouk, m. de Tlemcen, Tlemcen)
Explored by locals, mapped by Collignon and Pablo in 1981.

8. **rhar es Skhoun** ....1750 m
(Azerou el Kbir, Bibans, Bordj Bou Arreridj)
(*Karstologia*, 1983 (2) plan).

9. **anou Ifflis** ....1600 m
(Ras Timedouine, Djurdjura Range, Bouira)
(*Spéléologie algérienne*, 1982-1983, plan).

10. **rhar Medjraba** ....1500 m
(Azerou el Kbir, Bibans, Bordj Bou Arreridj)
Mapped from 1973 to 1975 by Coiffait and Quinif (*Karstologia*, 1983(2) plan).

11. **"source N° 4 de" Misserghin** ....1200 m
(Misserghin, Oran)
Mapped in 1951 and 1963 by Larat.

12. **rhar el Djemaa** (djebel Taya, Guelma) ....1100 m

13. **rhar Bou Akouss** ....1050 m
(djebel Doukane, plateau de Cheria, Tebessa)
Explored in 1946 (Trelaun) and from 1978 to 1980 (Coiffait, Quinif *et al.*, *Spelunca* 1980(3) plan).

ANGOLA

The karst areas of this country are poorly known. Since the time of the Portuguese colonization, a few people have become interested in the mountains located around Humpata (in western Huila province) and Manlunqa. The longest cave found to date appears to be the **Gruta do Rio N'Gunza** (Novo Rodondo), approximately 500 m long.

BOTSWANA

Politically independant, Botswana has been partially explored by cavers from South Africa (notably the region of Ngamiland in 1972-1974). Only one cave, **Drotsky** (Kwinabe), is presently known to be longer than one kilometer (1200 m more precisely). It was explored in the 1930's and mapped in 1969-1970 by Falcon *et al.* (Geoqr. Jl., 1975, 141 (3) plan). We should also mention **Lobatse N° 2** (Lobatse) with a vertical extent of 132 m and **Aha** (Xai Xai, Ngamiland) with a vertical extent of 55 m (*Botswana Notes & Records*, 1974 (6) and 1975 (7)).

CHAD

There are several small sandstone caves in Tibestri (D. Gavrilovic mapped 39 m in **Kéchou** and 30 m in **Yangar Nié**, near Bardai, Actes V Congr. Intern. Spél., Stuttgart, 1969, II). An interesting non-karstic phenomena, also in the Tibesti area (pic Toussidé), is the **trou au Natron**, a volcanic crater at an altitude of 2700 m that is 1160 m deep. It can be easily descended due to steps formed by crumbling rock. Another crater 430 m deep has been noted nearby.

CONGO

While it was still a colony, the Congo was visited by several biospeleologists from France in 1948 and from Switzerland in 1957 (Aellen and Strinati). New explorations were made in 1984 by the S. C. d'Albi.

LONG CAVES:

1. **grotte de Kila-Ntari** (Mouyondzi)...approx. 1500 m
Explored by the S.C.A. Well known for its large room of 400 by 50 by 30 m.

2. **perte de Mboumba** (Kimponzi) .... 1200 m
Explored in 1984 by the S.C.A. (*Spelunca*, 1986 (21) plan).

3. **grotte de Meya-Nzourari** (Bangou) .... 988 m
In 1961 by J.P. Adam and R. Caron and in 1983 by F. Vincent (*Ann. Spél.*, 1966(3): 717, plan).

EQUATORIAL GUINEA

Since we do not have any information on the caves known of Bioko Island, we will simply mention the existence of the granite cave **cueva Caracas** (Ebebiyin, Rio Muni), 36 m long and located near the Cameroon border.

ETHIOPIA

ITYOPIYA

Although essentially volcanic, Ethiopia contains sedimentary rocks in which some caves have formed: essentially in the regions of Mekele (Tigre), Harrar, Fantale, Shalla, and Bale. If one does not count the biospeleological forays (F. de Zeltner, 1901; T. Monod 1971, for example), the only foreign expedition to the country was the 1972 British Speleological Expedition. The University of Addis Abeba mapped many of the deepest caves in 1976. Since this time our knowledge of Ethiopian caves has not progressed due to political difficulties and access problems (many of the caves with religious significance are closed to foreigners). For this reason the only reference publication available is the *Transactions of the Cave Research Group of Great Britain*, 1973, 15(3).

DEEP CAVES:

1. **enkoftu Mohu** (Bedenno) .... − 192 m
In 1976 by the Univ. of Addis Abeba.

2. **enkoftu Dideesa** .... − 80 m
In 1976 by the Univ. of Addis Abeba.

3. **enkoftu Hade Kure** (Bedenno) .... − 66 m
In 1976 by the Univ. of Addis Abeba.

4. **Tula kiliwisa N° 1** (Tula, Harrar) .... − 64 m
Explored in 1971 by Bill Morton *et al.*, (*op. cit.*, profile).

LONG CAVES:

1. *Sof Omar* **(Bale)** .... **15,100 m**
This famous cave (6° 55' N and 40° 45' E) has long been a tourist attraction. It is located near the little village of Sol Omar. It is an underground section of the Web river, which is lost at **Ayiew maco** (alt. 1345 m) and resurges 1200 m away at Holuca (alt. 1330 m). A maze has formed near the resurgence, giving the cave a total of 42 entrances and a length of 15,100 m.

The first known visit was that of Arthur Donaldson-Smith in 1897, followed by Italians in 1913 and 1938. The first through trip was made by Eric Robson, Chris Clapham (British) and Kabir Ahmed in 1966. They mapped 8000 m in eight days of surveying. In the spring of 1972, the length was brought to 15,100 m by the BSEE.

Map: after the BSEE map published in Trans. C.R.G. of G.B., 1973, 15 (3).

2. **Nur Mohaned** (Goro, Bale) .... 2500 m
Approx. alt. 2600 m. In 1972 by the BSSE (*op. cit.*, plan).

3. **Zayei beati** (Mekele, Tigre) .... 330 m
Mapped in 1962 by Dave Causer (*op. cit.*, plan).

GABON

The following information comes essentially from my personal explorations and from tidbits gathered here and there. I believe they reflect fairly accurately the caving situation in Gabon at present. I must first remark that cave exploration is poorly developed due to the difficulties in gaining access to certain areas. However, the karst potential is significant and it is quite likely that much larger caves will be found in the future than those mentioned here.

The caves of the Lastoursville area are formed in some of the oldest non-metamorphosed sedimentary rocks in the world (between 2 and 2.3 billion years old). The caves are limited to a horizontal strata of dolomite about forty meters thick that is interbedded with non-soluble rocks. The greatest depths (about 30 m) are found in the caves of Kessipougou, Lastoursville, and Paouen N° 1.

Gabon has caves in jasper (Franceville area), in sandstone, (Mount Ngagui and Mount Mouba, Franceville) and

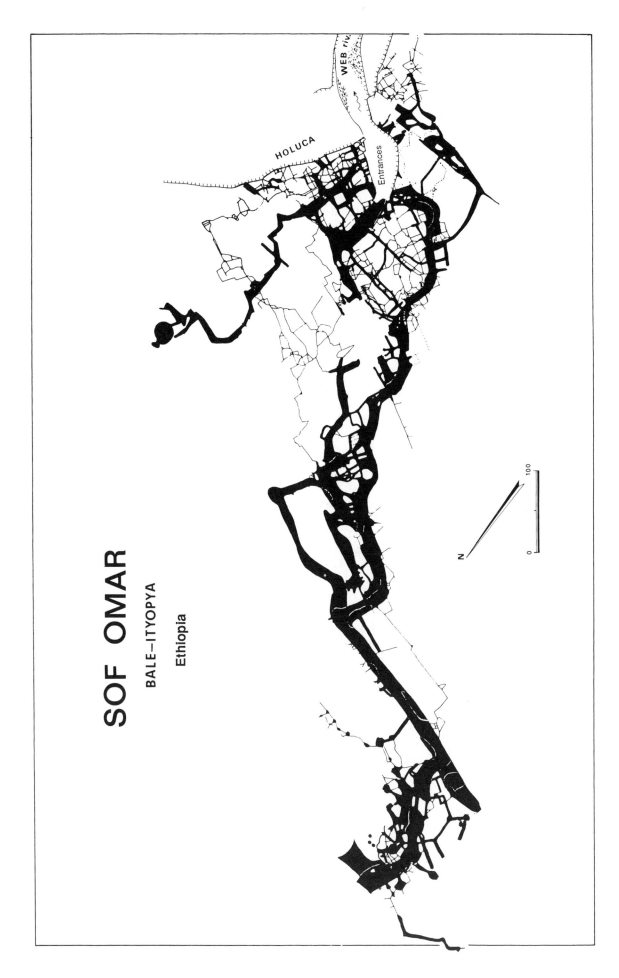

SOF OMAR

BALE–ITYOPYA

Ethiopia

HOLUCA

WEB riv.

Entrances

in oxides and hydroxides of iron (Belinga deposits, in the NE of the country). An example of the latter is the grotte du Faucon with a 10 m by 10 m room!

Gérard Delorme

LONG CAVES:

1. **grotte de Kessipougou** .... 1552 m
(Lastoursville, Ogooue-Lolo)
Approx. alt. 470 m, Franceville (dolomite). Completely explorable 1200 m long underground river segment, explored in 1976-1978 by G. Delorme (*Spelunca*, 1979 (4) plan). Estimated to be over 2000 m long.

2. **grotte de Lastoursville** .... 1352 m
(Lastoursville, Ogooue-Lolo)
Approx. alt. 400 m. Stream of 1000 m, explored in 1976-1978 by G. Delorme. Estimated length of 1440 m (*Spelunca*, 1979 (4) plan).

3. **grotte de Bongolo** .... 1150 m
(Lebamba, Ndende, Ngounie)
Ann. Spél., 1952 (2) plan.

4. **grotte de Paouen N° 1** .... 750 m
(Lastoursville, Ogooue-Lolo)
Approx. alt. 340 m. Stream passage explored in 1976-1978 by G. Delorme (*Spelunca*, 1979 (4) plan). Contains a room 170 m by 60 m.

5. **grotte de la rivière Ibembe** .... 641 m
(Lastoursville, Ogooue-Lolo)
Approx. alt. 360 m. Stream passage explored in 1976-1978 by G. Delorme (*Spelunca*, 1979 (4) plan). Estimated length of 940 m.

GUINEA
Guinée

It was in the West of the country, in Moyenne Guinée at the western flank of the Fouta-Djalon mountain range, where the principal Western African rivers of Senegal, Gambia, and Kroubal are born, that cavers J.L. Fantolil and R. Durand found several caves in 1983. These interesting caves are found in the high valleys and the plateaus near the villages of Kindia, Mamou, and Labé. Locally they are called *fomes*. They are distinguished by being formed in volcanic rocks such as dolerites, gabbrodunites, conglomerates, and sandstones. Large and extraordinary colonies of bats live in these strange and wild caves.

Jean-Louis Fantoli.

LONG CAVES:

1. **fomé Tiouki nord** (Gora Yamba, Labé) .... 400 m

2. **grotte de la Plantation** (Sougéta, Kindia) .... 400 m

3. **Touké fomé** (Seguaya, Kindia) .... 300 m

4. **grotte de Nianka** (Allya, Kindia) .... 220 m

5. **fomé Tiouki est** (Gora Yamba, Labé) .... 220 m

6. **fomé Bougohey** (Sabou, Mamou) .... 200 m

7. **fomé Goubambaya** (Goubambaya, Mamou) .. 150 m

8. **fomé Mili-Bili** (Goubambaya, Mamou) .... 120 m

9. **grotte de Kombétidé** (Seguaya, Kindia) .... 105 m

10. **fomé Frillé** (Sougéta, Kindia) .... 100 m

KENYA

The only caves known in Kenya at present are volcanic. They are being studied by the Cave Exploration Group of East Africa, created in 1964. They put out a periodic bulletin, recently renamed the *Speleophant*. **Leviathani Cave** is composed of several segments for a total depth of 465 m, of which 408 come from the Upper Leviathani cave segment (*Newsletter of C.E.G. of East Africa*, 1978 (2), schematic profile). Not having been able to consult all of the C.E.G. publications, the following list must be considered provisional.

LONG CAVES:

1. **Leviathani Cave** (Chyulu Mts., Kibwezi) .. 11,152 m
Alt. 1600 m. Has eleven entrances. Discovered in 1975, the cave was explored in 1976. The longest segment is 9152 m (*Speleophant*, 1982, 6).

2. **Mathaioni Cave** (Chyulu Mts., Kiboko) ...... 1700 m
Also called Mathalone Lava Tunnel. Show cave explored in 1966-1967.

3. **Kimikia Cave** (Chyulu Mts., Kibwezi) .... 1500 m
Explored in 1965.

4. **Cave 12 Mt. Suswa** .... 542 m
(*Bull. C.E.G. of E. A.*, 1972 (3) plan).

5. **Kangundo Cave** (Kangundo, Machakos) .... 228 m
(*Nl of C.E.G. of E. A.*, 1965 (3) plan).

LIBYA
LIBIYA

Several expeditions have explored the speleological potential of Libya (Circolo Speleologica Romano, 1959-1960, between Banghazi and Al-Bayda, Cyrenaica; Comex and Société des Eaus de Marseille, near Banghazi; and a Hungarian team led by Attila Kósa, 1979-1981, exploring the gypsum karst of Dir al Ghanam, Tripoli area). In spite of the climate, which is unfavorable for karstification, some remarkable caves have nonetheless developed. Few Libians seem interested in caving, and the sole publication dedicated to Libya, to which we owe a great debt, is the excellent report of Attila Kósa, *Bir al Ghanam Karst Study Project* (1981).

DEEP CAVES:

1. **Bukarma** (Banghazi, Cyrenaica) .... –101 m
    Alt. 25 m. 78 m are underwater. In 1975 by COMEX and S.E.M.

2. **Abraq** (Abraq, Cyrenaica) .... –92 m
    In 1983 by A. Kósa and Csernavölgyl.

3. **aïn Mizraq** (Nasmah, Tripolitainia) .... –86 m
    Alt. 200 m approx. Descended around 1936 (Kósa, *op cit.*, profile).

4. **"puits aux jumars"** (Banghazi, Cyrenaica)... –85 m
    In 1977 by COMEX and S.E.M. Has 60 m underwater.

5. **"gouffre sans nom"** .... –75 m
    (Jebel Akhdar, Banghazi, Cyrenaica)
    In 1959 by C.S.R. (*Notiziario C.S.R.*, 1969 (18-19) profile).

6. **Muntaqa Aqaranta** (al Marj, Cyrenaica) .... –60 m
    In 1983 by A. Kósa and Csernavölgyl. Depth estimated at 100 m.

7. **haua el Labrag** (Labrag, Cyrenaica) .... –53 m
    In 1959 by C.S.R. (*Not. C.S.R.*, 1969 (18-19) profile).

8. **Umm al Masabih** ....+52.5 m
    (Bir al Ghanam, Tripolitainia)

LONG CAVES:

1. **Umm al Masabih** ....3593 m
    (Bir al Ghanam, Tripolitainia)
    Twelve entrances. Gypsum. in 1981 (Kósa, *op. cit.*, plan).

2. **Bukarma-Habibi** (Banghazi, Cyrenaica) ......2256 m
    Part of the Zayana system. In 1976 and 1977 by COMEX and S.E.M. Total of 1454 m underwater. Map not drawn.

3. **Mirisi** (Banghazi, Cyrenaica).......................2117 m
    In 1976 by COMEX and S.E.M. Totally underwater. Part of the Zayana system. Map not drawn.

4. **Abu an Niran N° 1** ....858 m
    (Bir al Ghanam, Jefren, Tripolitaine)
    (Kósa, (*op. cit.*), plan.)

5. **aïn Debussia** (Banghazi, Cyrenaica) ....847 m
    Partly underwater (1977, COMEX and S.E.M.)

6. **"puits aux jumars"** (Banghazi, Cyrenaica) ...830 m
    Part of the Zayana system (1977, COMEX and S.E.M.).

7. **aïn Fasat** (Bir al Ghanam, Tripolitainia) ....618 m
    (Kósa, (*op. cit.*), plan.)

8. **Abu an Niran N° 3;** ....365 m
    (Bir al Ghanam, Jefren, Tripolitaine)
    (Kósa, (*op. cit.*), plan.)

MADAGASCAR
Madagasikara

In spite of a past closely linked with France, Madagascar remained largely unknown to cavers until 1981, when the first expeditions visited the country since its independence.

The karst of the country is all located along the west coast, extending from the south to the north ends of the island, and has a total surface area of 30,000 square kilometers. It is divided into seven distinct areas. The most interesting for cavers are the **Ankarana** (150 sq. km), in which over 100 km of passage has been mapped, some of it containing rivers with crocodiles, the **Mikoboka** (2000 sq. km), which although very difficult to get to offers the best depth potential, and the **Bemaraha** (4000 sq. km), which has seen little exploration to date. A reconnaisance to the karst of **Kelifery** (9000 sq. km) showed it to have little potential. The **Mahafaly** area (9000 sq. km) seems to be a very shallow submerged karst, also with little promise for cavers.

Although many of the caves have been explored and used by the locals, the first recorded explorations began in 1938 with the presence of the French. The 1960's were the most active years of this period, when J. de Saint-Ours did a lot of work. A summary of caving in Madagascar before Independence can be found in articles by R. Decary and A. Klener in *Annales de Spéléologie*, 1970: 409-440 and 1971: 31-46.

After the French left, the only explorations were those of J. Radofilao (University of Antsiranana) in the Ankarana range. Several surface studies by the geographers of the ERA 282 of the C.N.R.S. (G. Rossi, J. N. Salomon) were summarized in 1981 in the thesis of G. Rossi. This forms the basis of current explorations which are slowly answering many of the intriguing questions that are posed by the Madagascan karst.

Eric Gilli

DEEP CAVES:

1. **gouffre de Tolikisy** (Fiherenana, Toliara) .. –160 m
    Explored in 1966-1968 by J. Radofilao.

2. **aven de Lavaboro** (Mahafaly, Toliara) .... –115 m

3. **aven de Manamby** (Fiherenana, Toliara).... –105 m

4. **Andetobe** (Ankarana, Ambilobe) .... –105 m

LONG CAVES:

1. *Ambatoharanana* ....18,100 m
    (Ankarana, Ambilobe)
    This cave, once called the Underground River of Mananjeba, is located in the Ankarana mountain range, 10 km north of Ambilobe and about 75 km south of Antsiranana. The Mananjeba river traverses the cave from east to west, cutting through an isolated limestone butte on the south flank of the range before resurging into the Mozambique Canal.

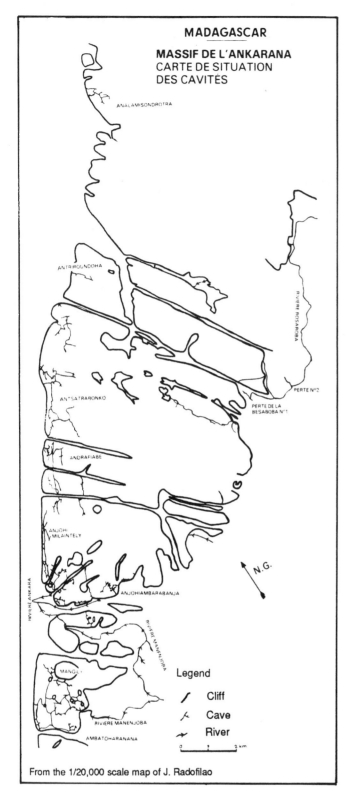

MADAGASCAR

MASSIF DE L'ANKARANA
CARTE DE SITUATION
DES CAVITÉS

Legend

/ Cliff

⟋ Cave

⤳ River

0   1   2 km

From the 1/20,000 scale map of J. Radofilao

1974, Jean Radofilao, with Philippe Andriambololona and Félix Raoelison, organized an expedition with the goal of discovering all the branches of the underground river. In three weeks, most of them were explored and mapped, bringing the total length to 18,100 m.

Map: from the summary of J. Radofilao in *Spelunca*, 1984 (13).

2. **Andrafiabe** (Ankarana, Ambilobe) .... 12,030 m
The initial explorations of J. de Saint-Ours in 1961 and 1963 to 2800 m were followed up by the A.S.U.M. (9900 m in 1964-1966, 11,200 m in 1971) and by J. Radofilao beginning in 1980.

3. **Ambatoanjahana** (Ankarana, Ambilobe) .. 10,810 m
Explored by the 1983 French expedition. Contains 3700 m of boating passage.

4. **Antsatrabonko** (Ankarana, Ambilobe) ..... 10,475 m
Explored to 5760 m in 1966-1970 by the A.S.U.M., then by J. Radofilao and French cavers in 1977-1978.

5. **Anjohimilaintety** (Ankarana, Ambilobe) .... 9005 m
Explored to 7500 m in 1966-1972 by the A.S.U.M., then by J. Radofilao and the 1980-84 French expeditions.

6. **Anjohiandranoboka** .... 5330 m
(Andranoboka, Mahajanga)
Explored in 1951 by J. de Saint-Ours, Paulian, and A. Ramahalimby (*Ann. Spél.*, 1959 (3-4) plan).

7. **Ampandriampanihy-Nord** .... 4480 m
(Ankarana, Ambilobe)
In 1977-1978 by J. Radofilao and French cavers.

8. **Andetobe** (Ankarana, Ambilobe) .... 4260 m

9. **"B"** (Ankarana, Ambilobe) .... 4080 m
By Radofilao, 1978-1879.

10. **Ambatomanjahana** (Ankarana, Ambilobe) .. 3850 m
1971, A.S.U.M., then French expedition of 1982.

11. **Anjohin'ny Voamboana** .... 3750 m
(Ankarana, Ambilobe)
1971, A.S.U.M., then French expedition of 1982.

12. **"Y"** (Ankarana, Ambilobe) .... 2560 m

13. **Ambarabanja** (Ankarana, Ambilobe) .... 2325 m
In 1972 by the A.S.U.M.

14. **Anjohiantsatrabonko** .... 2070 m
(Ankarana, Ambilobe)

15. **Anjohibefoka** (Ankarana, Ambilobe) .... 1940 m
In 1971 by the A.S.U.M.

16. **Analamisondrotra** (Ankarana, Ambilobe) ... 1900 m
(Madagascar, expeditions Spéléologiques 81, report in *Spéléologie*, Nice, N° 115, map)

Note: "B" and "Y" are provisional names.

It is a maze cave in which the main river can be followed by canoe for over 2700 m in large passages. A huge collapsed doline in the heart of the butte can be found by following several fossil passages.

The cave is formed in the reef limestones of the Middle Jurassic.

First explored for about 6 km by J. de Saint-Ours, the cave was mapped to 2700 m in 1972 by l'Association Sportive de l'Université de Madagascar (A.S.U.M.). In

MAURITANIA

Mauritanya

The Sahara desert contains sandstone outcrops in Mauritania, Algeria, Mali and Niger that contain small caves. They are difficult to access and have not attracted much interest. In Mauritania, the Frenchmen De Keyser and Villiers explored 84 m in the sandstone cave **grotte du Lapin** (Kanoal, Atar) (*Bull. Direction des Mines*, Kakar, 1952 (15) map).

MOROCCO

Maghreb

Morocco has a surface area of 724,734 sq. km, of which about 100,000 are limestone, making it a first class area for caving. The principle limestone areas are located in three distinct geologic zones, which are themselves divided by major tectonic accidents: the area of Anti-Atlas in southern Morocco, of Precambrian age, where the vertical relief attains 2000 to 3000 m, extending from east to west; the area of the Atlas, composed of a Caledonian to Alpine age range extending from SW to NE; and the Late Alpine age Rifin area in the North.

Caves have been known from prehistoric times, used as shelters for habitation and for religious ceremonies. Modern explorations began in 1925 with the arrival of the French. The rhar bel Hårdaïfa, south of Romani, was exploited for its guano. The exploration of Tasserakout, in the Oujda region, took place in 1927. Norbert Casteret visited the caves of the Central Atlas, south of Taza, in 1934. He discovered the entrance to the Friouato system in the bottom of the shaft, and explored the Chiker cave to its low point. The rhar Goran (Safi) was explored for the first time in 1937. The kef Aziza, at Tazzouguert (west of Er Rachidia) was also explored before WWII.

The first systematic studies began in 1947 with the creation of the Société Spéléologique du Maroc, followed by the Spéléo-Clubs of Rabat, Meknès, Fès, Oujda, Agadir, and Casablanca.

The first inventory of the country, published in 1953, cited over sixty caves and pits.

Several important discoveries were made after 1953, notably the kef Toghobeït in the Rif area and the underwater river of Wit Tamdoun, NE of Agadir. Sumps were dived in Chiker and Friouato. Others, such as the aïn Melghfi in the Middle Atlas of Beni Mellal area, the river of Chara (south of Taza), the ifri N'Taoui, or the grotte du Caïd (south of Azilal) in the High Atlas were explored, considerably broadening our knowledge of Moroccan caves.

The *Inventaire Spéléologique du Maroc*, by J. Camus and C. Lamouroux appeared in 1981. It describes 543 caves and is an essential reference for those planning expeditions to the country.

Camille Lamouroux.

DEEP CAVES:

1. ***kef Toghobeït*** (Rif, Bab Taza, Chaouene).. **–713 m**
    The kef Toghobeit is near the western extreme of the Rif, near the picturesque town of Chaouene. To get there,

one must pass through the village of Bab Taza, and take a trail through the forest in the direction of Djebel Laacheb. After 8 km, one arrives in a vast depression, where a spring and several buildings serving as shelter for the shepherds can be found. Shortly after this depression, the trail for Toghobeït leaves the main one. The cave is reached after half an hour of walking along this trail, and is at an altitude of 1700 m. (See the 1/50,000 map Bab Taza; x 523.8; y 498.8).

A dye trace done in 1980 has shown that the water which disappears in the sump at –395 m emerges in the oued Serafat, 850 m below the cave and 3500 m away. The resurgence for the sump at –700 m is presently unknown.

The cave was shown to members of Spéléo-Club de Rabat in 1959 by shepherds. That year, a team formed of Penot, Camus, Lamouroux, Serrano, and Renner reached a huge room at –135 m. In 1960, the S.C.R. was stopped at a sump at –380 m (originally erroneously measured as –544 m).

The exploration was taken up again in 1969 by a Belgian group from the Spéléo-Club de l'Université Catholique de Louvain. They drained the short sump but were stopped by another 15 m lower.

The Spéléo-Club de Blois, led by Michel Chassier, became interested in the cave in 1971: they discovered a new passage at –240 m, but were soon stopped by a tight spot. This was widened the following year, and the S.C.B. continued to –540 m, where they were stopped by lack of ropes. In 1973 a first terminus was reached at –677 m. In 1974, a fossil side passage was explored, again by the S.C.B., which led to a water-filled pit at –697 m (map of 1981). The length of the cave then reached 3600 m.

A dive by the Association pour la Recherche Spéléologique du Maroc in 1981 brought the total depth to –713 m and the length to 3725 m.

Map: based on the drawings of the S.C.B., provided by Michel Chassier.

Bibliography: Chassier (M.) - Le gouffre du Toghobeït (Maroc, Rif) *Spelunca*, 1975 (2): 29-32.

2. **kef Tikhoubaï** .... –310 m
    (djebel Tazekka, Middle Atlas, Taza)
    Alt. 1050 m. Explored by the S. C. Taza (–120 m, 1964), the S. C. Rabat (–155 m, 1967), and the S. C. Aix-en-Provence (–310 m, 1969) (Camus, Lamouroux, *op. cit.*, 1981, profile).

3. **Friouato** (djebel Messaoud, Middle Atlas)... –271 m
    Alt. 1450 m. Explored by the S. C. Taza (1930-1931), then by N. Casteret (1934). Bottomed by the S. C. Taza in 1948 (Camus, Lamouroux, *op. cit.*, 1981, profile).

4. **aïn Melghfi** ....(–115, + 136) 251 m
    (Middle Atlas, Azilal)
    Alt. 1350 m. By the S. S. Maroc (1948) and the S. C. Alpin Languedocien and the S. C. Rabat (1974).

5. **kef El Sao** .... –220 m
    (djebel Bou-Messaoud, Middle Atlas, Taza)
    Alt. 1510 m. Explored by Casteret (–144 m, 1934), S. C. Taza (–199 m, 1957) and G. S. Bagnois-Marcoule (–220 m, 1983) (*Spelunca*, 1984 (14) profile).

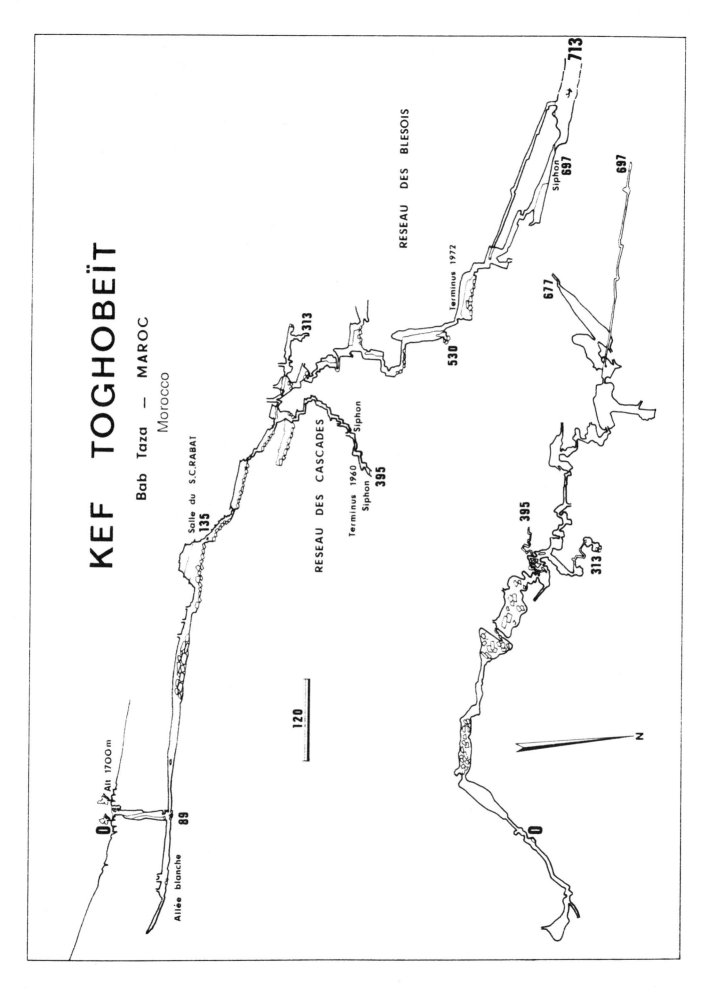

KEF TOGHOBEÏT

Bab Taza — MAROC

Morocco

Salle du S.C.RABAT

RESEAU DES BLESOIS

RESEAU DES CASCADES

Terminus 1972

Terminus 1960

Siphon

Siphon

Allée blanche

Alt 1700 m

120

N

713

697

697

677

530

395

313

395

313

135

89

0

0

6. **kef Anefid** .... −214 m
(Middle Atlas, Bab Bou Idir, Taza)
Alt. 1280 m. In 1968 by Sheffield U.S.S. (*S.U.S.S. Journal*, 1978, 2 (6) profile).

7. **kef Amafane** (Middle Atlas, Beni Mellal) .... −190 m
Alt. 1100 m. In 1971 by P. Courbon (Camus, Lamouroux, *op. cit.*, 1981, profile).

8. **kef Taounza** (Middle Atlas, Beni Mellal) ..... −165 m

9. **kef Rhachaba** .... −152 m
(Rif, Bab Taza, Chaouene)
Alt. 1580 m. Explored in 1960 by S.C. Rabat (Camus, Lamouroux, *op. cit.*, 1981, profile).

10. **rhar Chiker** (Middle Atlas, Taza) .... −146 m

LONG CAVES:

1. **Wit Tamdoun** .... 7550 m
(Tazroukht, Immouzer, Agadir)
Also called A Ouit n'Doun. Explored by the S.S.M. (1200 m, 1950), the S. C. Rabat (7550 m, 1957) and the E.R.E. Catalunya (1971, 1973) (Camus, Lamouroux, *op. cit.*, 1981, plan).

2. **rhar Chara** (Middle Atlas, Taza) .... 6200 m
Underground river explored in 1962-1963 by Chatelain and Matthieu, by the S. C. Périgueux in 1964-1965 and by the S. C. Aix-en-Provence in 1967 (Camus, Lamouroux, *op. cit.*, 1981, plan).

3. **kef Aziza** .... 3950 m
(Central Altas, Bou Denib, Errachidia)
Length given by the cave's Yugoslavian explorers (*Echo des Vulcains*, suppl. N° 41, 1981 map showing 1541 m).

4. **rhar Chiker** (Middle Atlas, Taza) .... 3865 m
Explored beginning in 1934 (N. Casteret). Length figure obtained in 1982 by the S. C. Nantua.

5. **kef Toghobeït** (Rif, Bab Taza, Chaouene) ... 3725 m

6. **ifri N'hamed N'Taouïa** .... 3600 m
(Atlas Central, Aït N'hamed, Azilal)
The S. S. Maroc explored 1400 m in 1950, length brought to 3600 m in 1981 by the Westminster S. G. (*Caves & Caving*, 1982 (16) plan).

7. **aïn Melghfi** (Middle Atlas, Azilal) .... 2732 m
In 1972 by the S.C.U.C. Louvain (Camus, Lamouroux, *op. cit.*, 1981, plan).

8. **rhar Isk N'Zouya** .... approx. 2300 m
(Essaouira, Agadir)
In 1960 by the S.S.M. (Camus, Lamouroux, *op. cit.*, 1981, plan).

9. **rhar Karkar** (Maroc Central, Safi) .... 2300 m
(Camus, Lamouroux, *op. cit.*, 1981, plan).

10. **kef Azar** (Middle Atlas, Merhaoua) . approx. 2000 m

NAMIBIA

Thanks to the reports of Pierre Strinati, we know that Namibia contains large scale karst phenomena, such as the **Guinas lake** (Otavi Bergland) 153 m deep, or the **Otjikoto lake**, 60 m deep for a diameter of 160 m. The divers of the South African Speleological Association reached − 110 m in **Harasib** (Otavi Bergland), a 153 m deep cavity. A cave named **Drachenrauchloch** (Dragon's Breath Hole) was discovered in 1986 by the SASA. It contains what may be the world's largest known underground lake, 207 m long by 107 m wide on the surface, and larger underwater. The lake was dived in 1987 by the SASA.

LONG CAVES:

1. **Ghaub** (Ghaub, Otavi, Bergland) .... 2190 m
Alt. 1550 m. Depth 31 m. Map showing 1070 m in Verin für Höhlenforschung, Windhoek, 1969 (6-7).

2. **Drachenrauchloch** .... approx. 500 m

3. **Nooitgedaghöhle** (Karibib) .... 390 m
Map: *ibid.* 1968 (4).

4. **Uhlenhöhle** (Karibib) .... 312 m
Map: *ibid.* 1968 (4).

5. **Nosib** (Otavi Bergland) .... 225 m
Map: *ibid.* 1968 (3).

6. **Aigamas** (Sissekab, Otavi Bergland) .... 120 m
Map: *ibid.* 1969 (5).

NIGERIA

The speleological potential of Nigeria would seem to be somewhat limited, but significant caves are nonetheless found in granite (in Kano province), sandstone (in Amambra province) and basalt (Bauchi province). They have been studied by Anglo-saxon cavers.

LONG CAVES:

1. **Oche** (Uwani Ono, Udi, Amambra) ..... approx. 90 m

2. **Ogbunike** (Ogbunike, Onitsha, Amambra) ...... 77 m
(*Studies in Speleology*, 1984, V, plan).

3. **Kassa** (Kassa, Bauchi) .... 45 m

4. **Birnin Kudu N° 2** (Kano) .... 17 m

RWANDA

Although the continent of Africa has a relatively poor speleological potential on the global scale, one does find interesting pseudokarst phenomena of great interest. Such is the case in Rwanda, where Barcelonean cavers explored two lava tubes in 1977. **Ubuwumo bwa Musanze** (Run-engeri) is 4560 m long and 210 deep, but is divided in several segments. **Ubuwumo bwa Nyirabadogo** (Bigowa) is appoximately 1500 m long.

SOMALIA
SOMALIYA

Small areas of limestone have been found along the Jubba river (Gedo province) by Italian biospeleologists. In 1984 Messana, Chelazzi and Baccetti mapped **Snowli Berdi** (Baardheere/Bardera) for a length of 675 m and estimated the length of nearby Mugdile at 200 m. The two caves appeared to be connected (*Monitore Zool. Italiano*, 1985 (15) maps).

SOUTH AFRICA

The karst areas most interesting for speleologists are in the Cape Province in the south and the Transvaal Province in the north. On the international scale, South African caves, which are mostly formed in lower Protozoic dolomites and cherts, are in modest dimensions. In contrast are some rather remarkable caves formed in quartzites and diabases, with over fourteen longer than 200 m.

Founded in 1954 the South African Speleological Association, divided into chapters according to province, has been responsible for the exploration of about two-thirds of the caves listed below, most of the rest having been explored by the Cave Research Organization of South Africa, the Free Cavers, or unaffiliated cavers in the past. The most celebrated example of the latter is the discovery and exploration of Cango Cave principally by a peasant named Van Zyl around 1780. This famous show cave contains the largest underground chamber in the country, measuring 220 by 35 by 35 m.

Of interest is the impact of gold mining in the Western Transvaal, which has lowered the local water table, permitting cavers to make extensive explorations in West-driefontein and other caves in the area.

DEEP CAVES:

1. **Westdriefontein** .... –183 m
(Far West Rand, Carletonville, Western Transvaal)
Alt. 1618 m. Explored in 1975 (*Proc. VIIth Intl. Congr. Spel.*, Sheffield, 1977, schematic profile).

2. **Wolkberg** .... –152 m
(Pietersburg, Northern Transvaal)
Alt. 1484 m.

3. **Chaos Cave** .... –112 m
(Carletonville, Western Transvaal)

4. **Wondergat** (Western Transvaal) .... –104 m

5. **Abyss** (Northern Transvaal) .... –96 m

6. **Bathole** (Northern Transvaal) .... –95 m
Contains a vertical pitch of 92 m.

7. **Nico's II Cave** (Western Transvaal) .... – 95 m

LONG CAVES:

1. *Apocalypse Pothole* ....**12,139 m**
or *Bobbejaansgat*
This maze cave is located near the town of Carleton-ville in the Far West Rand. The entrance pit has been known for a long time by the name of Bobbejaansgat and was descended in the 1960's by the owner, W. H. E. Kinnear, who used an oil drum hooked up to a tractor. He explored the Old World section solo. The SASA explored and mapped the Old World section from February to December 1973, finding about 3000 m. In March of 1974 a key discovery was made. This Stromatolite Passage led to the New World which was mostly explored the end of 1974. Additional pushing in 1975 brought the total cave length to 10,850 m. In 1982-1986, the discovery of additional galleries brought the length to 12,139 m and the depth to 85 m.

The cave is formed in Proterozoic dolomite and chert (Transvaal System, Malmani subgroup) and the passage formation was phreatic. The cave is completely dry, except in one place where the water table is reached.

Map: furnished by Jacques Martini. From S.A.S.A (Transvaal) 1973-1975.

Bibliography: Martini (J.), Kavalieris (I.), Verhulsel (J.). Apocalypse Pothole, *SASA Bull.*, 1975, pp. 4-9, plan.

2. **Wonderfontein** ....9420 m
(Carletonville, Western Transvaal)
First known in the 1840's. Eighty percent explored in 1955, with extensions through 1978. (Kent et *al.*, in *Geol. Surv. South Afr. Annals*, 1976, 11).

3. **Crystal Cave** ....7984 m
(Carletonville, Western Transvaal)
Explored begining in 1971.

4. **Westdriefontein** ....5785 m
(Carletonville, Western Transvaal)
Explored from 1955 to 1978 (*SASA Bull.*, 1979-1980, plan).

5. **Cango Cave** (Oudtshoorn, Cape) ....5275 m
Famous show cave. Extensions discovered begining in 1976 by SASA (Cape).

6. **Thabazimbi** ....4480 m
(Thabazimbi, Northern Transvaal)
Explored from 1959 to 1979 (*SASA Bull.*, 1979-1980, plan).

7. **Chaos Cave** ....4125 m
(Carletonville, Western Transvaal)
Explored begining in 1975.

8. **Empire Cave** (Western Transvaal) ....4010 m
Empire/Westminster/Rising Star Cave. Explored by SASA and Free Cavers.

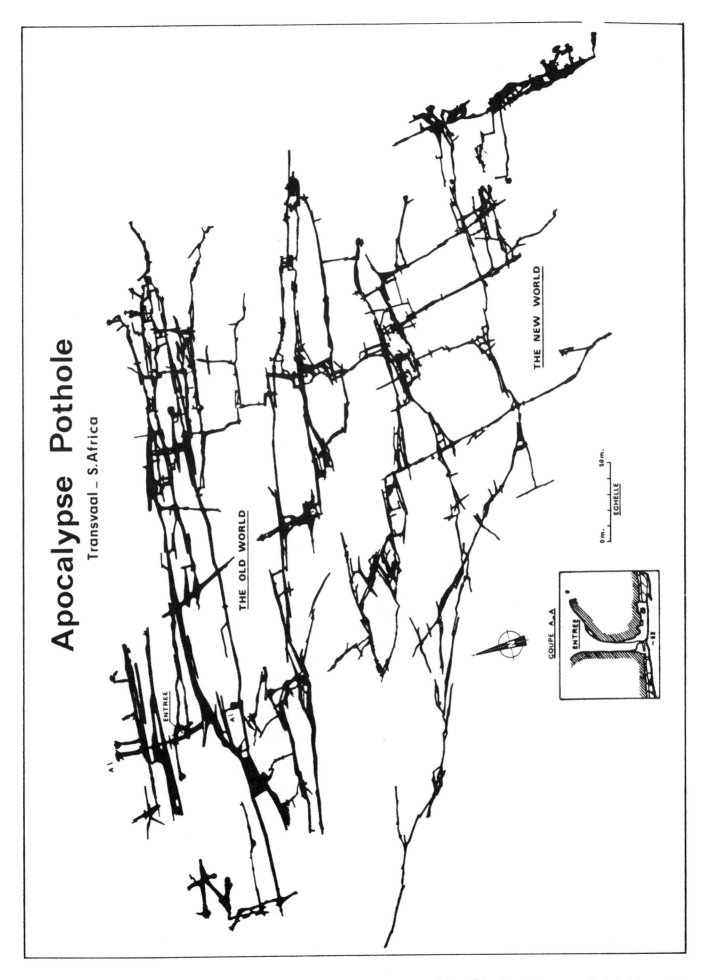

Apocalypse Pothole

Transvaal – S.Africa

THE NEW WORLD

THE OLD WORLD

ENTRÉE

A'

COUPE A–A

ENTRÉE

0 m.   50 m.

ECHELLE

9. **Boons Cave** ....3350 m
   (Carletonville, Western Transvaal)
   (*Geol. Surv. South Afr. Annals*, 12, plan).

10. **NH3 Cave** (Western Transvaal) ....3060 m
    Explored begining in 1978.

11. **Nico's II Cave** (Western Transvaal)...........2900 m

12. **Efflux Cave** (Cango Valley, Cape) ....2450 m

13. **Stroomwater Cave** ....2325 m
    (Cango Valley, Cape)

14. **Sterkfontein** (Western Transvaal) ....2210 m

15. **Jock's Cave** (Eastern Transvaal) ....1935 m

16. **Sudwala Cave** (Eastern Transvaal) ....1800 m

17. **New Cave** (Western Transvaal) ....1750 m

18. **Echo Cave** (Eastern Transvaal) ....1720 m

19. **Mogoto Cave** (Northern Transvaal)............1615 m
    Cave formed in quartzite.

20. **Van Rooy Cave** (Western Transvaal) ....1585 m

TANZANIA

There have been several archeological and paleontological investigations in Tanzania in which caves have played a peripheral role, but until now no cavers have become interested. In 1912, one cave, **Nduil** (Kibata), was supposedly explored for a length of 3500 m! On a smaller scale, Jeannel and Racovitza noted 200 m long **Kulumuzi C** (Kyomoni, Tanga) and 100 m long **Kulumuzi A** (*Arch. Zool. Expérim. et Générale*, 1914, 53 (7) sketches).

TUNISIA

Tunis

In spite of the fact that a large portion of the Tunisian landscape is composed of limestone (or dolomite), one finds only a few karstified ranges (djebels Zaghouan Oust, Ben Saïdane and Fkirine, for example). In addition, the layers are relatively thin, except in the djebels Serdj, Bargou, Taboursouk, or Cap Bon.

The Tunis section of the Club Alpin Français began several explorations from 1948 to 1958 when Tunisia was a French protectorate. Several caves located in mining areas were visited by miners. More recently (1969-1976), various French clubs (Société Spéléologique d'Avignon, Spéléo-Ragaï de Vedène) have undertaken caving explorations. Today, Tunisian caving is still developing, thanks especially to Lotfi Ghattas. A fairly complete summary of Tunisian caves can be found in a 164 page article in *Spéléo-Drack* N° 14 (Le Havre, 1980) by Jean-Jacques Lhopiteau.

DEEP CAVES:

1. *rhar Djebel Serdj* ....(–167, + 100) 267 m
   (Ousseltia, Kairouan)
   Alt. 950 m. Has three artificial entrances dug before 1913. Explored by the C.A.F. around 1948-1950. (–109, +70), then by the S. S. Avignon in 1969 and the Spéléo-Ragaï de Vedène in 1976 (*Spelunca*, 1978, (2) profile).

2. **rhar Aïn Et Tseb** ....+ 160 m
   (djebel Serdj, Ousseltia, Kairouan)

3. **rhar Kriz** .... –150 m
   (djebel Teboursouk, Aïn Younes, Khaled Nord)

4. **Mongass el Hamam** .... –100 m
   (djebel Ben Saïdane, Aïn Djougar, Pont du Fahs)
   (*Spéléo-Drack*, 1980 (14) profile).

5. **rhar Zaghouan** (Zaghouan, Tunis) .... –77 m
   Alt. 945 m (*Spéléo-Drack*, 1980 (14) profile).

6. **damous Ben Habib** .... –60 m
   (djebel Teboursouk, Nefza)
   Also called Khanget Ket Tout. Alt. 350 m (*Annales Spél.*, 1952 (2) profile).

7. **aïn Djougar N° 1** .... –55 m
   (djebel Ben Saïdane, Aïn Djougar, Pont du Fahs)
   (*Spéléo-Drack*, 1980 (14) profile).

8. **rhar Djebel Saïkra N° 1** .... –50 m
   (Matmata, Medenine)
   (*Spéléo-Drack*, 1980 (14) profile).

LONG CAVES:

1. **rhar Aïn Et Tsab** ....approx. 2600 m
   (djebel Serdj, Ousseltia, Kairouan)
   Alt. 940 m. In 1976 by the S. R. Vedène and in 1981 by the S. C. Périgueux (plan showing − 565 m in the report S.R.V., *Djebel Serdj, Tunisie*, 1977).

2. **rhar Djebel Serdj** (Ousseltia, Kairouan) ..... 1700 m
   In 1969 (1260 m) by the S.S.A., and in 1976 by the S.R.V. (*Spelunca*, 1978 (2) plan).

3. **rhar Kriz** ....1130 m
   (djebel Teboursouk, Aïn Younes, Khaled Nord)
   (*Spéléo-Drack*, 1980 (14) plan).

4. **rhar Aïssa N° 1** ....360 m
   (djebel Chambi, Aïn Touerf, Kasserine)
   (*Spéléo-Drack*, 1980 (14) plan).

5. **damous Ben Habib** ....335 m
   (djebel Teboursouk, Nefza)
   (*Annales Spél.*, 1952 (2) plan).

6. **rhar Tilist** (Tamerza-Midès) ....320 m
   Or rhar Tamerza (*Spéléo-Drack*, 1980 (14) plan).

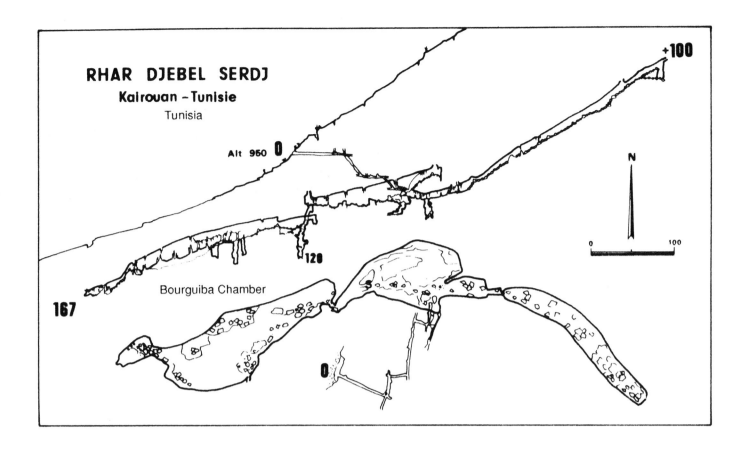

RHAR DJEBEL SERDJ
Kairouan – Tunisie
Tunisia

Alt 950 **0**

+100

120

Bourguiba Chamber

167

0

N
0   100

7. **rhar Aïssa N° 2** ....250 m
(djebel Chambi, Aïn Touerf, Kasserine)
(*Spéléo-Drack*, 1980 (14) plan).

ZAIRE

It has above all been Belgian biospeleologists who have revealed the speleological potential of Zaire. Among these were Geerts in 1912, Schwetz in 1936, and especially N. Leleup, who prospected from 1948 to 1954 in the west of the country (Mont Hoyo, Itombwe, Maniema, Shaba; Leleup, *La faune cavernicole du Congo Belge...*, Tervuren, 1956), and, with M. Heuts, in Lower Zaire between Kinshasa and the Angolan frontier (Heuts, Leleup, *La géographie et l'écologie des grottes du Bas-Congo*, Tervuren, 1954).

There has been little exploration since then except for two 1984 Belgian expeditions to Zaire, led by Yves Quinif, the "Kwilu 1984".

The limestones date to the Precambrian and contain horizontally developed caves with large passages and underground rivers. Those in the Shaba are known for their prehistoric art.

DEEP CAVES:

1. **tadi Gwango** (Kiasi-Mankala, Bas Zaïre) ...... –70 m
1984, "Kwilu 84" (*Spéléo-Flash*, 1985 (146) profile).

2. **tadi Kieza** (Mbanza Ngungu, Bas Zaïre) .... –55 m

LONG CAVES:

1. **grotte de Pitanshi** ....approx. 7000 m
(Kamina, Shaba)

2. **grotte de Ngovo** ....5860 m
(Mbanza Ngungu, Bas Zaïre)
Anonymous explorations, map at the Musee Royal d'Afrique Centrale in Tervuren.

3. **grotte de Kiwakishi** (Mitwaba, Shaba) ....2100 m
Explored in 1957-1958 by the S. S. Elisabethville and the S. C. Jadotville (*Karstologia*, 1984 (4) plan).

4. **grotte de Salanga** ....approx. 2000 m
(Gandajika, Lupata, Kasai)

5. **grande grotte de Ngungu**..........................1665 m
(Mbanza Ngungu, Bas Zaïre)
Show cave. Map at the Musee Royal d'Afrigue Centrale of Tervuren.

6. **grotte de Ndimba Dimba** ....986 m
(Mbanza Ngungu, Bas Zaïre)
1984, "Kwilu 84" (*Spéléo-Flash*, 1985 (146) plan).

7. **tadi Kieza** (Mbanza Ngungu, Bas Zaïre) .....1300 m
1984, "Kwilu 84" (*Spéléo-Flash*, 1985 (146) plan).

8. **"grotte supérieure Tordeur"** ....986 m
(Mbanza Ngungu, Bas Zaïre)
(Heuts, Leleup, *op. cit.*, plan).

9. **tadi Ndieka** (Kanka, Lovo, Bas Zaïre) ....635 m
1984, "Kwilu 84" (*Spéléo-Flash*, 1985 (146).

10. **grotte de Mbuzi** ....610 m
(Mbanza Ngungu, Bas Zaïre)
1984, "Kwilu 84" (*Spéléo-Flash*, 1985 (146).

ZIMBABWE

Chirorodziva is a famous show cave located 100 km NW of Harare, near Chinhoyi (formerly Sinoia). It was the scene of several tribal wars in the nineteenth century. Divers from the South African Normalair Underwater Club reached –103 m in 1969 and sounded the lake to − 172 m.

III
AMERICA

ANGUILLA

The small island of Anguilla, administered by Great Britain, has a 22 m deep, 65 m long cave named **The Fountain**. It is descibed by P. Wagemar Hummelinck in *De Grotten van de Nederlandse Antillen*, 1979 (map).

ANTIGUA and BARBUDA

Among the caves known on these two small independent islands are **Dark Cave** (Barbuda), 140 m long (P. W. Hummelinck, *De Grotten van de Nederlandse Antillen*, 1979, plan) and **Bat Cave** (Falmouth, Antigua), 75 m long, explored in 1961 by Spence and Gurnee (*International Speleologist*, 1961, 1 (1) plan).

ARGENTINA

Argentina has a variety of geological settings and several mountain ranges. Although some 500,000 km² contain rocks of speleological interest, the most promising area is the Patagonia range of the Andes, where one finds a band of schist mixed with clay, sandstone, and limestone.

Although over 1000 caves of a great variety have been explored, they generally are not very long or deep. Often times small caves are found to connect with each other to form systems hundreds of meters long.

The Grupo Espeleológico Argentino, one of the main local clubs, has published an up-to-date guidebook on the caves of Argentina. They also publish the newsletter *Salamanca*.

Victor H. Demaria Pesce.

DEEP CAVES:

1. **sima de Huitrin** (Neuquén) .... –55 m
   Bottomed in 1987 by the G.E.A. (*Bol. G.E.A.*, 1987 (4) profile).

2. **cueva de las Brujas** (Mendoza) .... (–21,+30) 51 m

LONG CAVES:

1. **sistema del Cuchillo Curá** .... 1503 m
   (Las Lajas, Picunches, Neuquén)
   System of four caves, including caverna del Gendarme. Length estimated at over 3500 m.

2. **cueva de las Brujas** .... 1200 m
   (Bardas Blancas, Malargüe, Mendoza)
   Alt. 1940 m. Mapped in 1981-1984 by the G.E.A. Length estimated at 3000 m.

3. **cueva de Doña Otilla** (Mendoza) .... 838 m
   Lava tube (*Bol. Soc. Venezolana Espel.*, 1973, 4 (2) plan).

4. **cueva del León** .... 631 m
   (Las Lajas, Picunches, Neuquén)
   Gypsum cave (*Aire y Sol*, 1984 (137) plan with 550 m). Contains rooms 44 by 63 by 40 m and 53 by 43 by 24 m.

5. **caverna Helada** .... 370 m
   (El Puesto, Pulién, La Pampa)
   Basalt cave.

BAHAMAS

This archipelago has unusual karst phenomena known as blue holes, first described by George Benjamin. These are submerged cave openings along the coast that lead to water-filled caves underneath the islands. Between 1967 and 1971, divers under the direction of Commander Cousteau made serious reconaissance trips. After 1978 it has mostly been American (D. Williams, G. Melton) and British (M. Farr, Palmer) divers who have made the most interesting discoveries.

DEEP CAVES:

1. **Normans Castle Sinkhole** .... approx. –70 m
   (Little Abaco island)

LONG CAVES:

1. **Lucayan Caverns** (Grand Bahama island) .. 9184 m
   Explored in 1980 and 1981 by American divers. Totally submerged system with seven entrances.

2. **Conch Blue Hole** (Andros island) ....900 m
    Submerged cave explored in 1981 by the British.

3. **Rat Cay Blue Hole** (Andros island) ....600 m
    Submerged cave explored by the British in 1981.

BARBADOS

This small independant island country was visited in 1945-1950 by Ken Pawson (from Calgary). His unpublished 53 page manuscipt describes over twenty caves. Because it is a show cave, Harrison's Cave is probably the best studied in the country.

LONG CAVES:

1. **Bowmanston Cave** (St. John Parish) .... 1646 m
    Cave 90 m deep explored in 1981 by R. Goddard *et al.*

2. **Coles Cave** ....approx. 1000 m
    Explored in 1948 by K. Pawson and in 1979 by M. Buck (*The Florida Entomologist*, 1981, 64 (4) plan).

3. **Harrison's Cave** (St. Tomas Parish) ....780 m
    The map by O. Knox, A. Torres *et al.* can be found in J. Gurnee's *A Study of Harrison's Cave*, 1978.

BELIZE

The small country of Belize (formally British Honduras) is located in the southeast of the Yucatan Peninsula. It has a central highland core (the Maya Mountains) of noncarbonate rock. Runoff from the plentiful rainfall (125-425 cm/year) pours radially outward into cockpit karsts developed on 3000 km² of Cretaceous carbonates fringing much of the mountains. These large streams often form massive, unitary trunk caves that emerge at low elevations onto alluvially-covered Quaternary limestones. These younger limestones are occasionally karstified, but are in areas of high water tables. The trunk conduits originating in the Mayas accumulate local inputs from the Cretaceous carbonates they pass through, but these passages are generally small and quickly terminate in sumps.

Considerable exploration of Belize caves occurred by the Maya Indians, primarily between 650-960 A.D., apparantly for ceremonial puposes. For this reasons, **all cave entry must be acknowledged by the Belize Department of Archeology.**

Deep water stopped Mayan penetration. Some entry of caves certainly occurred in the centuries following settlement by the British, but there is little or no record until about 1890. Brief accounts by adventurers such as Thomas Gann or Karl Sapper were published, but most were the result of local archeologists such as A. H. Anderson in the 1920's to 1930's.

Modern exploration can be divided into three phases. The first commenced in about 1958, consisting primarily of visits by expatriate Americans W. Ford Young and Frank Norris to many caves and pits. Barbara MacLeod and Dave Albert of the United States first visited Belize in about 1970. MacLeod later moved to Belize for several years, and was joined by Carol Jo Rushin. With the aid of visitors such as Tom Miller, Logan McNatt, and members of the Sligo Grotto (Maryland, USA), a small program of cave exploration and survey was begun, chiefly concentrating on the Caves Branch System.

The latest and most active phase began in 1976-1977 with the first of many field seasons by Tom Miller. With the aid of Logan McNatt and others, mapping of the major elements of the Caves Branch System was completed by 1979. Since 1979, projects have been undertaken throughout the country, with the most mapping completed in the Chiquibul Cave system during National Geographic sponsored expeditions directed by Miller. The Rio Grande Project, initially led by Percy Dougherty, began in 1984 in southern Belize.

References for Belize caves are: Young (W. F.), *International Speleologist* 1(1) 4-13; *N.S.S. News* 29(1) 6-9; Miller (T.) *Caving International Magazine* 11, 16-22; Rushin (C.J.) *Caving International Magazine* 12, 12-18; Dougherty (P.), *N.S.S. News* 43 (11) 329-334; Miller (T.) *Karst Hydrology and Geomorphology of Belize*, *Proc. 9th Int. Cong. of Spel.*, Barcelona, Spain, 23-24.

Tom Miller.

DEEP CAVES:

1. **actun Lubul Ha** ....approx. –120 m

2. **Tunichil Muknal** ....approx. –105 m

3. **Kabal Group** ....approx. –95 m

LONG CAVES:

1. ***Chiquibul Cave System***
    (Cave District, Belize and Guatemala)
    The system is made up four segments. The three located in Belize are:
    a. **Cebada Cave** ....approx. 15,000 m
    b. **actun Tun Kul** ....approx. 12,200 m
    c. **Kabal Group** ....approx. 12,000 m
    The fourth segment, Xibalba, is located entirely in Guatemala (see that country). Access to the system is by long hikes through the jungle. The system is formed by the Chiquibul River, which sinks just upstream of the Kabal Group, and resurges in Guatemala some 10 km distant as the crow flies. Exploration and mapping have taken place in 1982, 1984, and 1986 in expeditions directed by Tom Miller and partially funded by the National Geographic Society. Participants included cavers from several areas in Canada, the U.S.A., and Belize. During the 1986 expeditions, Cebada was explored for 8 km in March, and pushed to 15 km in April and May. Discoveries included the Mind Broddler/Prodder passage, which is up to 80 m wide for a considerable length, and a large room just 200 m from the sump at the end of actun Tun Kul. This segment contains the Belize Chamber, one of the largest in the world with dimensions of 300 by 150 by 65 m. The system also contains the Chiquibul Chamber, which measures 250 by 150 m.
    Map: provided by Tom Miller.
    Bibliography: *Canadian Caver* 18(2).

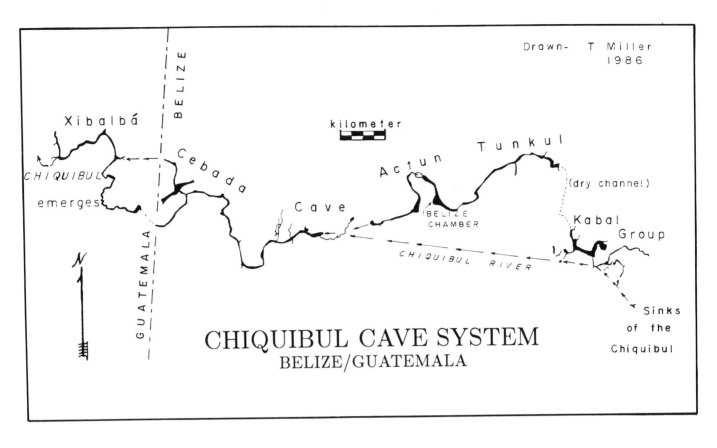

CHIQUIBUL CAVE SYSTEM
BELIZE/GUATEMALA

Drawn- T Miller
1986

2. **Caves Branch System** (Cave District)

This is another river cave system, formed by the following segments:

   a. **"Perdido" Cave** ....approx. 6000 m
   b. **St. Herman's** ....approx. 5000 m
   c. **actun Tzimin** ....approx. 3000 m
   d. **Lower Nab Nohol Group** ....approx. 10,000 m

Explored and mapped by W.F. Young, B. MacLeod, C.J. Rushin, T. Miller, L. McNatt, the Sligo Grotto, and others from 1958 to present.

3. **actun Chek** (Cave District) ....approx. 8000 m

Mapped by T. Miller, L. McNatt *et al.* in 1976-1979.

4. **Hokeb Ha System** ....approx. 5000 m
(Toledo District)

Mapped by T. Miller, J. Wyeth, L. McNatt, and cavers from the McMaster University (Canada).

5. **actun Lubul Ha** (Cave District) ......approx. 4000 m

Mapped by T. Miller, L. McNatt, and cavers from the McMaster University.

6. **Tunichil Muknal** (Cave District) ....approx. 3500 m

Mapped by T. Miller, S. Engler, D. Coons, and M. Saul in 1986.

7. **Hich Tulz** (Toledo District) ....approx. 3000 m

Or Tiger Cave. Explored by MacLeod, Rushin, Topsey, and Sosa in 1974. Mapped by Dougherty and the Rio Grande Project, 1984-1986.

8. **"Pat Cartwright's Cave"** ....approx 3000 m
(Cave District)

Explored by B. MacLeod, P. Cartwright, *et al.* Mapped by T. Miller, P. Cartwright, P. Mothes, R. Jameson, and A. Stone.

9. **actun Yax Te Ha** (Toledo District) ....2950 m

Mapped by T. Miller, D. Coons, and M. Saul in 1986.

BERMUDA

Bermuda consists of a small group of Eolian limestone islands situated atop a volcanic seamount in the NW Atlantic Ocean about 1000 km east of the United States. Bermuda's limestone caves are believed to have formed during low sea level stands of the Pleistocene Ice Ages. As post-glacial sea levels rose, much of the former extent of Bermuda's caves were drowned in sea water such that today many of the caves have considerable extensions below present sea level accessible only through cave diving. Over 150 caves are known in Bermuda with many of the inland caves containing tidal sea level pools.

Thomas M. Iliffe.

DEEP CAVES:

1. **Church Cave** (Hamilton Parish) .... –67 m
Of which 23 m are below sea level.

2. **Crystal Cave** (Hamilton Parish) .... –62 m
Of which 26 m are below sea level.

3. **Tucker's Town Cave** (St-George's Parish) .... –46 m
25 m are below sea level.

4. **Admiral's Cave** (Hamilton Parish) .... –43 m
Of which 13 m are below sea level.

5. **Windgate's Cave** .... –42 m

LONG CAVES:

1. **Green Bay Cave** (Hamilton Parish) .... 1970 m
A second entrance is called Cliff Pool entrance. Explored by the Bermuda Cave Diving Association. Completely submerged cave (map with 1532 m in *Proceedings 8th Internat. Cong. Spel.*, 1981: 163).

2. **Walsingham Cave** (Hamilton Parish) .... 700 m
Of which 650 m are submerged.

3. **Palm Cave** .... 500 m
Completely submerged.

4. **Crystal Cave** (Hamilton Parish) .... 500 m
Contains 350 m of submerged passage.

5. **Admiral's Cave** (Hamilton Parish) .... 500 m
Contains 50 m of submerged passage.

BOLIVIA

Bolivia has not attracted the attention of many cavers, perhaps because the karst areas are dispersed and shallow, hard to get to, and the political climate is often unstable. In 1966, the Spéléo-Club de Paris, with Henry Saavedra Coca, explored the **gruta de Umalajanta**, a swallow located at Torotoro, in Charca Province, at an altitude of 2850 m (depth –130 m approx., length 1620 m; *Grottes & Gouffres*, 1967 (39) plan). In 1968 the ephemeral activity of the Club Andino Boliviano led to the exploration of 660 m long **gruta de San Pedro** (Sorata, Larecaja Province, La Paz), at an altitude of 2590 m. We have not heard of any caving activity since that date.

BRAZIL

Brazilian caves were visited in the 17th and 18th centuries mostly by naturalists. Beginning in 1834, the Danish paleontologist Wilhelm Lund made excavations in Lagoa Santa (Minas Gerais). The real pioneer of Brazilian caving was the German geographer Richard Krone, who from 1895 to 1906 studied the caves of Iporanga (Río Ribeira valley, São Paulo). He made both archaeological and paleontological discoveries, describing 41 caves.

The Sociedade Excursionista e Espeleológica was formed on October 12, 1937 under the auspices of the Mining School of Ouro Preto (Minas Gerais). They explored numerous caves (Minas Gerais, São Paulo, Bania, ...) and are still active today. A first Sociedade Brasileira de Espeleologia (S.B.E.) was founded on August, 14, 1958. On June, 26, 1959, the Clube Alpino Paulista decided to create a caving section; this creation coincided with the arrival of the Frenchman Michel Le Bret who led the exploration of many of the large caves in the state of São Paulo (Iporanga). At the end of the 4th national congress, on November 1, 1969, a new S.B.E. was formed which today represents 18 local clubs and publishes *Espeleo-Tema*. Starting in 1970, cavers began exploring the great underground river systems of the Golas, to the north of Brasilia.

Among the 18 large provinces or speleological areas, we cite the vale do Ribeira, to the south of São Paulo (Precambrian limestones), where the vertical potential seems the best, the Bambuí, covering the eastern and south-eastern part of Goiás, the north west part of the Minas Gerais and the west of Bahía (Proterozoic limestones where large horizontal systems form, notably to the south of São Domingos), the Serra de Bodoquena (Cambrian and Precambrian), the Alto Pargual (Cambrian), the Chapada da Ibiapada (to the NW of the Ceará, Upper Cambrian) and the Río Grande do Sul (Precambrian).

Brazil contains many caves formed in material other than limestone, including sandstone, granite, gneiss, schist, mica schist, quartzite, and basalt. The caves and rock shelters of the Minas Gerais are known for their engravings and prehistoric paintings.

DEEP CAVES:

1. **abismo do Juvenal** .... –252 m
(Iporanga, São Paulo)
Explored in 1977 by the Centro Excursionista Universitário de São Paulo (*Espeleo-Tema*, 1978 (11) profile).

2. **cavernas Ribeirãozinho II e III** .... –220 m
(Iporanga/Apiai, São Paulo)
In 1978 by CAMIN and in 1979 by the C.E.U.

3. **gruta do Corrego Fundo** .... –195 m
(Iporanga, São Paulo)
In 1974 by the Clube Alpino Paulista.

4. **gruta Ouro Grosso** .... –192 m
(Iporanga, São Paulo)
Three entrances connected in 1973 by the C.E.U. (*Espeleo-Tema*, 1976 (10) profile).

5. **gruta da Casa de Pedra** .... (–90, +70) 160 m
(Iporanga, São Paulo)
Alt. 170 m (swallow), 80 m (resurgence). The cliff of this swallow of the Río Maximiano is 173 m tall!

6. **abismo de Gurutuva** .... –154 m
(Iporanga, São Paulo)
In 1969 by Le Bret and G. Collet.

7. **abismo de Hipotenusa** .... –153 m
(Iporanga, São Paulo)
In 1974 by the C.E.U.

8. **gruta da Agua Suja** .... +153 m
(Iporanga, São Paulo)

9. **gruta Engenho do Farto** .... –150 m
(Iporanga, São Paulo)

10. **abismo Tobias** (Iporanga, São Paulo).........−146 m
In 1969 by Le Bret and F. Valla.

11. **abismo da Lagoa Grande** ....−140 m
(Iporanga, São Paulo)

12. **gruta da Tapagem** (Eldorado, São Paulo) .. −140 m

LONG CAVES:

1. *Conjunto São Mateus-Imbira*...............**20,540 m**
(São Domingos, Goiás)

This hydrologic system is formed by the underground confluence of the São Mateus and Imbira (or Cascudeira) rivers. It is located about 50 km south of São Domingos and about 6 km west of São Joao, in the heart of the sierra do Calcáro, on the lands of the Cedral and Terra Ronca fasendas (farms). (13°39'S and 46°23'W).

The existence of the system was noted in 1970 by the Sociedade Excursionista Espeleológica de Ouro Preto and the access doline was discovered in 1973 during an expedition led by G. Collet with members of the C.E.U. de São Paulo. That year, 4300 m of passage were explored and mapped.

In 1974 G. Collet and the C.E.U. mounted another expedition which brought the length of over 13,500 m. Further explorations in 1976, 1978, 1979, and 1980 brought the length to 20,540 m, still without succeeding in a connection with the resurgência do São Mateus, due to the presence of sumps.

Map: after the plan quoted below. Work of G.L.N. Gusso, I. Karmann, C. F. Iino, L. E. Sánchez, J. -C. Setúbal and C. F. Zilio.

Bibliography: Karmann (I.), Setúbal (J. C.) - Conjunto espeleológico São Mateus-Imbira: principais aspectos físicos e histórico da exploraçâo, *Espeleo-Tema*, 1984 (14): 43-53, map.

2. **lapa dos Brejões** ....7750 m
(Morro do Chapeu/Ireçê, Bahía)

Has seven entrances. Known since 1877 (J. de Vasconcelos), explored and mapped in July, 1967 by the SEE of the Federal University of Ouro Preto.

3. **lapa do Angelica** ....6390 m
(São Domingos, Goiás)

Explored and mapped from 1972 to 1985 (P. Martin, R. Avari *et al.*).

4. **gruta Olhos d'Agua** ....6300 m
(Itacarambi, Minas Gerais)

Explored and mapped in July, 1985 by the Bambui de Belo Horizonte group.

5. **lapa do Convento** (or Salitre) ....5670 m
(Campo Formoso, Bahía)

Explored and mapped by the SEE d'Ouro Preto in July, 1970.

6. **gruta Canabrava** ....approx. 5500 m
(Santana, Bahía)

Reconnaissance and preliminary exploration in January, 1986 by the Bambui de Belo Horizonte group.

7. **caverna de Sant'Anna** ....5180 m
(Iporanga, São Paul)

Entrance described by R. Krone in 1898. Explored by several clubs from 1964 to 1975. New map by Sergio Beck in 1984/85.

CONJUNTO SÃO MATEUS - IMBIRA

GOIÁS - BRAZIL

8. **lapa de Terra Ronca** ....4850 m
(São Domingos, Goiás)
Explored and mapped by the SEE d'Ouro Preto in December, 1970.

9. **gruta da Tapagem** (or caverna do Diabo) ...4800 m
(Eldorado, São Paulo)
Known to R. Krone in 1898, first traversed in November, 1964 by the Clube Alpino Paulista, and mapped in 1965 by M. Le Bret and in 1967 by the SEE d'Ouro Preto.

10. **lapa Nova** (Vazante, Minas Gerais) ....4550 m
Explored by the SEE d'Ouro Preto in 1967. New map by the same group in July, 1985.

11. **gruta da Cabana** (Iporanga, São Paulo) ......4185 m
Explored by the Clube Alpino Paulista and mapped by P. Slavec from 1979 to 1984.

12. **gruta do Areado Grande II** ....3400 m
(Iporanga, São Paulo)
Explored by the Clube Alpino Paulista. Mapped by Peter Slavec and Bruno Sellmer from 1975 to 1983.

13. **gruta das Areias I** (Iporanga, São Paulo) ....3260 m
Explored by M. Le Bret and the Clube Alpino Paulista in 1961. New map by the SEE d'Ouro Preto in 1968.

14. **lapa de Mangabeira** (Ituaçu, Bahía) ....3230 m
Known to Joseph Mawson in 1886 as a pilgrim site, explored by Le Bret and Collet in 1968. New maps by the SEE d'Ouro Preto (July, 1969) and Coelba (1985).

15. **lapa do Janelão** (Januária, Minas Gerais) ..3020 m
Explored and mapped by the SEE d'Ouro Preto in July, 1975.

16. **lapa do Bezerra** (São Domingos, Goiás)......3010 m
Explored and mapped from 1973 to 1975 (P. Martin *et al.*)

17. **lapa de São Vicente II** ....2920 m
(São Domingos, Goiás)
Exploration in progress by the Clube Alpino Paulista. Mapped by P. Slavec (1974/78).

18. **gruta dos Paiva** (Iporanga, São Paulo) ....2880 m
Explored and mapped by the Centro Excursionista Universitário (São Paulo) in December, 1972.

19. **gruta da Agua Suja**......................approx. 2695 m
(Iporanga, São Paulo)
Noted by R. Krone in 1898. Explored and mapped for 1800 m in October, 1965 by Le Bret and Martin. New exploration by the Clube Alpino Paulista (1971/76) and mapped for 2100 m by the SEE d'Ouro Preto in December, 1968.

20. **gruta Buenos I** (Iporanga, São Paulo).........2580 m
Explored and mapped by the CAMIN club in 1978.

21. **lapa Grande** ....2200 m
(Montes Claros, Minas Gerais)
Explored and mapped by the SEE d'Ouro Preto in September, 1965.

22. **lapa de São Vicente I** ....2,155 m
(São Domingos, Goiás)
Explored by the Clube Alpino Paulista from 1974 to 1984.

23. **lapa Encantada** ....approx. 2000 m
(Montes Claros, Minas Gerais)
Explored by the Bambui de Belo Horizonte group in October, 1985.

24. **lapa de Maria Cobra** ....approx. 2000 m
(Coração de Jesus, Minas Gerais)
Explored in 1984 by the Bambui de Belo Horizonte group. Mapping in progress.

25. **gruta Jabuticaba** (Formosa, Goiás) approx. 2000 m
Explored by various clubs from Brasilia. Mapping in progress.

Pierre Martin, March ,1986.

CANADA

The documentation of Canadian caves began in the 19th century with the publication of descriptive articles, including George Gibb's "On Canadian Caves" published in 1861. Most early cave investigations occurred in eastern and central Canada, although in the West mountain travelers recorded some discoveries and the Nakimu Caves enjoyed a commercial history as early as 1904. The 1960's saw the beginning of intensive exploration and the formation of the principal caving groups across the country.

Canada's longest and deepest caves have been found in the Rocky Mountains. Despite the wide extent of the limestone deposits, caves in the Rockies are relatively rare due to the destructive nature of successive periods of glaciation. Those that are known have often been left in topographically illogical locations or have had passages plugged by inwashed glacial debris. The pace of new discoveries has been slowed by the eight month snow cover in the high country and the lack of roads or major population centers in the mountains. Happily, this also means that many potential caving areas remain unexplored and within recent years at least one major new discovery has made the 'top ten' list every season.

It is on Vancouver Island, one of the wettest and warmest regions in Canada, that one finds the greatest concentration of caves. Over a thousand caves have been located since systematic searches began in 1962; of these, four are over 300 m deep. Here, as in the Rockies, local cavers and visitors from across Canada or from other countries work together on explorations.

Several urban centers in Ontario and Quebec host caving clubs although the local caves are usually small. Minor caves occur in the eastern maritimes, and the Nahanni River Valley contains the only significant caves known in the northern territories.

Today over 1500 cavers belong to some 25 clubs or regional associations such as the Société Québecoise de Spéléologie, McMaster University Caving and Climbing Club, Toronto Caving Group, Alberta Speleological Society and the British Columbia Speleological Federation. Many publish their own newsletters, but the most important and comprehensive journal is *The Canadian Caver*, an independent publication with editorship shared across the country, started in 1969.

Ian McKenzie and Paul Griffiths.

DEEP CAVES

1.   *Arctomys Cave*........................(–523, +13) 536 m
(Mount Robson, Rocky Mountains, B.C.)

This cave is located at an altitude of about 1950 m in the hanging Arctomys Valley in Mount Robson Provincial Park, about 50 km west of Jasper, Alberta. It is developed in a 100 m thick band of Mural limestone, a massive carbonate of the Lower Cambrian group, dipped at about 30 degrees and sandwiched between impermeable rocks. The terminal sump is close to valley level and resurges in pools near the Moose River. The cave is accessed via a 25 km hike. The upper half of the cave contains innumerable free-climbs ("The Endless Climb") that gradually pick up water. Below "The Refresher" the cave is generally flatter with shallow pools and cascades. Only five pitches require tackle, the deepest being 15 m. Seasonally high water makes the cave more sporting.

1/50,000 map 83 E/2 West. Coordinates: 727815.

The first reported visit was by well-known mountaineers A. O. Wheeler, Conrad Kain, Byron Harmon, George Kinney and Donald Phillips in 1911 when they explored the cave to about –76 m, recorded in the 1912 edition of the *Canadian Alpine Journal.* Mike Goodchild confirmed the report in 1971 and the following August the cave was pushed by Canadian and British cavers to a depth of 297 m where exploration was halted by wet conditions near "The Thousand Foot Waterfall". In August 1973 members of the Alberta Speleological Society (ASS), McMaster University Climbing and Caving Club (MUCCC), and the Guelph University Caving Club (GUCC) mapped 2414 m of passage to the terminal sump at –522 m, making Arctomys the deepest known cave north of Mexico, a record that still stands. Higher passages were mapped by the ASS in 1977 and by the Anglo-Canadian Rocky Mountain Speleological Expedition (ACRMSE) in 1984 resulting in an overall vertical range of 536 m and a length of 3496 m.

Map: from the work of the ASS, the MUCCC, sent by Mike Shawcross and from *The Canadian Caver*, 1984, 16 (1).

Bibliography: Tracey (G.) "A New Anglo-American Depth Record: Arctomys Cave", *The Canadian Caver*, 1973, vol. 5 (2): 3-9; Pollack, *The Canadian Caver* vol. 9 (2).

2.   Castleguard Cave ....(–12, +375) 387 m
See below.

3.   Yorkshire Pot...–384 m
See below.

4.   Thanksgiving Cave .... –356 m
(Tahsis, Vancouver Island, B.C.)

Located at an altitude of 560 m on Vancouver Island and formed in Quatsino limestone (1/50,000 map 92 E/15). Two resurgences are known nearby but have not been positively linked with the cave. The Main Entrance was discovered in 1975 by the Vancouver Island Cave Explora-

tion Group (VICEG), and primarily explored and surveyed by them to 1978. The Staircase Entrance was found in March 1978 from surface surveys. Pillar Pit enabled easy access to the far end of the cave after its discovery by digging in August 1983. The depth was increased from –219 m to –364 m in September 1985 with the surveying of the Final Option by the the VICEG and the British Columbia Speleological Federation (BCSF). The present length is 6458 m (*The Canadian Caver*, 1978, vol. 10 (1); 1984, 16 (2); 1986, vol. 18 (1)).

5.   Glory 'ole .... –313 m
(Port McNeill, Vancouver Island, B.C.)

Located at an altitude of 760 m on Vancouver Island. Discovered and explored in 1982 by VCC, ASS, MUCCC, and BCSF cavers to a depth of 293 m and a length of 1472 m. Creekside Entrance was pushed and connected in 1983 by the VCC and ASS for a length of 1628 m and a depth of –313 m (*The Canadian Caver*, 1982, vol. 14 ; 1983, vol. 15 (2)).

6.   Arch Cave .... –302 m
(Port McNeill, Vancouver Island, B.C.)

Another Vancouver Island cave at an altitude of 665 m, discovered and explored in 1982 by BCSF, VCC, ASS, and MUCCC cavers to –299 m for a length of 982 m. It was extended by them in 1983 to a depth of –302 m and a length of 3770 m (*The Canadian Caver*, 1982, vol. 14; 1985, vol. 15 (2)).

7.   Q5 .... –301 m
(White Ridges, Vancouver Island, B.C.)

Altitude 1192 m. Discovered and explored in 1975 by the BCSF and extended to a terminal sump at –301 m in 1978 by BCSF, ASS, MUCCC, and Toronto Caving Group (TCG) cavers. Extensions by the SQS and various persons in 1984 and 1985 brought the length to 2066 m. If hoped-for connections to Windy Link Pot and Quatsino Cave were realized, a system 630 m deep and 7 km long would be formed (*The Canadian Caver*, Vol. 9 (2); Vol. 10 (2); Vol. 16 (2)).

8.   Close To The Edge .... –294 m
(Dezaiko Range, Rocky Mountains)

The 15 x 20 m entrance to Close To The Edge is located at 1580 m in a ridgetop cliff, but is concealed behind a low parapet carved by solution on one side and glaciation on the other. It is only visible from the air, and was in fact discovered during a brief reconnaissance with the helicopter supporting the 1985 Dezaiko Cave exploration. Subsequent investigations by the ASS led to the first bottoming trip by ASS and VCC cavers in 1986. The cave includes a remarkable 244 m deep shaft which is the deepest known north of Mexico, and is blocked at the base of the second pitch at – 294 m with a surveyed length of only 329 m (*The Canadian Caver*, 1987, vol. 18 (2)).

9.   Gargantua ....(–271, +15) 286 m
(Crowsnest Pass, Rocky Mountains, B.C.)

Alt. 2530 and 2520 m. Discovered in 1970 and explored to – 271 m for a length of 2440 m by the MUCCC. They discovered the Boggle Alley, the largest known room in Canada with dimensions of 30 m wide by 200 m long. Explorations in 1975 culminated in the discovery of the GB

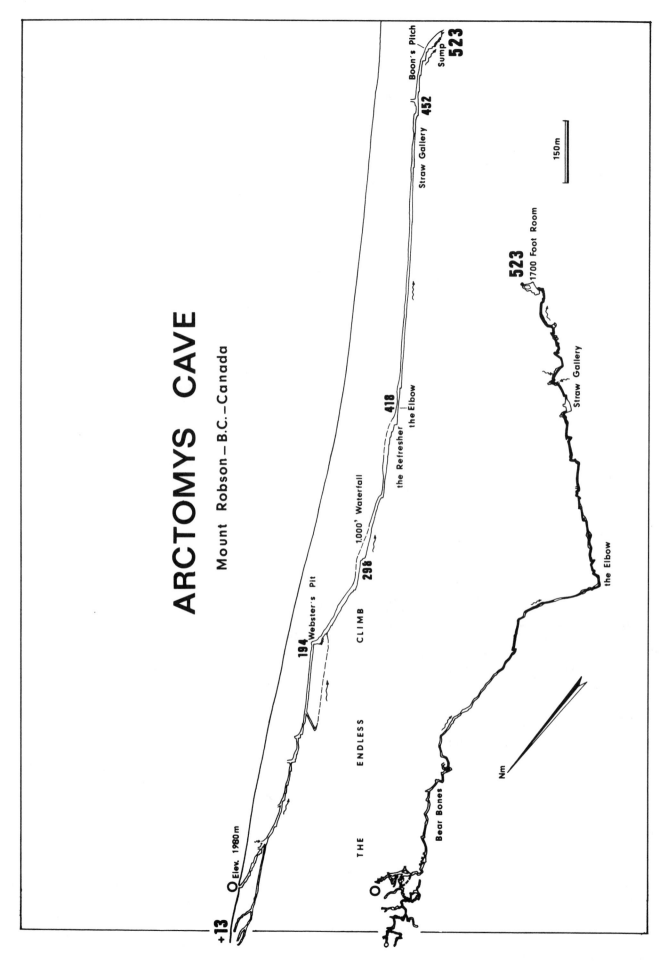

ARCTOMYS CAVE

Mount Robson—B.C.—Canada

Elev. 1980 m

+13

194 Webster's Pit

THE ENDLESS CLIMB

298

1,000' Waterfall

418

the Refresher

the Elbow

Straw Gallery

452

Boon's Pitch

Sump

523

523

1700 Foot Room

Straw Gallery

the Elbow

Bear Bones

Nm

150m

Passage, increasing the length to 5266 m. Two lower entrances were discovered by the ASS in 1985 and 1986, respectively. The current length is 5940 m (*The Canadian Caver*, N° 3; N° 5; Vol. 6 (2); Vol. 18 (1)).

10. **Nakimu Caves** .... –270 m
(Cougar Valley, Selkirk Mountains, B.C.)
Alt. 1710 m. These caves were explored as early as 1902 by D. Woolsey and W. Scott, and were commercialized and publicized by C. Deutschman in 1904. The first survey was begun in 1905 by Deutschman, Wheeler, and Agres and much of the cave was explored at that time. Commercialization ceased in 1935. A group of NSS cavers surveyed and explored additional passages in the 1950's. In 1965 another resurvey and further explorations commenced by the MUCCC. Nakimu was closed to the public by Parks Canada in the 1970's (*The Canadian Caver*, N° 4; N° 5; Vol. 8 (1)).

11. **White Hole** .... –253 m
(Bocock Peak, Rocky Mountains, B.C.)
Alt. 1600 m. Investigated by the ASS and explored by the ACRMSE in 1984 with a surveyed length of 1320 m (*The Canadian Caver*, Vol. 17).

12. **Dezaiko Cave** .... –253 m
(Dezaiko Range, Rocky Mountains, B.C.)
Alt. 1950 m. Explored and mapped by the ASS and VCC in 1983 to –95 m and a length of 308 m, and various persons in 1984 to –253 for a length of 710 m (*The Canadian Caver*, 1984, Vol. 16 (1); Vol. 18 (1).

13. **Fang Cave** .... (–143, +95) 238 m
(McGregor Range, Rocky Mountains, B.C.)
Explored in 1981 by ASS cavers and friends from Prince George to –143 m and a length of about 1 km. A major extension was discovered by the ASS in 1983 and surveyed to +95 m to reach an overall length of 2977 m in 1987. (*The Canadian Caver*, 1983, Vol. 15 (1); 1984, Vol. 16 (1)).

14. **Cadomin Cave** .... –220 m
(Nikanassin Range, Rocky Mountains, Alberta)
Alt. 1890 m. Original exploration unknown, but probably occurred in the early 1900's. Mapped by Biggs and Taylor in 1959 and by the ASS in 1972. Major extensions by the ASS brought the length to 1704 m (depth –171 m) in 1977, to about 2000 m in 1978 (depth –220 m), to 2468 m in 1982, and finally to 2791 m by 1985 (*The Canadian Caver*, Vol 4 (1); Vol. 10 (1); Vol. 14).

15. **Windy Link Pot** .... –209 m
(White Ridges, Vancouver Island, B.C.)
Alt. 887 m. Discovered in 1983 by the VCC, with a second entrance discovered in 1985. Explored by various cavers to –209 m and 3310 m long. See Q5 cave. (*The Canadian Caver*, Vol. 16 (2); Vol. 18 (1); Vol. 18 (2)).

16. **Grueling Cave** .... –153 m
(Tahsis, Vancouver Island, B.C.)
Explored by the BCSF in 1980 and 1984, and the BCSF and ASS in 1985. Total length is 697 m (*The Canadian Caver*, Vol. 17).

17. **Wapiabi Cave** .... –151 m
(Bighorn Range, Rocky Mountains, Alberta)
Discovered in 1967 by G. Hutchenson and explored in 1970 by the Youth For Christ, and in 1971 by the ASS to –130 m. Extensions by B. B. Maranda, M. Labrie, P. Lemaire and the ASS in 1976 and 1977 brought the depth to –151 m and the length to 540 m (*The Canadian Caver*, Vol. 4 (2); Vol. 9 (1)).

18. **Quatsino Cave** .... +151 m
(White Ridges, Vancouver Island, B.C.)
Explored by the BCSF and others from 1975 to the present. Part of the same hydrological system as Q5 and Windy Link Pot (*The Canadian Caver*, Vol. 9 (2); Vol. 16 (2)).

19. **Crackpot** .... –148 m
(White Ridges, Vancouver Island, B.C.)
Explored by the BCSF in 1977 and mapped in 1985 (*The Canadian Caver*, Vol. 9 (2)).

20. **Sawtooth Cave** .... –146 m
(Kootenay Lake, Purcell Mountains, B. C.)
Explored in 1977 and 1978 by the VICEG and the TCG (*The Canadian Caver*, Vol. 10 (2)).

21. **Moon Valley Cave** .... –134 m
(Park Ranges, Rocky Mountains)
Alt. 1920 m. Explored by the ASS in 1982 (*The Canadian Caver*, Vol. 15 (1)).

22. **Pinto Lake Caves** .... –123 m
(White Goat, Rocky Mountains)
Explored by the ASS in 1979 (*The Canadian Caver*, Vol. 11 (2)).

23. **Lizard Pot** .... –120 m
(Lizard Range, Rocky Mountains)
Explored in 1982 by the VICEG and Army Caving Association of Britain (*The Canadian Caver*, Vol. 15 (1)).

24. **Moon River Cave** .... +120 m
(Park Ranges, Rocky Mountains)
Alt. 1860 m. Explored by the ASS in 1982 and 1983. Exploration incomplete (*The Canadian Caver*, Vol. 15 (1); Vol. 15 (2); Vol. 16 (1)).

LONG CAVES:

1. ***Castleguard Cave*** .... **19,500 m**
(Columbia Icefield, Rocky Mountains, Alberta)
Castleguard Cave is located in the gently dipping Cathedral Formation Cambrian limestone adjacent to and beneath the Columbia Icefield, near the Alberta-British Columbia border in the north end of Banff National Park. The cave resurges mainly at Big Spring (alt. 1706 m) into the Castleguard River with mid-summer discharges of about 10 m³/sec. Sudden and unpredictable summer melts cause the main cave to flood in the entrance section, so explorations are confirmed to mid or late winter. Access is via a 20 km ski to the entrance, located in a pine forest at the south end of Castleguard Meadows, at an altitude of about 1950 m.

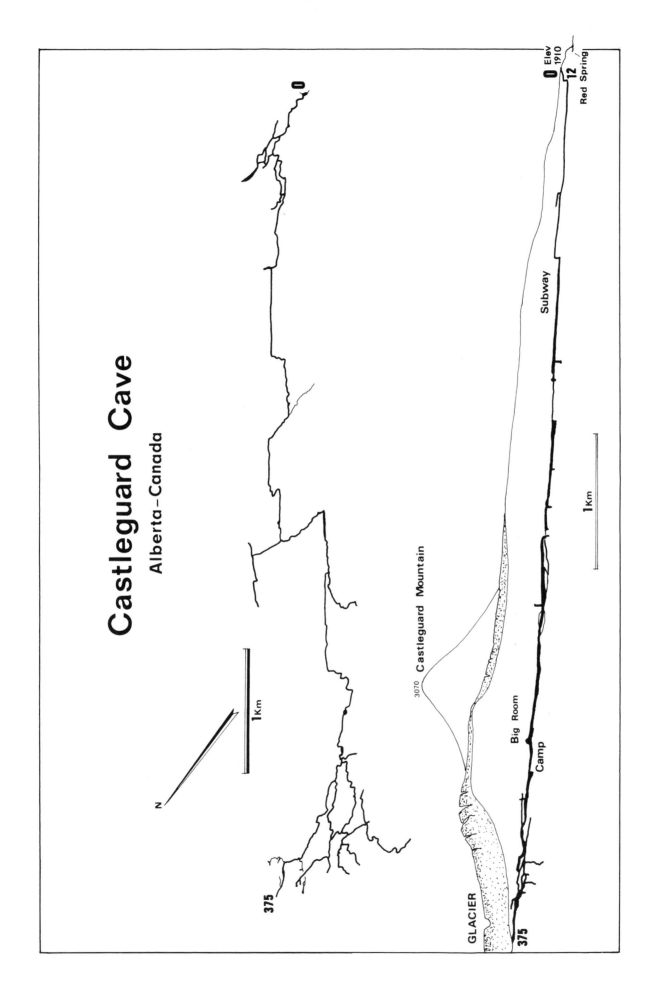

Castleguard Cave

Alberta–Canada

Castleguard is the only known cave in the world located beneath an existing icefield, with passages blocked by glacial ice extruded in from the sole of the ice sheet. Its most remote surveyed passage is 9.4 km from the only entrance by the shortest route, so that a trip to the back of the cave normally requires two underground camps.

The first recorded visit was by members of W. O. Field's mountaineering expedition in 1924, reported by L. R. Freeman in a 1925 edition of National Geographic Magazine. It is unlikely, however, that any exploration beyond the 8 m drop near the entrance was attempted prior to the 1967 investigations by Dr. Derek Ford and the MUCCC. Since that time Castleguard has been explored by cavers from Canada and around the world, with most trips organized by the MUCCC and more recently by the ASS.

Explorations between 1967 and 1970 by the MUCCC mapped over 10 km of passage. The famous Ice Plug was discovered by Mike Boon in 1970 during his celebrated solo Phantom trip, a remarkable feat considering the cave's remote location and length of passage traversed. Surveying by the MUCCC in 1973 established a length of 10,960 m and an elevation of 310 m (later corrected to 368 m) for the Ice Plug, then the highest point in the cave. Exploration in 1974 established a new length of 13,196 m during the filming of "The Longest Cave". The cave was gated and access was restricted by Parks Canada in 1975. Major extensions were explored by the Société Québecoise de Spéléologie (SQS) in 1979 (Boulevard du Quebec) and in 1983 by the ASS and the Vancouver Caving Club (VCC) (Jasper Avenue) establishing a new high point of 375 m. The MUCCC explored the Bon Appetit area. In 1987 John Pollack and Keith Sawatzky dived the sump at Boon's Blunder to a depth of 21 m and laid an additional 115 m of line upstream without reaching an end. The current surveyed length is approx. 19,500 m.

Map: from drawings by S. Worthington *et al.*, sent by S. Worthington.

Bibliography: Thompson (P.) "A Brief History of the Exploration of Castleguard Cave":, *The Canadian Caver*, 1970, N° 2, 26:35; Waltham (A.) "Castleguard Cave, Canada", *B.C.R.A. Bull.* 1974 N° 5, 18-28; Ford (D.) "Notes on the Speleogenesis of Castleguard Cave", *The Canadian Caver*, 1980, Vol. 12 (1): 9-15; Yonge (C.) *et al.*, "The 1984/1985 Castleguard Expeditions", *The Canadian Caver*, Vol. 17.

2. *Yorkshire Pot* ....**9413 m**
(Crowsnest Pass, Rocky Mountains, B.C.)
The main entrance to Yorkshire Pot is located on the Andy Good Plateau, south of Crowsnest Pass, in the Flathead Range, at an altitude of about 2400 m. The system has four entrances, Yorkshire Pot, Mendip Cave, The Backdoor, and Shorty's Cave. It is formed in the Livingstone and Banff Formation limestones.

The cave was discovered in 1969 by the MUCCC, who descended the first five pitches to a depth of − 165 m. After the final (6th) pitch of the entrance series was descended the following year, an extensive sub-horizontal network of passages was explored in ensuing summers and on one winter trip in 1972 to reach lengths of 1765 m in 1970, 2847 m in 1971, 3800 m in 1972, 5032 m in 1973, and 5544 m in 1976, for a depth of − 384 m.

Mendips Cave was discovered in 1969 by the MUCCC and explored by the MUCCC and ASS in 1970. The entrance has been blocked by ice since 1971.

The Backdoor was discovered by the ASS in 1979, and mostly explored in 1980, with connections to Mendips Cave in July and Yorkshire Pot in October.

Shorty's Cave was discovered by the ASS in 1973, and four pitches were descended to a bottom. In 1985, a traverse over the first pitch led to a second series of pitches and a connection in 1986 with Yorkshire Pot, bringing the total length of the system to 9413 m.

Map: from plots sent by S. Worthington.

Bibliography: Cave Exploration in Canada, 1976, *The Canadian Caver*, N° 1; N° 3; N° 5; Vol. 4 (2); Vol. 5 (1); Vol. 6 (2); Vol. 13 (1); Vol. 18 (1); Vol. 18 (2).

3. **Thanksgiving Cave** ....6458 m
See above.

4. **Gargantua** ....5940 m
See above.

5. **Nakimu Caves** ....5790 m
See above.

6. **Arch Cave**..3770 m
See above.

7. **Arctomys Cave** ....3496 m
See above

8. **Windy Link Pot** ....3310 m
See above.

9. **Grotte de Boischatel**....................approx. 3000 m
(Québec City, P.Q.)
Located in the town of Boischatel. Excavation work in 1979 by the SQS resulted in a surveyed length of 1540 m that year. By 1982 the length had been increased to about 2500 m. Exploration continues (*The Canadian Caver*, Vol. 12 (2); *Spéléo-Québec*, Vol. 7-9).

10. **Fang Cave** ....2977 m
See above.

11. **Cadomin Cave** ....2791 m
See above.

12. **Minigill Cave** ....2400 m
(Port McNeill, Vancouver Island)
Explored in 1977 by the BCSF (*The Canadian Caver*, Vol. 10 (1)).

13. **Grotte Mickey** ....2270 m
(South Nahanni River, NWT)
Explored in 1970 by the SQS and friends (*The Canadian Caver*, Vol. 5 (1)); *Spéléo-Québec*, Vol. 5-6).

14. **Q5** .... 2066 m
See above.

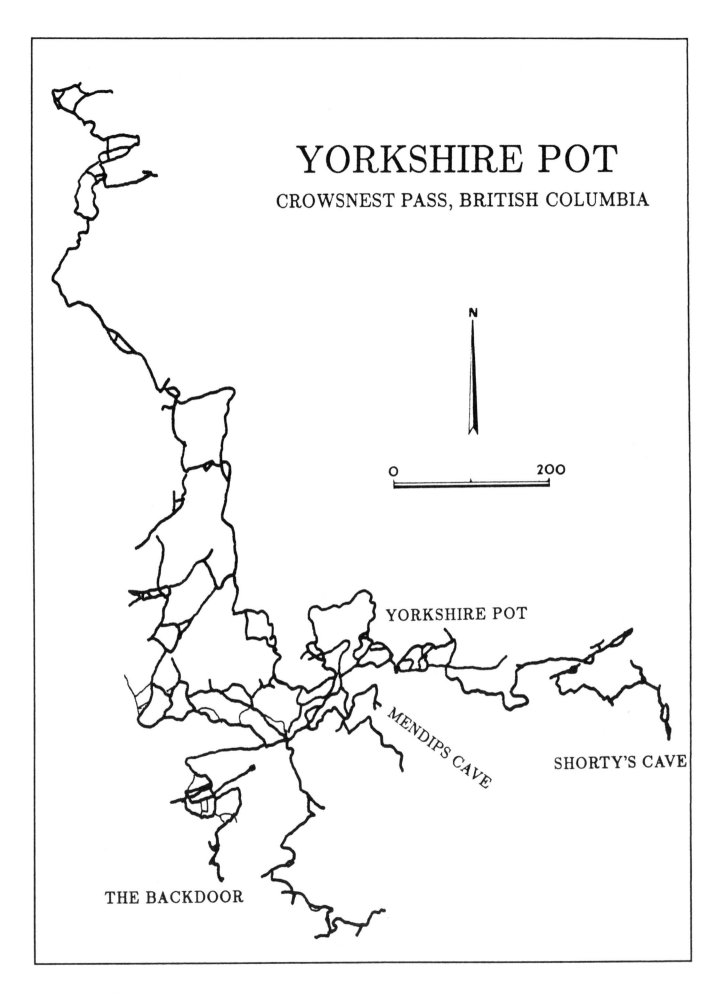

YORKSHIRE POT

CROWSNEST PASS, BRITISH COLUMBIA

N

0   200

YORKSHIRE POT

MENDIPS CAVE

SHORTY'S CAVE

THE BACKDOOR

15. **Grotte Valerie** .... 1900 m
    (South Nahanni River, NWT)
    Explored by the SQS and the MUCCC in 1971 (*The Canadian Caver*, N° 5; vol. 5 (1)).

CHILE

The country of Chile has few known caves. The principal ones are on Easter Island, where cavers from Bonsecours mapped several lava tubes in 1979. Among them are **Ana te Pahu** (Ahu Akivi, length 800 m) and **Ana Oke Ke** (Anu Akivi, length 450 m, vertical extent 40 m). The famous **cueva del Mylodon** (Puerto Natales, Magallanes) is 192 m long (*El Guácharo*, 1975-1976, 8-9 (1-2) plan).

COLOMBIA

Although Colombia has been known for its hoyo del Aire (see below) since the 19th century, it was not until the 1950's that modern exploration began. Dr. Eugenio Salas explored the Cueva de los Guácharos, and Luis Cuervo explored many other caves. Also in the 1950's Dr. Cabrera Ortiz began an extensive series of publications. The 1973 and 1975 trips of Tom Miller did much to reveal the speleological potential of the country. Interesting discoveries have been made by Polish cavers in 1975, French cavers from the G.S. Nice in 1977, and again French cavers from the A. S. Montreuil in 1980 and 1985. A valuable reference work was written in 1977 by B. Hof, *Recherches Spéléologiques en Colombie*.

DEEP CAVES:

1. **hoyo del Aire** (La Paz, Santander) .... −241 m
   Alt. 1763 m. In 1851, the Révérend Père Romualdo Cuervo was lowered to the bottom of the immense entrance pit (160 by 80 m), which is 115 m deep from the low side of the opening. The bottom was reached in 1975 and given a depth of − 270 m (figure depends on datum point chosen, Hof, *op. cit.*, profile). The depth was increased slightly in 1980 by the A. S. Montreuil.

2. **sistema Hermosura** .... −193 m
   (La Hermosura, Santander)
   System formed by the connection of the hoyo del Aguila (1977 French expedition, B. Hof, *op. cit.*, profile) and the hoyo Hermosura by the A. S. Montreuil in 1985.

3. **hoyo de los Pajaros** (Santander) .... −184 m

4. **cueva de Cunday** (Cunday, Tolima) .... −160 m
   Stream cave in sandstone (alt. 900 and 740 m) explored in 1953 by Wenceslao Cabrera Ortiz.

5. **hoyo de Colombia** (La Paz, Santander) ..... −150 m
   Alt. 1775 m. Bottomed in 1975 (B. Hof, *op. cit.*, profile).

6. **hoyo de las Flechas** .... −144 m
   (La Hermosura, Santander)
   1985, A.S.M., French and Colombians.

7. **hoyo del Niño** .... −140 m
   (La Hermosura, Santander)

8. **cueva de Gedania** (La Paz, Santander) ...... −110 m
   Alt. 1800 m. Polish exploration (B. Hof, *op. cit.*, profile).

9. **cueva de la Cuchara** .... −100 m
   (La Paz, Santander)
   Alt. 1900 m. Explored in 1975 and 1977 (B. Hof, *op. cit.*, profile).

LONG CAVES:

1. **sistema Hermosura** .... 4926 m
   (La Hermosura, Santander)
   Explored in 1980 (3000 m) and 1985.

2. **cueva del Indio** (Pitalito, Huila) .... 1966 m
   Explored in 1973 and around 1974 (Hof, *op. cit.*, plan).

3. **cavernas del Paramo** .... 1280 m
   (Paramo, Santander)
   Explored by Yugoslavians in 1984.

4. **cueva Gdynia-Paramo** .... 1240 m
   (Paramo, Santander)
   Alt. 1200 m. In 1975 (Hof, *op. cit.*, plan).

5. **cueva de los Guácharos** (Pitalito, Huila) .... 1000 m
   Alt. 1950 m. Explored in 1876, 1899, and 1973 (Hof, *op. cit.*, plan).

6. **cueva Antigua** .... 990 m
   (San Gil, Santander del Norte)
   (Hof, *op. cit.*, plan).

7. **cueva del Yeso** ....891 m
(San Gil, Santander del Norte)
Gypsum cave explored in 1975 (Hof, *op. cit.*, plan).

8. **cueva Danta** (?) ....873 m
In 1983 by Szentes (*Brit. Caver*, 1983 (89) plan).

9. **hoyo de Colombia** (La Paz, Santander) ....840 m
(Hof, *op. cit.*, plan).

10. **cueva de Cunday** (Cunday, Tolima) ....850 m
(Hof, *op. cit.*, plan).'

COSTA RICA

There has been little cave exploration in Costa Rica until recently. In 1973, the Cave Research Foundation studied the karst resouces of the Barra Honda region (*C.R.F. Annual Report*, 1974), but found only small caves. The Circolo Speleologico Romano visited the same area in 1974. Local cavers have also been active, and recently formed the Asociación Espeleológica Costaricense, which is actively exploring large new caves.

DEEP CAVES:

1. **sima Santa Ana** .... –170 m
(Barra Honda, Nicoya, Guanacaste)
In 1971 by the Club Montañeros (*El Guácharo*, Caracas, 1974, 7 (1) profile). Alt. 320 m.

2. **sima 110** .... –110 m
(Barra Honda, Nicoya, Guanacaste)
By the C.S.R. (*Notizario C.S.R.*, 1974 (1-2) profile).

3. **pozo La Trampa** (Barra Honda, Nicoya)..... –106 m
By the C.S.R. (*ibid.*, profile).

LONG CAVES:

1. **caverna del Cabinarraca** ....1500 m
(Venado, Alajuela)
(*ibid.*, plan).

2. **caverna de Damas** (Quepos, Puntarenas) ....390 m
(*ibid*, plan).

3. **sima Santa Ana** (Nicoya, Guanacaste) ....292 m

CUBA

Antonio Núñez Jiménez, pioneer of Cuban speleology (he founded the Sociedad Espeleológica de Cuba starting in 1940), with his numerous publications (*Clasificación genética de las cuevas de Cuba*, 1967; *Cuba: dibujos rupestres*, 1975; *40 años explorando a Cuba*, 1980) has made known to the world the activities of Cuban speleologists and the great varety of tropical karst to be found on the island. Seven types are karst are distinguished,

including the famous cone karst or mogotes in the sierra de los Organos (Pinar del Río), in which the largest Cuban caves are found, and the raised marine terrace karsts, notably found in the Oriente area.

Caves have been visited for a long time in Cuba (Guayabo Blanco Indians in prehistoric times, the Ciboney and Taíno Indians in historic times, etc.) and some are famous for their cave art (cuevas de Punta del Este in the island of Pinos). Systematic cave exploration began in the 19th century, but it was not until 1940 that the S.E.C. began finding and exploring large cave systems.

DEEP CAVES:

1. **cueva Cubamagyar** (Santi Spiritus) .... –396 m ?

2. **cueva Jibara** (Santiago-de-Cuba, Oriente) ..–242 m
or sumidero del Río La Palera. Alt. 580 m. Explored in 1961 by the Expedición Cubanó-Polaca and the Departamento de Espeleológia de la Academia de Ciencias (D.E.A.C.).

3. **pozo Prieto** (Santiago-de-Cuba, Oreinte) ....–225 m
In 1961 by the Expedición Cubanó-Polaca (*W Sentach I Lodach Sviata*, 1971, IV: 333, profile).

4. **furnia de Pipe** .... –165 m
(Jiguaní, Santiago-de-Cuba, Oriente)
Alt. 620 m. Contains 145 m deep pit dropped by the D.E.A.C. in 1966 (*Serie Espel. y Carsol.*, 1973 (51) profile).

5. **cueva del Cura** (Juruco, La Habana) .... –152 m

6. **cueva Martin** .... –150 m
(Topas de Collantes, Guamuhaya)

7. **cueva Rollando** (Camaguey)...................... –112 m

LONG CAVES:

1. **sistema cavernario de los Perdidos** ......26,000 m
(Rancho Mundito, Sierra del Rosario, Pinar del Río)
System explored beginning in 1969 by the G. E. Marcel Loubens.

2. **gran caverna de Santo Tomas** ....25,000 m
(Sierra de Quemadas, Pinar del Río)
Explored from 1954 to 1970.

3. **sistema cavernario Majaguas-Cantera** ..23,500 m
(Sierra de San Carlos, Pinar del Río)
System explored from 1962 by the Grupo Martel de Cuba, with four entrances. Part of the Cuyaguateje system.

4. **cueva Fuentes** ....12,000 m
(Sierra de Mesa, Pinar del Río)
Explored from 1961 by the D.E.A.C.

5. **cueva del Gato Jibaro** ....11,000 m
(Matanzas, Matanzas)
Explored from 1960 by the D.E.A.C. and the G. E. Carlos de la Torre.

6. **sistema cavernario de la Amistad** ....7460 m
(Sierra de Sumidero, Pinar del Río)
Explored from 1961 by the G. Martel de Cuba.

7. **cueva grande de Santa Catalina** ....7000 m
(Carboneras, Matanzas)

8. **caverna de los Majaes** ....4800 m
(terrazas marinas de Siboney, Santiago-de-Cuba, Oriente)

9. **cueva Mejico** (Camaguey) ....4000 m

10. **caverna de Moa** (Holquin) ....4000 m
Explored in 1955 by the G.E.A. Martel de Camaguey.

11. **sistema cuevas el Boqueron** (Las Villas) ...3357 m

12. **cueva grande de Caguenes** ....3236 m
(Cayo Caguanes, Mayajigua, Las Villas)

13. **cueva de Bellamar** ....3225 m
(Bahia de Matanzas, Matanzas)

14. **cueva de los Cuatrocientos Rozas** ....3000 m
(Banas, Holquin)

DOMINICAN REPUBLIC
Republica Dominicana

The Dominican Republic, located next to Haïti, has five major karst regions, including the sierras de Neiba, sierras de Baoruca, Bahia de Samana, and the plateau de Higuey. The mountainous karst terrains are in Cretaceous and Tertiary limestones, and much of this is uninhabited cone karst. The low lying areas where most cave exploration has taken place are Pleistocene reef limestones.

DEEP CAVES:

1. **sumidero del Río Los Bolos** .... –95 m
(Sierra de Neiba)
Alt. 1086 m. Explored in 1986 by the McMaster Univ. C.C.C. and the U.S. Peace Corps (*Canadian Caver*, 19 (1), 1987, plan).

LONG CAVES:

1. **cueva Fun Fun** or **Boca del Infierno** ..... 10,100 m
(Hato Major)
A large resurgence in Ordovician limestone. The first 1 km of passage is regularly visited by local inhabitants. In 1986, 7000 m of stream passage was mapped by Circolo Speleologico e Idrologico Friulano (*Speleologia* 15, 1986, plan). In 1987, the cave was extended by the MUCCC and the Peace Corps (*Canadian Caver*, 19 (1), 1987, plan).

2. **cueva Misterio** (Boca de Chavon) ....2000 m
In Pleistocene limestone. Explored by the Sociedad Dominicana de Espelogia and CSIF (*Speleologia* 15, 1986, plan).

3. **las cuevas de Camili** (Boca Chica) approx. 2000 m
Explored by the CSIF (*Speleologia* 15, 1986).

4. **cueva de las Maravillas**.............................800 m
(Soco, San Pedro de Macoris)
In Pleistocene limestone. Explored by the SDE and the CSIF (*Speleologia* 15, 1986, plan).

5. **los Manantiales** (La Caleta) ....500 m
Almost totally underwater cave in Pleistocene limestone. Explored by Parker and Palmer (*Bull. BCRA*, 21, 1983, plan).

6. **sumidero del Río Los Bolos** ....500 m
(Sierra de Neiba)
See above.

DUTCH ANTILLES
Nederlandse Antillen

This semi-independant group of islands is made up of Aruba, Curaçao, and Bonaire. The most notable karst phenomena are found on Aruba and Curaçao. The best reference, which includes descriptions and maps of all caves listed below, is the work of P. Wagemar Hummelinck, *De Grotten van de Nederlandse Antillen*, 1979.

LONG CAVES:

1. **Lago Colony Cave** (Aruba) ....480 m

2. **Tunnel of Love** (Aruba) ....200 m

3. **cueba di Noordkant** (Curaçao) ....160 m

4. **Huliba Cave** (Aruba) ....150 m

5. **cueba di Quadirikiri** (Aruba) ....150 m

6. **cueba di van Hato** (Curaçao) ....125 m

7. **cueba di Barcadera** (Bonaire) ....120 m

8. **cueba Bosa** (Curaçao) ... .120 m

9. **cueba di Jetchi** (Curaçao) ....110 m

10. **cueba di Ratón** (Curaçao) ....110 m

ECUADOR

The large caves in Ecuador are almost all located in the western part of the Andes Cordillera. Some lava tubes are found in the Galápagos archipelago. In 1969 the well-known tropical karst explorer, the Hungarian Dénes Balázs, told the world about the Jumandi caves. In 1976 the British took an interest in the famous cueva de Los Tayos de Coangos. Much of the recent explorations have been by

French cavers (Société Spéléologique et Préhistorique des Pyrénées-Occidentales, Horde Spéléologique de Néanderthal de Lyon in 1982 and 1984); "Equatoriales" expedition in 1985-1986.

DEEP CAVES:

1. **cueva de Los Tayos de Coangos** .... –201 m
    (Santiago de Mendez, Morona-Santiago)
    Alt. 800 m approx. (*Caving Internat. Magazine*, 1978 (1) profile).

2. **cueva de Shimpiz** .... –88 m
    (Logrono, Morona-Santiago)
    In 1983-1984 by the H.S.N. (*Bull. de la H.S.N.*, 1985 (2) profile).

3. **cueva de Gallardo** .... –81 m
    (Puerto Ayora, Santa Cruz, Galápagos)

4. **Nagembaimi** (Logrono, Morona-Santiago) .... –66 m

5. **Eturco uctu** (Archidona, Napo) .... –64 m
    In 1984 by the S.S.P.P.O.

6. **caverna de Bucanero No. 1** .... –57 m
    (Puerto Ayora, San Salvador, Galápagos)
    In 1978 by the Yugoslavians of the J.K. Crni Galeb (SSPPO, *Ecuador 82*, profile). In basalt.

7. **bahia de Bucanero No. 2** .... – 57 m
    (Puerto Ayora, San Salvador, Galápagos)
    Cave in basalt.

LONG CAVES:

1. **cueva de Los Tayos de Coangos** ....4800 m
    (Santiago de Mendez, Morona-Santiago)
    Known to indians, noted by Prono in 1860, explored by gold seekers and military persons (1969), this cave was completely mapped by the British in 1976 (*B.C.R.A., Bull.*, 1976 (14) plan).

2. **cueva de San Bernardo** ....2460 m
    (Archidona, Napo)
    In 1984 by the SSPPO and HSN.

3. **cueva de Gallardo** ....2331 m
    (Puerto Ayora, Santa Cruz, Galápagos)
    Cave in basalt in four segments 68, 23, 669 and 1315 m long (SSPPO, *Ecuador 82*, plan).

4. **cueva de Shimpiz** ....2305 m
    (Logrono, Morona-Santiago)
    In 1984 by the H.S.N. (*Bull. de la H.S.N.*, 1985 (2) plan).

5. **cueva de Lagarto** (Cotundo, Napo).............2228 m
    In 1983-1984 by the H.S.N. (*Spéléologie-Dossiers*, 1984 (18) plan with 1602 m; sequel in *Bull. H.S.N.*, 1985 (2)).

6. **Eturco uctu** (Archidona, Napo) ....2060 m
    In 1982 and 1984 by the S.S.P.P.O.

7. **Amaron uctu** or **cavernas de Jumandi** .....1900 m
    (Archidona, Napo)
    In 1979 by Southampton Univ. (S.S.P.P.O., *Ecuador 82*, plan).

8. **Huana Ucto** (Archidona, Napo) ....1767 m

9. **cueva de Mariposa Negra** ....1592 m
    (San Francisco, Archidona, Napo)
    In 1983 by the H.S.N. (*Spéléologie-Dossiers*, 1984 (18) plan).

10. **cueva Espin** (Santa Cruz, Galapagos) ....1557 m

11. **cueva de Los Tayos de Chinganaza** ....1250 m
    (Yaupi, Morona-Santiago)

12. **cueva Arias** (Santa Cruz, Galapagos) ....1225 m
    Explored by "Equatoriales" expedition in 1985-1986.

13. **cueva Sonia** (Santa Cruz, Galapagos) ....1011 m
    Explored by "Equatoriales" expedition in 1985-1986.

14. **Agua yacu** (Archidona, Napo) ....1000 m
    Explored by "Equatoriales" expedition in 1985-1986.

15. **cueva de Kubler** ....902 m
    (Puerto Ayora, Santa Cruz, Galápagos)
    Cave in basalt explored in 1970 by D. Balázs (*N.S.S. Bull.*, 1975, 37 (1) plan).

EL SALVADOR

There are no known karst areas in El Salvador, but some lava tubes have been found, such as **cueva Nanarita** (Santa Ana), 179 m long and explored in 1953 by W. H. Grebe (*Die Höhle*, 1956 (4) plan).

GUATEMALA

The karst areas of Guatemala are localized in Alta Verapaz, Quiché, Péten, and Huehuetenango. They are interesting from both the hydrological and archeological points of view. The caves were intensively visited by the Mayans, beginning around 600 B.C. They left many artifacts, unfortunately often now pillaged. Modern exploration began in the 19th century, with the investigations of the German geologist K. Sapper in the Alta Verapaz and Lanquin cave (1894-1899). In 1902, E. Seller uncovered Mayan artifacts in the Quen Santo cave.

After the second world war, interest in the caves of the country increased. Investigations were made by, among others, Americans R. Gurnee (from 1958), D. MacKenzie (1967), Frenchmen R. Vergnes (1956), D. Dreux (from

1968), and M. Siffre (from 1974) and S. Knutson (from 1980). Many caves were explored by cavers from McMaster University (M.U.C.C.C., Canada) and the Expédition Spéléologique Française (E.S.F.) of 1974-1975. More recently, the political situation has discouraged visits, but expeditions led by S. Knutson have returned in 1986, 1987 and 1988, and renewed interest has been shown by other cavers as well.

It is difficult to classify caves by length in a tropical country like Guatemala, where river systems are frequently segmented, often for very short distances. Thus we have counted the Río Candeleria river complex as one entity composed of several sections. The same problem arises in classifying the long caves of Belize.

DEEP CAVES:

1. **Jul Mas Nim** (Alta Verapaz) ....-294 m
    Explored in 1988 by a team led by S. Knutson (*The California* Caver, Vol. 38 N° 4).

2. **el ojo "Grande" de Mal País** ....approx. –240 m
    (Barrillas, Huehuetenango)
    In 1976 by the MUCCC (*Canadian Caver*, 1976, 8 (1) profile).

3. **el ojo "Chiquito" de Mal País** .... –210 m
    (Barrillas, Huehuetenango)
    See above.

4. **Xibalba** .... –187 m
    Part of the Chiquibul Cave Complex (see Belize). Explored and mapped in May, 1986 by an expedition directed by Tom Miller and partially funded by the National Geographic Society. Others in the exploration party included S. Grundy, J. Ganter, J. Rollins, G. Dunkley, G. Veni, and C. Yonge. The main river passage is over 4 km long and very large, up to 100 m wide. The river ends in a sump. The fossil section of the cave was called actun Zactun. It was explored in January and March of 1986, and connected with Xibalba in May of that year (*Canadian Caver* 18(2)).

5. **sótano del Cimarron** (Huehuetenango) ..... –183 m
    Pit 183 m deep explored by the Guatemalans Lopez and Piedra Santa and the MUCCC in 1970 (*Canadian Caver*, 1971 (4) profile).

6. **sistema de Seamay-Sejul** .... –170 m
    (Finca Seamay, Alta Verapraz)
    In 1976 by the MUCCC.

7. **sótano de Chinamá No. 4** .... approx. -165 m
    (Senahú, Alta Verapaz)
    In 1972 by D. Dreux.

LONG CAVES:

1. **sistema del Río Candeleria** ....(21,980 m)
    (Raxruja, Chisec, Alta Verapaz)
    Alt 170 to 160 m. Explored in 1974-1975 by the E.S.F. It is made up of the following segments, from upstream to downstream (*Spelunca*, spécial No. 1, 1976, plan).

    a) **cueva Veronica**, 18 entrances! ....7900 m
    b) **cueva del Venado No. 2** ....1730 m
        800 m from a), 6 entrances
    c) **tributario del Venado** ....2050 m
        between b) and d), 6 entrances
    d) **cueva del Venado No. 1** ....1150 m
        (800 m from b)
    e) **cueva del Mico** ....4300 m
        180 m from d), 12 entrances
    f) **cueva de los Nacimientos** ....3800 m
        800 m from e), 12 entrances
    g) **cueva fosile de los Nacimientos** ....1050 m
        3 entrances

2. **Jul Mas Nim** (Alta Verapaz) ....8021 m
    See above.

3. **Xibalba** ....approx. 7000 m
    See above.

4. **sumidero del río San Antonio** ....3700 m
    (Raxjura, Chisec, Alta Verapaz)
    In 1975 by the E.S.F. (*Spelunca*, spécial No. 1, 1976, plan).

5. **cueva de Agua Escondida** ....3463 m
    (La Democracia, Huehuetenango)
    Explored from 1968 to 1974 by the Potomac S.C., then the MUCCC (*Canadian Caver*, 1974, 6 (2) plan).

6. **sistema de Seamay-Sejul** ....approx. 3000 m
    Explored in 1975 (Gurnee, Storek), then in 1971 and 1976 by the MUCCC (*Canadian Caver*, 1971 (4) plan).

7. **grutas Lanquin** ....2298 m
    Cave mapped by S. Knuston, S. Linn, C. Ream, R. Rigg, R. Spahl, P. Mothes, and P. Hill in February and April, 1987. According to legend, Prof. Sapper traveled upstream 75 km by boat in this cave long ago!

8. **Naj Tunich** ....approx. 2000 m
    Cave explored in 1981.

9. **Jul'ik'** (Chisec, Alta Verapaz) ....1750 m
    In 1972 by D. Dreux.

HAITI
Haïti

Located on the western half of the island of Santo-Domingo, the Republic of Haiti has karst areas throughout most of its surface area, notably in the Hotte range in the south. In fact, "haïti" is the local term for cone karst. Although caves have been long known in the country, and were often used for religious purposes, most of the exploration and documentation of the caves of Haiti comes from the 1980 Expédition Spéléologique Française (E.S.F.), led by Alan Oddou (*Bull. Centre Méditerranéen de Spél.*, 1980, 2: "Contribution à la connaissance spéléologique de la République d'Haïti", 33 p.). Their work was facilitated be the previous geologic studies of J. Butterlin.

DEEP CAVES:

1. **bim Séjourné** (vallée de Jacmel) .... −167 m
    Alt. 710 m (*op. cit.*, profile).

2. **trouin Sene** (massif de la Selle, Seguin) ...... −92 m

3. **trouin Seguin** (massif de la Selle, Seguin) ... −72 m
    (*op. cit.*, profile).

4. **trou Maïo 1** or **Nan Cadet** .... −65 m
    (massif de la Selle, Seguin)
    (*op. cit.*, profile).

5. **trouin de la Scierie** .... −63 m
    (massif de la Selle, Seguin)
    (*op. cit.*, profile).

LONG CAVES:

1. **trouin Sene** (massif de la Selle, Seguin) ..... 1668 m
    (*op. cit.*, plan).

2. **grotte de Port à Piment** ....approx. 1000 m
    (massif du Macaya, Port à Piment)

3. **grotte de Conoubois** ....approx. 900 m
    (Camp Perrin)

HONDURAS

The karst areas of Honduras (see for example the map published in *Grottes & Gouffres*, 1974, No. 53) have until recently been seldom visited. Expeditions led by R. Finch beginning in 1974 have explored caves in the Santa Barbara area and scouted much of the rest of the country. Expeditions led by S. Knutson beginning in 1980 explored caves on Mt. Santa Barbara and at Atima. Good potential exists for finding more caves in this country.

DEEP CAVES:

1. **sumidero de Maigual** (Santa Barbara) ...... −420 m
    Explored in 1980 by S. Knutson , S. Burkig, L. Wilt, M. Cavanaugh and E. Garza. (*Caving Internat. Magazine*, 1981 (10) profile). This may be the deepest explored cave in the Western hemisphere south of Mexico.

2. **sumidero del Río San Jose de Atima** ...... −190 m
    (Santa Barbara)

LONG CAVES:

1. **cueva del Río Talgua** (Guanaja, Olancho)..3090 m
    Surveyed by L. Cohen *et. al.* in 1984 (*N.S.S. News*, 1986 (5) plan).

2. **sumidero del Río San Jose de Atima** ....2450 m
    (Santa Barbara)
    Explored in 1981 and 1985-'87 by S. Knutson *et. al.*

(*N.S.S. News*, 1986 (5) map with 863 m). Difficult river cave requiring the use of over 1 km of rope in 53 riggings to make the traverse from the entrance to resurgence.

3. **cueva Masical** .... 1330 m
    (San Nicolas, Santa Barbara)
    Explored in 1981 and mapped by R. Finch *et al.* (*Speleonews*, 1982, XXVI (2) plan).

JAMAICA

Over two thirds of the exposed rocks of Jamaica consists of limestones of Cenozoic age, and numerous caves are found throughout the island. In the high country one finds the famous cockpit karst, a classic form of tropical karst.

Cave exploration has been going on ever since the island was colonized, first by the Spanish, then by the British. Archaeological studies have shown that caves were used in some areas by the first known natives, the American Indian "Arawaks", as burial sites, and possibly for habitation or refuge. In the time of slavery, the caves were often used by fugitive slaves as hiding places.

The scientific exploration of caves in Jamaica dates from the early 1800's with the first description of the geology of the island. During the years of the second world war, a systematic survey of caves was undertaken in order to locate bat guano deposits which might be used to supplement the dwindling supplies of agricultural fertilizers.

The Jamaican Caving Club (J.C.C.) was formed in 1958 and members continue to explore and record the caves of the island. Apart from the work of the J.C.C., a major contribution to the speleological knowledge of the country has been made, over the years, by visiting parties of British and American explorers.

At the present time, about 1000 cave sites have been explored, and the number increases slowly year by year. It appears that most of the easily accessible caves have been found and explored, but good potential for new caves exists in the remote mountain areas.

Alain G. Fincham.

DEEP CAVES:

1. **Dunn's Hole** (Stewart Town, St. Ann) .... −229 m
    Alt. 299 m. In 1957 by the NSS (A. Fincham, Jamaica Underground, 1977, profile). Datum point ambiguous.

2. **Morgan's Pond Hole** .... −186 m
    (Cobblers, Manchester)
    Alt. 762 m. Explored in 1975 by the British Speleological Expedition (*B.C.R.A. Bull.*, 1975 (10) profile).

3. **Thatchfield Great Cave** .... −177 m
    (Browns Town, St. Ann)
    Alt. 442 and 414 m. In 1977 by the Jamaican Caving Club (Fincham, *op. cit.*, profile).

4. **Volcano Hole** (Norwood, St. Ann) .... −158 m
    Alt. 670 m. In 1966 by the English Karst Hydrological Expedition (Fincham, *op. cit.*, profile).

5. **New Hall Cave** .... –151 m
(Georges Valley, Manchester)
Alt. 595 m. In 1975 by the J.C.C. (Fincham, *op. cit.*,
profile).

6. **Asuno Hole** (Grant Bailey, St. Ann) .... –137 m
Alt. 640 m. In 1966 by the E.K.H.E. (Fincham, *op. cit.*,
profile). Contains 116 m deep pit.

7. **Sploosh Pot** (Quashies River, Trelawny) .... –125 m
Alt. 427 m. In 1966 by the E.K.H.E.

8. **Wright's Hole** (Cobblers, Manchester) .... –119 m
Alt. 731 m. In 1975 by the B.S.E.

9. **Everlasting Hole** (Allison, Manchester) ..... –116 m
Alt. 670 m. In 1975 by the B.S.E.

10. **Quashies River Cave** .... –111 m
(Sink Quashies River, Trelawny)
Alt. 405 m. In 1966 by the E.K.H.E. (Fincham, *op. cit.*,
profile).

LONG CAVES:

1. **Still Waters Cave** ....3597 m
(Tyne River Sink, St. Elizabeth)
Explored in 1977 by the Liverpool University P. C.
(*Trans. of B.C.R.A.*, 1980, 7 (3) plan).

2. **Gourie Cave** (Dump, Manchester) ....3505 m
Explored from 1972 to 1977 by the J.C.C. (Fincham,
op. cit., plan).

3. **Jackson's Bay Cave** ....3354 m
(Jackson's Bay, Clarendon)
Alt. 15 m. Has 8 entrances. Explored in 1964 by the
J.C.C. (Fincham, *op. cit.*, plan).

4. **Printed Circuit Cave** ....3220 m
(Rock Spring, Trelawny)
In 1966 by the E.K.H.E. (Fincham, *op. cit.*, plan). Has
ten entrances.

5. **Mouth Maze Cave** ....3190 m
(Mouth River, Trelawny)
Has five entrances. Explored in 1966 by the E.K.H.E.
(Fincham, *op. cit.*, plan).

6. **Windsor Cave** (Windsor, Trelawny) ....2980 m
(Fincham, *op. cit.*, plan).

7. **Saint-Clair-Lemon Ridge Cave** ....2896 m
(Dolly Ground, St. Catherine)
In 1954 by the Geol. Survey Dept. (Fincham, *op. cit.*,
plan).

8. **Coffee River Cave** ....2802 m
(Auchtembeddie, Manchester)
In 1958 by the J.C.C. (Fincham, *op. cit.*, plan).

9. **Rock Spring Caverns** ....2590 m
(Pear Tree Grove, St. Mary)

In 1963 by the Univ. Leeds Hydrol. Survey Exped.
(Fincham, *op. cit.*, plan).

10. **Noisy Water Cave** (Norwood, St. Ann) ....2475 m
In 1966 by the E.K.H.E. (Fincham, *op. cit.*, plan).

11. **Riverhead Cave** ....2440 m
(Black River, St. Catherine)
In 1976 by the J.C.C. (Fincham, *op. cit.*, plan).

12. **Bristol Cave** (Quashies River, Trelawny) ....2260 m
In 1966 by the E.K.H.E. (Fincham, *op. cit.*, plan).

MEXICO

The karst of Mexico has lived up to the great hopes placed in it in the 1960's: it presently has two very long and deep systems, one of which became the third deepest in the world in April, 1987. Limestone areas are found throughout Mexico, alternating with volcanic and metamorphic rocks, and flat plains. From the Sierra Madre Oriental, which begins in the north in Tamaulipas to the southern karsts of Yucatan or Chiapas, and from sea level to the high mountains, one finds major caves in nine of the 31 states which make up this large country: Oaxaca, Guerrero, Querétaro, San Luis Potosí, Tamaulipas, Puebla, Chiapas, Santa Cruz, and the Yucatan. Known for its deep pits, such as the sótano de las Golondrinas bottomed in 1967, Mexico is also a country with large volume caves, such as the gruta de Cacahuamilpa (Guerrero) and the gruta de Palmito (Nuevo Leon). The latter was the subject of speleological investigations in the 19th century (from 1835 to 1875). The history of cave exploration goes back much further, to the Mayans, who penetrated deep into many caves hundreds of years in the past and often left artifacts. These were the subject of archeological investigation by the American H. C. Mercer in 1895, who documented his results in *The Hill Caves of Yucatan*, 1896.

In the 1930's, Mexican cavers began cave exploration in the modern sense; in 1937 they made the first traverse of the underground river of Chontalcoatlán and in 1940 made that of the San Jéronimo. Both these systems are in Guerrero and resurge at the famous Dos Bocas.

Biologists became active in the 1950's. C. Bolivar and especially Federico Bonet published information on caves in the Xilitla area (San Luis Potosí) and in Guerrero in 1953 and 1971.

The 1960's marked a turning point in Mexican speleology. The Grupo Espeleológico Mexicano was formed around 1960 under the direction of Jorge de Urquijo Tovar, and in 1962 cavers from Texas formed the Association for Mexican Cave Studies (A.M.C.S.), which began a historic series of explorations. They reconnoitered the Sierra Madre Oriental from north to south, then the states of Oaxaca, Vera Cruz, Guerrero, Puebla, Yucatan, sometimes helped by Canadian cavers from McMaster University. The first generation (including Terry Raines, William Russell, James Reddell, John Fish, etc.) spawned a second even hardier generation, including Bill Stone, Bill Steele,

Jim Smith, and Peter Sprouse. To the ever increasing activities of the A.M.C.S. were added the efforts of the French (beginning in 1972), the Australians (beginning in 1978) and the Belgians (1980's). The Mexican cavers have also become more active, forming the Sociedad Mexicana de Exploraciones Subterráneas, which under the direction of Carlos Lazcano has been making many new discoveries, especially in more remote areas which are difficult to get to for foreigners.

In spite of the steady increase in exploration activity, the caving potential of Mexico is still enormous. It is a country with a great variety of karst forms and cave types - there is something to satisfy every caver's dreams.

The A.M.C.S. publishes the *A.M.C.S. Newsletter* and the *A.M.C.S. Activities Newsletter*, as well as Bulletins on selected areas or topics. These publications summarize the activities of the S.M.E.S., which now publishes its own bulletin. The A.M.C.S. keeps an up-to-date library on all caves and caving activities in Mexico.

DEEP CAVES:

1. *sistema Huautla* .... –1353 m
(Huautla de Jimenez, Oaxaca)

An impressive limestone range stretches 80 km to the south of Orizaba, in the Sierra Madre Oriental. At the south end of this limestone, 5 km to the east of Huautla de Jimenez, is a region of swallows draining the plateau at altitudes between 1600 m and 1800 m. Many of the caves go deep, and among them are four that that have been connected to form the sistema Huautla: sótano de San Agustín, Li nita, nita Nanta, and La Grieta. The entrances are near the villages of San Agustín and San Andres Hidalgo.

The plateau under which the system formed is 1500 m above the río San Domingo and the dammed lake of Miguel-Aleman. The presumed resurgence of the system is the cueva de la Peña Colorada, near Mazatlan.

The first reconnaissance trips to the Huautla plateau were made in 1965 by the Texan William Russell. Starting in 1966 the Association for Mexican Cave Studies (A.M.C.S.) based in Austin began exploration of the sótano de San Agustín (alt. 1750 m approx.), stopping at –280 m. In 1967 and 1968, with the help of Canadians from the McMaster University Caving and Climbing Club (M.U.C.C.C.), the depths of –494 m and –612 m were reached. They stopped at a sump and reached a length of 1860 m. A side passage at –536 m (Route 68) was explored for 500 m to a breakdown pile.

Incidents in which ropes were cut or stolen and other hostile acts by the local Mazatec Indians forced interruption of exploration until 1976.

Using Route 68, the depth of –760 m was reached on Christmas 1976. In the spring of the following year, two sumps were found at –806 and –859 m, bringing the length to 5900 m.

A nearby cave named Li nita was discovered at an altitude of approx. 2110 m on December 29, 1979, and explored to a depth of –162 m. The same expedition brought the length of San Agustín to 12,364 m. The connection with Li nita was made on May 9, 1980, bringing the depth to –1121 m and the length to 21,300 m. Pushing brought the length in 1981 to 24,025 m and the terminal sump was dove for a depth of 25 m without re-emerging into airspace. The mapping of a side passage in 1985 increased the vertical extent of the system by 6 m.

Cueva Grieta (alt. 1860 m) was located in 1965, and explored to –95 m in 1968, then to –401 m in 1976. In May, 1977, a camp was established at the –520 m level and the depth of –665 m was reached (by diving the 15 m long Pato Mojado sump at the –520 m level). By the end of 1977, the depth of –734 m was reached at an inpenetrable swallow, while the length reached 8782 m. Much effort was expended in searching for a connection with San Agustin, but success was not achieved until April 24, 1985, bringing the length of the system to 33,789 m.

Nita Nanta (alt. 2237 m) is the highest cave on the Huautla plateau and has over ten entrances, including nita Sa, nita Zan, nita Mazateca, and nita Lajoa. Exploration by the A.M.C.S. (led by Bill Stone, Bill Steele, Steve Zeeman, Mark Minton, and others) began in 1980, when the depth of –445 m was reached. In 1981, the depth was increased to –927 m. In 1982 and 1983 many of the entrances were connected together and the cave became the second on the American continent to pass the –1000 m depth mark, with –1026 m. Attempts to connect with sistema Huautla were unsuccessful in late 1983, but the length increased to 10,769 m, then 11,655 m. The connection was finally made on March 26, 1987 when Jim Smith dived through a 10 m long, 1.5 m deep sump. This brought the depth of the system to –1353 m, the third deepest in the world. The length of the system was increased to 52,111 m with the discovery of a 3 km trunk passage leading from the Sala Grande (near the entrance of San Agustín), and two new entrances, Bernardo's cave, and nita Ina, which connected together at –330 m and connected to nita Nanta at –660 m. This brought the total number of entrances to the system to seventeen.

It should be noted that sistema Huautla offers one of the most interesting through trips in the world. It is possible to enter nita Nanta, drop 1240 m to dive a sump, then ascend 1100 m to the Li nita entrance, all without ever retracing one's steps.

In parallel, Bill Stone led the exploration of the resurgence, cueva de la Peña Colorada, beginning in 1981. They dived the first 524 m long sump, only to later find a bypass using another entrance. An expedition in 1984 dived eleven more sumps, using two camps, and reached a length of 7793 m, of which 1350 m are submerged. In spite of this tremendous progress, a substantial distance remains for a connection with the system. The present terminus is sump No. 7, which is at the bottom of a 54 m deep pit. It was dived for 165 m to a depth of 55 m. A return expedition which will certainly push the technological limits of caving to its limits is being planned.

Map: from computer plots sent by Mark Minton.
Bibliography: *The Canadian Caver*, 1970 (3); *A.M.C.S. Newsletter*, 1966 (2); *A.M.C.S. Activities Newsletter*, 1978 (2); 1979 (10); 1980 (11); 1982 (12); 1983 (13); 1984 (14); 1985 (15); 1987 (16); *N.S.S. News*, 1978, 36 (4); 1983 (6); 1987 (12).

2. **Akemati** (Puebla, Mexico) .... approx. –1130 m
Explored by GSAB in early 1988. Located in the same region as sistema Ocotempa.

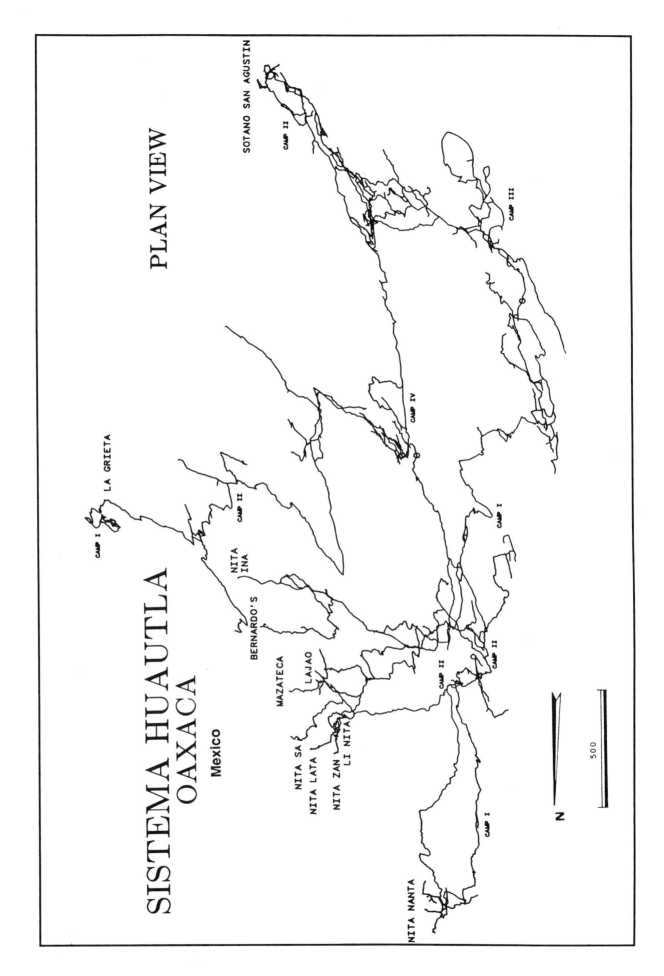

SISTEMA HUAUTLA
OAXACA

Mexico

PLAN VIEW

LA GRIETA

CAMP I

NITA
INA

CAMP II

BERNARDO'S

MAZATECA

LAJAO

NITA SA

NITA LATA

NITA ZAN

LI NITA

CAMP II

CAMP II

NITA NANTA

CAMP I

SOTANO SAN AGUSTIN

CAMP II

CAMP III

CAMP IV

CAMP I

N

500

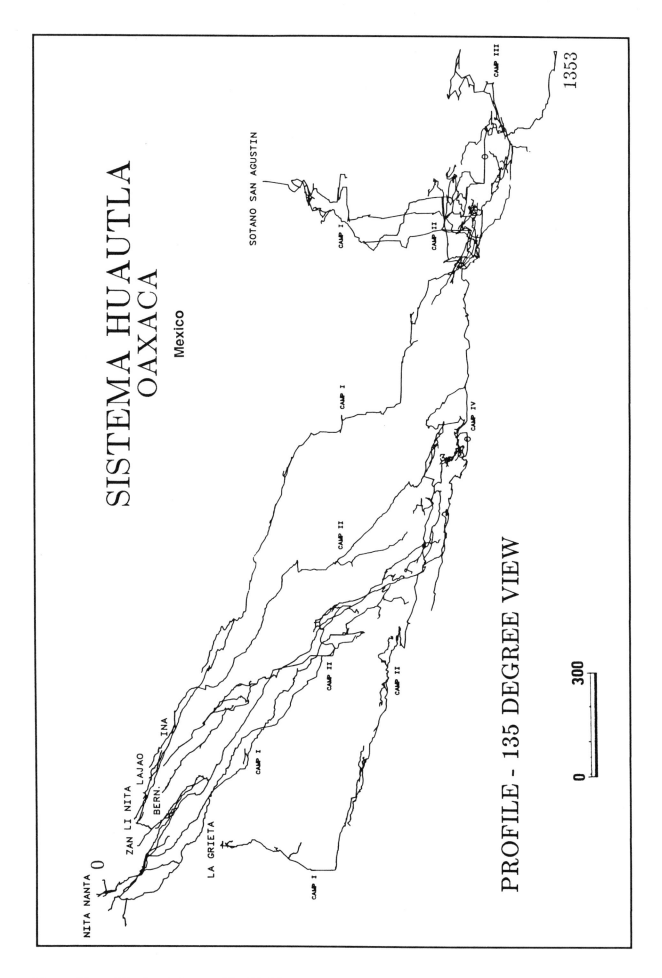

SISTEMA HUAUTLA
OAXACA

Mexico

PROFILE - 135 DEGREE VIEW

0   300

3.   sistema Ocotempa ....approx. –1063m
(Ocotempa, Puebla)

This cave is located at an altitude of approx. 1700 m near the village of Ocotempa, which is about 20 km SE of Tehuacán. It was explored in 1985 by a Belgian expedition organized by the G.S.A.B. They reported the first drop to be 380 m deep (or a 340 m free drop from a natural bridge at –40 m, bypassing a ledge at –306 m). A group of sixteen cavers from Texas, Tennessee, Quebec, and Mexico City returned on January 1, 1987 to do the long free rappell, only to find that pozo Verde actually consists of two pits, the first 221 m deep, the second pit dropping 82 m from the –228 m level to the –306 m level. Their survey was done with a highly accurate laser rangefinder. The same group, which included Terry Raines and Marion O'Smith, also discovered a way on by following a strong breeze through an obscure crawlway. This led almost immediately to the third drop.

Part of the same group of cavers returned in February and pushed down a series of drops to –647 m, where they stopped for lack of rope at the top of yet another pit. At this point they had only traveled 48 m horizontally from the entrance, making it one of the most vertical caves in Mexico. The passage was clean and pleasant and the amount of water relatively small except when it rained.

A return trip by mostly the same cavers was made in May, 1987. In a series of three trips they pushed the survey down to –773 m, and explored to about –850 m.

Terry Raines led a return trip of American, Mexican and Belgian cavers in December, 1987, and a sump was reached at –1041 m. In early 1988 Belgian cavers linked a slightly higher entrance to the cave at around –800 m.

Map: sent by Mark Minton, showing part of cave mapped by the Americans.

Bibliography: Spéléo-Flash No. 148, 1985; A.M.C.S. Activities Newsletter 1987 (16).

4.   sistema Cuicateca .... –1230 m
(Cuicatlan, Oaxaca)

The large main entrance has long been known to local Indians. Systematic exploration began in December, 1986 when Bill Farr and Carol Vesely (from California) explored about 1 km for a depth of –100 m in the large rooms and passages close to the entrance. They returned in March, 1987 to descend to –194 m for a length of 1460 m. Five U.S.A. cavers joined them for a 3-week trip in January, 1988. The Giant's Staircase and 150 m deep Saknusum's Well were discovered, leading back to the main stream passage, for a depth of –736 m and a length of 4100 m. Farr and Vesely led an expedition "Proyecto Papalo '88" of about 15 U.S.A. cavers in March, 1988 during which the main stream was followed downwards for a distance of several kilometers to a breakdown choke. A higher entrance was connected to the system, bringing the number of entrances to five. At the end of the expedition the depth was –1038 m and the length 9180 m.

Map: sent by Bill Farr.

5.   Guixani Ndia Guinjao .... –940 m
(Zongolica-Chilchotla, Oaxaca)

The cave is located near the village of Zongolica (near Maria Luisa, in the NW corner of the Sierra Mazateca) in the same massif as the sistema Huautla. The name means "You're About to Get Married Cave" in the local language! The cave was explored in November and December, 1985 by a group of cavers from Australia (the Chilchotla '85 expedition). Gentle streamways in the upper section of the cave gradually give way to a pitch series and a sand choke at –940 m. A climbing lead was not pushed due to the unpleasant nature of the cave at the bottom. Possibilities exist for connections with other nearby caves, such at nita Xonga. Total length 1950 m.

Map: A.M.C.S. Activities Newsletter, 1987 (16).

Bibliography: N.S.S. News, 1986 (10); A.M.C.S. Activities Newsletter, 1987 (16).

6.   sistema Purificación .... –895 m
(Villa Hidalgo, Tamaulipas)

This system was formed by the connection of cueva del Brinco (alt. 1900 m) and cueva del Infiernillo (alt. 1098 m), both located in the limestone massifs of the Sierra Madre Oriental, about 50 km west of Ciudad Victoria in the municipio de Villa Hidalgo. Access takes a 5-7 hour drive on logging roads using four-wheel drive vehicles to reach Conrado Castillo, where the Brinco entrance is located. Other entrances to the system include la cueva del Vapor (connected November 25, 1978), the cueva de los Franceses (alt. 1950 m approx.), connected by digging on April 13, 1979, the cueva del Oso, and the six entrances of the sumidero de Oyamel, connected on April 11, 1980.

The discovery of the sistema Purificación was due to the intensive reconnaissance trips of cavers in the 1970's, among whom Charles Fromen was particularly active. Subsequent exploration and mapping of the system has been led by Peter Sprouse and Terri Treacy of the A.M.C.S.

The cueva del Brinco, previously known and used by locals, was explored to –105 m in 1973 (length 1500 m), –115 m in 1976, and –197 m in 1977 (for a length of 5,200 m). In April 1976, the Infiernillo entrance was located by Charles Fromen. Opening in the side of a cliff, it requires a climb of 35 m to get inside. The cave acts as an overflow, the permanent resurgences being located lower down in the the Purificación valley. The cave was pushed to a length of 1158 m in 1976, and 4100 m in 1977, for a vertical extent of 216 m.

Connection fever ran high as cavers followed the air currents upward in Infernillo. After intensive effort, the connection was made on July 14, 1978, bringing the vertical extent to 836 m (–802, +34) and the length to 20,086 m. Over a hundred leads had been passed up during the connection effort, which were pushed in the following years to give a length of 27,962 m in 1979, 36,795 m in 1980, 45,468 m in 1981, 51,170 m in 1983, 55,078 m in 1985, and 67,599 m in 1987. The spectacular increase in length has being made possible by numerous 7 to 10 day long camps established in various parts of the system and an influx of new cavers. The vertical extent was increased to 895 m with the exploration of upper level passages.

The potential for increasing the length remains high as the cave is extended ever further under the mountain towards potential connections with caves such as cueva del Tecolote (over 11 km long) or cueva de la Llorona (over 3 km long). The sistema has a feature that is rare in the world: its 895 m of vertical extent can be traversed without climbing up any pits, and requires only a few rappels down

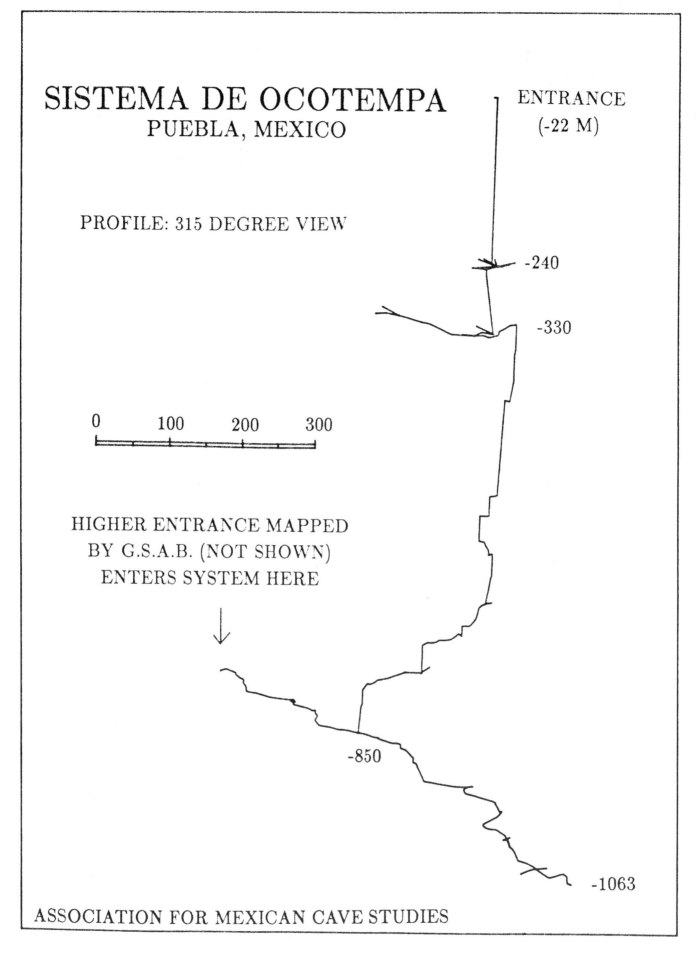

SISTEMA DE OCOTEMPA
PUEBLA, MEXICO

PROFILE: 315 DEGREE VIEW

ENTRANCE
(-22 M)

-240

-330

0   100   200   300

HIGHER ENTRANCE MAPPED
BY G.S.A.B. (NOT SHOWN)
ENTERS SYSTEM HERE

-850

-1063

ASSOCIATION FOR MEXICAN CAVE STUDIES

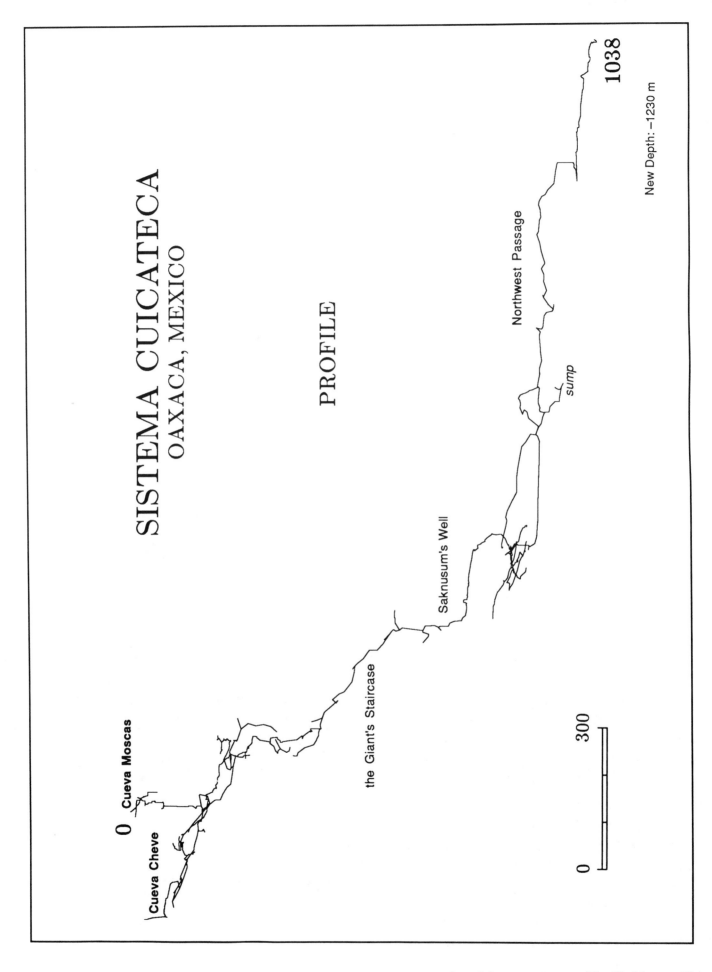

SISTEMA CUICATECA
OAXACA, MEXICO

PROFILE

0 Cueva Moscas

Cueva Cheve

the Giant's Staircase

Saknusum's Well

Northwest Passage

sump

1038

New Depth: –1230 m

0   300

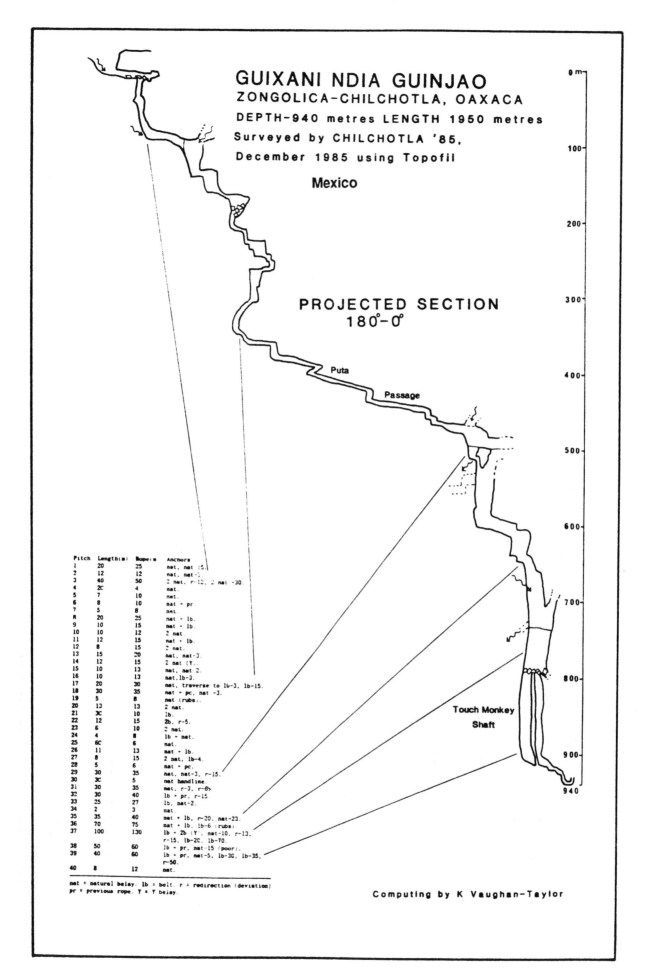

GUIXANI NDIA GUINJAO

ZONGOLICA-CHILCHOTLA, OAXACA

DEPTH-940 metres LENGTH 1950 metres

Surveyed by CHILCHOTLA '85,

December 1985 using Topofil

Mexico

PROJECTED SECTION
180°-0°

Puta

Passage

Touch Monkey
Shaft

Pitch	Length(m)	Rope(m)	Anchors
1	20	25	nat, nat :5.
2	12	12	nat, nat-5.
3	40	50	2 nat, r-12, 2 nat -30.
4	2C	4	nat.
5	7	10	nat.
6	8	10	nat + pr.
7	5	8	nat
8	20	25	nat + lb.
9	10	15	nat + lb.
10	10	12	2 nat
11	12	15	nat + lb.
12	8	15	2 nat.
13	15	20	nat, nat-3.
14	12	15	2 nat (Y.
15	10	13	nat, nat 2.
16	10	13	nat, lb-3.
17	20	30	nat, traverse to lb-3, lb-15.
18	30	35	nat + pc, nat -3.
19	5	8	nat (rubs).
20	13	13	2 nat.
21	3C	10	lb.
22	12	15	2b, r-5.
23	6	10	2 nat.
24	4	8	lb + nat.
25	6C	6	nat.
26	11	13	nat + lb.
27	8	15	2 nat, lb-4.
28	5	6	nat + pc.
29	30	35	nat, nat-3, r-15.
30	3C	5	nat handline.
31	30	35	nat, r-3, r-6.
32	30	40	lb + pr, r-15
33	25	27	lb, nat-2.
34	2	3	nat.
35	35	40	nat + lb, r-20, nat-23.
36	70	75	nat + lb, lb-6 (rubs).
37	100	130	lb + 2b (Y), nat-10, r-13, r-15, lb-2C, lb-70.
38	50	60	lb + pr, nat-15 (poor).
39	40	60	lb + pr, nat-5, lb-30, lb-35, r-50.
40	8	12	nat.

nat = natural belay. lb = bolt. r = redirection (deviation)
pr = previous rope. Y = Y belay.

Computing by K Vaughan-Taylor

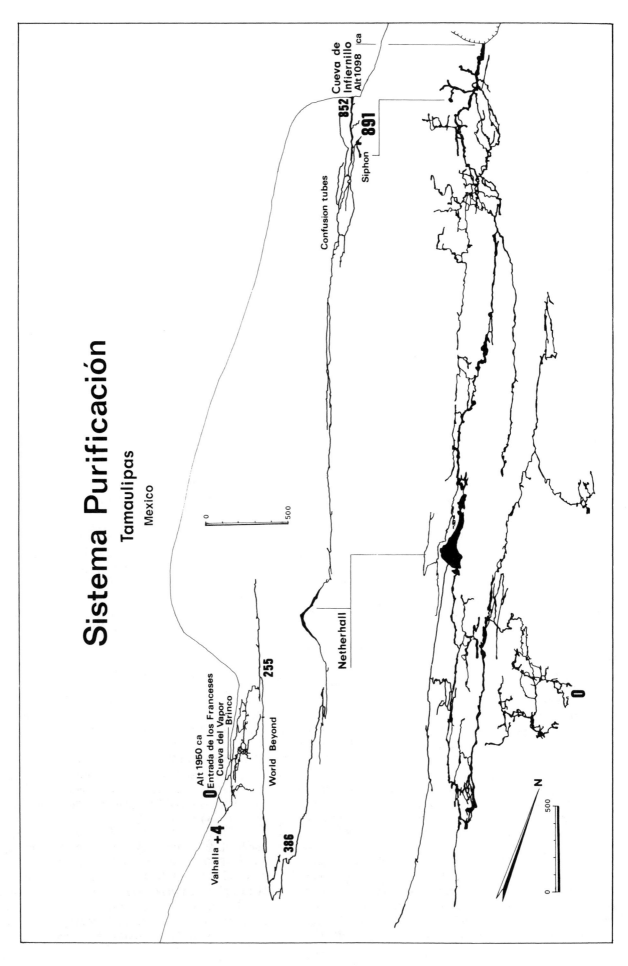

Sistema Purificación

Tamaulipas

Mexico

Valhalla **+4**

Alt 1950 ca
0 Entrada de los Franceses
Cueva del Vapor
Brinco

255

World Beyond

386

Confusion tubes

Cueva de
Infiernillo
Alt 1098

852

891
Siphon
ca

Netherhall

0

N

500

0

0

500

pits. This makes it one of the finest traverses in the world.

Map: from the work of Peter Sprouse (*Proyecto Espeleológico Purificación*), sent by P. Sprouse.

Bibliography: *A.M.C.S. Activities Newsletter*, 1978 (8); 1979 (10; 11), 1982 (12); 1983 (13); 1985 (15); 1987 (16).

7. *sótano de Agua de Carrizo* .... –836 m
(Huautla de Jimenez)

This is another cave on the Huautla plateau. It is located near San Andres Hidalgo, at an altitude of about 1960 m, 213 m higher than San Agustín and 120 m higher than La Grieta. It is part of the same hydrological system as sistema Huautla.

The entrance was located in 1977 by cavers from the A.M.C.S. including Bill Stone and Bill Steele. Exploration the following January was spearheaded by A.M.C.S. cavers, and a depth of –778 m was reached. A sump at –836 m was reached in May and June of the same year, stopping progress.

Map: from the maps of the A.M.C.S., sent by Bill Stone.

Bibliography: *A.M.C.S. Activities Newsletter*, 1978 (8); 1979 (9).

8. *sótano de Trinidad* .... –834 m
(Xilitla, San Luis Potosi)

The Xilitla area was one of the first Mexican karst areas to be investigated. It was visited in the 1940's and 1950's by the biologist Federico Bonet, then by the A.M.C.S. in the 1960's.

Sótano de Trinidad was located on December 30, 1977 by a group of Canadian cavers from McMaster University. The approach is via a trail from El Balcon, to the west of Xilitla (alt. approx. 1150 m). The trail goes north to arrive at Ejido Trinidad, which is a polje. The opening is an altitude of 1880 m, to the west of a depression which is 4 km to the east of Ejido Trinidad.

Exploration began December 31, 1977, and continued in January and March 1978. The Canadians were stopped by a sump at –559 m, a figure which was revised to –583 m in 1981. In that year exploration of new passages allowed them to reach the depth of –827 m. In December, 1982, A.M.C.S. cavers explored the cave and added 7 m to the vertical extent.

Map: from the M.U.C.C.C. (1981), sent by P. Sprouse and C. Lazcano.

Bibliography: Fish (J.) - The Xilitla Plateau and sótano de Trinidad, *The Canadian Caver*, 1978, 10 (2): 39-53 and 1979, 11 (1): 3-23. *A.M.C.S. Activities Newsletter*, 1983 (13): 11.

9. *Sonyance* ....–745 m
(Zongolica-Chilchotla, Oaxaca)

Located 200 m south of Guixani Ndia Guinjao (see above). The cave was explored in December, 1985 by the Australian "Chilchotla '85" expedition. Name means "Place of the Yance Tree". This is a clean cave which collects many streams. Explored in two weeks, interrupted by a flood. The water can be followed the entire distance except for one bypass through breakdown. The breeze was lost 50 m before the bottom. The length of the cave is 1785 m.

Map: from *A.M.C.S. Activities Newsletter* 1987 (16).

Bibliography: *N.S.S. News* 1985 (10); *A.M.C.S. Activities Newsletter* 1987 (16).

10. *nita Xonga* .... –740 m
(Zongolica-Chilchotla, Oaxaca)

Located 600 m of Guixani Ndia Guinjao (see above). The cave was located by Australian cavers in April 1985 at the end of an expedition to the area. On the last day of their trip they dropped a 55 m pitch only to be stopped by a deeper one at –430 m. The return "Chilchotla '85" expedition bottomed the cave within a week of their arrival in November, 1985. The pitch at –430 m was measured to be 310 m deep and took three days to rig and survey. They found a giant chamber floored with boulders at the bottom. The only lead was a too-tight blowing slot. A parallel shaft system was found to re-connect with the main one. The name means Little Stream Cave in the local language. The length is 1550 m.

Map: *A.M.C.S. Activities Newsletter* 1987 (16).

Bibliography: *N.S.S. News* 1985 (10); *A.M.C.S. Activities Newsletter* 1987 (16).

11. *Aztotempa* (Ajalpa, Puebla) .... –700 m
This cave in the municipio de Ajalpa is at an altitude of 1400 m, an hour's walk from the hamlet of Huitzmalloc, not far from the village of Alcomongua.

The entrance, known to residents of Huitzmalloc, was shown to a member of the Belgian expedition of March-April, 1985, organized by the Groupe Spéléo-Alpin Belge. The exploration was immediately undertaken and the depth of –700 m reached.

This cave has a complex development including shafts mixed with horizontal passages. An area of squeezes at –400 m leads to a series of wet pits, then the terminal sump at –700 m. The cave contains a large room 250 m long by 150 m wide. The length is about 4 km.

The presence of an air current and many un-pushed leads indicates a good potential for future increases to the length and depth of the cave. The presumed resurgence is Coyalatl.

Map: from the map in *Spéléo-Flash*, 1985 (148).

Bibliography: *Spéléo-Flash*, Bruxelles, December, 1985, No. 148, pp. 47-56.

12. *sótano de los Planos* .... –694 m
(Tlacotepec de Diaz, Puebla)

Explored by Société Québécoise de Spéléologie (S.Q.S) cavers in January, 1988. It has a 300 m entrance pit.

Map: from the S.Q.S., sent by Steve Worthington.

13. *sótano de Tilaco* .... –649 m
(Landa de Matamoros, Querétaro)

The entrance (alt. 1050 m) is located in a large 5 km long polje, about 500 m from the village of Tilaco, in the NE section of Querétaro.

It was located in 1975 by Sergio Zambrano of the Asociación Alpina de México (A.A.M.) who explored it to about –250 m. At Easter 1977 the A.A.M. returned and descended an additional 100 m. In November, 1978, the Mexican cavers were stopped for lack of rope at about –500 m. On May 19, 1980, the A.A.M. and the Grupo Expedicionario Xaman-Ek reached a sump at the bottom of the cave at –649 m. The survey was made in December, 1980,

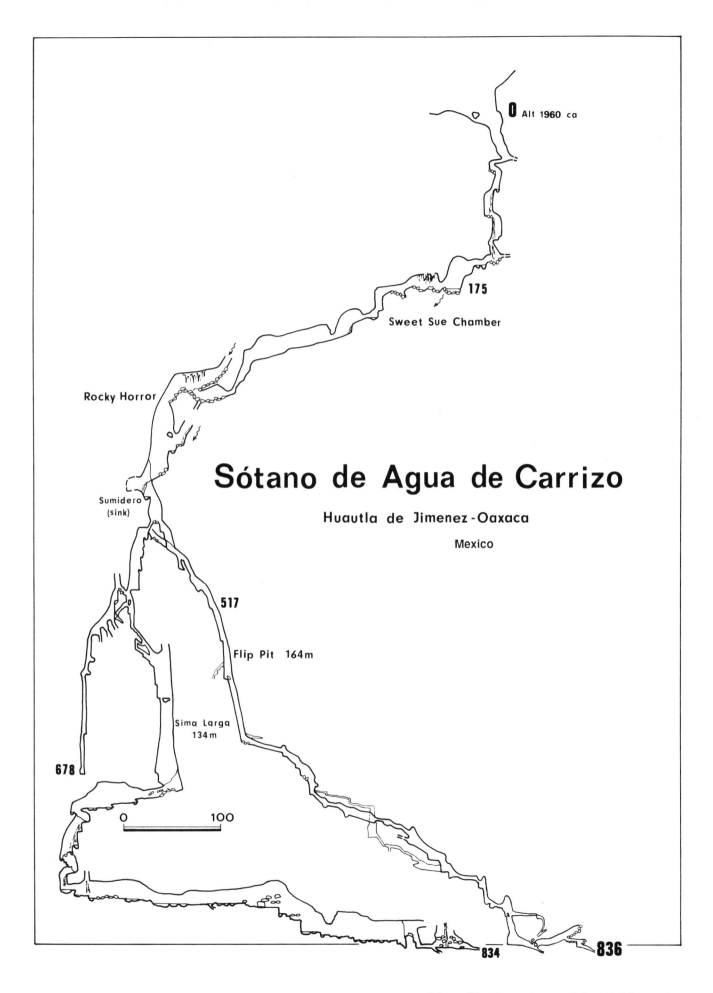

0 Alt 1960 ca

175

Sweet Sue Chamber

Rocky Horror

Sótano de Agua de Carrizo

Huautla de Jimenez-Oaxaca

Mexico

Sumidero
(sink)

517

Flip Pit 164m

Sima Larga
134m

678

0   100

834   **836**

Sótano de Trinidad

Xilitla-SLP

Mexico

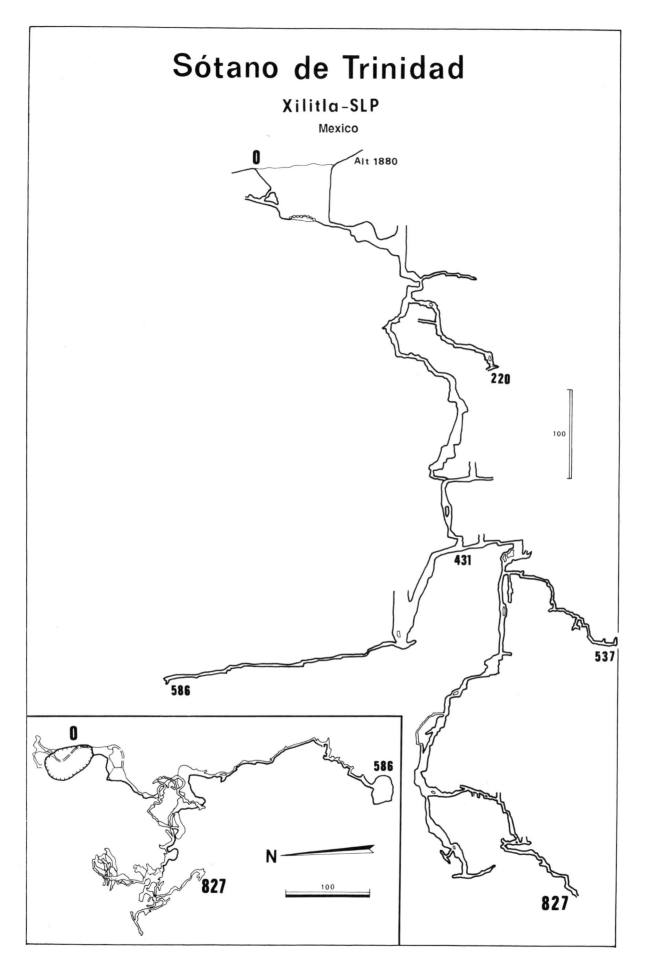

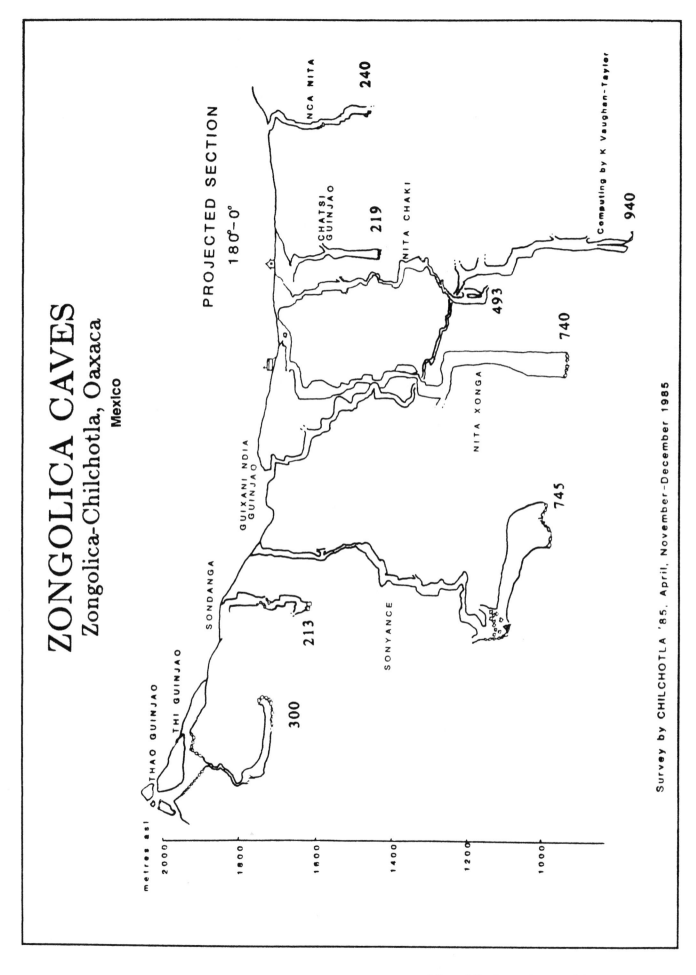

ZONGOLICA CAVES
Zongolica-Chilchotla, Oaxaca
Mexico

PROJECTED SECTION
180°-0°

THAO GUINJAO
THI GUINJAO
SONDANGA
300
213
SONYANCE
GUIXANI NDIA GUINJAO
745
NITA XONGA
740
NITA CHAKI
493
CHATSI GUINJAO
219
NCA NITA
240
Computing by K Vaughan-Taylor
940

metres asl
2000
1800
1600
1400
1200
1000

Survey by CHILCHOTLA '85, April, November-December 1985

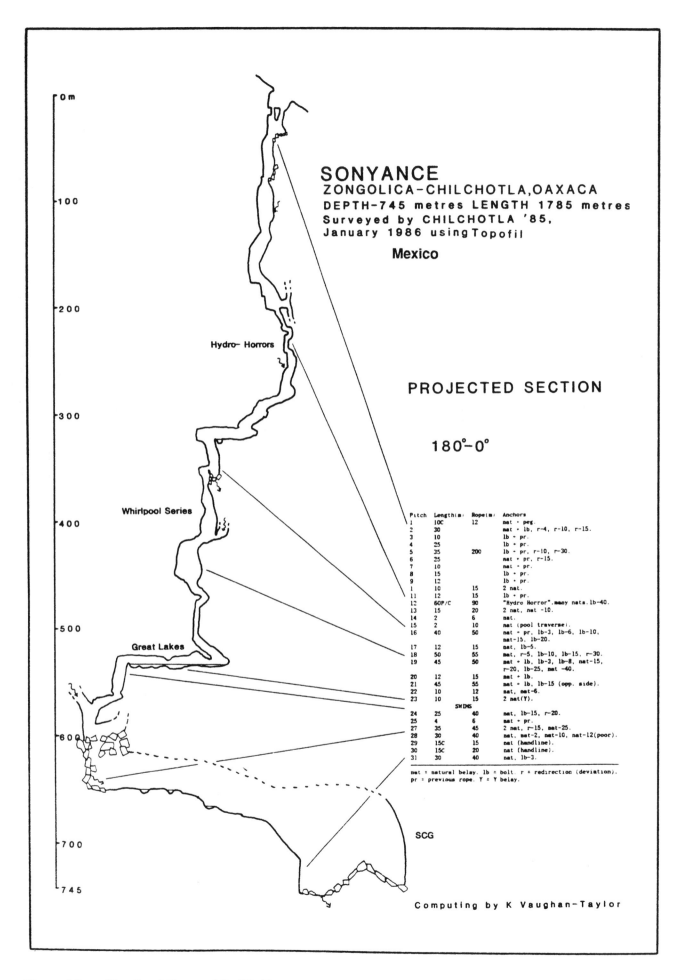

SONYANCE
ZONGOLICA-CHILCHOTLA,OAXACA
DEPTH-745 metres LENGTH 1785 metres
Surveyed by CHILCHOTLA '85,
January 1986 using Topofil

Mexico

PROJECTED SECTION

180°-0°

Hydro-Horrors

Whirlpool Series

Great Lakes

SCG

Pitch	Length(m)	Rope(m)	Anchors
1	10C	12	nat + peg.
2	30		nat + 1b, r-4, r-10, r-15.
3	10		1b + pr.
4	25		1b + pr.
5	35	200	1b + pr, r-10, r-30.
6	25		nat + pr, r-15.
7	10		nat + pr.
8	15		1b + pr.
9	12		1b + pr.
1	10	15	2 nat.
11	12	15	1b + pr.
12	60P/C	90	"Hydro Horror".many nats.1b-40.
13	15	20	2 nat, nat -10.
14	2	6	nat.
15	2	10	nat (pool traverse).
16	40	50	nat + pr, 1b-3, 1b-6, 1b-10, nat-15, 1b-20.
17	12	15	nat, 1b-5.
18	50	55	nat, r-5, 1b-10, 1b-15, r-30.
19	45	50	nat + 1b, 1b-3, 1b-8, nat-15, r-20, 1b-25, nat -40.
20	12	15	nat + 1b.
21	45	55	nat + 1b, 1b-15 (opp. side).
22	10	12	nat, nat-6.
23	10	15	2 nat(Y).
		SWDMS	
24	25	40	nat, 1b-15, r-20.
25	4	6	nat + pr.
27	35	45	2 nat, r-15, nat-25.
28	30	40	nat, nat-2, nat-10, nat-12(poor).
29	15C	15	nat (handline).
30	15C	20	nat (handline).
31	30	40	nat, 1b-3.

nat = natural belay. 1b = bolt. r = redirection (deviation).
pr = previous rope. Y = Y belay.

Computing by K Vaughan-Taylor

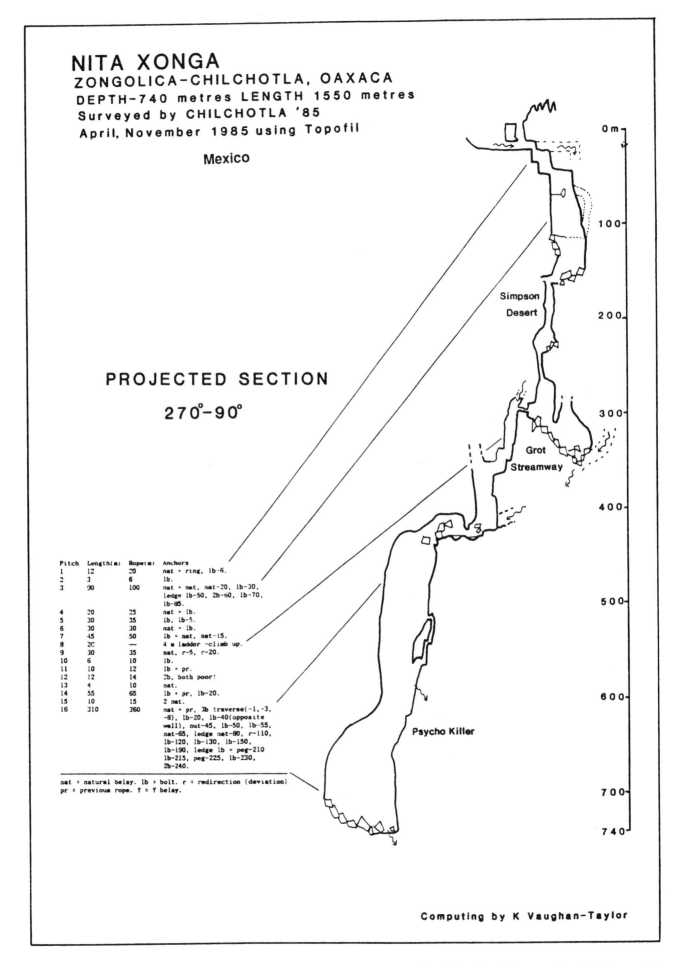

NITA XONGA
ZONGOLICA-CHILCHOTLA, OAXACA
DEPTH-740 metres LENGTH 1550 metres
Surveyed by CHILCHOTLA '85
April, November 1985 using Topofil

Mexico

PROJECTED SECTION
270°-90°

Simpson
Desert

Grot
Streamway

Psycho Killer

Pitch	Length(m)	Rope(m)	Anchors
1	12	20	nat + ring, lb-6.
2	3	6	lb.
3	90	100	nat + nat, nat-20, lb-30, ledge lb-50, 2b-60, lb-70, lb-85.
4	20	25	nat + lb.
5	30	35	lb, lb-5.
6	30	30	nat + lb.
7	45	50	lb + nat, nat-15.
8	20	—	4 m ladder -climb up.
9	30	35	nat, r-5, r-20.
10	6	10	lb.
11	10	12	lb + pr.
12	12	14	2b, both poor!
13	4	10	nat.
14	55	65	lb + pr, lb-20.
15	10	15	2 nat.
16	310	360	nat + pr, 3b traverse(-1,-3, -8), lb-20, lb-40(opposite wall), nut-45, lb-50, lb-55, nat-65, ledge nat-80, r-110, lb-120, lb-130, lb-150, lb-190, ledge lb + peg-210 lb-215, peg-225, lb-230, 2b-240.

nat = natural belay. lb = bolt. r = redirection (deviation)
pr = previous rope. Y = Y belay.

Computing by K Vaughan-Taylor

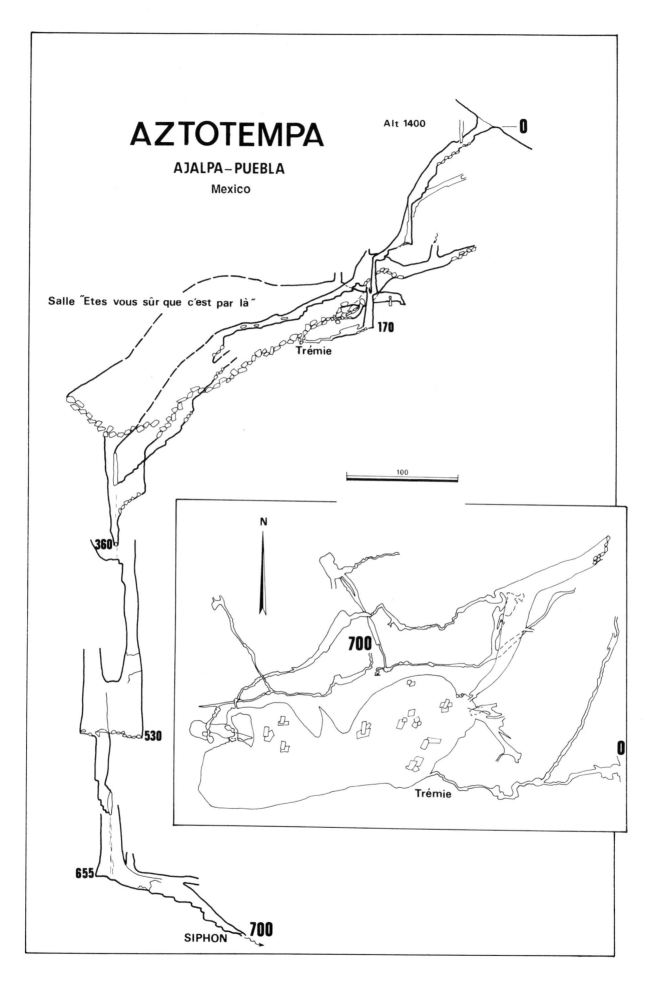

AZTOTEMPA

AJALPA–PUEBLA

Mexico

Alt 1400

0

Salle "Etes vous sûr que c'est par là"

170

Trémie

100

N

700

0

Trémie

360

530

655

SIPHON 700

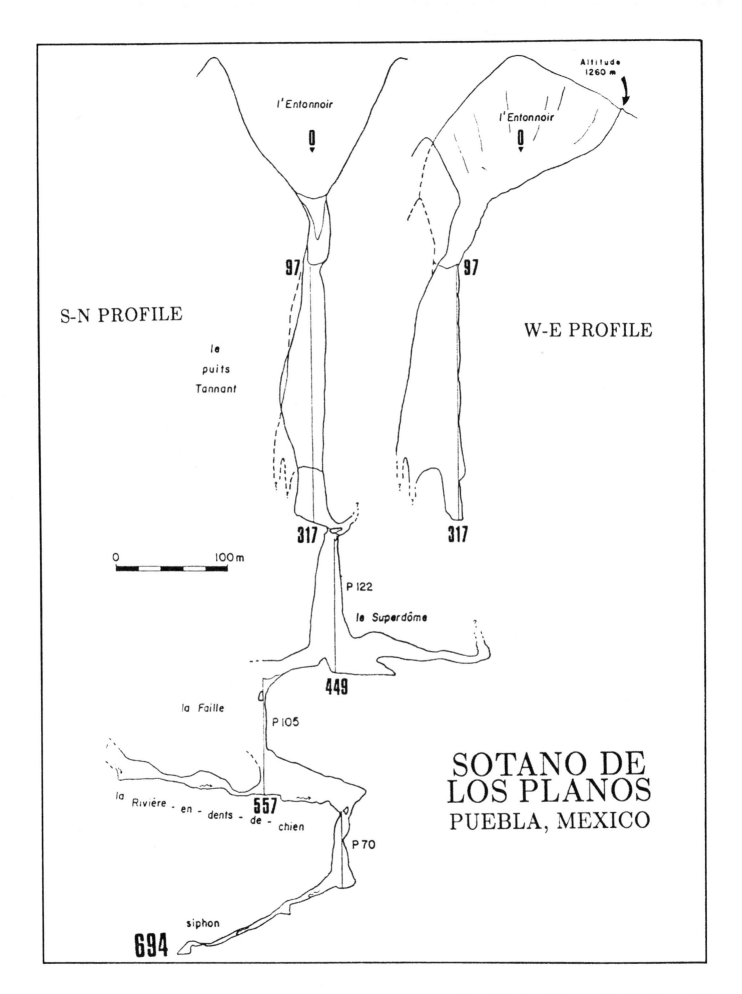

S-N PROFILE

W-E PROFILE

l'Entonnoir

l'Entonnoir

Altitude
1260 m

0

0

97

97

le
puits
Tannant

317

317

0   100 m

P 122

le Superdôme

449

la Faille

P 105

SOTANO DE
LOS PLANOS
PUEBLA, MEXICO

la Rivière - en - dents - de - chien

557

P 70

siphon

694

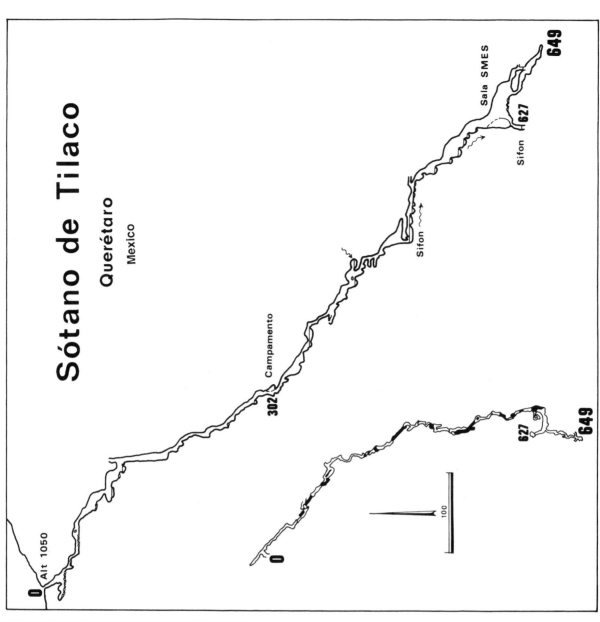

Sótano de Tilaco

Querétaro

Mexico

Alt 1050

0

302 Campamento

Sifon

Sifon

Sala SMES

627

649

100

0

627

649

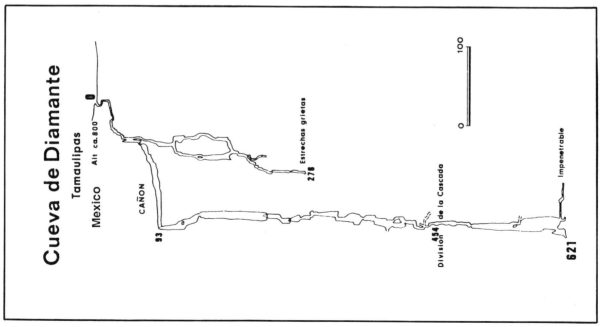

Cueva de Diamante

Tamaulipas

Mexico   Alt ca. 800

0

CAÑON

93

Estrechas grietas

278

454

Division   de la Cascada

Impenetrable

621

100

0

by the Sociedad Mexicana de Exploraciones Subterráneas, led by Carlos Lazcano.

Map: from the maps of the S.M.E.S., sent by Carlos Lazcano.

Bibliography: *A.M.C.S. Activities Newsletter*, 1980 (11); 1982 (12).

14. **nita Nashi** .... –641 m
(Huautla de Jimenez, Oaxaca)
Nita Nashi is the northernmost cave of the Huautla hydrological system. The entrance was located in 1980, but not explored by the A.M.C.S. until 1982. With a depth of –641 m, it is the third deepest on the plateau. Heading to the south, the cave comes very close to La Grieta and Li nita, but is stopped by the same inpenetrable layer which limits the depths of other caves on the plateau. (*A.M.C.S. Activities Newsletter*, 1983 (13), see sistema Huautla area map).

15. *cueva de Diamante* .... **–621 m**
(sierra de El Abra, Laguna del Mante, Tamaulipas)
This cave drains a 1000 m long, 500 m wide, and 76 m deep polje located on the crest of the sierra de El Abra, at an altitude of 800 m, 120 km to the west of Tampico and 5 km north of the state line between Tamaulipas and San Luis Potosí.

The cave was located in 1974 by William Russell who, with Andy Grubbs, explored to about –90 m. The same year another group of A.M.C.S. cavers explored down to –120 m. In 1975 progress was stopped at –276 m by impenetrable fissures. Another series of pits was pushed in 1976 to the top of a wet passage at –454 m. In late 1977 the explorers were stopped at –621 m by a set of two tight horizontal fissures not far from base level.

Map: from the maps of Mark Minton and the A.M.C.S. (1974-1977).

Bibliography: *A.M.C.S. Activities Newsletter*, 1975 (2); 1978 (8).

16. **nita He** (Huautla de Jimenez, Oaxaca) .... –594 m
Another deep cave on the Huautla plateau. Explored in 1980 by the A.M.C.S. (A.M.C.S. Activities Newsletter, 1983 (13) profile).

17. **sistema Cuetzalan** (Cuetzalan, Puebla) ..... –587 m
Located a few minutes walk from the small town of Cuetzalan. Large system with seven entrances, formed by the connection of the sumidero de Chichicasapan and the sumidero de Atischalla in March, 1978. Exploration began in 1977, when the depth of –261 m was reached. American and British cavers continued exploration into the early 1980's, when they were joined by Belgian cavers (*A.M.C.S. Activities Newsletter* 10). With 22,432 m of surveyed passage, this is the third longest cave in Mexico, but a good map of the system has yet to be drawn.

18. **sótano de las Coyotas** .... –581 m
(Tortugas, Xichu, Guanajuato)
Explored by Mexican cavers from the S.M.E.S. and the G.E.U.M. in 1982 (unpublished map).

19. **sótano Arriba Suyo** .... –563 m
(Xilitla, San Luis Potosí)
Explored by British cavers in 1985-1986.

20. **sótano del Río Iglesia** .... –531 m
(Huautla de Jimenez, Oaxaca)
Swallet cave located 500 m south of the village of San Agustin, 1 km to the south of the sótano of the same name, at an altitude of 1600 m. Located in 1965 by William Russell and explored in 1966 by the A.M.C.S. Bottomed in late 1967 by M.U.C.C.C. cavers (Courbon, *Atlas des Grands Gouffres du Monde*, 1979, pp. 36, 46).

21. **sótano de Nogal** .... –529 m
(San Juan, Querétaro)
This cave is located 5 km to the north of the hoya de las Conchas (see below), at an altitude of 1300 m not far from the rancho San José. Located by A.M.C.S. cavers in 1975, and explored the following year. In March, 1975, A.M.C.S. cavers Tracey Johnson, Bill Stone, Roy Jameson, and Steve Ward descended to –247 m. In May, the bottom was reached by T. Johnson, B. Stone, Gary Stiles and Larry O'Loane (*AMCS Bull.*, 1977,7; Courbon, *op. cit.*, 1979, pp. 37, 47).

22. **sótano de Ahuihuitzcapa** .... –515 m
(Zongolica, Vera Cruz)
Explored in 1983 by the S.M.E.S., then by a (non-authorized!) group of French cavers (*Spelunca*, 1984 (13) profile).

23. *sótano de Las Golondrinas* .... **–512 m**
(Aquismón, San Luis Potosí)
Famous pit whose huge opening is located in the Sierra Madre Oriental, in the SE corner of the state of San Luis Potosí. Access is from Aquismón, to the west of route No. 85. The entrance is 4 km to the NW of the village of Tamapatz, not far from the trail to San Rafael, at an altitude of 740 m. A new road has shortened the former long approach march.

Coordinates: 21°36'N; 99°06'W. The cave is formed in limestones of the El Doctor formation, of Cretaceous age. The presumed resurgence is the nacimiento del Río Coy, 15 km distant, at an altitude of 50 m.

The cave has long been known to the Huasteco Indians. The entrance was visited by French and Mexican mountaineers in 1957, and again in December 1966 by T. R. Evans, Charles Borland, and Ronald Stearns. T. R. Evans returned in April 1967 with other A.M.C.S. cavers to finally drop the 333 m deep pit (see "The Great Pits").

In late 1968, T. Wilson and T. Petranoff discovered the opening to a second drop at the bottom of the first one. In late 1969, a team from the Bloomington Indiana Grotto descended this new pit to a depth of –512 m. A strong breeze indicates a good potential for more cave passage. The official depth of the cave is either –512 m or –488 m, depending on which side of the entrance is chosen as the zero point.

Map: from the maps of the A.M.C.S., sent by Steven Bittinger.

Bibliography: Raines (T.) (ed.) - "Sótano de las Golondrinas", *Bull. A.M.C.S.*, 1968 (2): 20 p.

24. **hoya de las Conchas** .... –508 m
(San Juan, Querétaro)
This cave is at an altitude of 1300 m and is located 10 km to the north of Jalpan, not far from la Purisima and 3

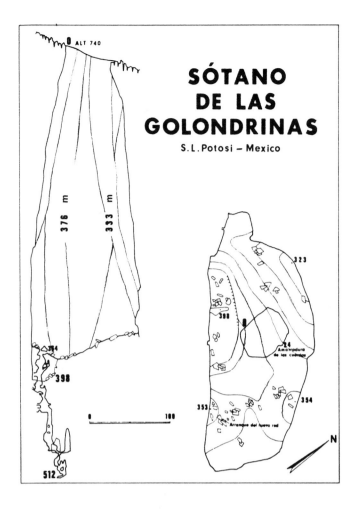

SÓTANO DE LAS GOLONDRINAS

S.L. Potosi – Mexico

km to the north of ejido San Juan, at the bottom of a large doline.

The entrance was located in November 1975 by A.M.C.S. cavers Bill Stone and T. Johnson who descended to –65 m. In January 1976 they were assisted by Australians J. James and N. Montgomery, then by M. Farr to reach the depth of –470 m. In March 1976 a group of 33 cavers returned to explore the cave, but were stopped at a sump at –508 m (*AMCS Bull.*, 1977, 7; Courbon, *op. cit.*, 1979, pp. 37-38, 49).

25. **sótano del Buque** (El Pino, Querétaro) ...... –506 m
This cave is located to the north of Ahuacatlán, near La Ciénega, in the municipio de El Pino. It was located in the summer of 1972 by A.M.C.S. cavers who descended to –471 m that same year and reached the terminal sump at –506 m in 1974. The loss of the survey notes obliged an A.M.C.S. team to resurvey the cave in January 1979 (R. Jefferys, sótano del Buque, *AMCS Activities Newsletter*, 1979 (1), 71-74).

26. **gruta de San Cristobal**.................approx. –500 m
(Rancho Nuevo, Chiapas)
Extended in January, 1987 by Italian cavers.

27. **nita Chaki** .... –493 m
(Zongolica - Chilchotla, Oaxaca)

The cave is located very close to nita Xonga (see above) and explored in November 1985 by the Australian "Chilchotla 85" expedition. The name means Cricket Cave. The cave starts off small and clean, but the rock quality deteriorates near the bottom of the cave, where an impenetrable bedding plane squeeze stopped exploration (*N.S.S. News* 44 (10)); *A.M.C.S. Activities Newsletter* 1987 (16).

28. **hoya de las Guaguas** .... –478 m
(Tampaxal, Aquismón, San Luis Potosí)
The entrance is a vertical bell-shaped pit 262 m deep on one side, and 147 m on the other side. Explored in 1977. The terminal sump was dove for a distance of 13 m (*AMCS Activities Newsletter*, 1979 (9) profile).

29. **cueva de San Augustín** .... –461 m
(Huautla de Jimenez, Oaxaca)
Explored in 1970 (*The Canadian Caver*, 1971 (4) profile).

30. **sótano del Barro** (Jalpan, Querétaro) .... –455 m
See "The Great Pits".

31. **hoyo de San Miguel** (Taxco, Guerrero) ...... –455 m
In 1984 by the S.M.E.S. (*AMCS Activities Newsletter*, 1984 (14), profile).

32. **sótano Itamo** (Vera Cruz) .... –454 m

33. **cueva de la Peña** .... –448 m
(Rancho La Presa, San Luis Potosí)
Explored in 1978 by A.M.C.S. cavers (*AMCS Activities Newsletter*, 1978 (8) profile).

34. **sótano de Tlamaya** .... –447 m
(Xilitla, San Luis Potosí)
Explored in 1964-1965 by the A.M.C.S. (*A.M.C.S. Bull.*, 1967 (1), profile). The loss of the survey notes required a resurvey by the A.M.C.S. in December, 1984.

35. **cueva de la Llorona** .... –412 m
(Villa Hidalgo, Tamaulipas)
Cave near the sistema Purificación explored by A.M.C.S. cavers.

36. **cueva del Reefer Madness** .... –411 m
(Xilitla, San Luis Potosí)
Explored by British cavers in 1985-1986.

37. **sótano del Alhuastle** .... –410 m
(Tlacotepac de Diaz, Puebla)
Explored by S.Q.S. cavers in January, 1980. The last drop is 329 m deep.

38. **cueva Tan-go-jo** .... –405 m
(Xilitla, San Luis Potosí)
Explored by British cavers in 1985-1986.

39. **Kajahe Xuntua** (Oaxaca) .... –400 m

40. **sumidero de Santa Elena** .... –400 m
(Xochitlan, Puebla)
Long through river cave requiring over 30 drops to

traverse, explored in 1980, 1981 and 1983 by groups led by S. Knutson (*AMCS Activities Newsletter*, 1983 (13) profile; *N.S.S. News* 1982 (9) plan with 6664 m).

41. **atepolihuit de San Miguel** .... –399 m
(Cuetzalan, Puebla)

42. **Veshtucoc** (Chiapas) .... –380 m

43. **sótano de la Joya de Salas** .... –376 m
(Jaumave, Tamaulipas)

44. **hoya del Poleo** (Guanajuato) .... –375 m

45. **sótano Tomasa-Kiahua** .... –374 m
(San Jose Independencia, Vera Cruz)
See "The Great Pits".

46. **sótano de la Virgen** .... –352 m
(Landa de Matamoros, Querétaro)

47. **sótano del Perro Vivo** (Hidalgo) .... –350 m

48. **sistema El Chorreadero** (Chiapas) .... –345 m

49. **cueva de Los Hornos** (San Luis Potosí) ..... –341 m

50. **cueva de Xocotlat** .... –339 m
(San Pablo Zoquitlán, Puebla)

51. **sótano del Río Coyomeapan** .... –337 m
(San Pablo Zoquitlán, Puebla)

52. **sótano de Los Herrandez** (Querétaro) .... –330 m

53. **cueva de Santa Cruz** .... –327 m
(Huautla de Jimenez, Oaxaca)

54. **sumidero del Río Xocotlat** .... (–308, +15) 323 m
(San Pablo Zoquitlán, Puebla)

55. **sótano de Seis Segundos** .... –323 m
(Huautla de Jimenez, Oaxaca)

56. **sótano de Ahuacatlán** .... –320 m
(Pinal de Amoles, Querétaro)

57. **hoya de Zimapán** .... –320 m
(Tamuín, San Luis Potosí)

58. **sumidero de Atikpak** .... –319 m
(Zongolica, Vera Cruz)

59. **cueva del Arroyo de Tenejapa** .... –317 m
(Chiapas)

60. **nita Ntau - nita Nido** (Oaxaca) .... –310 m

61. **sótano de Jabalí** .... –308 m
(El Quirino, Querétaro)

62. **Thao/Thi Guinjao** (Oaxaca) .... –300 m

LONG CAVES:

1. **sistema Purificación** ....71,583 m
(Villa Hidalgo, Tamaulipas)
See above.

2. **sistema Huautla** ....52,111 m
(Huautla de Jimenez, Oaxaca)
See above.

3. **sistema Cuetzalan** ....22,432 m
(Cuetzalan, Puebla)
See above.

4. **cueva Coyalatl** ....approx. 19,000 m
(Tlacotepec de Diaz, Puebla)
Explored in 1985 by the G.S.A.B.

5. **cueva de Tecolote** ....11,084 m
(Villa Hidalgo, Tamaulipas)
Located near sistema Purificación and explored from 1984 to present by A.M.C.S. cavers. Contains large trunk passages.

6. **sistema Cuicateca** ....9300 m
(Cuicatlan, Oaxaca)
See above.

7. **gruta de San Cristobal** ....9000 m
(Rancho Nuevo, Chiapas)
See above.

8. **sumidero de Santa Elena** ....7884 m
(Xochitlán, Puebla)
See above.

9. **cueva de la Peña Colorada** ....7793 m
(Huautla de Jimenez, Oaxaca)
See sistema Huautla under "Deep Caves".

10. **atepolihuit de San Miguel**) ....7700 m
(Cuetzalan, Puebla)

11. **sótano del Arroyo** ....7200 m
(Los Sabinos, Ciudad Valles, San Luis Potosí)
Explored from 1967 to 1973 by U.S. and Canadian cavers. Contains large horizontal passages subject to sudden flooding (*The Canadian Caver*, 1974, 6(1) plan).

12. **actún de Kaua** (Yucatán) ....6700 m

13. **sumidero de Jonotla** (Jonotla, Puebla) ......6381 m

14. **sótano de las Calenturas** ....6032 m
(Yerbabuena, Tamaulipas)
Explored from 1979 to 1984 by the A.M.C.S. (*AMCS Activities Newsletter*, 1983 (13) plan with 5877 m).

15. **gruta del Río Chontalcoatlán** ....5827 m
(Tetipac, Guerrero)
(*The Canadian Caver*, 1976, 8(1) plan; *Grottes et Gouffres*, 1980 (76)).

16. **gruta del Río San Jerónimo** ....5600 m
(Tetipac, Guerrero)
(*The Canadian Caver*, 1976, 8(1) plan).

17. **grutas de Juxtlahuaca** ....5098 m
(Colotlipa, Guerrero)
In 1971 by the A.M.C.S. (*AMCS Newsletter*, 1974, V(1) plan).

18. **Veshtucoc** (Acala, Chiapas) ....4900 m

19. **cueva del Nacimiento del Río San Antonio** (Acatlan, Oaxaca) ....4570 m

20. **sótano de la Tinaja** ....4502 m
(Ciudad Valles, San Luis Potosí)
Part of the same hydrological system as sótano del Arroyo.

21. **sótano del Japonés** ....approx. 4500 m
(Yerbaniz, San Luis Potosí)

22. **sistema San Andres** (Puebla) ....4471 m

23. **sótano del Río Iglesia** ....4206 m
(Huautla de Jimenez, Oaxaca)

24. **sistema Zoquiapan** (Cuetzalan, Puebla) ....4107 m

25. **sima del Borrego** ....4087 m
(Chilpancingo, Guerrero)

26. **cueva Aztotempa** ....approx. 4000 m
(Ajalpa, Puebla)

27. **sumidero de San Bernardo** ....3931 m
(Xochitlán, Puebla)
River cave explored by S. Knsuton *et al.*

28. **sumidero de Pecho Blanco No. 2** ....3790 m
(Chiapas)

29. **sótano de Agua de Carrizo** ....3748 m
(Huautla de Jimenez, Oaxaca)

30. **nita Nashí**...3524 m
(Huautla de Jimenez, Oaxaca)

31. **cueva del Río Jalpan** ....3440 m
(Jalpan, Querétaro)

32. **actún Xpukil** (Yucatan) ....3353 m

33. **cueva de la Laguna Verde** ....3350 m
(Acatlan, Oaxaca)

34. **sumidero Yochib** (Tenejapa, Chiapas)........3316 m
Difficult river cave. (Steele, C. W. - *Yochib: The River Cave*, Cave Books, St. Louis, 1985, 164 p.).

35. **sistema El Chorreadero** (Chiapas)............3280 m

36. **sumidero La Joya** (Guerrero) ....3245 m

37. **cueva de la Llorona** (Tamaulipas).............3136 m

38. **Atepolihuit de Nauzontla** (Puebla) ....3066 m

39. **sótano de Tlamaya** ....3057 m
(Xilitla, San Luis Potosí)

40. **sistema de Montecillos** ....3022 m
(Los Sabinos, San Luis Potosí)

41. **resumidero de Toxim** (Jalisco) ....3005 m

42. **sótano de Huitzmolotitla** ....3002 m
(Xilitla, San Luis Potosí)

43. **sumidero de Atliliakan** (Guerrero) ....3000 m

44. **sótano del Río Coyomeapan** (Puebla) ....3000 m

45. **Tamazcalco** (Puebla) ....3000 m

46. **sótano del Tigre** (San Luis Potosí) ....3000 m

PANAMA
Panamá

Panamanian caves have been studied by American biologists since 1912. Among those explored are the **cuevas de Chilibrillo** (Buenos-Aires, Prov. Panama) which has three segments 340, 94, and 30 m long (see S. Peck, *Annales de Spéléologie*, 1971, 26(2)).

PARAGUAY

Little is known of the speleological potential of this country. Two small caves in the limestone hills in the NE part of the country were explored by French cavers from the Spéléo-Club d l'Aube in 1982: **gruta-abismo de la Fraternidad** (Santa Maria, Concepción), 66 m long and 24 m deep, and **gruta-abismo Santa Maria** (*Ibid.*), 50 m long (Lo Bramavenc, *Calcassonne*, 1982 (5) maps).

PERU
Perú

Caves in Peru have been known for a long time: the naturalists Humboldt (in 1802) and Raimondi (in 1868) visited the gruta de Uscopisco and the comte de Castelnau Sansonmachay (in 1846). Starting in 1932 Cesar Garcia Rosell began studying several caves and published his useful inventory *Cavernas, grutas y cuevas del Perú* in 1965, year in which the Sociedad Peruana de Espeleológia was formed. Explorations by foreign cavers began with the Polish in 1971 and continue to the present. The most significant finds have been made by the Imperial College Karst Research Expedition to the Peruvian Andes (ICK-REPA) in 1972 and by the Groupe Spéléologique de Bagnois-Marcoule in 1979.

DEEP CAVES:

1. *millpu de Kaukiran*...............**(–402, +5) 407 m**
    (Palcamayo, Tama)

    Also called sima de Racas Marca. To the north of Huagapo, the Pucará limestones are notched by quebrada de Ushto. At the head of this deep gorge, about 2000 m from the río Shaca, a little valley is found, where the Hualli Hurah stream sinks near the village of Kaukiran. The entrance to the cave is 50 m lower, downstream from the insurgence.

    x 8756.64 ; x 413.51: alt. 3992 m.

    Polish cavers from the Klub Wysokogorski explored the cave to − 60 m in 1971. The following year Modesto Castro, a guide from Huagapo, reached –120 m. The same year, the I.C.K.R.E.P.A. reached the terminal sump at –402 m, about 20 m in elevation above the resurgence.

    The cave is formed in a regularly sloping gallery heading to the NE and interrupted by drops between 5 and 11 m deep. The sump is about 1800 m from the cueva de Guagapo (alt. 3572 m) which is the presumed resurgence. From here the water travels to the río Shaca.

Map: from *Cave Science*, 1973 (52).

Bibliography: *Cave Science, Journal of British Spel. Assoc.*, 1973, No. 52.

2. **tragadero de San Andres** .... **–334 m**
    (Parque National, Cutervo, Cajamarca)

    In 1977 the Club Excursionista de Catalunya explored down to –125 m. The bottom, where the stream sinks into clay flooring a room, was reached in 1979 by the G.S.B.M. (Y. Sammartino, G. Staccioli, J.-D. Klein - *Pérou 79. Expédition du G.S. Bagnols- Marcoule*, report, 183 p., profile).

3. **cueva de los Guácharos** (Cajamarca) .... **–180 m**
    In 1976 by the E.R.E. Catalunya.

4. **gruta de San Andres**............................. **–117 m**
    (Cutervo, Cajamarca)

    Explored in 1977 (S. Vilchez Murga), 1977 (E.R.E. Barcelona) and 1979 (G.S.B.M.) (Y. Sammartino *et al., op. cit.*).

5. **red de las grutas** ....**(–65, +29) 94 m**
    (Cutervo, Cajamarca)
    (Y. Sammartino *et al., op. cit.*, profile).

LONG CAVES:

1. **millpu de Kaukiran** (Palcamayo, Tarma)....2141 m
    See above.

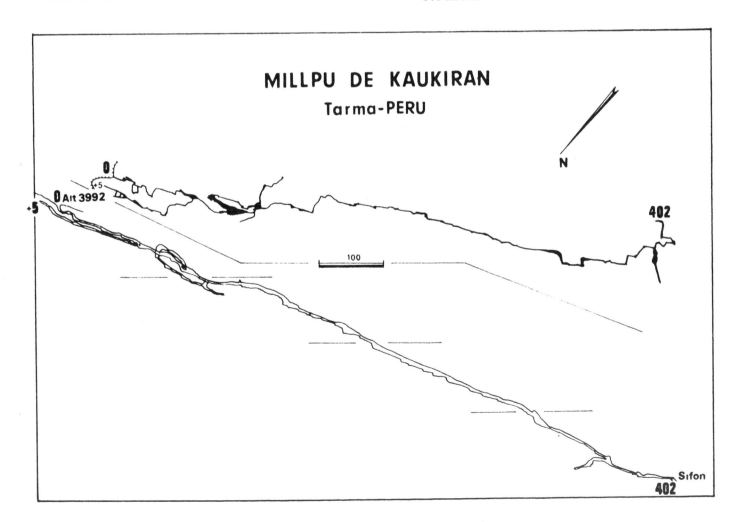

MILLPU DE KAUKIRAN
Tarma-PERU

2. **cueva de Guagapo** (Palcamayo, Tarma) ...... 1883 m
Also written as Huagapo. Explored in 1969 (C. Morales Arnao), 1971 (K. Wysokogorski), 1972 (I.C.K.R.E.P.A.) and 1976 (Centre Aixois E.S.) (*Cave Science*, 1973 (52) plan; Y. Sammartino et al., *op. cit.*, plan).

3. **cueva de los Guácharos** (Cajamarca) .... 1334 m

4. **gruta de Uscopisco** .... 1100 m
(Ninabamba, Santa Cruz, Cajamarca)
Has five entrances. Explored in 1973 by the G.E.S. Barcelona (Y. Sammartino *et al., op. cit.*, plan).

5. **gruta de San Andres** .... 1097 m
(Cutervo, Cajamarca)
See above.

6. **quebrada Churos** .... 1440 m
In 1982 by the G.S.B.M.

7. **Limbo tocco Yorac Casa** .... 835 m
(Karañahue, Yanoaca, Cuzco)
Alt. 3891 m. In 1981 by the E.R.E. Barcelona.

8. **cueva de Pacu Huayan** .... 800 m
(San Pedro de Cajar, Tarma)
In 1979 by the G.S.B.M. (Y. Sammartino *et al., op. cit.*, plan).

PUERTO RICO

The caves of Puerto Rico (or Borinquen) are not known for their exceptional length, but often have large passage dimensions and major rivers flowing through them. Abundant guano deposits have been the cause of numerous cases of histoplasmosis suffered by cave explorers. The rich fauna found in Puerto Rican caves has been the subject of much biospeleological study.

Most of the caves are formed in Cretaceous and Tertiary age limestones, but the largest ones are formed in Oligocene age limestone. The caves are often visited by local inhabitants, and the bottoms of dolines and depressions are frequently cultivated. Most systematic exploration has been by cavers from the National Speleological Society (U.S.) and more recently by the Sociedad Espeleológica de Puerto Rico.

Valuable references can be found in *Discovery at the Río Camuy* by R. Gurnee and J. Gurnee (1974), The Karst Landforms of Puerto Rico by W. Monroe (*U.S. Geol. Survey Prof. Paper*, 1976, 69 p.) and "Some Preliminary notes on Caves in Puerto Rico" by D. St Pierre (*C.R.G. of G. B. Newsletter*, 1971 (125): 10-20).

David St. Pierre.

DEEP CAVES:

1. **sistema del Río Encantado** .... approx. –250 m
(Florida)
See below.

2. **cueva del Río Camuy** ....approx. –180 m
(Río Camuy, Bayaney)
See below.

3. **cueva de Empalme** .... –127 m
(Río Camuy, Bayaney)
See below.

LONG CAVES:

1. *sistema del Río Encantado* .... 16,910 m
(Barceloneta, Florida)
This system has five entrances: the Río Encantado, alt. 175 m, Escalera, 200 m, Vidrios, 200m, cueva Juan Nieves or Tito, and manantial de las Aguas Frias. Exploration began in 1974 (Basco, Garrison, and Quires, to a length of 2500 m) and was pursued in 1977 (Basco, Miller, Herceg, to about 5000 m). In 1983, Basco, Troester and the McMaster Univ. C.C.C. mapped a total of 7060 m. In 1984, N.S.S. cavers, the M.U.C.C.C. and the S.E.P.R. Inc. connected Encantado with Escalera to bring the length to 9100 m. The connection with Vidrios brought the length to 9332 m in 1985. Meanwhile, the exploration of the cueva Juan Nieves began in 1983, and was connected to Aguas Frias by a sump dive by Skiles and Nicholson. This cave was pushed to 3500 m in 1984 by N.S.S. and S.E.P.R.I. cavers. In 1985 the upstream sump was dove to bring the length to 5357 m.

The connection of Juan Nieves with the Encantado system by N.S.S. and S.E.P.R.I. cavers in February 1986 brought the length to 16,910 m, argueably making this the longest continually traversable underground river in the world (*N.S.S. News* 44 (9)).
Map: sent by Kevin Downey.

2. **sistema del Río Camuy** (Bayaney) .... 8900 m
The insurgence is at an altitude of 270 m, while the resurgence is at 150 m. The system has three segments, listed from upstream to downstream:
   a. **cueva del Huimo-**
     cueva de los Angeles .... 3550 m
    Explored in 1958 and from 1970 to 1974 by the N.S.S. and Puerto Rican, Australian, and British cavers. Connection on January 5, 1972 (*N.S.S.News*, 1975, 33 (2), map with 3237 m).
   b. **cueva del Río Camuy** .... 3600 m
    This segment has four entrances: sumidero de los Tres Pueblos, sima Espiral, cueva La Ventosa and sumidero de los Angeles. Explored by the N.S.S. from 1962 to 1966 (*N.S.S. Bull.*, 1967, 29 (2), map with 3200 m). Extended to 3600 m in 1983. Separated from the preceding cave by a doline 80 m wide and 150 m long.
   c. **cueva de Empalme** .... 1000 m
    A show cave with four entrances, including the sumidero de los Tres Pueblos, connected in 1966. Separated from the preceding cave by a 140 m diameter doline.
   d. other segments including cueva del Agua Evaporada (400 m), the resurgence (200 m), and upstream Empalme (150 m) (*N.S.S. News* 1987 45 (5)).

3. **cuevas de Aguas Buenas** (Caguas) .... 2800 m
Cave with twelve entrances explored in 1899 by M. Dinwiddie, and in 1958, 1968, and 1976 by N.S.S. cavers (*N.S.S. Bull.*, 1976 (1) map with 2500 m).

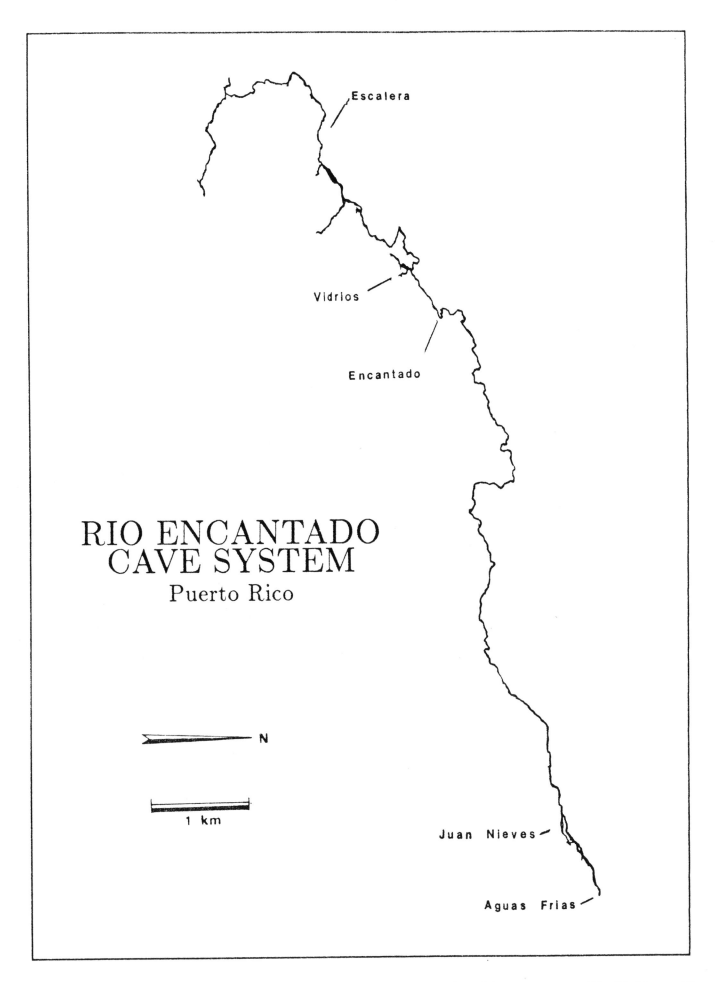

RIO ENCANTADO
CAVE SYSTEM
Puerto Rico

Escalera

Vidrios

Encantado

Juan Nieves

Aguas Frias

N

1 km

4. **cuevas del Viento** (Rosario) .... approx. 2000 m (*Nittany Grotto News*, 1983, 30 (2)). Has eight entrances.

5. **Quebrada No. 8** (Camuy) .... 1494 m
Explored in 1975. Length potential estimated at 1600 m (*Decateur Caver*, 1975, 4 (6)).

6. **el Zumbo** (Hato Viejo, Ciales) .... 1247 m
Cave upstream from the sistema del Río Encantado, explored in 1984 and 1985.

7. **cueva Mantilla** (Juncal, San Sebastian) ..... 1000 m
Explored in 1983 by the M.U.C.C.C. and the S.E.P.R.I.

TRINIDAD AND TOBAGO

As for Barbados, it is the publication of Ken Pawson describing exploration done in 1945-1950 that is the most valuable reference for this island country located to the north of Venezuela. **Aripo No. 1** is reported to be a large cave full of bats (*Canadian Caver*, 1974, 6(1); 21-27, length and depth not stated). In 1978 the Bloomington Indiana Grotto explored 221 m in **Oropouche Cumaca** (*B.I.G. Newsletter*, 1979, 14 (2) plan).

UNITED STATES OF AMERICA

Approximately 10% of the land surface of the United States consists of karst topography developed in carbonate rocks or gypsum. These areas are distributed throughout many different geologic regions, of which the following are the most important:

(1) The Appalachian Mountains contain numerous Paleozoic limestone and dolomite formations that grade from metamorphic rock in the east, through a band of folding and faulting in the center, to high plateaus of only slightly deformed and nearly horizontal rocks in the west.

Marble from the metamorphism of Ordovician limestone and dolomite is exposed mainly in western New England and the Adirondack Mountains of New York, although a few smaller exposures occur in the mid-Atlantic states. Marble caves in these areas are small but intricate, and many have walls brightly banded with multi-colored rock.

Carbonate rocks up to 6000 m thick occur in the folded Appalachians, but unfortunately the thickest are exposed in low valley regions where deep or long caves are unable to develop. However, the lowland regions contain such well-decorated caves as Luray. Most of the large caves are located along the flanks of resistant ridges. Their overall patterns are most typically long and narrow, extending along the strike of the beds, which is also the trend of the ridges and valleys. The longest cave in the folded

Appalachians is the Butler-Sinking Creek System in Virginia.

The plateaus to the west of the fold belt contain the broadest exposures of limestone in the Appalachians. With few restrictions on their lateral extent, caves are able to achieve much greater length than in the folded rocks to the east. This region contains a greater number of caves on the present long cave list than any other of comparable size in the world. The longest are Friars Hole and the Organ Cave System in West Virginia. Along the Cumberland escarpment, the westernmost border of the Appalachian plateaus, are Cumberland Caverns and Xanadu Cave in Tennessee and Sloan's Valley Cave System in Kentucky. The deepest caves in the East are located in the southern Appalachian plateaus of Tennessee, Alabama, and Georgia. Ellison's Cave in Georgia is not only the deepest of these but also contains the deepest shaft in the U.S.A. (180 m deep Fantastic Pit, which also has access through a lower passage to a balcony 156 m above the floor).

(2) Florida, in the southeastern coastal plain, consists mainly of Eocene limestone containing extensive caves, most of which are below the water table. Some of these were flooded because of a rise in sea level since the late Pleistocene, but drilling has revealed others that extend hundreds of meters below sea level and must always have been water filled. Florida contains by far the longest subaqueous caves in the country, and probably the longest in the world. Access to most of them is gained through shallow water-filled sinkholes. Most caves consist of tubular passages as much as 30 m in diameter that interconnect in a crude dendritic pattern. Their great size and, in places, high water velocities have led to many deaths among ill-prepared divers. The Cathedral-Falmouth and Peacock Springs Systems are presently the longest subaqueous caves, although with recent advances in diving technique their present length records will not hold for long. The largest air-filled caves in Florida, such as Warren's Cave, are complex networks that appear to have formed, or to have been enlarged, by the mixing of waters of different chemical character.

(3) Low plateaus of Paleozoic carbonate rocks extend throughout large areas of Kentucky, Tennessee, Indiana, Missouri, and Arkansas. Here the cavernous rocks are typically only 100 to 200 m thick, but the very small dip of only 5 to 10 m/km causes them to be exposed over a very large area. The low relief prevents deep caves from forming, but many of the world's longest caves are located in this region, including 500+ km long Mammoth Cave in Kentucky. In view of the known density of passages in the area, the extent of available limestone with underground drainage, and the many levels that allow connections with other caves, future exploration should easily extend the known length of Mammoth cave by several hundred kilometers.

At Mammoth Cave, Fisher Ridge Cave, and Cumberland Caverns a protective cap of insoluble rocks has preserved many of the higher passages from destruction by surface erosion. In comparison, those areas in which the protective cap-rock is absent are restricted to one or two levels beneath sinkhole plains. These include Crevice Cave in Missouri, the Hidden River Complex (or Hicks) in Kentucky, and Blue Spring Cave and Binkley's Cave in Indiana.

To the north, particularly in Iowa and Minnesota, limestone and dolomite are exposed in only a few places. Glacial deposits cover many of the soluble rocks that were once at the surface in the north-central states, although their low relief probably limited the amount of cave development even before glaciation. Mystery Cave in Minnesota (a network maze) is the largest in the region. Many caves in Iowa are mineralized and may owe their origin to hydrothermal processes.

(4) The southern rim of the Great Plains in Texas is a vast plateau of nearly horizontal Cretaceous limestone. Karst is concentrated along the fault-controlled Balcones Escarpment, which extends through the east-central and south-central parts of the state. This region contains many large and well-decorated caves, such as Natural Bridge Caverns and Caverns of Sonora, as well as some that are long, wet, and gruelling, such as Powell's Cave and Honey Creek Cave. Many of the caves in the drier parts of the state seem to be relics from deep-seated cave development (similar to that of Carlsbad) or from a former more humid climate. To the north, in Oklahoma and Kansas, gypsum is prevalent and contains some of the largest gypsum caves in the country, such as Jester Cave.

(5) The Black Hills of South Dakota, although geologically part of the Rocky Mountains, merit consideration as a separate speleogenetic region by virtue of the unique caves they contain. Mississippian limestone is exposed around the rim of the uplift, and several large caves have been formed in this limestone by a combination of rising thermal water and surface recharge. These include Jewel Cave and Wind Cave, which are perhaps the most complex network maze caves in the world. Many of the present passages were enlarged from paleo-caves, which formed soon after the limestone was deposited. Their potential for additional discoveries is enormous. Known caves occupy a tiny fraction of the favorable limestone, and exploration is hindered mainly by breakdown along fault zones and by the fact that large complex sections of caves are connected to each other by a few narrow passages. Air movement is controlled precisely by barometric pressure changes at the surface, creating winds up to 60 km/hr in certain narrow passages during sudden periods of low pressure. Calculations of cave volume from barometric winds indicate that less than 2% of the volume has been discovered — although much of what remains is probably inaccessible.

(6) The southwestern U. S. consists mostly of arid plateaus and mountains, many of which contain Paleozoic carbonate rocks. Gypsum is common in some of the lower basins. Although surface karst is not well developed because of the dry climate, there are many caves. The Guadalupe Mountains of southeastern New Mexico consist of Permian reef and back-reef limestones containing some of the world's finest and most unusual caves. Carlsbad Cavern, one of the largest, fits perfectly the popular image of a cave: huge, well-decorated rooms that interconnect in a vast three-dimensional labyrinth. Most of the Guadalupe caves appear to have been formed by rising hydrogen sulfide, which oxidized to sulfuric acid at the water table.

Broad, high plateaus of limestone occur in many parts of Arizona, New Mexico, and Utah. Owing to the arid climate and rugged terrain, caves are sparse and usually difficult of access. The longest in this region is Fort Stanton Cave in New Mexico. Although few achieve significant depth or length, many have spectacular settings. Most caves open into the walls of canyons, including the Grand Canyon.

(7) The mountain ranges of the western U. S. contain hundreds of small exposures of Paleozoic limestone and dolomite. The limestone is not very thick and does not rise in high karst massifs like those in the Pyrénées and Alps, but instead is limited mainly to narrow bands along the flanks of mountains. The total relief of the carbonate rocks is close to 1000 m in places, but recharge areas are rather small and large caves are rare. The U. S. depth record was broken three times between 1978 and 1981 and now stands at − 472 m in Columbine Crawl in the Teton Range of Wyoming. The entrance of Columbine is at an elevation of 2990 m, making it one of the highest major caves in the world. Potential for new depth records is very good throughout the Rocky Mountains, particularly in Wyoming and Montana.

Paleokarst is well developed at the top of Mississippian limestones throughout much of the West. Caves associated with it are predominantly mazes similar to those of the Black Hills. Their origin is partly due to exhumation of paleo-caves, most of which were filled with Pennsylvanian sediment. The Bighorn-Horsethief System, which straddles the Montana-Wyoming border, is one of the largest.

The westernmost mountain ranges in the U. S. are composed mainly of igneous rocks and insoluble metamorphic rocks. However, isolated belts of Paleozoic and Mesozoic marble occur throughout the region, particularly in California and Oregon. The Bigfoot-Meatgrinder Cave System in northern California is currently the longest in the state and the fourth deepest in the country. Lilburn Cave, also in marble, is a stream cave with a complex network of overflow passages. Banded marble in this cave makes it one of the most attractive in the country.

(8) The northwestern states contain vast plateaus of Cenozoic volcanic rocks. The Cascade Range also contains prominent volcanic peaks, some of which are still active. Lava caves are common in the youngest flows. One of the largest in the continental U.S., Ape Cave, is located on the flank of Mt. St. Helens and survived the recent explosive eruption.

(9) Some of the world's largest lava caves are located in the Hawaiian Islands, particularly on the largest island, which is still active. Ainahou Ranch Cave and Kazamura Cave are the deepest lava caves in the U. S. The Hawaiian peaks rise to more than 4000 m above sea level and have been built up by lava that erupted in a more fluid state than that of most other volcanic mountain areas. These conditions are favorable to the origin of lava caves. Hawaii may very well be the future home of the U. S. depth record.

Sea caves are found all along the West coast of the U. S. and on the Hawaiian islands. The longest of them are over 300 m long.

Cave exploration in the U. S. is principally carried out by members of the National Speleological Society (organized into local chapters called Grottos) and the Cave Research Foundation.

Arthur N. Palmer.

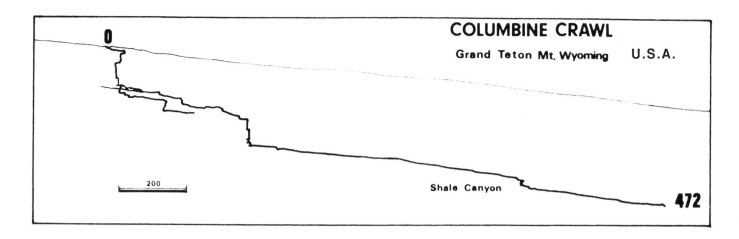

COLUMBINE CRAWL

Grand Teton Mt. Wyoming   U.S.A.

0

200

Shale Canyon

472

DEEP CAVES:

1. *Columbine Crawl Cave* .... –472 m
(Teton County, Wyoming)

Columbine Crawl is located in Darby Canyon, on the western slopes of the Teton Mountains, at an altitude of 2990 m. It is formed in the Madison limestone. The entrance is 850 m above the lowest resurgence in the area. The entrance is a small hole 1 m by 0.7 m at the base of a small cliff bank in the midst of rugged alpine karst. It is covered by snow for up to ten months of the year.

Mike Beer was the first to enter the cave when he explored 70 m at the end of 1980. The depth of –453 m was reached in 1981 by Jed Blakeley, Jean Jancewicz and Bob Benedict, for a length of 2900 m. The cave consists of a 70 m entrance crawl followed by 2 series of drops through two layers of limestone. The first series is characterized by rotten rock and tight squeezes on rope. The rock of the second series is solid; there are bigger shafts and lots of water. The cave drops steeply through the Madison limestone into a soft shale layer at the top of the Darby formation. There are some 100 m long sewer pipes requiring nearly total immersion in 1° C water to get through. At the end of a low air space belly crawl is a small chamber at the bottom of a breakdown filled canyon that marks the end of the cave.

One significant side passage was found to contain several torrential waterfalls and a howling wind that ends in a shaft whose floor is covered with gravel and a 0.5 m deep lake.

Map: from Rick Rigg and B. Benedict showing 3701 m of passage.

Bibliography: *1984 N.S.S. Convention Abstracts.*

2. *Lechuguilla Cave* .... –458 m
(Guadalupe Mountains, Eddy County, New Mexico)

Located within Carlsbad National Park, about 7 km from Carlsbad Cavern. The entrance area has been known since 1914 when it was mined for guano. With the incentive of winds blowing up to 60 km/hr, mining rubble was removed from the back of the cave in 1985 and 1986, giving access to a series of large rooms and passages. Exploration in 1986 reached a depth of –283 m for a length of over 2200 m. Many more large passages containing unusual speleothems were discovered in 1987 in a series of expeditions bringing together cavers from all over the western U. S. Passage size generally averages 3 to 6 m wide

and 2 to 6 m high. The Western Borehole averages 15 to 45 m wide and 6 to 25 m high for its 1200 m length. The largest room in the cave is currently 90 x 180 m and the deepest pit is 45 m. The most remarkable fact about Lechuguilla is not its tremendous passage length and size, but the fact that there is virtually no mud or organic debris anywhere in the cave, which averages 19.4° C. Lechuguilla is one of the most heavily decorated caves in the U. S.A. with vast expanses of snow-white gypsum and aragonite, numerous pools and flowstone formation areas. The depth reached –430 m in March, 1988 for a length of over 26 km with a linear extent of over 2 km. The length had increased to over 53 km by March, 1989. This is one of the most significant cave discoveries in the U. S. in the 1980's. Exploration in progress. (*N.S.S. News*, 1986, 44 (9); *N.S.S. News*, 1988, 46 (2); 46(10) and 46(11)).

Map: from Lechuguilla Cave Project plots.

Rick Bridges.

3. **Great EXpectations Cave** .... –430 m
(Bighorn Mountain, Washakie County, Wyoming)

A trip from the upper entrance (Great X, alt. 2590 m), where the cold waters of the N. and S. Trapper creeks converge and sink at the dolomite-granite contact, to the lower entrance (Great Exit, 2160 m) is one of the nicest through trips in the U. S. The cave essentially consists of a single 12 km long stream passage that narrows down to the Grim Crawl of Death near the lower entrance. It was first explored in 1976 by cavers from Sheridan, Wyoming, then from 1976 to 1980 by cavers from Wyoming, S. Dakota, Colorado, Idaho, and Montana. The connection to Great Exit was made from this dug entrance in August, 1980 by T. Miller and P. Shifflett (*N.S.S. News*, May 1981), but was not mapped until 1985 (*N.S.S. News*, June 1986).

4. **Bigfoot Cave** .... –367 m
(Marble Mountains, Siskiyou County, California)

This alpine cave (air 3° C, water 1° C) is located in the Marble Valley cirque on the east slope of the Marble Mountains. Along with other caves in the area, it is one of the most challenging in the west coast. Discovered by organized cavers in 1974, the depth of –326 m was reached in July, 1975. This was increased to –340 m in August, 1976. In September, 1977 S. Knutson and D. Cowan made the connection between the bottom of Meatgrinder Cave and Bigfoot Cave to bring the depth to –367 m. This was the deepest cave in the U. S. until August, 1987. (*N.S.S. News*, 1978, 36 (5); *N.S.S. News*, 1985, 45 (4)).

5. **Neffs Canyon Cave** .... –357 m
    (Wasatch Mountain, Salt Lake County, Utah)
    Located east of Salt Lake City. First explored by organized cavers from Salt Lake Grotto in 1952. Bottom of cave (stream sinking into gravel) reached in 1956. Developed along a fault with a dip of about 50 degrees, with no side leads and few speleothems. The 9 rope drops up to 50 m deep and the 5° C temperature make this a strenuous cave to explore. (*N.S.S. News*, 1957, 15 (12)).

6. **Ainahou Ranch Cave** .... –352 m
    (Kilauea, Hawaii)
    Lava tube explored in 1977 by J. and L. Neiland and 1979 by British expedition. (*Cave Science*, 1981, 8 (3)).

7. **Ellison's Cave** .... –324 m
    (Walker County, Georgia)
    Located about 10 km west of Lafayette on Pidgeon Mountain, which is a north-south spur off the east side of Lookout Mountain. The resurgence (Blue Springs or Blue Hole) is on the east side of the mountain. The Historic, Dug, Lord's, and Gross entrances are on the west side, while Staircase entrance is on the east side of the mountain. The cave is formed in the entire thickness of the Mississippian-age limestone unit.

    The Historic Entrance has been long known to the Ellison family. Most of the exploration of the cave occurred from fall 1968 to summer 1969, when over 15 km of passages were mapped by cavers from Huntsville, Smoky Mountain, and Dogwood City grottoes and other cavers from many areas, led by R. Schreiber and D. McGuffin. The discoveries included 170 m deep Fantastic Pit (mostly free) and 134 m deep Incredible Pit. In 1974-1976 another 5 km were mapped with the discovery of the Attic area and extensions downstream. (Smith (M.) *The History of Ellison's Cave"*, (book), c. 1974).

8. **Silvertip Cave System** .... –321 m
    (Flathead County, Montana)
    Nearly all the known caves on Silvertip Mountain were explored and mapped by the Association for Mexican Caves Studies under the direction of Mike McEachern, Steve Zeman, and Bill Steele, 1974-81. This cave system has about a dozen entrances and about 8 km of passages extending throughout much of the northern cirque of Silvertip Mountain. The uppermost entrance is Getout Cave, which is a steeply descending canyon passage leading to shafts and lower-level stream passages. Bell and Rainbow Caves consist of dry upper-level canyons and tubes that drop through shafts to lower stream levels. The main passage of the system is a large tubular stream passage that terminates in breakdown just upstream from the stratigraphically perched main spring (*Alpine Karst*, Summer 1978).

9. **Carlsbad Cavern** .... –313 m
    (Guadalupe Mountains, Eddy County, New Mexico)
    See below.

10. **Big Brush Creek Cave** .... –262 m
    (Uintah County, Utah)
    Developed in Madison Limestone and Humbug Formation (Mississippian). Early exploration and mapping of entrance sections by Salt Lake Grotto in the 1950's. Explored and mapped to 4600 m and –122 m by A. N. and M. V. Palmer, 1968-1971 (*N.S.S. Bulletin*, 37, p. 68). The length and depth have been doubled since then by Salt Lake Grotto under the direction of Dale Green. A large entrance receiving seasonal overflow from Big Brush Creek, which sinks just upstream, leads to a wide, breakdown-floored passage that terminates after several hundred meters in a boulder choke. Extensive mazes of floodwater passages extend beyond and resolve themselves into several convergent stream passages interrupted by short drops with deep plunge pools. Contains high concentrations of carbon dioxide and perennial ice.

11. **Kazamura Cave** (Kilauea, Hawaii) .... –261 m
    This is the depth of the longest segment. With a length of 11,713 m, this is one of the longest lava tubes in the world. It was explored in 1979 by a British expedition (*Cave Science*, 1981, 8 (3)).

12. **Papoose Cave** (Idaho County, Idaho) .... –251 m
    This cave located about 10 km SW of Riggins, in the Nez Perce National Forest, and has two entrances (alt. 1540, 1555 m) both located near Papoose Creek. Known to locals since at least 1959, it was explored and mapped from 1965 to 1972 by cavers from the Cascade, Gem State, Great Basin, and Xanadu Grottoes and the EWSCSC (Huppert (G.) "Speleography of Papoose Cave", Idaho Bureau of Mines and Geology, *Info. Cir. 35*, October 1981, 25 p).

13. **Meanderbelt Cave** .... –246 m
    (Flathead County, Montana)
    Explored and mapped by the AMCS, 1975-1977. Known alternatively as "Dismal Dip" by the explorers, this is one of the most unpleasant caves on Silvertip Mountain. It consists of two divergent passages that extend approximately down the local 35° dip of the limestone as narrow canyons and end in constrictions. Length 1164 m (*Alpine Karst*, Summer 1978).

14. **Sunray Cave** .... –244 m
    (Flathead County, Montana)
    Explored and mapped by the AMCS, 1976-1977. The main passage is an inactive tube that extends directly down the 40° dip of the rocks, then turns along the strike and terminates at a second entrance (*Alpine Karst*, Summer, 1978).

15. **Spanish Cave** (Colorado) .... –230 m
    Located at an elevation of approx. 4000 m.

16. **Bull Cave** (Blount County, Tennessee) .... –225 m
    Alt. 560 m. Stream cave 2245 m long formed along the Great Smokies Fault in the Jonesboro Limestone (Barr (T.) *Caves of Tennessee*, 1961). Remapped in 1969 by R. Schreiber (*Speleotype 4* (3), 1969).

17. **Lost Creek Siphon** (Montana) .... –225 m

18. **Virgin Cave** (Eddy County, New Mexico) .... –221 m
    Includes a drop requiring 200 m of rope (The Four O'Clock Staircase). (*1986 N.S.S. Convention Guidebook*).

19. **Ape Cave** (Washington) .... –214 m
Lava tube explored in 1963 by W. Halliday *et al.*

20. **Bobcat Cave** .... –213 m
(Beth/Highia Counties, Virginia)
(See below).

21. **Simmons Mingo - My Cave System** .... –207 m
(Randolph/Pocahontas Counties, West Virginia)
Cave with five entrances: Simmons-Mingo, Simmons Pot, Stans Blowing Rock, Oildrum Falls, and My Cave. Developed along lineament extending under Mingo Knob, with a linear extent over 4,200 m. Explored by the Potomac Speleological Club beginning in the mid 1960's and presently being resurveyed.

22. **Dorton Knob Smokehole** (Tennessee) .... –201 m
First explored by Bill Cuddington in 1957, and mapped by Chris Kerr (unpublished). Cave 1280 m long with several short drops.

LONG CAVES:

1. *Mammoth Cave System* .... 530,000 m
(Edmonson, Barren, and Hart Counties, Kentucky)
Mammoth Cave National Park, which contains the greater part of the Mammoth Cave System, is located halfway between the cities of Louisville and Nashville, to the NW of Cave City and Park City, through which passes Interstate Highway 65. The Mammoth Cave System is by far the longest known cave in the world. The system has developed on the south bank of the Green River, in the Pennyroyal Plateau. Most of the system lies under Toohey Ridge, Joppa Ridge, Flint Ridge, and Mammoth Cave Ridge. The last two are separated by Houchins Valley.

The passages are formed in Paleozoic age Mississippian limestones, including the St. Louis, Ste. Genevieve, and Girkin formations, corresponding to the three principal levels in the system. The limestone dips at 0.5° to the NW and the cave occupies a relief of about 90 m. It has twenty-one entrances, of which only three were not dug open.

Water enters the system through infeeders in the Pennyroyal Plateau and numerous sinkholes south of Park City, and resurges at numerous springs along the Green River. These include (from upstream to downstream) River Styx Spring and Echo River Spring (draining Mammoth Cave Ridge) and Turnhole Spring (the largest draining Joppa Ridge, at 400 l/sec). In times of flood, the Green River floods part of the cave system.

The history of exploration begins some 2240 years ago when Indians explored for great distances in Salts Cave and Mammoth Cave, going many kilometers from the entrances using cane torches. They appear to have been true explorers, not motivated solely by utilitarian or religious goals, as the careful research of Dr. P. Watson and others has shown.

Modern exploration began in Mammoth Cave ("Historic Entrance"). Although first explored by prehistoric Indians, the first recorded visit was that of a bear hunter named Houchins in 1797. It was "rediscovered" in 1809.

As with many other caves in the area, it was intensively mined for saltpeter during the war of 1812. The great extent of the cave was soon recognized, and a first map was made by E. F. Lee in 1835 showing 12,874 m of passages. The cave was visited by tourists from 1838 to 1841. Their guide, Stephen Bishop, considerably extended the cave by exploring the passages beyond the Bottomless Pit. Other notable explorers in the 19th century were F. Demunbrun in 1858 and later the Lee brothers. By the beginning of the 20th century 48 km had been mapped, notably by Hovey, Call, and Martel, who forecast the connections that occurred in the latter half of the century. The now famous map of the German Max Kaemper was made in 1908 and showed 56,300 m of passage. In 1909, M. Kaemper and E. Bishop reached 70,810 m (a figure which cannot be exactly confirmed today). They also discovered the Violet City Entrance, which was connected to the system in 1931.

The cave wars among land owners for access to commercial exploitation of the cave began in 1916 and led to the opening of many new entrances: the Cox and Morrison Entrances were dug open by Cox and Morrison in 1916, the New Entrance and Frozen Niagara Entrance were dug open by Morrison in 1921 and 1924 respectively, Cathedral Dome Entrance was opened in 1930, Carmichael Entrance was opened in 1931, and finally the New Discovery Entrance was dug open in 1940.

A new map by H. D. Walker showed 33,800 m of passage in 1936, of which 5600 m were not on Kaemper's map, giving an official length of 61,900 m to the system. Hanson and Hunt discovered several passages in 1938 that were forgotten and rediscovered in 1972 from Flint Ridge. The 1956 map of R. Nelson showed 52,300 m, giving the system a length of 71,614 m if one counts the additional passages shown on the 1908 and 1936 maps.

The Cave Research Foundation began a new survey in 1969, bringing the length to 78,696 m in 1971. By August 1972, just before the historic connection with Flint Ridge Cave System, the length of Mammoth Cave was 93,200 m.

The history of the exploration of Flint Ridge Cave system is considerably more complicated due to the numerous connections which took place to form it:

(1.) **Colossal Cave** was discovered on July 15, 1895 by L.H. Lee who explored it in 1896 and made the connection with the Bedquilt Entrance, discovered in 1871. The other two entrances to this cave are Woodson/Adair (1890) and Hazen Entrance (1897). Exploration continued by L. Hazen in 1897 and Vaughan, Marshall, and Armstrong in 1902, for a length of 4270 m. Colossal Cave was connected to Salts Cave on August 22, 1960 by the C.R.F.

(2.) **Salts Cave** was first explored by prehistoric Indians, then by Turner and Floyd Collins in 1912. It has a second entrance called the Pike Chapman Entrance. Exploration continued by J. Lehrberger and Reccius in 1952 and J. Lehrberger and W. Austin in 1954. Salts was connected with Colossal on August 22, 1960, and with Unknown/Crystal Cave on August 21, 1961.

(3.) **Crystal Cave** (or Great Crystal Cave, or Floyd Collins' Crystal Cave) was discovered on December 17, 1917 by Floyd Collins who explored it up to 1922. Dyer, Miller and Austin resumed exploration in 1948. The N.S.S. organized a large expedition in 1954 that marked the beginning of systematic exploration under Flint Ridge. J.

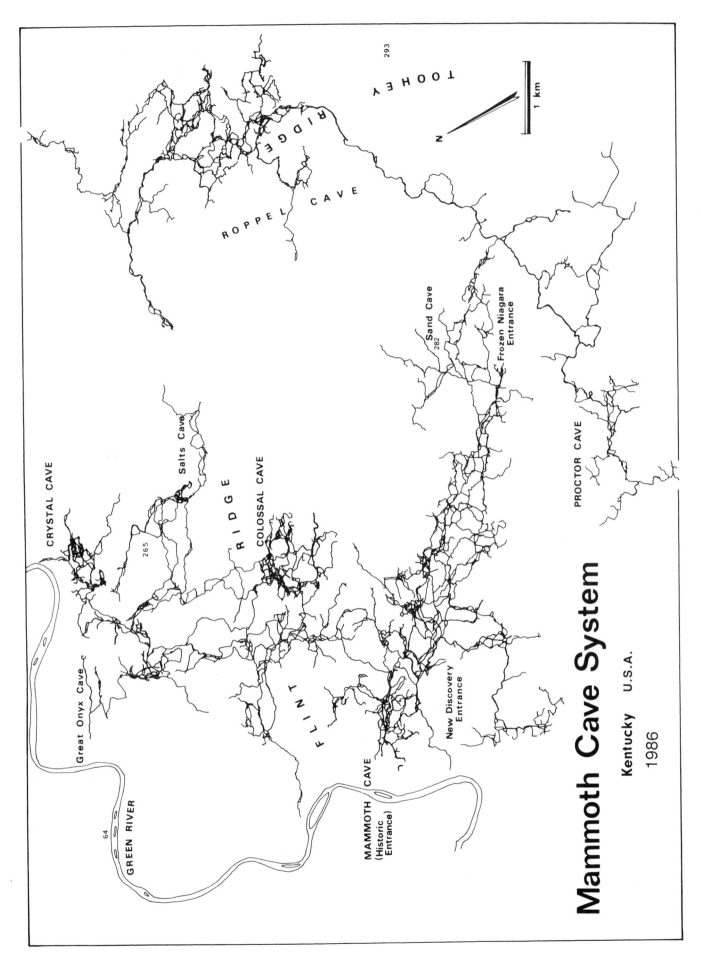

Mammoth Cave System

Kentucky   U.S.A.

1986

CRYSTAL CAVE

Great Onyx Cave

Salts Cave

265

R I D G E

COLOSSAL CAVE

F L I N T

MAMMOTH CAVE (Historic Entrance)

New Discovery Entrance

Sand Cave
282

Frozen Niagara Entrance

PROCTOR CAVE

R I D G E

ROPPEL CAVE

T O O H E Y

293

GREEN RIVER

64

N

1 km

Lehrberger and W. Austin connected Crystal to Unknown Cave in September 1955.

(4.) **Unknown Cave** is thought to have been discovered in 1903 by M. Charlet and explored for 2090 m the same year. In 1954 and 1955, J. Lehrberger and W. Austin continued exploration, leading to the connection with Crystal Cave.

In May 1956, thanks to a feat of precision surveying, the Austin Entrance was dug open to facilitate exploration in Crystal-Unknown Caves. Another important date is 1957, when the Cave Research Foundation (CRF) was created by Philip Smith, E.R. Pohl, and others. Their goal was to connect all the known caves into a single system. By applying scientific principles and always surveying new passages as they were discovered, rapid progress was soon made.

Thus with the connection of Salts/Colossal to Unknown/Crystal on August 21, 1961, the Flint Ridge Cave System was formed with a length of 33,956 m. The long quest to pass under Houchins valley and connect with Mammoth Cave then began. The dry chronology of lengths achieved reads as follows: 45,222 m in 1963, 52,946 m in 1964, 59,384 m in 1965, 68,718 m in 1966, 80,466 m in 1967, 90,122 m in 1968, 99,939 m in 1969, 118,607 m in 1970, and 132,930 m in 1971.

The long-sought-after connection with Mammoth Cave was made on September 9, 1972, by a team of six CRF cavers led by J. Wilcox (the actual connection was made on August 30, but not mapped). This brought the length of the system to an astounding 232,500 m, assuring it the rank of the world's longest cave for the foreseeable future.

This connection did not slow down the steady increase in length, i.e. 258,295 m in 1973, 268,917 m in 1974, 284,366 m in 1975, 297,080 m in 1976, 306,949 m in 1977, and 312,240 m in 1978.

A connection with Proctor Cave on June 28, 1979 caused the length to jump to 345,227 m. This Joppa Ridge cave was discovered in 1863 and explored beginning in 1972 to a length of 16,254 m at the time of the connection. The next years saw the length of the entire system grow to 361,620 m in 1980, 367,890 m in 1981, and 379,140 m in 1982.

Going back for a moment, it was in April 1976 that J. Borden and J. Currens discovered Roppel Cave, located on Toohey Ridge, outside of the National Park. They formed the Central Kentucky Karst Coalition (CKKC) and explored 3272 m in 1977 and 8046 m in 1978. A breakthrough in the "S Survey" to the Hobbit trail opened up the cave, and the length jumped to 18,137 m in 1979, 35,985 m in 1980, and 42,980 m in 1981. The connection of Roppel Cave with Mammoth Cave System on September 10, 1983 brought the length of the entire system to 473,680 m!

Although a substantial percentage of the mapping involves re-surveying passages to improve the map quality, new discoveries continue at a steady pace, for a total length of 484,300 m in 1984, 500,506 m in 1985, 523,023 m in 1986, and who knows how much by the year 2000?

Map: produced by the Cave Research Foundation and the Central Kentucky Karst Coalition (1986), sent by Richard Zopf.

Bibliography: Palmer (A.) - *A Geological Guide to Mammoth Cave National Park*, Zephyrus Press, Teaneck, 1981, XIV-196 p.

Watson, Patty J. (ed.) - *Archeology of the Mammoth Cave Area*, Academic Press, New York, 1974, XX-255 p.
Brucker (R.), Watson (R.) - *The Longest Cave*, A. Knopf, New York, 1976, XX-316-XIV p.

2. *Jewel Cave* 123,771m
(Custer County, South Dakota)

Jewel Cave National Monument is located in the southern Black Hills, 23 km to the west of Custer, on US Highway 16. It is found in the upper half of the Pahasapa Limestone.

There is no certain record of the original discovery of the cave, but a mining claim was filed on the site in 1900 by Felix, Frank, and Albert Michaud, and Charles Bush. Though the cave became a National Monument in 1908, the Park Service did not assume responsibility until 1933. Perhaps 3 km was discovered between 1900 and 1959, but only 1.3 km were surveyed. Dwight Deal, along with Herb and Jan Conn, began a systematic exploration of the cave in 1959. The Conns continued the effort from 1960 to 1980, when Mike Wiles and Ken Allgier took over.

Jewel Cave is dry, with little evidence of past or present stream activity. The known cave is located under about 5 km² of surface area. Passages occur at four fairly distinct levels. The vertical relief at any one point is about 80 m, while the total vertical extent of the cave is 135 m. The largest vertical drop is 30 m deep. The cave has one natural entrance near the northwest extremity which was originally too small for human entry. An elevator and portal entrance now exist near the center of the known system, greatly aiding exploration. Breezes occur throughout the cave in response to barometric pressure changes, with wind speeds in some tight spots up to 51 km/hour. Except for the highest levels, the passages are coated with a layer of calcite crystals up to 17 cm thick. Deposits of manganese oxides are found throughout the cave, with some as much as 20 cm thick.

Map: from the 1975 map of H. Conn and documents sent by Bob Gulden.

Bibliography: Conn (H.) - "Barometric Wind in Wind and Jewel Caves, South Dakota", *N.S.S. Bull.*, 1966, 28(2): 53-79.
Deal (D.) *Geology of Jewel Cave National Monument*, thesis from the Univ. of Wyoming, 1962, 183 p.
Conn (H. and J.) -*The Jewel Cave Adventure*, Zephyrus Press, Teaneck, 1977, 239 p.
Conn (H.) - *N.S.S. News*, December, 1982.

Mike Wiles.

3. *Wind Cave* ....82,074 m
(Custer County, South Dakota)

The entrance to Wind Cave lies in the floor of Wind Cave Canyon, 100 m north of the National Park Visitor Center, in the southern Black Hills of South Dakota.

According to stories, the cave was discovered in 1881 by Tom and Jesse Bingham. Little happened for 10 years until in 1890 the South Dakota Mining Company hired J. D. McDonald to manage a mining operation. From notes in a diary kept by his son Alvin, it has been ascertained that possibly as much as 13 km of cave were explored by them. An official survey was done in 1902 by the U. S. Government.

Systematic exploration of the cave began in 1959, the Colorado Grotto mapped 4800 m. In 1963-1964 Herb and

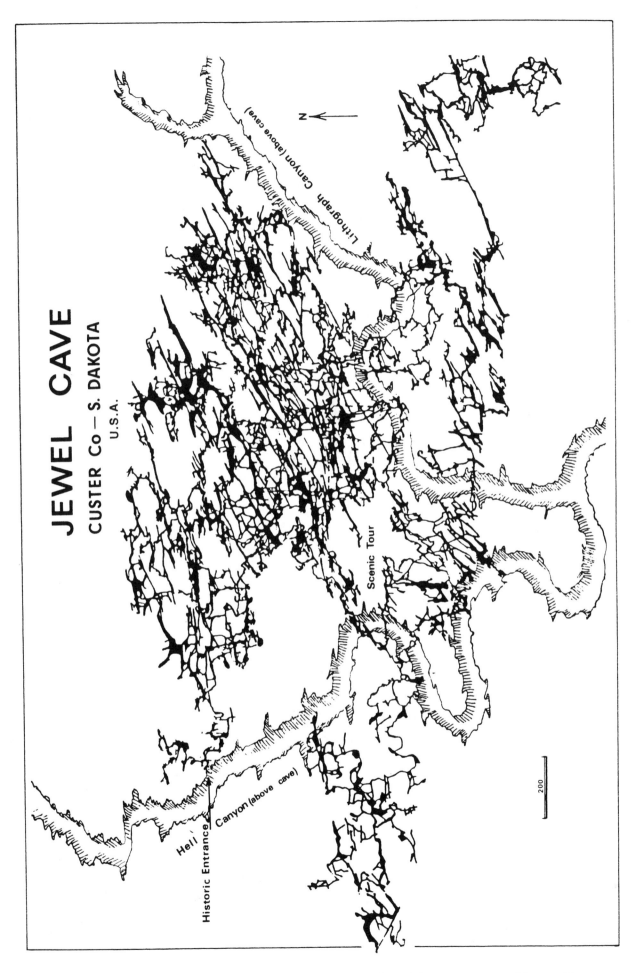

JEWEL CAVE
CUSTER Co—S. DAKOTA
U.S.A.

Lithograph Canyon (above cave)

Scenic Tour

Hell Canyon (above cave)

Historic Entrance

N

200

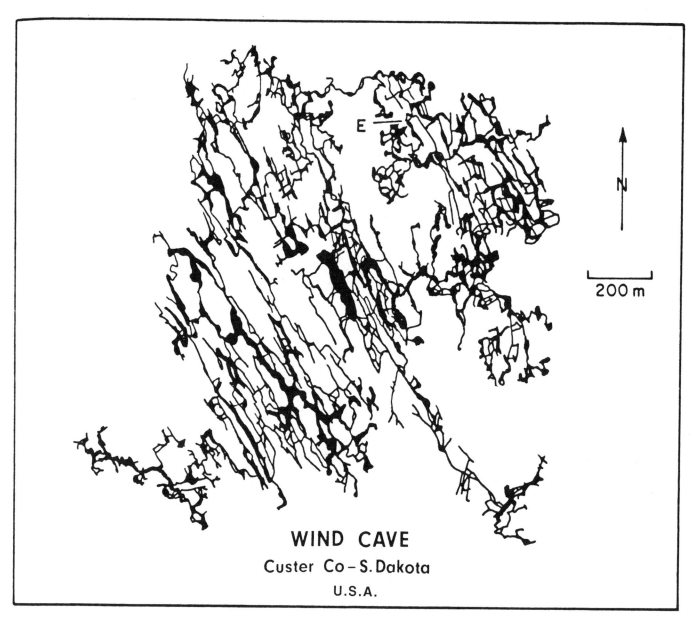

WIND CAVE
Custer Co – S. Dakota
U.S.A.

200 m

E

N

Jan Conn mapped an additional 4800 m. The length of the system was increased to 32,614 m by the end of 1973, and 49,150 m in 1977. John Scheltens has been organizing the survey effort by various groups and Park Service employees since 1979. Total surveyed lengths were 56,327 m in 1981, 64,000 m in 1984, 70,039 m in 1985, and 77,262 m in 1986. The 50-mile mark (80,467 m) was passed in 1987. Over 2000 leads remain to be checked as exploration continues.

The natural entrance is a 35 cm diameter hole. Over 2 km of passages have been developed for commercial tours. An artificial entrance with an elevator shaft was constructed during the 1930's. The cave is a complex maze underlying less than 3 km² of surface area. It has a vertical relief of 172 m. At the lowest point the cave has intersected the water table at Windy City Lake, Calcite Lake and Land of 10,000 Lakes. The cave is best known for its boxwork formations which are found throughout the cave.

Wind Cave is formed in the Madison limestone of Mississippian age, which dips to the south and east. The cave cuts all of the major stratigraphic sections of the limestone.

Map: map of John Scheltens sent by A. Palmer.
Bibliography: Palmer (A.) *The Geology of Wind Cave*, Wind Cave Natural History Association, 1981.
Annual Reports of the Windy City Grotto for 1970, 1971, 1972, and 1973, and *Annual Reports* of John Scheltens for 1979, 1980, 1981, and 1982.
N.S.S. News, January and February, 1988.

4. *Fisher Ridge Cave System* ....71,500 m
(Hart County, Kentucky)

This system lies under a portion of the dissected Fisher Ridge, about 3 km SW of the small community of Northtown on the Horse Cave 7.5 minute quadrangle. Most of the cave is located at an altitude between 150 m and 200 m.

The system has three entrances. The Historic entrance was first noted as a blowing hole at the sandstone/limestone contact and was dug open in February 1981. A multiple drop shaft complex led to a multilevel complex of subparallel passages that was 60 m in vertical extent. In April of that year Splash Cave was found and explored as

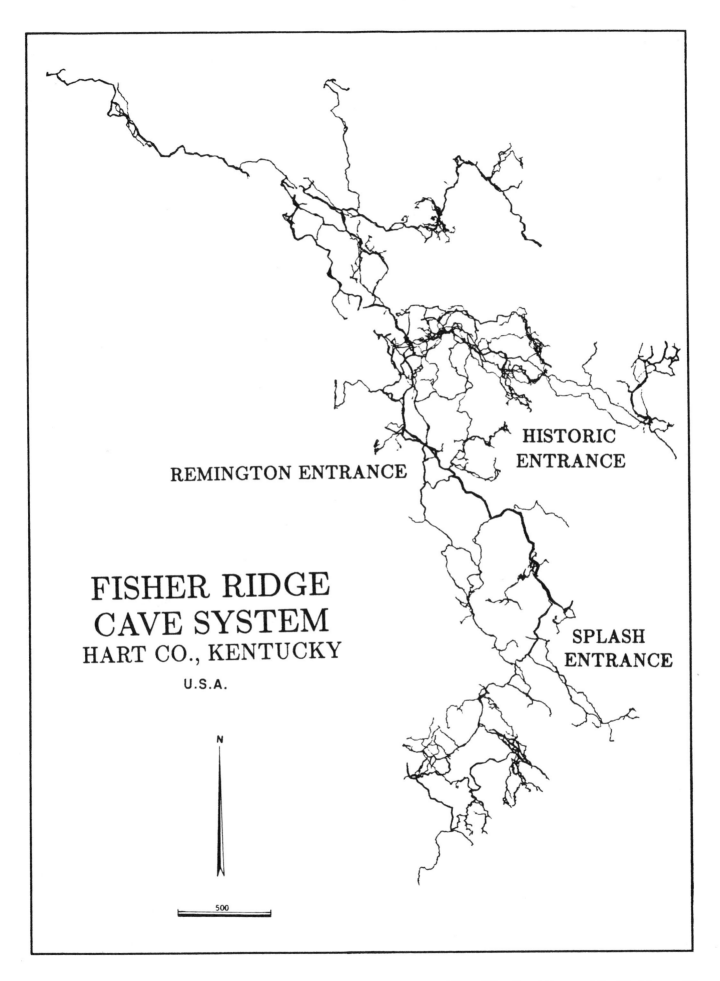

FISHER RIDGE CAVE SYSTEM
HART CO., KENTUCKY
U.S.A.

REMINGTON ENTRANCE

HISTORIC ENTRANCE

SPLASH ENTRANCE

N

500

a 200 m long cave with no obvious continuation. In May a major abandoned trunk passage was found that led nearly 2 km to the south where in June it intersected Splash Cave, giving Fisher Ridge a second entrance. The trunk passage led also to the north, into Ice Cave Ridge, after amorphous flowstone blocking the passage was cleared away. Several major related base level stream segments were later discovered. A 1983 radio location confirmed that the Remington Room was directly under a major sinkhole, and digging subsequently opened the cave's third entrance here. In 1984 a difficult climb led to Bob's Borehole, a major trunk with no apparent hydrologic relation to the rest of the system. A flood in May 1984 resulted in the opening of two new passages. They were followed in 1985 to over 10 km of new passage under Eudora Ridge, under which portions of the Mammoth Cave System can also be found. About 5 km of new passage was found in 1986, over 6 km were found in 1987 and exploration still continues. The mapping and study of the cave has been a project of the Detroit Urban Grotto of the NSS.

Evidence of aborigine visitation was discovered in 1982. The remains discovered have been tentatively dated at 800-1200 B.C. by Dr. P. Watson.

The base level streams resurge at Lawler Blue Spring, 4 km to the north. Fisher Avenue flows to the NNW, probably to a much different discharge point on the Green River. There are numerous sumps at base level, which has slowed exploration. Fisher Ridge Cave System is only 250 m east of Mammoth Cave System and 500 m west of Crump Spring Cave (see below).

The system is formed in Girkin, Ste. Genevieve, and upper St. Louis limestones, all of Mississippian age.

Map: from the 1984 map sent by Bob Gulden and the 1985 map by Chip Hopper.

Bibliography: Saunders (J.) Kentucky's Fisher Ridge Cave System. *Wisc. Speleologist* 1983, 18 (3): 14-17. Monthly issues of the *DUG SCOOPS*, the publication of the Detroit Urban Grotto (N.S.S.).
N.S.S. News, March, 1982.

Joe Saunders and Keith Ortiz.

5. *Friars Hole Cave System* ....**68,824 m**
(Pocahontas and Greenbrier Counties, West Virginia)
Friars Hole Cave System is located in and near the Friars Hole Valley on the west side of Droop Mountain in northern Greenbrier and southern Pocahontas Counties. The area is about 2 km to the north of upper Spring Creek.

The system has ten entrances, at altitudes between 680 and 760 m: Snedegar Stream, Crookshank Pit, Snedegar Saltpetre, Rubber Chicken, Friars Hole Cave, Staircase, Snedegar Maze, Canadian Hole, Toothpick Cave and Radio Pit. The first three have been known since the middle of the the 19th century. The seven others were discovered or dug open between 1960 and 1983. The Snedegar-Crookshank Cave System was connected to the Rubber Chicken-Friars Hole system in August, 1977, Rubber Chicken and Friars Hole Cave having been connected the previous year. Canadian Hole was integrated into the system in September, 1977, and Toothpick Cave was connected in 1978, and Radio Pit was connected in 1987.

While parts of the system were surveyed as separate caves in the early 1960's, a complete resurvey was begun in 1976 in by N.S.S. cavers led by Doug Medville, when a length of 24,080 m was quickly reached. Meanwhile, cavers from McMaster University (Hamilton, Canada) mapped the northern parts of the system, Canadian Hole in particular. As connections were made and new passages found, the length of the system increased to 33,470 m in 1977, 43,854 m in 1978, 48,270 m in 1979, 56,674 m in 1980, 61,629 m in 1981, 68,122 m in 1984 and finally 68,824 m in 1987, for a total vertical extent of 188 m.

The system contains three almost-separate drainages. The southernmost of these carries local water sinking in the Friars Hole Valley southwest and along the strike. The system's central drainage, carrying water from several of the caves entrances (Snedegar Stream, Toothpick, Crookshank Pit, Snedegar Staircase and Rubber Chicken) flows to the north, finally terminating at a sump at the system's lowest point. A clearly defined drainage divide separates the southern and central drainages in the system. The northern part contains a stream which sinks several km to the north (Hills Creek) and which rises from breakdown at the system's northern end. It flows to the south then is lost in rockfall along a fault, but pirates water from the central and southern drainages.

The system also contains several large paleo-trunks, one of which (The Highway) extends for over 4 km. It also contains some large chambers, such as Monster Cavern, over 100 m across and 60 m high.

The system is developed in the Mississippian Greenbrier Group of limestones. All of the streams in the cave have been traced to a spring along Spring Creek, 11 km south of the southernmost part of the system.

Map: from plots sent by S. Worthington and D. Medville.

Bibliography: Medville (D.) "Geography of the Friars Hole Cave System, U.S.A.', *Proc. 8th Inter. Congr. Spel.*, *Bowling Green, 1981*: 412-414.
Worthington (S.)*The Paleodrainage of an Appalachian Fluviokarst: Friars Hole, West VA*, thesis, McMaster Univ., Ontario, 1984, 218 p.
N.S.S. News, February, 1979; October, 1979; February 1981.

6. *Organ Cave System* ....**60,510 m**
(Greenbrier County, West Virginia)
The Organ Cave System is located in Greenbriar County below the community of Organ Cave. It is one of 56 caves which have been found on the Organ Cave Plateau. This plateau covers a surface area of about 40 km², and ranges between altitudes of 120 and 200 m. It is bounded on the north and the west by the Greenbrier River, on the south by Second Creek, and on the east by White Rock Mountain.

While accounts of the cave date back to the late 1700's, and the cave was mined for saltpetre during the 1800's, modern exploration did not begin until 1948, the same year that saw the connection of Organ Cave and Hedricks Cave. Much of the early exploration of the Organ Cave System was conducted by members of the Charleston Grotto of the N.S.S. and the West Virginia Association for Cave Studies (WVACS). Lipps Cave was connected into the system in 1958. Members of the District of Columbia Grotto (DCG) of the NSS continued the exploration and

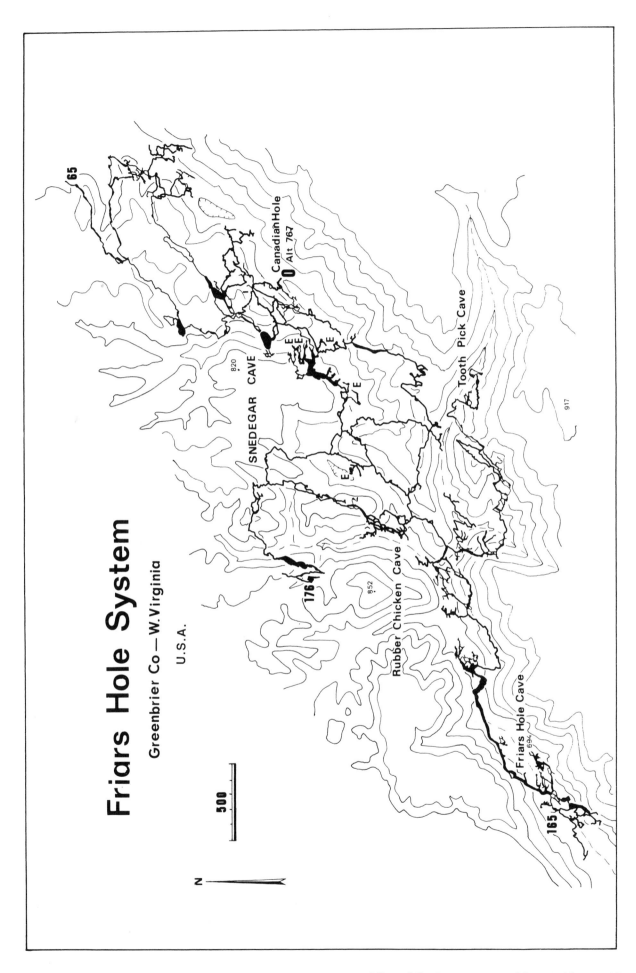

Friars Hole System

Greenbrier Co — W.Virginia

U.S.A.

500

N

SNEDEGAR CAVE

Canadian Hole
Alt 767

Tooth Pick Cave

Rubber Chicken Cave

Friars Hole Cave

65

176

165

820

852

694

917

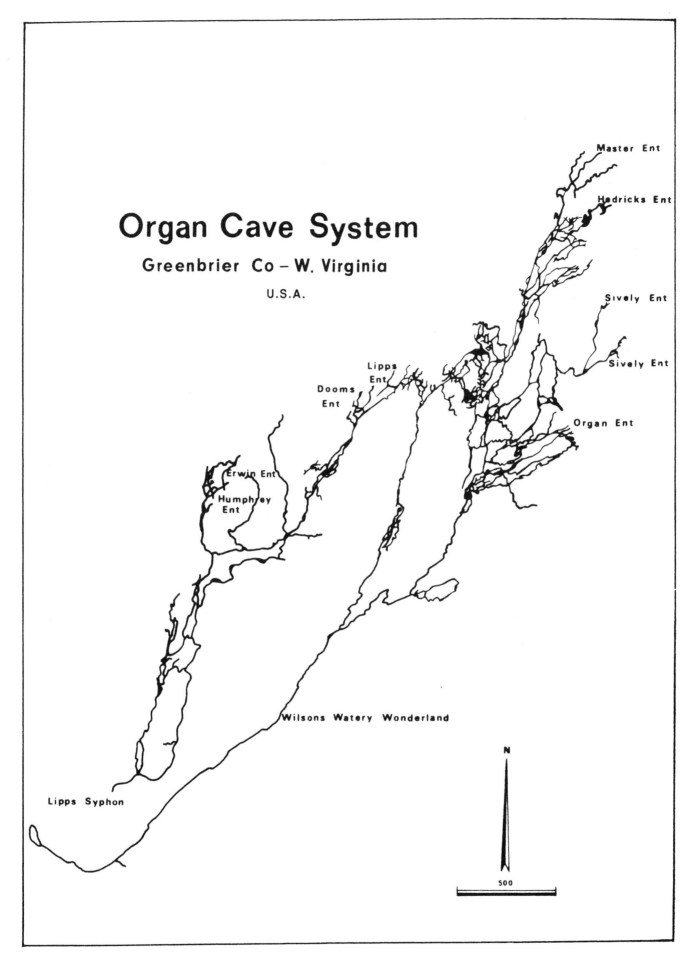

Organ Cave System

Greenbrier Co – W. Virginia

U.S.A.

Master Ent

Hedricks Ent

Sively Ent

Sively Ent

Organ Ent

Lipps Ent

Dooms Ent

Erwin Ent

Humphrey Ent

Wilsons Watery Wonderland

Lipps Syphon

N

500

survey of the system during the 1970-1982 period, reaching a length of 51,490 m in 1976 (replacing the 70 km announced earlier) and a length of 59,546 m in 1980. Since then members of both DCG and WVACS have worked together to further the project. As a result of these efforts, detailed maps of the caves were produced. The Organ Cave System presently has ten entrances and a surveyed length of 60,510 m.

The cave is formed in the Hillsdale and Sinks Grove limestone members of the Middle Mississippian Group. The main stream passages of the Organ Cave System are formed along the Caldwell Syncline, which plunges to the southwest toward Second Creek with a dip of about 5°. The cave streams resurge at a series of springs along Second Creek and the Greenbrier River.

Map: 1980 version published in the *D.C. Speleograph* (1981), sent by Paul Stevens.

Bibliography: *D. C. Speleograph*, 1981, 37 (2).

Paul Stevens..

7. *Lechuguilla Cave* ....**53,000 m**
(Eddy County, New Mexico)
See above.

8. *Crevice Cave* (Perry County, Missouri) ...**45,385 m**
Crevice Cave is a complex system of passages located in Perry County, near the town of Perryville. It has three entrances: the Historic Entrance, Echo Pit, and the Pipistrell Entrance. The four major stream passages are the First and Second Miles, the Bends, the Nile, and Merlins. The normally placid streams rapidly change to torrents in periods of heavy thunderstorms. The cave is formed in Champlainian Series limestone of Ordovician age.

Exploration began with the efforts of Stanley Sides in 1960. The length was brought to 1,900 m in 1965 with the explorations of Dennis Drum. The length reached 4,055 m in 1967. In 1968 Paul Hauck and John Schneider reached a length of 6,035 m, then 8,288 m in 1972. Echo Pit was opened in 1974 and Pipistrell Entrance was discovered in 1975, allowing the pace of discovery to accelerate rapidly. Total lengths of 17,702 m and 36,692 m were soon reached. After that, new discoveries became harder to find, and the length reached 39 km in 1976 (SEMO Grotto), then 41,399 m in 1978, 43,895 m in 1983, and 45,385 m in March 1986.

Connections are being sought with Moore Cave System (see below) and Mertz Cave.

Map: from the 1979 map of Paul Hauck, sent by A. Palmer and J. Vineyard.

Bibliography: Hauck (P.) Crevice Cave, *The Southeast Caver*, 1976, 2 (3).

from text by Paul Hauck.

9. *Cumberland Caverns* ....**44,444 m**
(Warren County, Tennessee)
Cumberland Caverns is located on the SE flank of Cardwell Mountain, on the Cumberland plateau, near McMinnville. It has three entrances: Historic Entrance, Onyx Curtain Entrance, and Henshaw Cave. It is formed in the Mississippian age Monteagle limestone.

Formerly known as Higgenbotham Cave, the name was changed to Cumberland Caverns when it was opened as a show cave in 1956.

The discoverer was thought to have been Aaron Higgenbotham in 1810. He explored about 250 m of passage before his light failed and he had to wait for three days before being rescued! Shortly afterwards, locals explored about 1500 m. Nearby Henshaw Cave was used to mine saltpetre during the war of 1812, and was later explored by Shelah Waters and Collins around 1869.

While a description of Higgenbotham Cave was published in 1918 by Thomas Bailey, its systematic exploration did not begin until the 1940's when Tank Gorin, Tom Barr, Bent Denton and other N.S.S. cavers discovered the great potential of the cave. In the 1950's they were joined by other active explorers including Roy Davis and Kentuckians C. Fort, B. Lawson, J. Reccius, G. Welsh, and J. Lehrberger. The Onyx Curtain Entrance was discovered from inside the cave around 1950 by Milford Garner and Duane Fertig. The connection with Henshaw Cave was made in April 1953 by T. Barr, T. Gorin, and D. Smith.

The Nashville Grotto (P. Zawislak, J. Smyre J. Hodson) began a new survey in 1971, reaching 29,040 m in 1972, 32,648 m in 1973, 38,400 m in 1975, 43,768 m in 1977, and finally 44,444 m in 1984.

The owner of the cave is known for his generosity in hosting events in the cave, including a memorable banquet for the 1000 participants in the 1981 8th International Congress of Speleology!

Map: from the map of R. Zawislak and J. Smyre, sent by A. Palmer.

Bibliography: Davis (R.) My Cave!, *Inside Earth*, 1976 (4):16-24.
Crawford (N.), Vineyard (J.) - *Guidebook to the Karst and Caves of Tennessee and Missouri*, 1981, 176 p.

10. *Sloan's Valley Cave System* ....**39,640 m**
(Pulaski County, Kentucky)
This system is located in southern Pulaski county and to the east of Cumberland Lake, which sometimes floods the lower levels. Access is via US Highway 27. The system has at least sixteen entrances: Garbage Pit, Post Office, Hughes, Great Rock Sink West and East, Scowling Tom's, Railroad Tunnel North and South, Screaming Willy, Mainton Main, Right and Left, Dry Weather, Martin Creek North, South and River.

Some of these entrances were known to seekers of prehistoric artifacts. During the war of 1812, the two Great Rock Sink entrances were exploited for saltpetre mining. These two swallets are now flooded by Cumberland Lake, but were open to tourism in the 1920's or 1930's, under the names of Crystal Cave or Cumberland Caverns.

Mapping of the Sloan's Valley Cave System began in 1940-1942 when Malott and Malott surveyed 13 km. Jillson made a map in 1952. Al Geiser from the Greater Cincinnati Grotto mapped about 25 km (unpublished) from 1964 to 1966. His plans to purchase the system to make a show cave were unsuccessful.

The survey shown in this book was begun in 1970 by Dewe Beiter and the Bluegrass Grotto and pursued by the Central Ohio Grotto in 1971 for a length of 27,925 m. The map was updated in 1976 to show a length of 36,200 m. A few additional discoveries brought the length to 39,640 m in 1982.

Map: from the map of the Central Ohio and Bluegrass Grottos (1980), sent by Angelo George.

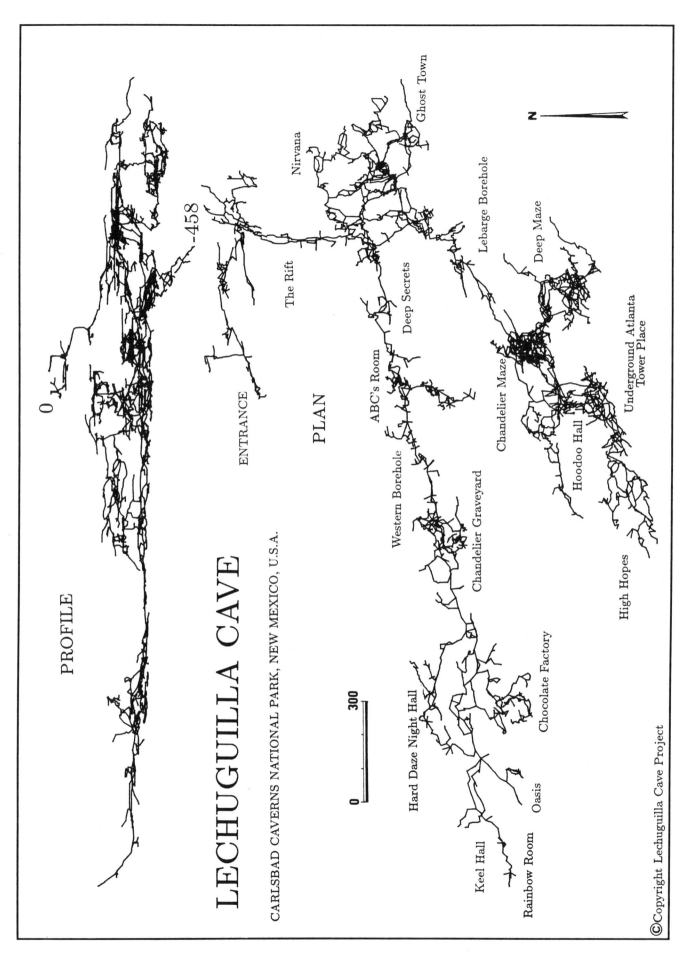

LECHUGUILLA CAVE

CARLSBAD CAVERNS NATIONAL PARK, NEW MEXICO, U.S.A.

PROFILE

0

-458

ENTRANCE

PLAN

The Rift

Nirvana

Ghost Town

Deep Secrets

Lebarge Borehole

ABC's Room

Deep Maze

Chandelier Maze

Western Borehole

Underground Atlanta
Tower Place

Chandelier Graveyard

Hoodoo Hall

High Hopes

Hard Daze Night Hall

Chocolate Factory

Keel Hall

Oasis

Rainbow Room

300

0

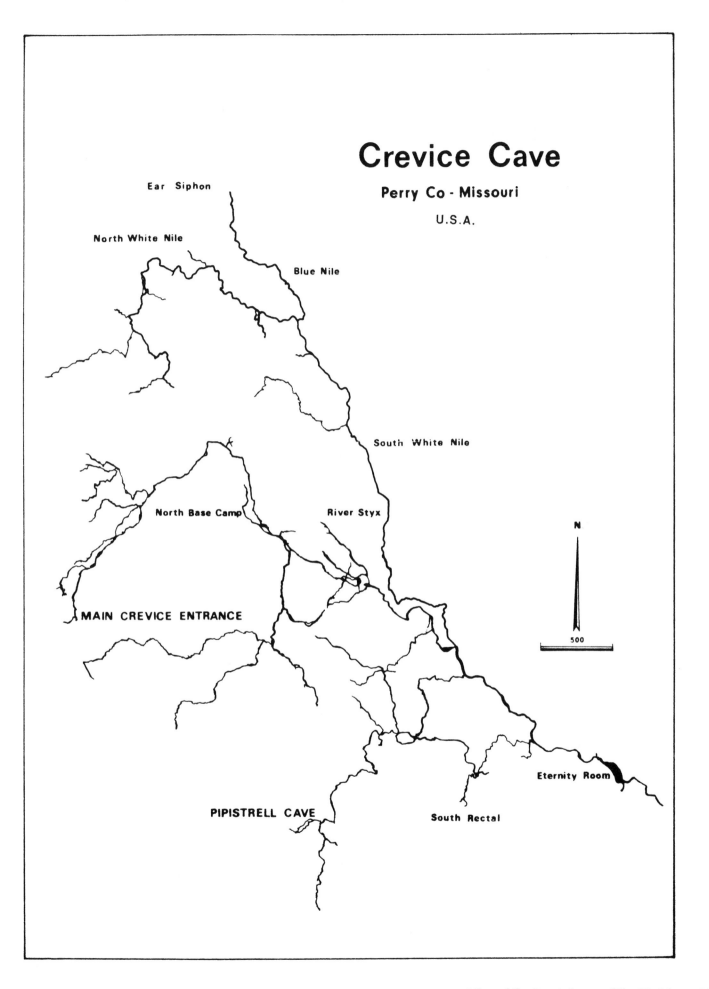

Crevice Cave

Perry Co - Missouri

U.S.A.

Ear Siphon

North White Nile

Blue Nile

South White Nile

North Base Camp

River Styx

N

500

MAIN CREVICE ENTRANCE

Eternity Room

PIPISTRELL CAVE

South Rectal

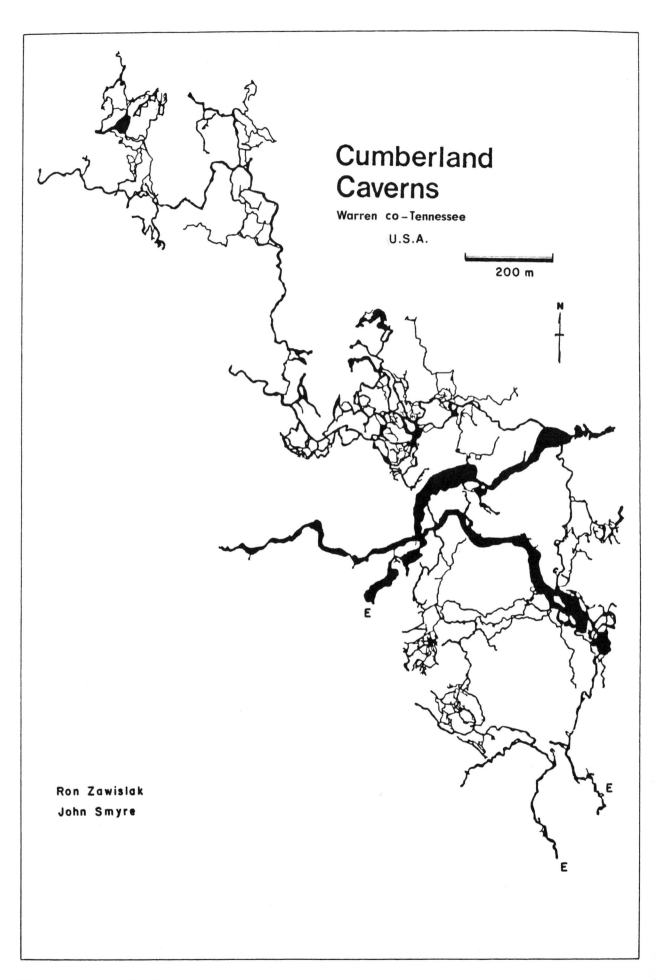

Cumberland Caverns

Warren co–Tennessee

U.S.A.

200 m

N

E

E

E

Ron Zawislak

John Smyre

Sloan's Valley Cave System

Pulaski Co - Kentucky

U.S.A.

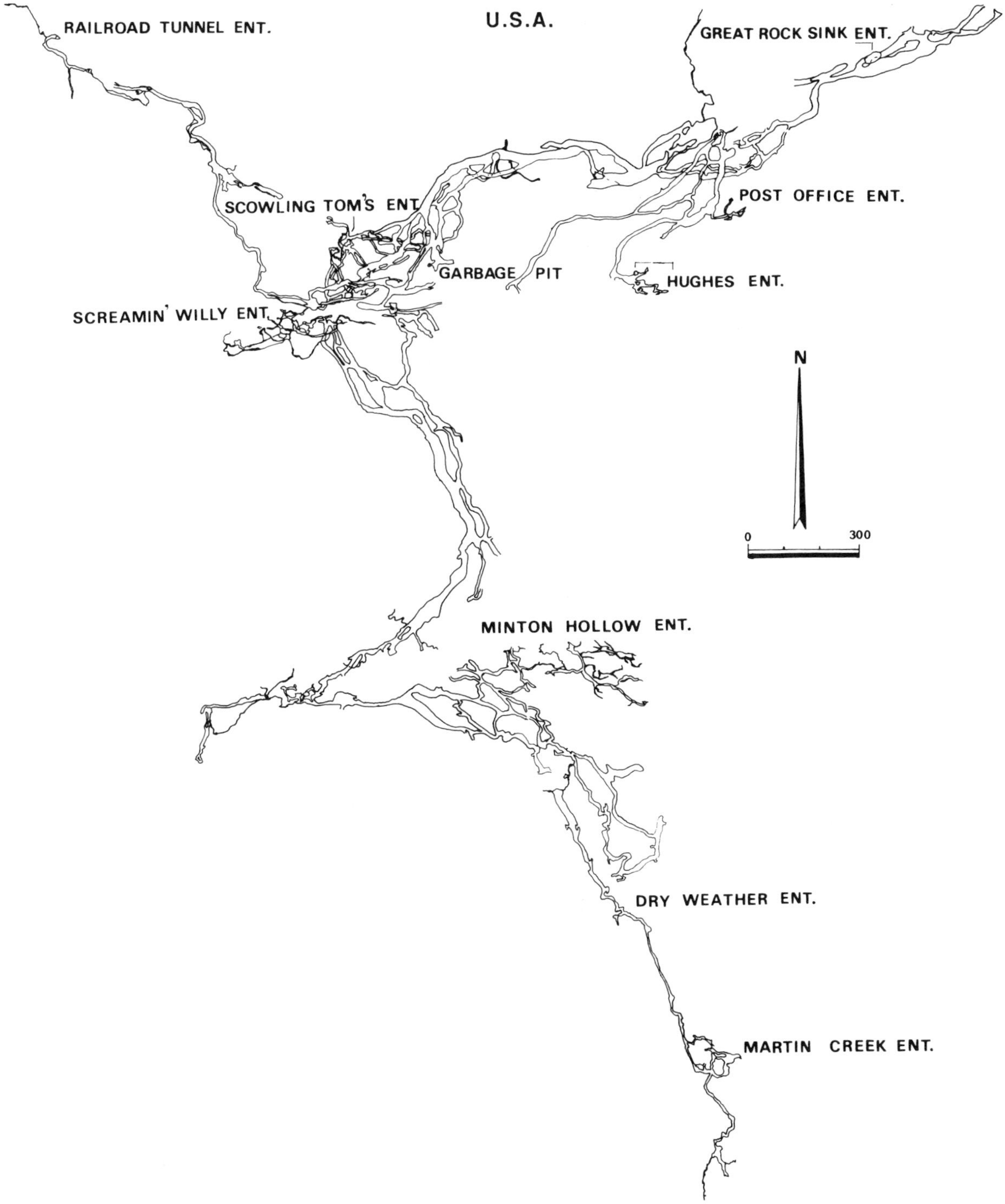

RAILROAD TUNNEL ENT.

GREAT ROCK SINK ENT.

SCOWLING TOM'S ENT.

POST OFFICE ENT.

GARBAGE PIT

HUGHES ENT.

SCREAMIN' WILLY ENT.

N

0   300

MINTON HOLLOW ENT.

DRY WEATHER ENT.

MARTIN CREEK ENT.

Bibliography: Simpson (L.) - Sloan's Valley Cave System, *Guidebook to the 1980 Kentucky Speleo-Fest*, 1980: 3-17.
N.S.S. News, March, 1981.

11. *Xanadu Cave System* ....38,623 m
(Fentress County, Tennessee)

Xanadu Cave is located southwest of Jamestown, on the east slopes of the Cumberland Plateau, in the gorge of the East Fork of the Obey River.

The three main entrances were discovered on March 26, 1977 by Ray Lewis, Sid Jones and Pat Sullivan. The cave was not known to locals, although one name was found in the entrance passage. It read "Pat Stevens, 1927. This is as far as we got". Ray and crew explored several kilometers of virgin passage, including the main trunk (Sandhill Passage), the connection between the upper Xanadu section and the lower Zoroaster section and the lower Zoroaster Maze. Ray turned the exploration over to the mappers, since he felt it was a large cave. In July 1977, the survey was begun by Jeff Sims, Charles Clark and members of the Smoky Mountain and Shelah Waters Grottos of the NSS. The survey crews discovered the remainder of the passages while mapping. Significant discoveries include the finding of Cumberland Avenue by Clark, Sims and Martha Clark, Fort Sanders by Chris Kerr, Jim Nicholls, and John Yust, and the discovery of Middle Alph, 747 Passage, Fifth Avenue, Sunday City and the Evil Frog Section.

The total lengths were 13,792 m in 1978, 16,383 m in 1979, 25,762 m in 1980, 26,431 m in 1981, 37,390 m in 1984, and 38,623 m in 1986. Exploration is still in progress.

Xanadu Cave has six entrances: the upper Xanadu Entrance, the Alph Entrance and the Lower Zoroaster Entrance area, which includes four entrances. The main trunk in the cave is called Sandhill Passage, and extends for over 1 km with passages up to 30 m wide and 25 m tall, with 15 large sandhills, some over 15 m high. The cave can be divided into about five levels, the upper three of which are dry while the lower two are damp, muddy and sometimes have active streams.

The cave is formed in limestone of the Monteagle limestone formation of Mississippian age in the level strata of the Cumberland plateau.

Map: from the map of Jeff Sims, sent by Bob Gulden.
Bibliography: Clark (C. and M.), Sims (J.) - Xanadu, *N.S.S. News*, 1981, 39 (8).

Jeff Sims.

12. *The Hole* ....36,838 m
(Greenbrier County, West Virginia)

The system has three entrances: the Boggs entrance (through which the original discoveries were made), and the Perkins and Gibbs entrances which were discovered later and excavated. The system is located near the town of Frankford, not far from Lewisburg.

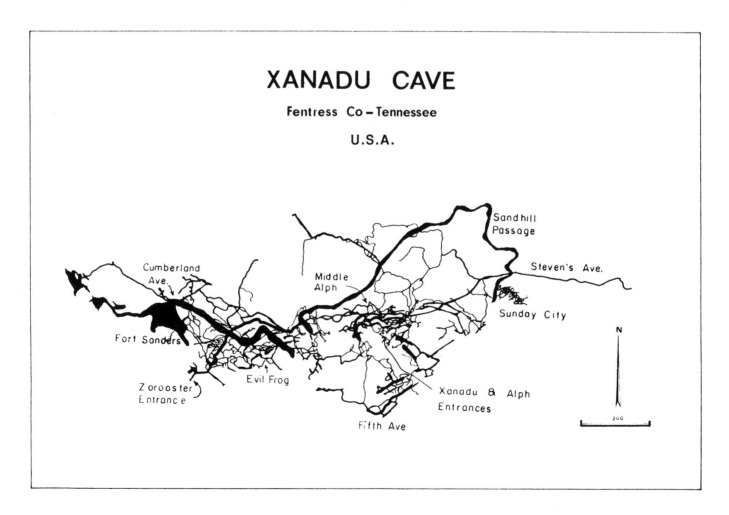

XANADU CAVE

Fentress Co – Tennessee

U.S.A.

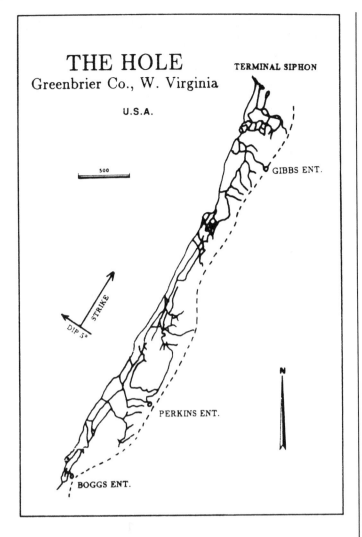

THE HOLE
Greenbrier Co., W. Virginia
U.S.A.

TERMINAL SIPHON

500

GIBBS ENT.

STRIKE
DIP 5°

N

PERKINS ENT.

BOGGS ENT.

The cave is formed at the contact of the Greenbrier Limestone and the Maccrady Shale, of Mississipian age. The long axis of this cave system lies parallel to the regional geologic structure. Surface water flowing on the shale sinks at the contact and continues to follow components of the dip until reaching the master stream, Shale River. This stream in turn delivers water along the strike to a major resurgence along Spring Creek, a tributary of the Greenbrier River.

The cave was discovered by members of the West Virginia Association for Cave Studies (WVACS) in the middle 1960's. Approximately 25,000 m were surveyed in the initial phase under the direction of John Rutherford and Charlie Maus before interest waned. In the mid 1970's the Pittsburgh Grotto (headed by Charlie Williams and Don Schleicher) undertook the resurvey of the system which was later incorporated into another WVACS attempt to finish the system. The survey continues under the direction of Bill Balfour.

Map: C. Williams & D. Schleicher 1981 (WVACS), sent by Bill Balfour.

Bibliography: WVA Geol. Surv. Bull., 1973, 36, map, 25,300 m.

Bill Balfour.

13. *Whigpistle Cave* ....**36,209 m**
(Edmonson County, Kentucky)

The main entrance to Whigpistle Cave, a floodwater swallet near the southwest corner of Mammoth Cave National Park, was discovered in 1976. The cave has been mapped by karst research teams directed by the National Park Service. It was discovered as a result of a search for the cave system that was believed to exist between two karst windows, Mill Hole and Cedar Sink.

The cave consists chiefly of long, dry trunk passages, some of the most continuous in the Mammoth Cave area, which are accessible only via small, wet, muddy, often-flooded crawlways. The long, low crawls through active stream passage, plus the tendency of the entrance to be sealed by floods, make exploration without judicious study of weather forecasts extremely unwise.

Whigpistle Cave is significant for several reasons. It has one of the larger rooms in any North American cave (the Big Womb, 55 m x 240 m and 15 m high). A dye-trace run in 1983 showed that a newly-discovered stream segment in Whigpistle is the continuation of the trunk stream in Proctor Cave, 6 km to the east, thus proving a hydrologic connection to the Mammoth Cave System. Also, it is part of a groundwater drainage system in which floodwaters are diverted to adjacent groundwater basins — thus complicating the design of a reliable system for monitoring groundwater quality.

Map: after a compliation by Don Coons.

Bibliography: N.S.S. News, February, 1979; August, 1979.

Quinlan, J. F. - Special problems of groundwater monitoring in karst terranes, in Neilsen, D. M. and J. F. Quinlan, eds., 1988, *Symposium on Standards Development for Ground Water and Vadose Zone Monitoring Investigations.* ASTM Special Tech. Paper, American Society for Testing and Materials, Philadelphia, PA (in press).

Quinlan, J. F., R. Ewers, J. Ray, R. Powell and N. Krothe - Groundwater hydrology and geomorphology of the Mammoth Cave Region, Kentucky, and of the Mitchell Plain, Indiana, in Shaver, R. H. and J. A. Sunderman, eds., 1983, *Field Trips in Midwestern Geology. Geol. Soc. Am. and Indiana Geol. Survey, Bloomington, Indiana,* v. 2, p.. 1-85.

Qunlan, J. F. and J. A. Ray - Groundwater Basins in the Mammoth Cave Region, Kentucky: *Friends of the Karst, Occasional Publication N° 1* (map), 1981.

J. F. Quinlan and Don Coons.

14. *Culverson Creek System* ....**33,507 m**
(Greenbrier County, West Virginia)

Culverson Creek Cave has eleven entrances. The Wild Cat Entrance is the most often used because of its ease of access to the main Culverson trunk, overflow passages, and the trunk passage in the McLaughlin Unus section of the cave. The main entrance is a 20 m by 20 m opening at the base of a 30 m cliff. It swallows Culverson Creek, a large surface stream which drains a basin with a surface area of 71 km². The drainage for the entire cave system totals 109 km². The resurgences are four springs along Spring Creek, 9 km to the east.

This system was one of the earliest caves entered by WVACS. The Fuller section, a 3 km long canyon and trunk which carries Culverson Creek, was connected to the main Culverson section in 1969. The McLaughlin Unus section (a tributary stream trunk and a section of the main Culverson Stream) were connected later. The total surveyed lengths have been 8,400 m in 1973, 17,700 m in 1975, 27,359 m in 1979, and 33,507 m in 1985. The present downstream end of the cave is Dream Siphon, 104 m below the main entrance.

The cave is developed in the Union Limestone of the Greenbrier Group. The passages in the system are typically large stream canyons which can sometimes flood rapidly and unpredictably. In many sections, recent sediment deposits reach the ceiling. Trees up to 15 m in length have been carried over 2.4 km into the cave and wedged 10 m above the canyon floor.

Map: from the WVACS, sent by P. Lucas.

Bibliography: Jones (W. K.) and Lucas (P. C.) - Proceedings of the 1977 NSS Annual Convention, *West Virginia Speleological Survey*, page 31-33.

Phil Lucas.

15. *Carlsbad Cavern* ....**33,490 m**
(Carlsbad, Eddy County, New Mexico)

Carlsbad Cavern is a world-renowned cave which has been a part of the U. S. National Park system since 1923. It is located in southeastern New Mexico, in Eddy County, about 25 km west of the town of Carlsbad (U.S.G.S. 7.5 minute Carlsbad Caverns quadrangle). The cave has one natural entrance, at an altitude of 1318 m. Two man-made shafts also enter the cave: the 229 m deep elevator shaft, providing visitor access to the Big Room area, and an abandoned 52 m deep bat guano mining shaft entering near the east end of the Bat Cave section.

Pictographs and artifacts in the natural entrance show that Indians knew of the cave, and records indicate that local ranchers knew its location in the late 1800's. The famous evening flight of Mexican Freetail bats (colony of

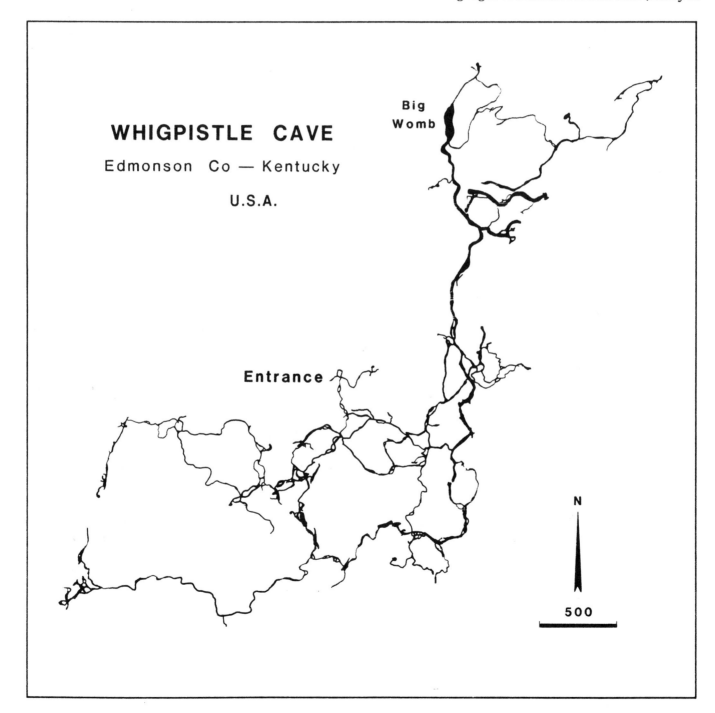

WHIGPISTLE CAVE

Edmonson Co — Kentucky

U.S.A.

Big Womb

Entrance

N

500

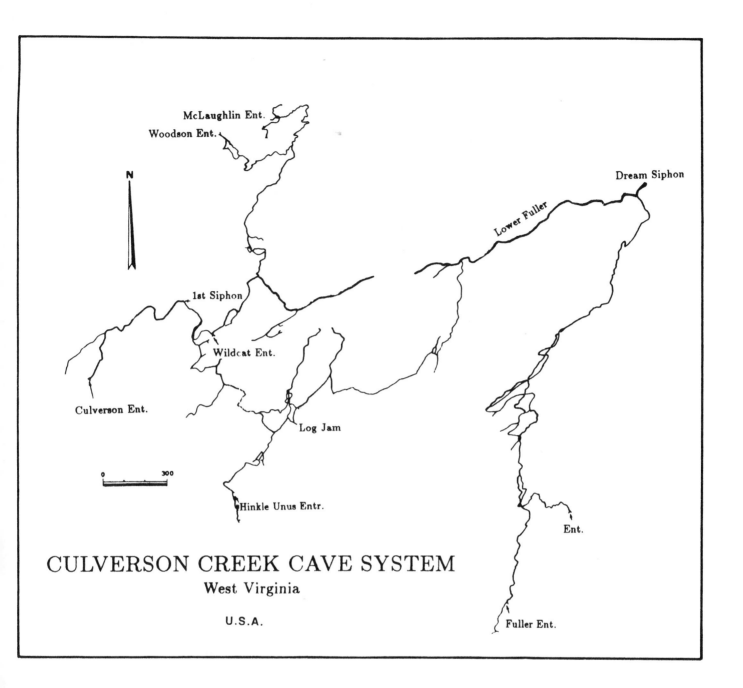

CULVERSON CREEK CAVE SYSTEM
West Virginia

U.S.A.

Map labels: McLaughlin Ent., Woodson Ent., N, Dream Siphon, Lower Fuller, 1st Siphon, Wildcat Ent., Culverson Ent., Log Jam, Hinkle Unus Entr., Ent., Fuller Ent.

0   300

300,000 in recent years) undoubtedly led people to the entrance from the earliest times. The first documented penetration of the cave beyond the 25 m entrance pit occurred in 1883 when a local boy, Roth Sublett, was lowered in by his father.

In 1903, Abijah Long filed a claim for the rights to mine bat guano from the cave. Mining operations soon began and continued for 20 years. During this period Jim White, one of the guano miners, became fascinated with the cave and was the first person to begin systematic exploration beyond the entrance mining area. His stories of the magnificent cave led others to join on his trips, leading eventually to an official government reconnaissance of Carlsbad Cavern by Robert Holley who surveyed 5 km of passage from the Bat Cave shaft into the Big Room. The significance of the cave was immediately recognized

and it was designated a National Monument in October, 1923, and subsequently, a National Park in May, 1930. Investigations and a survey by Willis T. Lee and R. H. Runyon in the 1920's began the process of study and surveying that has continued to the present with the support of the National Park Service. Organized group efforts have occurred under the auspices of the Guadalupe Cave Survey in the 1960's and the Cave Research Foundation from the 1970's to the present.

Carlsbad Cavern is formed primarily in the massive member of the Capitan Limestone, a reef limestone of Permian age. The entrance is in the Tansil formation and parts of the cave extend into a breccia member (reef talus) of the Capitan. The cave is known for its abundant displays of large speleothems. In addition, a wide variety of rare speleothems and speleogens have been identified. The

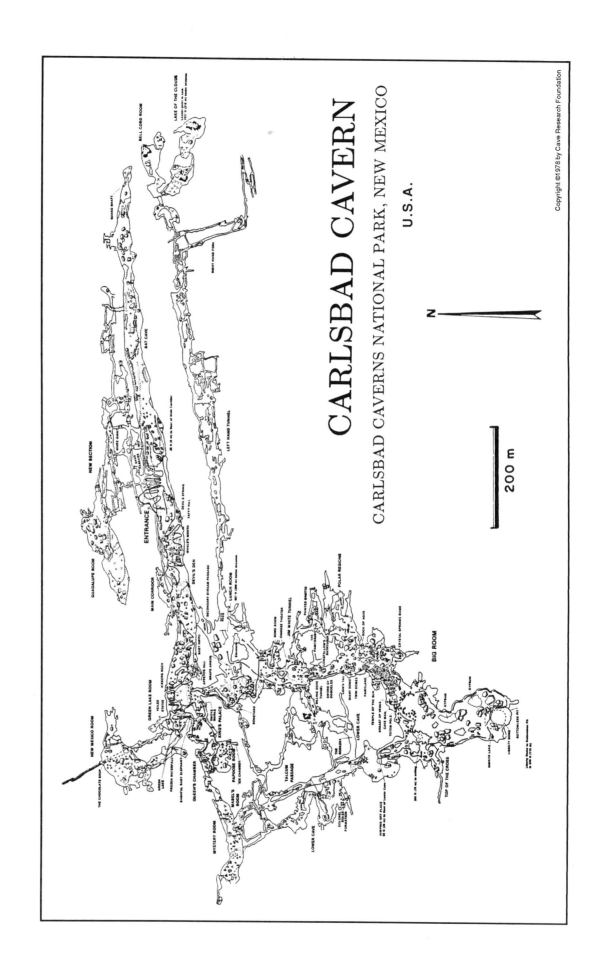

CARLSBAD CAVERN

CARLSBAD CAVERNS NATIONAL PARK, NEW MEXICO

U.S.A.

N

200 m

occurrence of extensive deposits of massive gypsum in the cave, along with the incidence of related elemental sulfur, has led to the development of a theory that solution by sulfuric acid (from hydrothermal waters of pyrite oxidation) played a significant role in the formation of Carlsbad Cavern.

Map: from the 1979 map by the C.R.F.

Bibliography: Jagnow (D.) - *Cavern Development in the Guadalupe Mountains*, Jan. 1979, Ph.D. Dissertation, C.R.F.

Barnett (J.) - *Carlsbad Caverns: Silent Chambers, Timeless Beauty*, 1981. Carlsbad Caverns Nat. Hist. Assn.

Hill (C.) - *Geology of Carlsbad Cavern...*, Bulletin 117, N. M. Bureau of Mines and Mineral Resources, 1987.

Lee (W. T.) - 1924, A Visit to Carlsbad Cavern, *Nat. Geog. Mag.* 45 (1):1-40; c1925, New Discoveries in Carlsbad Cavern, *Nat. Geog. Mag.* 48: 301-320.

Sutherland (M) and E. Helen - Carlsbad Cavern in Color, *Nat. Geog. Mag.* 54 (4): 433-368.

Alan Williams.

16. *Blue Spring Cave* ....32,251 m
(Lawrence County, Indiana)

Explored and mapped by A. and M. Palmer, D. J. and D. O. Chase, J. Richards and others in 1964-1975. Only the main stream passage was known prior to 1964, when four major extensions were made in rapid succession. The cave is developed beneath the Mitchell sinkhole plain in prominently jointed Salem Limestone and bedded St. Louis Limestone (Mississippian). It is an extensive branchwork cave with 55 tributary stream passages that converge into a single large passage navigable by boat for 4 km. Passages occupy three distinct levels, but with very few unconnected crossings. Most of the passages are high joint-controlled canyons and fissures. Breakdown midway in the main stream passage has resulted in a floodwater diversion maze containing 2 km of passages.

The Colglazier Entrance is open to tourists as Bluespring Caverns. Several hundred meters of the main passage upstream from the entrance are toured by boat. All other entrances are currently closed by sinkhole collapse or by land-owner edict.

Map: from the map of A. Palmer *et al.*, sent by A. Palmer

Bibliography: *N.S.S. News*, January 1969.

Arthur N. Palmer.

17. *Butler-Sinking Creek System* ....32,187 m
(Bath County, Virginia)

Butler Cave, the longest in Virginia, is located in Bath County, near the town of Burnsville, at an altitude of 773 m.

Location: 38° 11' 17" N, 79° 39' 00" W. Burnsville 7.5 minute USGS Quadrangle.

The cave was discovered by systematic ridgewalking on May 30, 1958 by I. Kennedy (Ike) Nicholson. Discovery of the Sizeable Trunk Channel later that summer quickly led to exploration of much of the cave. The early exploration and mapping were managed by the Nittany Grotto (Pennsylvania State University) of the N.S.S. By the end of 1960, 8,500 m had been surveyed and the present depth of 190 m had been established. In 1963, a small stream canyon near the downstream end of the cave was followed for 200 m to a 12 m drop. This drop led to the new section of the cave called Marlboro County, which yielded over 5 km of new passage. In 1964 blasting of a stream crawl led

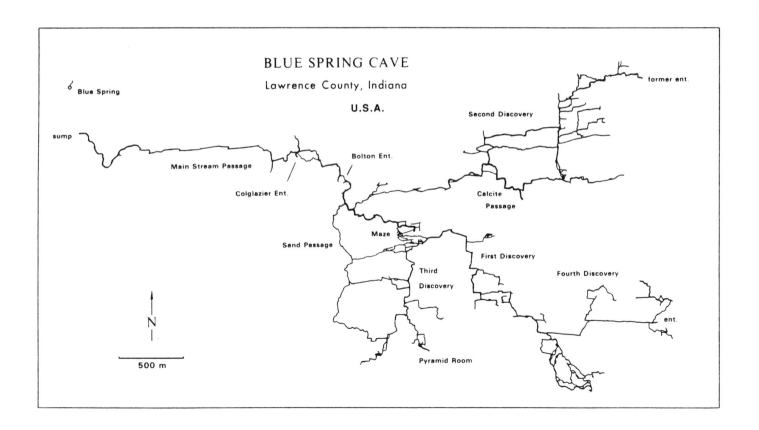

BLUE SPRING CAVE

Lawrence County, Indiana

U.S.A.

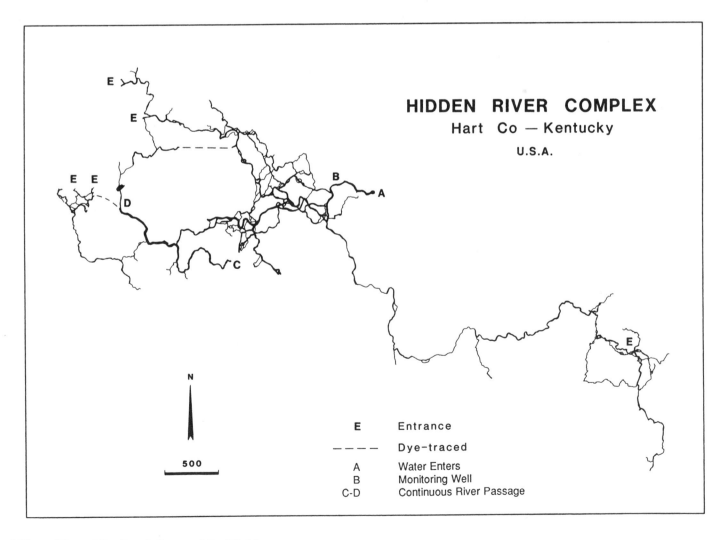

BUTLER CAVE - SINKING CREEK SYSTEM
BATH COUNTY, VIRGINIA
U.S.A.

HUNTLEY'S CAVE SECTION

BUTLER CAVE SECTION

ENTRANCE

MARLBORO COUNTRY

DYNAMITE SECTION

SNEAKY CREEK

DOWNSTREAM LOOOP

SAND CANYON SECTION

SINKING CREEK

POOL ROOM SECTION

BEYOND THE LAKE

N

250

HIDDEN RIVER COMPLEX
Hart Co — Kentucky
U.S.A.

E
E

E E

D

B
A

C

E

N

500

E	Entrance
– – –	Dye-traced
A	Water Enters
B	Monitoring Well
C-D	Continuous River Passage

to the discovery of the Dynamite Section, adding about 2 km to the known cave length. Since 1964 several small discoveries of never more than 2000 m have pushed the surveyed length of the cave to over 30 km.

Butler Cave is owned by Butler Cave Conservation Society which was formed in 1970 to manage the cave.

The system is formed along a synclinal valley that gently plunges to the NE. The cave is formed in the lower and middle Keyser limestone of Silurian-Devonian age.

Several permanent streams are present in the system, with the largest generally flowing to the NE and terminating in sumps. Dye tracing has shown that the resurgence is Mill Run (Agua) Spring. Other caves in the hydrologic system are Boundless Cave, Breathing Cave, Better Forgotten Cave, and the northern end of Bobcat Cave.

Map: from the 1985 map of the Butler Cave Conservation Society.

Bibliography: Wefer (F. L.) and I. K. Nicholson - Mapping and Exploration of the Sinking Creek System, *N.S.S. Bull.*, 1982, 44 (3).
White (W. B.) and Hess (J.) - Geomorphology of Burnsville Cove and the Butler Cave-Sinking Creek System, *N.S.S. Bull.*, 1982, 44 (3).
Wefer (F. L.) - Characteristics of the Butler Cave-Sinking Creek System, *B.C.C.S. Newsletter*, 1986, 12.

John Rosenfeld.

18. **Hidden River Complex** ....**31,817 m**
or **Hicks Cave** (Hart County, Kentucky)
The major entrances to this cave are along the south bank of Green River.

This cave was first explored in 1975 as a result of digging at high-level springs after interpreting the chemistry of waters in several low-level springs. It was mapped through the heroic efforts of several karst research teams organized by the National Park Service. From 1975 until 1978, 29,223 m were mapped. An entrance shaft was completed in 1986. Since then, 2,594 m have been mapped in the eastern end of the cave.with more surveying still to be done.

The cave is significant because it is one of the few mapped underground distributaries; it is part of the Hidden River distributary (3.5 km wide along an 8 km reach of Green River) which is part of the Bear Wallow distributary (13.5 km wide along a 19 km reach of Green River).

Map: after a compilation by Tom Ahlers.
Bibliography: Quinlan, J. F. and J. A. Ray - *Groundwater Basins in the Mammoth Cave Region*, Kentucky, Friends of the Karst, Occasional Publication n° 1, map.

J. F. Quinlan.

19. **Binkley's Cave System** ....**30,738 m**
(Harrison County, Indiana)
This cave was explored and mapped by the Bloomington Indiana Grotto in the early 1960's to a length of 10 km, and continued to its present length by the Indiana Speleological Survey during the early 1970's. It is a dendritic cave with several major stream passages and high-flow crossovers, developed beneath the Mitchell sinkhole plain in the Salem and St. Louis Limestones (Mississippian). Passages occur in several closely spaced levels with very few unconnected crossings. Most of the cave consists of

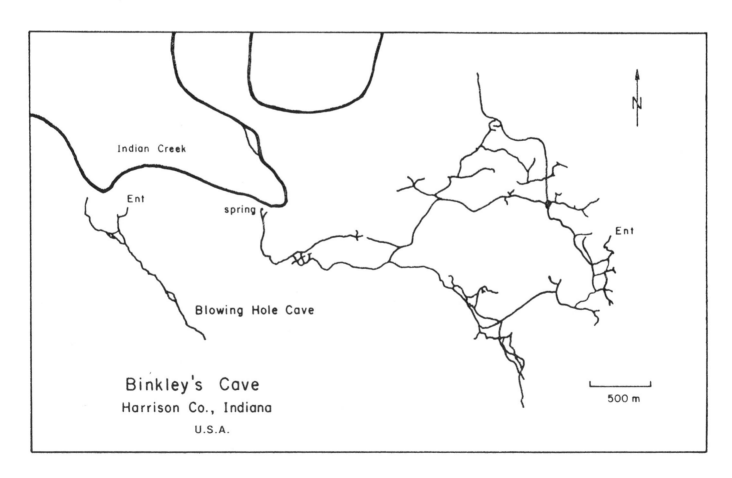

Indian Creek

Ent

spring

Blowing Hole Cave

N

Ent

500 m

Binkley's Cave
Harrison Co., Indiana
U.S.A.

wide tubes, including several major stream passages. Breakdown terminations are common, and the entrance is located in a very unstable collapse sinkhole. A nearby cave, Blowing Hole Cave, which also contains a major stream, has no known connection with Binkley's.

Map: from the map of the Indiana Speleological Survey, sent by A. Palmer.

Bibliography: *Guidebook to the 1982 Kentucky Speleofest.*

Arthur N. Palmer.

20. **Windymouth Cave** ....28,968 m
(Greenbrier County, West Virginia)

This cave has only one known entrance located in a cliff face in the gorge of the Greenbrier River. The resurgence is not far downstream from the entrance. The cave was first mentioned in Davies' *Caves of West Virginia* in 1949. Since the 1960's, exploration and mapping (still in progress) has been by numerous groups under the direction of Jim Hixson. Formed in Greenbrier limestone (Mississippian age).

21. **Mountain's Eye Cave System** ....28,967 m
(Fentress, Tennessee)

The system was formed with the connection of Cobb Creek Saltpetre Cave and Lost Deane Cave. The system (alt. 310 m) has nine entrances: Neff, East Eye, West Eye, Lost Dean, Spring, Sorcerers Grotto, Nargathrond, Nightshade, and Equulei Crack. The cave is developed in the Monteagle Formation. The Cobb Creek section was extensively mined for saltpetre and contains many artifacts. Parts of the system were found by Ray Lewis in 1977. Mapping by Chris Kerr and the Smokey Mountain Grotto is still incomplete, and presently unpublished (*SERA Cave Carnival Guidebook*)

22. **Mystery Cave** ....27,356 m
(Perry County, Missouri)

Mystery Cave is a large, integrated stream system developed in Ordovician dolomites and limestones beneath the southern Perryville karst plain. The cave was discovered in the early 1960's by members of the Southeast Missouri Grotto but was thoroughly explored and mapped in the late 1960's and early 1970's by the Little Egypt Student Grotto. The Historic Entrance is subject to flooding and an artificial entrance was dug in the late 1960's. Early in history of its exploration the cave was connected to nearby Spider Pit Cave and Harrington Cave, both of which had been known for many years. Over 4 km had been mapped in Lost and Found Cave before it was connected to Mystery, giving the system its present length. The cave stream resurges at several springs along Cinque Hommes Creek. (*The Southeast Caver*, 2 (3), 1976; *N.S.S. News*, May, 1969).

Scott House.

23. **Berome Moore Cave** ....27,077 m
(Perry County, Missouri)

The system is developed in Ordovician dolomitic limestone and consists of two connected sections: the original part (Tom Moore Cave) was locally well-known and contains about 4 km of stream passages, while the downstream section (Berome Moore) was discovered by T. Yokum, L. Brod, S. Sides and others on October 29, 1961. Over 22 km of varied passages including a large stream and paleo trunks were quickly explored. Mapping by the Missouri Speleological Survey is still in progress. The cave is known for well preserved tracks of a very large cat, probably saber-tooth tiger, which extend for over 800 m.

The cave lies beneath the Perryville karst plain and its waters resurge in a series of springs, of which the most famous is Ball Mill Resurgence (8th Int. Congr. of Spel., *Guidebook to the Karst and Caves of Tennessee and Missouri*, Vineyard and Crawford, 1981).

Scott House.

24. **Thornhill Cave** ....26,920 m
(Breckenridge Co., Kentucky)

Cave with two entrances (Thornhill and Lower Big Spring) formed in Mississipian age limestone (Ste. Genevieve formation) and located west of Fort Knox. Known since the early 1800's and used for saltpeter mining. Explored beginning in the early 1950's by the Bartel Research Assoc., which later became the Sinking Creek Research Assoc. Mapped from 1966 to present (issues of *Karst Window*, the newsletter of the Louisville Grotto and various Kentucky Speleofest guidebooks).

25. **McClung Cave System** ....26,393 m
(Greenbrier County, West Virginia)

The McClung entrance has been known since the area was first settled in the 1790's and until 1986 was the only known permanent entrance to the cave. In early 1986 a pit entrance was found by Bill Balfour that subsequently led into the system in a remote section of the cave.

It is formed at the contact of the Greenbrier Limestone and the Maccrady Shale (Mississippian). The resurgence is Davis Spring (the largest spring in West Virginia) along the Greenbrier River, a straight line distance of 25 km.

The cave is being surveyed by the WVACS under the direction of Bob Amundson.

Bill Balfour.

26. **Honey Creek Cave** ....25,740 m
(Comal and Kendall Counties, Texas)

Longest cave in Texas. The construction of a 45 m deep artificial shaft entrance in 1985 has led to the discovery of much new passage. Virtually all the cave passages are wet, and the through trip between entrances involves about 3 km of swimming with full wet suits and flippers.

27. **Fern Cave** (Jackson, Alabama) ....25,153 m

Cave with seven entrances located near Paint Rock Valley. Discovered by B. Torode and three others in 1961. They quickly came to 133 m deep Surprise Pit (the deepest free fall pit in the U.S. until 1969) which was first descended by Bill Cuddington. The length of the cave was only about 300 m until 1969 when the discovery of the Johnson Entrance gave easy access to a system of large passages. Bill Torode led the survey of over 18 km in the next six months. Occasional discoveries have extended the length to 25,153 m. The resurgence into Paint Rock River is the lower entrance to the cave. SCUBA is required to make a through trip. (Myrick (D.) *Fern Cave*, 1972, 106 p.; *N.S.S. News*, March, 1981).

28. **Cave Creek Cave System** ....24,150 m
(Pulaski County, Kentucky)
Explored principally by the Central Ohio Grotto in 1972 (*Guidebook to the 1980 Kentucky Speleo Fest*).

29. **Benedicts Cave** ....23,900 m
(Greenbrier County, West Virginia)
Cave with two entrances (Benedict's and Persinger's) known for at least 80 years. Initially explored by R. Baroody and others for the WVACS in the early 1960's. In the mid 1970's Bill Douty undertook the resurvey of the system which was completed in 1978. Formed in the Greenbrier limestone (Mississippian age). Resurgence at Davis Spring and part of the same hydrological system as McClung's Cave (see above).

30. **Scott Hollow Cave** ....22,933 m
(Monroe County, West Virginia)
Cave formed in the Hillsdale and Sinks Grove Limestones, with a single man-made entrance shaft dug open in November 1984 by Mike Dore and Martin Deleggi. The stream in the Mystic River trunk passage (which averages 25 m by 25 m over a length of 5 km) flows north towards the Greenbrier River. Exploration is being actively pursued.

31. **Powell's Cave System** ....22,851 m
(Menard County, Texas)
Located in central Texas, Powell's Cave is a multi-level maze cave explored starting in 1962 by members of the Texas Speleological Association (T.S.A.) A sinkhole entrance joins the complex maze level 15 m beneath the flat surface. The downstream end of the Crevice Passage, a 3 km long dry streambed cutting through the maze levels, ends at the Water Passage (on the 28 m level). The Water Passage extends 2 km upstream to within a few hundred meters of Silver Mine Cave, and 1 km downstream through several siphons connecting with Neel's Cave.
Pete Lindsley.

32. **Bone-Norman Cave System** ....22,724 m
(Greenbrier County, West Virginia)
The cave has two entrances on either side of the system. A large stream flows through the cave and resurges near the Bone entrance. Serious exploration started in the early 1960's and in 1963 John Davis discovered a way through a constriction that quickly led to the connection of the two caves. A survey by WVACS cavers under the direction of Bill Douty began in 1974 and was completed in 1977. Formed in lower Greenbrier limestone (Mississippian).

33. **Rimstone River Cave** ....22,589 m
(Perry County, Missouri)
This cave is a large wet system developed in Ordovician limestones and dolomites under the southern Perryville karst. The system has several entrances, including Rimstone River (Schaupert Pit) and Flaming River Cave. The cave parallels Mystery Cave and comes within 200 m of connecting. It was discovered around 1968 and mostly mapped by the MSS over the next eight years (*N.S.S. News*, February 1972). Resurgence at a spring on Cinque Hommes Creek.

34. **Mystery Cave** (Fillmore, Minnesota) ....20,597 m
Cave with two entrances (Mystery I and II) located 10 km SE of Spring Valley. The first entrance was dug open in 1939 by J. Paddy. The cave was explored and developed as a show cave 1939-1948 by C. "Slim" Prohastas. Mystery II was dug open in the late 1950's and also commercialized. A third section of the system was found in the 1960's. A third generation mapping effort by the Minnesota Spel. Survey began in the 1970's and is still in progress. The cave is a joint-controlled flood-water maze, formed in the Ordovician Age Galina Formation; resurgences occur along Forestville Creek.

35. **Big Bat Cave** ....20,400 m
(Breckenridge County, Kentucky)
Big Bat has a natural entrance in the bottom of a large sink 750 m long by 300 m wide, and drains an area of about 3 km². Located at an altitude of 200 m and formed in Paoli and Ste. Genevieve limestone. A new entrance (Butler or Mushroom Entrance) was dug in August, 1977 to facilitate exploration in remote sections of the cave (1983 Kentucky Speleofest Guide, *N.S.S. News*, May, 1977).

36. **Anvil Cave** (Morgan, Alabama) ....20,385 m
An extensive network maze developed in the Gasper Limestone (Mississippian) beneath the Hartselle Sandstone. It consists of mainly walking-sized fissure passages that intersect in a rectangular grid. A few small breakdown rooms are located at complex junctions.

37. **Gradys Cave System** ....19,840 m
(Hart County, Kentucky)
Explored and mapped mainly during the 1970's by the Hart Attack Team led by Joe Saunders (12,067 m in 1973, 18,504 m in 1978, 19,264 m in 1982). The cave is developed beneath the Pennyroyal sinkhole plain in the St. Louis Limestone (Mississippian). It consists of a few long tubular stream passages requiring swimming or rafting for part of their length. Most of the larger passages end in breakdown.

38. **Ellisons Cave** (Walker, Georgia) ....18,878 m
See above.

39. **Carroll Cave** (Camden, Missouri) ....18,105 m
Located on the northwestern side of the Ozark plateau and developed in massive dolomite of Ordovician Age. Contains two main cave streams, one of which pirates the other, with a resurgence at Toronto Spring on Wet Glaize Creek. Exploration and mapping by the Missouri Speleological Survey began in the mid-1950's and is still in progress due to the difficult nature of exploration (*N.S.S. Bulletin* 27 (1), 1965; *Missouri Speleology* 23 (1-4), 1983).

40. **Bigfoot Cave** ....18,088 m
(Siskiyou County, California)
See above.

41. **Goochland-Poplar Cave Complex** ....18,000 m
(Rockcastle County, Kentucky)

42. **Coldwater Cave** ....17,719 m
(Winnishiek County, Iowa)

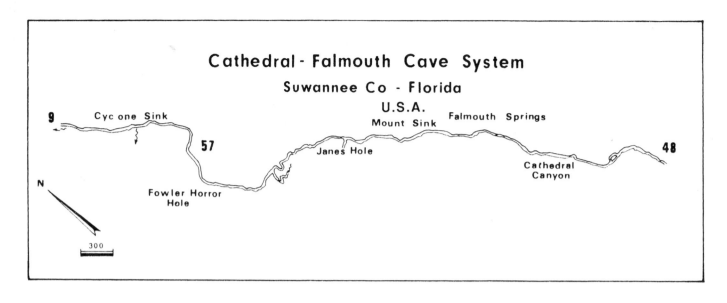

Cathedral - Falmouth Cave System

Suwannee Co - Florida

U.S.A.

43. **The Maria Angela Grotto** .... 17,678 m
(Grundy, Tennessee)

Explored in the early 1980's when J. Buckner, D. Parr, and D. Freeman opened up 3 m long Blowing Hole, long known to locals for its strong breeze. Most of the mapping by the Nashville Grotto took place in the following year before the entrance was closed by the landowner. The cave contains mainly dry walking passage, with good potential for future discoveries (issues of the Nashville Grotto *Speleonews*).

44. **Crumps Spring Cave** .... 17,600 m
(Hart County, Kentucky)

Explored and mapped during the 1960's by the Mid-Illinois Grotto and during the 1970's by the Hart Attack Team led by Joe Saunders. It has gained a certain notoriety as a tangle of nasty, tight canyons and crawls, without the large trunk passages of the large nearby caves such as Mammoth, Fisher Ridge, and Hicks. Formed in St. Louis, Ste. Genevieve and Girkin limestones of Mississippian age.

45. **Coral Cave System** .... approx. 17,000 m
(Pulaski, Kentucky)

46. **Lilburn Cave** .... 16,573 m
(Tulare County, California)

Cave formed in banded marble and located in Redwood Canyon, in Kings Canyon National Park, at an elevation of 1550 m. First known to miners (1910), its location was lost, then the cave was rediscovered, dug open and partially explored by G. Lilburn (1940). California cavers explored and mapped (E. Hedlund) much of the cave from 1967-1975. Research has been emphasized since 1968 (coordinated by the Cave Research Foundation since 1977). A new survey by the C.R.F. was begun in 1980 (P. Bosted). The resurgence is an ebb and flow spring named Big Spring. (*The California Caver*, 1987, 37(1); CRF annual reports 1977-1988).

47. **James - Thousand Domes Cave** .... 16,500 m
(Edmonson County, Kentucky)

Explored and mapped since the 1960's by a group led by Glen Merrill. The passage density is perhaps greater than that of any other long cave in the world. Formed in

St. Louis and Girkin limestones on the southern edge of the Chester upland south of Mammoth Cave.

48. **Wells Cave** (Pulaski, Kentucky) .... 16,300 m

Explored and mapped by the Dayton Area Spel. Society and the Miami Valley Grotto (*DASS Johnhouse News*, July Sept. 1983).

49. **Bighorn - Horsethief Cave System** .... 16,285 m
(Bighorn County, Wyoming and Montana)

Originally surveyed to this length in the 1970's (*N.S.S. News*, February 1979). New survey and scientific studies in progress by the Bighorn Research Project (7000 m in 1985, 10,600 m in 1986) in cooperation with the National Park Service (*N.S.S. News*, September, 1986).

50. **Perkins Cave** .... 16,154 m
(Washington County, Virginia)

51. **Sullivan Cave** .... 15,498 m
(Lawrence County, Indiana)

Explored and mapped during the late 1950's and late 1960's by the Central Indiana Grotto. Formed in the Crawford Upland and Ste. Genevieve Limestone (Mississippian). Small upper-level passages lead to a large lower-level stream passage with overflow cutarounds and a spring entrance (*1973 NSS Convention Guidebook*).

52. **Maxwelton Cave** .... 15,498 m
(Greenbrier County, West Virginia)

Cave with no entrances (one had a parking lot built over it and the other silted shut)! The cave was dug open by Pete Williams in 1971 and all of the surveying that took place was done in the next two years before the entrance silted shut again (*N.S.S. News*, 1974). Formed in the Greenbrier limestone (Mississippian age). Resurgence at Davis Spring (see McCluny's).

53. **Big Bone Cave** .... 15,494 m
(Van Buren County, Tennessee)

Large fossil cave mined for saltpeter during the Civil War. Of the two entrances, one is now closed. Mapping by the Nashville Grotto in the 1960's by led by John Smyre.

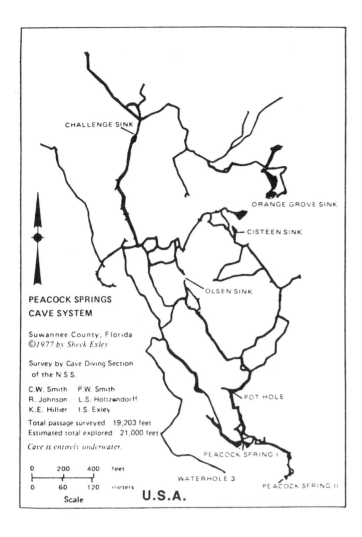

PEACOCK SPRINGS CAVE SYSTEM

Suwannee County, Florida
©1977 by Sheck Exley

Survey by Cave Diving Section
of the N.S.S.

C.W. Smith   P.W. Smith
R. Johnson   L.S. Holtzendorff
K.E. Hillier   I.S. Exley

Total passage surveyed 19,203 feet
Estimated total explored 21,000 feet

Cave is entirely underwater.

Scale

U.S.A.

The Arch Cave branch is mazy, while the Bone Cave section has long dry passages (Barr (T.), *Caves of Tennessee*, 1961; and issues of *The Speleonews*).

54. **Bobcat Cave** .... 14,661 m
(Bath and Highland Counties, Virginia)
This is a recently discovered cave in Burnsville Cove, Virginia. Mapping and exploration are currently proceeding via multi-day in-cave camps.

55. **Snail Shell Cave System** .... 14,598 m
(Rutherford County, Tennessee)
Cave with four entrances (Barr, Clark, Echo, Snail Shell) containing over 9 km of river passage requiring flotation to explore. Formed in Ridley limestone (N.S.S. News, August, 1982).

56. **Foglepole Cave** .... 14,500 m
(Monroe County, Illinois)
Cave with five entrances (Northwest, Southeast, Keller, Twin Pit, and Main Entrances), two main stream passages, and a large trunk passage. Surveying over the last 15 years by the Windy City Grotto has been coordinated by Father Paul Wightman (N.S.S. News 41 (2), 1983).

57. **Rice Cave** .... 13,824 m
(Van Buren County, Tennessee)

58. **Cuyler Cave** .... 13,132 m
(Sevier County, Tennessee)

59. **Fitton Cave** .... 13,106 m
(Newton County, Arkansas)
Fitton Cave (also known as Beauty Cave) is located in the Ozark Mountains of Arkansas. The Fitton entrance leads both to the East Passage and the Crystal Passage. Trunk passages are occasionally partially blocked by breakdown, and lower level steam passage traverses the length of the cave. A second entrance, Bat Cave, leads to a 14 m waterfall and connects to the East Passage. The cave is known for unique calcite and gypsum speleothems. The Cave Research Foundation is presently resurveying the cave in cooperation with the National Park Service.
Pete Lindsley.

60. **Dunbar Cave - Roy Woodard System** .... 13,000 m
(Montgomery County, Tennessee).

61. **Fort Stanton Cave** .... 12,875 m
(Lincoln County, New Mexico)

62. **Grassy Cove Saltpetre Cave** .... 12,703 m
(Cumberland County, Tennessee)
(*N.S.S. News*, July, 1977, map).

63. **Wolf River Cave** .... 12,698 m
(Fentress County, Tennessee)

64. **Great EXpectations Cave** .... 12,640 m
(Washakie County, Wyoming)
See above.

65. **Lee Cave** .... 12,580 m
(Edmonson County, Kentucky)

66. **Climax Cave** (Decatur County, Georgia) .. 12,440 m

67. **Lisanby Cave** .... 11,530 m
(Caldwell County, Kentucky)

68. **Webster System** .... 11,500 m
(Breckenridge County, Kentucky)

69. **Russell Cave** .... 11,330 m
(Jackson County, Alabama)

70. **Zarathustra Cave** .... 11,314 m
(Fentress County, Tennessee)
(*N.S.S. News*, November, 1977)

71. **Blanchard Springs Cave** .... 11,265 m
(Stone County, Arkansas)

72. **Paul Penley's Cave** .... 11,265 m
(Bland County, Virginia)
See above.

73. **Unthanks Cave** (Lee County, Virginia) .... 10,816 m

74. **Simmons Mingo - My Cave System** ...... 10,783 m
(Randolf and Pocahontas Counties, W. Va)
See above.

75. **Cassell Cave System** .... 10,670 m
(Pocohontas County, West Virginia)
Surveyed in the mid-1960's by D. B. Williamson, D. Medville and cavers from the West Virginia University, the Philadelphia Grotto, and the Potomac Speleological Club.

76. **Fallen Rock Cave** .... 10,284 m
(Tazewell County, Virginia)

77. **Groaning Cave** .... 10,270 m
(Garfield County, Colorado)

78. *Cathedral-Falmouth Cave System* ......**10,222 m**
(Suwannee County, Florida)
This is one of the longest explored underwater cave in the world. It is a gallery with seven entrances explored by the Cave Diving Section of the N.S.S., led by Scheck Exley. Around 1975 Falmouth Cave was dived for a distance 3810 m. In 1985, U.S. divers brought the surveyed length to 10,222 m!
Map: from the map of Sheck Exley (1985).

79. **Blissits Cave** .... 10,150 m
(Breckenridge County, Kentucky)

80. **Lost River Cave System** .... 9960 m
(Warren County, Kentucky)

81. **Jester Cave System** .... 9672 m
(Greer County, Oklahoma)
Longest cave formed in gypsum in the U.S.

82. **Vinegar Ridge Cave** .... 9636 m
(Hart County, Kentucky)

83. **Guess Creek Cave** .... 9553 m
(Jackson County, Alabama)

84. **Clarks Cave** (Bath County, Virginia) .... 9060 m

85. **Illinois Caverns** .... 9050 m
(Monroe County, Illinois)

86. **Precinct Eleven Cave** .... 9000 m
(Rockcastle County, Kentucky)

87. **Ludingtons Cave** .... 8948 m
(Greenbrier County, West Virginia)

88. **Byer's Cave** (Dale County, Georgia) .... 8851 m

89. **Acme Mine / Quarry Cave** .... 8690 m
(Greenbrier County, West Virginia)

90. **McFails Cave** .... 8687 m
(Schoharie County, New York)

91. **Wyandotte Cave** .... 8626 m
(Crawford County, Indiana)

92. **Devil's Icebox** .... 8625 m
(Boone County, Missouri)

93. **Hell Hole** .... 8610 m
(Pendleton County, West Virginia)

94. **Blowing Springs Cave** .... 8127 m
(Coffee County County, Tennessee)

95. **Spring Hollow Cave** .... 8127 m
(Bland County, Virginia)

96. **Abrogast - Cave Hollow System** .... 8047 m
(Tucker County, West Virginia)

97. **Pattyjohns Cave** .... 7998 m
(Walker County, Georgia)

98. **Blowhole (Nichols) Cave** .... 7986 m
(Cannon County, Tennessee)

99. **Dry Cave** (Eddy County, New Mexico) .... 7931 m

100. **Yellow Jacket Cave** .... 7925 m
(Eddy County, New Mexico)

———

126. *Peacock Springs Cave System* .... 6507 m
(Suwannee County, Florida)
Along with Cathedral-Falmouth Cave System and Lucayan Caverns (see Bahamas), Peacock Springs Cave System is one of the longest underwater caves in the world. It has eight entrances: Challenge Sink, Orange Grove Sink, Waterhole 3, Cisteen Sink, Olsen Sink, Pot Hole, and Peacock Springs I and II. Most of the passages are between -15 m and -18 m and are formed in the Ocala group. The water continues at depth to an unknown resurgence along the Suwannee River.
Exploration began in 1956 with the efforts of Vasco Murray in Peacock Spring I and Orange Grove Sink. The first connection (between Peacock Spring I and Pot Hole) was made by George Krasle, Howard Lilly and Dick Olsen in 1965. Next Cisteen Sink and Olsen Sink were connected by Rick Wright and Howard Bradbeer. Olsen and Challenge Sinks were connected around 1970 by Tom Mount and Frank Martz. The most significant connection was that of Orange Grove and Challenge Sinks in June 1970 by John Harper, Randy Halton and F. Martz.
The exploration of Waterhole 3 began in 1973 with Court Smith, Lewis Holtzendorff, and Sheck Exley. On July 7, 1973, David Fisk, Dana Turner and S. Exley emerged in Peacock Spring I. On September 3, 1973 C. Smith, L. Holtzendorff and S. Exley connected Waterhole 3 with Olsen Sink, establishing the Peacock Springs Cave System. Mapping began in 1975, with 5,833 m mapped by 1977 and the length of 6,507 m reached in 1980.
Map: by the Cave Diving Section of the N.S.S. (1977) showing 5853 m, sent by Arthur Palmer, published courtesy of Sheck Exley.
Bibliography: Exley (S.) - The Peacock Springs Cave System, *N.S.S. News*, 1978, 36 (3):43-44.

———

359. **Sinnett - Thorn Mt. Cave System** .... 3000 m
(Pendleton County, West Virginia)

URUGUAY

The longest cave known in this small country is **cueva de Arequita** (Minas), 41 m long, in granite.

VENEZUELA

The first explorations of Venezuelan caves were by indigenous groups who adventured into the more accessible sections of several caves, especially in the western part of the country. During the colonial period, from 1540 to 1799, about twenty speleological sites were investigated, but it was the visit of Alexander von Humboldt to the cueva del Guácharo in 1799 that marked the beginning of scientific studies. Numerous naturalists studied various caves in the 19th century, including Vicente Marano who explored more than 30 caves between 1883 and 1890 while studying guano deposits for potential agricultural use.

Modern speleology in Venezuela began in 1952 with the formation of the Sociedad Venezolana de Ciencias Naturales (S.V.C.N.) under the direction of Eugenio de Bellard-Pietri. The more enthusiastic members of this club formed the Sociedad Venezolana de Espeleológia in 1967, which has published an excellent bulletin, the *Catastro Espeleológico Nacional*, ever since. Some of the prime movers of this era were Juan Tronchoni, Omar Linares, and Franco Urbani. The Centro de Exploraciones Espeleológicas de la Universidad Simon Bolívar was formed in 1972, while the Grupo Espeleológico was formed in 1973. Venezuela hosted its first foreign expedition in 1973 (British Karst Research Expedition), and a second in 1976 (Polish Federation of Alpinism).

The limestone ranges are of Jurassic and Tertiary age and are of moderate extent, often fragmented by non-karstifiable rocks. This accounts for the relatively few number of long caves. The largest karst areas are the western part of the Serranía de l'Interior (Cretaceous age), the zone de Mata Mango in the state of Monagas, the central region of the Serranía Costera, and the Falconiana region in the Sierra de San Luis (Oligocene age).

The Guyana region, in the southern part of the country, has proven to contain some interesting caves formed in quartzites from the Roraima group, such as Autana, Urutany, Sarisariñama, Aonda, and others. This region, explored since 1974, is difficult to access, but is covered with huge pits (up to 400 m in diameter and up to 300 m deep) and labyrinths of fissures. It provides a challange for those interested in speleogenesis.

The parts of the country that border on Columbia, Brazil, and Guyana appear to have the best potential for future cave discoveries in Venezuela.

Carlos Galan and Franco Urbani

DEEP CAVES:

1. ***sima Aonda*** (Auyantepuy, Bolivar) .... **–362 m**
The sima Aonda has been known for many years from aerial reconnaissance, but it was not first explored until January, 1983 by members of the Sociedad Venezolana de Espeleológia (S.V.E.) led by Carlos Galan. The expedition was transported by helicopter to the top of the mountain.

The cave is located to the NW of Auyantepuy, at an altitude of 1600 m (6°2'N, 62°37'W). (See "The Great Pits").

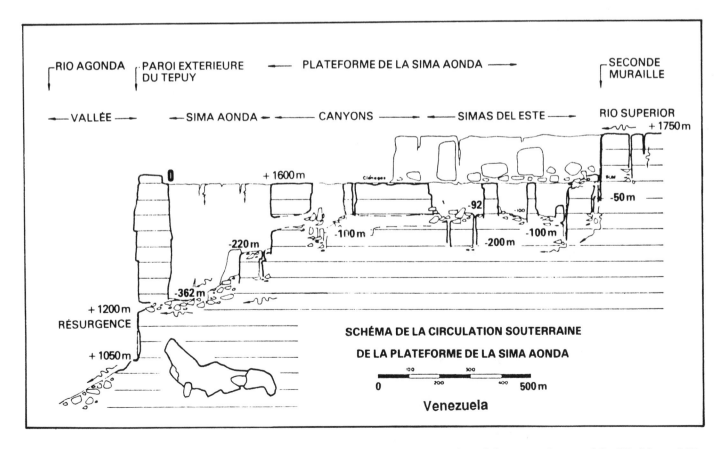

MESETA DE SARISARIÑAMA
Venezuela

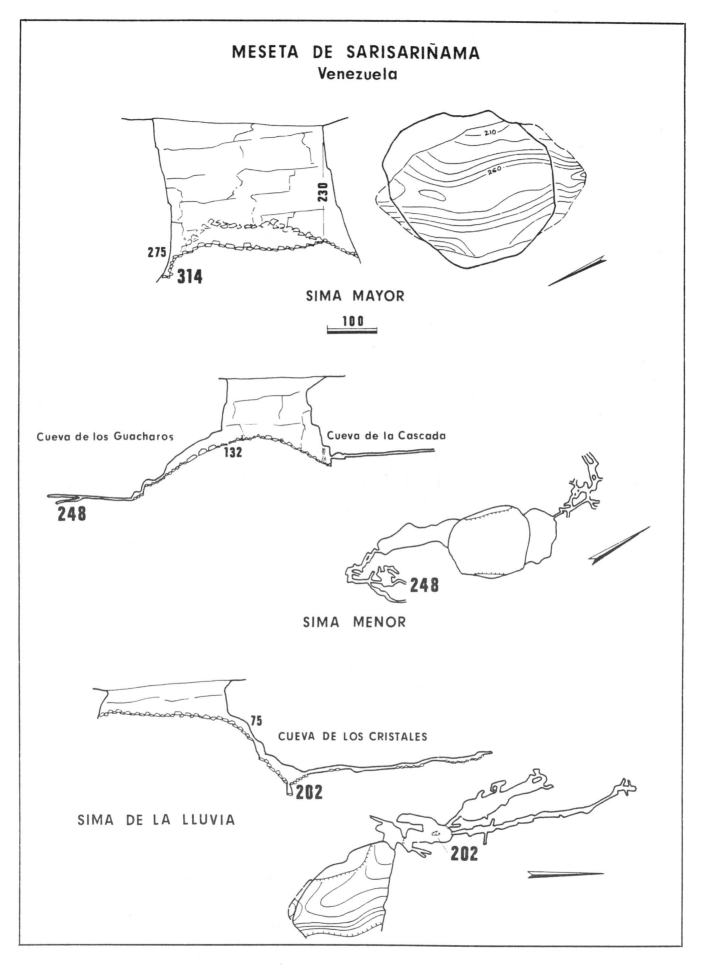

SIMA MAYOR

100

Cueva de los Guacharos   Cueva de la Cascada

SIMA MENOR

CUEVA DE LOS CRISTALES

SIMA DE LA LLUVIA

The cave is formed in quartzite from the Roraima group, of Precambrian age, and has been stable for the last 1.5 billion years. Its formation is linked to a major system of fractures bordering the tall table mountain Auyantepuy. This mountain is known for a large 970 m tall waterfall called Churún Merú (or Salto Angel), located 10 km from the cave. The sima Aonda is the deepest cave in the world not formed in limestone.

Map: taken from Spelunca, 1984 (14).

Bibliography: *Bol. de la Sociedad Venezolana de Espeleológia*, 1984 (20): 43-45.

Galan (C.) - La sima Aonda, *Spelunca*, 1984 (14): 14-17.

Galan (C.) - Sima Aonda, *S.V.E.* ed., 1983, 24 p., 15 color photos.

2. **sima Auyantepuy Norte** .... –320 m
(Auyantepuy, Bolivar)
Another deep cave formed in Precambrian quartzite explored in 1986 by the S.V.E.

3. *sima Major de Sarisariñama* .... **–314 m**
(Sarisariñama Jidi, Bolivar)
This impressive pit has been known for over twenty year by airplane pilots. It was first dropped in 1974 by an expedition of the Sociedad Venezolana de Ciencias Naturales led by Charles Brewer and David Nott. Another expedition took place in 1976 in which the S.V.E. was joined by Polish cavers. They studied and mapped the cave, measuring the depth with a theodolite (see 'The Great Pits').

As for the two preceding caves, this one is formed in the Precambrian quartzites of the Roraima group, at an altitude of 1430 m. The interior of the pit contains silica speleothems (opal) and traces of hydrothermalism which may have weakened the rock, explaining the large dimensions of the cavity.

Bibliography: Nesatas de Jaua, Guanacoco y Sarisariñama, *Soc. Venezolana Ci.* Nat. ed., 1973, 32 p., 17 fig. *Bol. S.V.E.*, 1976 (13):86-88.

4. **haitón del Guarataro** .... –305 m
(Sierra de San Luis, Curimagua, Falcón)
Alt. 1200 m. Explored in 1973 by the British Karst Research Expedition to Venezuela (B.K.R.E.) (*El Guácharo*, 1973, 6 (1-2)).

5. **sima Aonda Este 2** .... –295 m
(Auyantepuy, Bolivar)
Cave formed in quartzite descended in 1986 by the S.V.E.

6. **sima Aonda Sur 1** .... –290 m
(Auyantepuy, Bolivar)
Cave formed in quartzite explored in 1986 by the S.V.E.

7. **haitón de Sabana Grande No. 1** .... –288 m
(Sierra de San Luis, San Luis, Falcón)
Alt. 1040 m. Explored starting in 1973 by the G. E. Universidad Simon Bolivar (*Bol. S.V.E.*, 1983 (20)).

8. **sima del Cacao** .... –260 m
(Mata de Mango, Monagas)
Alt. 600 m. Explored in 1982 by the S.V.E. (*Bol. S.V.E.*, 1982 (10)).

9. **sima Menor de Sarisariñama** .... –248 m
(Sarisariñama, Bolivar)
Alt. 1430 m. Cave in Precambrian quartzite explored in 1974 (S.V.C.N.) and in 1976 (S.V.E. and Polish cavers; *Bol. S.V.E.*, 1976 (13)).

10. **sima de San Lorenzo** .... –232 m
(Macuquita, Falcón)
Explored in 1973 by the B.K.R.E. (*Bol. S.V.E.*, 1983 (20)).

11. **sima del Chorro** .... –220 m
(Mata de Mango, Monagas)
Alt. 1100 m. Explored in 1979 by the S.V.E. (*Bol. S.V.E.*, 1982 (19)).

12. **sima del Danto** .... –208 m
(Mata de Mango, Monagas)
Alt. 600 m. Explored in 1981 by the S.V.E. (*Bol. S.V.E.*, 1982 (19)).

13. **sima de la Lluvia** .... –202 m
(Sarisariñama, Bolivar)
Alt. 1400 m. Cave in quartzite explored in 1976 by the S.V.E. and Polish cavers (*Bol. S.V.E.*, 1976 (13)).

14. **sima Los González** .... –200 m
(Mata de Mango, Monagas)
Alt. 680 m. Explored in 1979 by the S.V.E. (*Bol. S.V.E.*, 1982 (19)).

15. **sima Aonda Norte** ....approx. –200 m
(Auyantepuy, Bolivar)
Cave in quartzite.

16. **sima La Bandera** (Curimagua, Falcón) ...... –198 m
Explored in 1973 by the B.K.R.E.

17. **cueva Segunda Cascada** .... –190 m
(Humocaro Alto, Lara)
Alt. 1235 m. In 1974 by the G.E. Inter Venezolano (Bol. S.V.E., 1975 (11)).

18. **sima Bastimento No. 1** ....(+7, –163) 170 m
(Mata de Mango, Monagas)
Alt. 720 m. In 1974 and 1977 by the S.V.E. (*Bol. S.V.E.*, 1977 (16)).

19. **sima Aonda Sur 2** .... –168 m
(Auyantepuy, Bolivar)
Cave in quartzite explored by the S.V.E. in 1986.

20. **sima Domingo-Hilario** .... –162 m
(Mata de Mango, Monagas)
Alt. 1000 m. In 1979 and 1983 by the S.V.E. (*Bol. S.V.E.*, 1982 (19) and 1984 (21)).

21. **haitón La Maleta** .... –161 m
(Sierra de San Luis, Curimagua, Falcón)
Explored in 1973 by the B.K.R.E.

22. **sima La Quebrada** .... –157 m
(Mata de Mango, Monagas)
Alt. 950 m. Explored by the S.V.E. in 1979 (*Bol. S.V.E.*, 1982 (19)).

LONG CAVES:

1. **cueva del Guácharo** (Caripe, Monagas)...10,200 m
This famous show cave was used by local indians as long as 3000 years ago. Written descriptions date back to 1662 and were made by priests. Alexander von Humboldt was the first to describe the cave in detail, in 1799, but he only explored the first 422 m. He also described the bird *Steatornis caripensis* (guácharo) which lives in the cave and uses echo-location in a manner similar to bats.

It was visited and described by numerous European naturalists in the 19th century, including A. Codazzi who explored 1074 m in 1835.

Modern exploration began in 1952, culminating in a mapped length of 10,200 m in 1972 (S.V.C.N. until 1965, then the S.V.E.).

The cueva del Guácharo is formed in Cretaceous age limestone (Aptien-Albien). It is basically a horizontal cave with several levels and a total vertical extent of about 50 m, located at an altitude of 1065 m. It has an unusual variety of formations, including carbonate, sulphate, and phosphate minerals. (*Bol. S.V.E.*, 1967 1 (1): 97-107, map with tourist route; 1972, 3(2):116-131, map of wild section of cave).

2. **cueva Alfredo-Jahn** (Birongo, Miranda) ..... 4292 m
Formed in marble. Explored from 1952 to 1956 by the S.V.C.N. (*Bol. S.V.E.*, 1973, 4(1)).

3. **cueva Segunda Cascada** ....3014 m
(Humocaro Alto, Lara)
Explored in 1974 by the G.E.I.V. (*Bol. S.V.E.*, 1975, 6 (11)).

4. **cueva La Peonía** (Barbacoas, Lara) ....2514 m
Explored in 1973 by the S.V.E. (*Bol. S.V.E.*, 1974, 5 (1)).

5. **cueva Grande Antón-Goering** ....2340 m
(Mata de Mango, Monagas)
Discovered in 1866 by A. Goering. Explored in 1973 and 1975 by the S.V.E. (*Bol. S.V.E.*, 1975 (12) plan). Contains guácharos birds.

6. **cueva Francisco-Zea** ....2000 m
(Río Guasare, Zulia)
In 1973 by the S.V.E. (*Bol. S.V.E.*, 1973, 4 (1) plan).

7. **haitón de Sabana Grande No. 1** ....1870 m
(Curimagua, Falcón)
From 1974 to 1983 by the C. E. de la U.S.B. (*Bol. S.V.E.*, 1983 (20) plan).

8. **cueva de La Quebrada del Toro** ....1602 m
(La Taza, Falcón)

Formed in quartzite. Contains guácharos birds. Explored in 1974, 1979, and 1980 by the S.V.E. (*Bol. S.V.E.*, 1983, (20) plan).

9. **cueva Grande**...........................approx. 1600 m
(Buenos Aires, Lara)
Explored in 1983 by the G.E.I.V.

10. **sima Los González** ....1526 m
(Mata de Mango, Monagas)
In 1974 and 1979 by the S.V.E. (*Bol. S.V.E.*, 1982, (19) plan).

11. **cueva del Agua** (Guanta, Anzoátegui) ....1367 m
(*Catastro Espel. Venezolano*, 1973, 1, plan showing 985 m.)

12. **sima de la Lluvia** ....1352 m
(Sarisariñama, Bolivar)
(*Bol. S.V.E.*, 1976 (13) plan). Contains guácharo birds.

13. **cueva Cruxent** (Birongo, Miranda) ....1310 m
Formed in marble. Explored in 1973 by the S.V.E. (*Bol. S.V.E.*, 1973, 4 (2) plan).

14. **cueva La Milagrosa** ....1282 m
(Mundo Nueva, Monagas)
Explored in 1963 by the S.V.C.N. and in 1973 and 1976 by the S.V.E. (*Bol. S.V.E.*, 1982 (19) plan).

15. **cueva de la Azulita** (La Azulita, Mérida) .... 1240 m
(*Bol. S.V.E.*, 1974, 5 (1) plan).

16. **sima Menor de Sarisariñama**...................1179 m
(Sarisariñama, Bolivar)
(*Bol. S.V.E.*, 1976 (13) plan).

17. **cueva Walter-Dupouy**.............................1172 m
(Capaya, Miranda)
Marble cave (*Bol. S.V.E.*, 1975 (12) plan).

18. **cueva Hueque No. 3** (Hueque, Falcón) ....1140 m
(*Bol. S.V.E.*, 1983 (20) plan).

19. **cueva El Sarao** ....approx. 1125 m
(Cuicas, Trujillo)

20. **sima de Simón** ....1120 m
(Mata de Mango, Monagas)
(*Bol. S.V.E.*, 1984 (21) plan).

21. **cueva Coy-Coy de Uria**............................1100 m
(Curimagua, Falcón)
(*Trans. B.C.R.A.*, 1981, 8 (1) plan).

22. **cueva Cagigal** (Birongo, Miranda) ....1085 m
In marble (*Bol. S.V.E.*, 1973, 4 (2) plan).

23. **cueva Zárraga** (Curimagua, Falcón) ....1028 m
(*Bol. S.V.E.*, 1983 (20) plan).

IV ASIA

AFGHANISTAN

Expeditions to this high mountain country (contains Hindu Kush, 7730 m high) have been undertaken by the Swedish biospeleologist Knut Lindberg (1957-1958), the A.R.I.A.S. de Rouen (1974), and the Franco-Spanish expedition of 1975 (S. C. Paris and E.R.E. Catalunya de Barcelona). No major caves were found. Although there are huge limestone deposits, climatic conditions have not allowed them to become karstified. The present list is from 1975 (*Mém. du S. C. Paris*, 1975, 1), but has not changed much in the meantime due to political developments in this country.

LONG CAVES:

1. **Ab Bar Amada** .... 1120 m
    (Salang massif, Qalatak, Parwan)
    Alt. 2240 m. Explored in 1974 and 1975 (*op. cit.*, map).

2. **ghar Bōlān Bābā** .... 730 m
    (Qalat, Zabul)
    Cave with religious usage explored in 1958 and 1975. Cave depth is − 33 m (*op. cit.*, map).

3. **ghar I-Djon** .... 326 m
    (Salang, Qalatak, Parwan)
    Alt. 2180 m. Mapped in 1975 (*op. cit.*, map).

CAMBODIA
Kâmpŭchea Prâcheathipâtéyy

Cambodian karst, noted in 1913 but not thought to be promising, is located in the provinces of Kâmpôt (Kâmpôt, Kâmpóng Trach, Tuk Meas) and Bătdâmbâng (to the west of this city) and contains many explorables caves. Biospeleological research in 1968-1970 by Frenchmen including Claude Boutin took place in over fifty caves. Among these was **roung Thmâr Sâr** (Kâmpông Trach, Kâmpôt), 73 m long.

CHINA
Zhōngguó

China has over a million square kilometers of cavernous limestone karst, with the principal areas being in the southern provinces of Guangxi, Hunan, Guizhou and Yunnan. Cave exploration in China has a long history, but it has mostly been for scientific rather than sporting reasons.

Caves and cave hydrology in North China were described in *The Mountain Scripture*, a book written over 2200 years ago. But the real pioneer of Chinese speleology was Xu Xiake (1587-1641), who explored over 300 caves. Chinese cave exploration nowadays is principally related to water supply and dam-building projects, and many kilometers of cave have been mapped during such projects. National recognition for karst studies came in 1976 with the establishment of the Institute of Karst Geology at Guilin; the Institute now employs several hundred people. Cave diving commenced in the Seventies.

Since 1984 there have been many foreign expeditions to China (Japanese, Canadian, British, French, New Zealand U.S.A.), with most of the exploration being in Guangxi and Guizhou. Cave statistics quoted in Chinese publications commonly refer to the length of cave systems demonstrated by dye-tracing rather than the length of explored passage in a single cave. The following list has applied the latter criteria, and thus unfortunately omits many long caves, details of which have not been published in the west. A report on tourist caves in China, by Mo Zhong Da, gives 24 karst tourist caves longer than 1 km, with the longest being Tenglongdong (length approx. 30 km, Lichuan, Hubei), and the longest tourist lava tube being Erdaobeihe (Jilin), with a length of 4 km.

DEEP CAVES:

1. **Wu Jia dong** (Guizhou) .... −436 m
    Explored in 1987 by British cavers (*Caves and Caving*, 3a)

2. **Baishui dong** (Jiangkou, Guizhou) .... −310 m
    Mapped by the Guizhou Geological Bureau.

3. **Danghai Cave** (Guizhou) .... −301 m
    Explored in 1986 by French cavers.

4. **Shen dong** (Qingzhen, Guizhou)................ −280 m
    Contains a vertical pit 275 m deep, descended by "China Caves 85".

5. **Da Guala dong** (Luota, Hunan) .... –258 m
(*Carsologica Sinica*, 1983, 2(1) profile).

LONG CAVES:

1. **Soliao dong** (Bama, Guangxi) ....7600 m
Underground river mapped by the Guilin Karst Institute.

2. **Sandang system** (Guizhou) ....6200 m
Explored by French cavers in 1986.

2. **Wan Hua dong** (Luota, Hunan) ..... approx. 5000 m
Fossil cave mapped by the I.K.G.

3. **Chuan yan** (Nanxu, Guangxi) ....3860 m
Underground river mapped by "China Caves 85".

4. **Guan yan** (Caoping, Guangxi) ....3830 m
Swallow cave mapped by "China Caves 85".

5. **Shinban Qiao dong** (Haiyang, Guangxi) .....2700 m
Underground river mapped by "China Caves 85".

6. **Xiaoheli yan** (Caoping, Guangxi) ....2690 m
Resurgence mapped by "China Caves 85".

7. **Xizhen yan** (Nanxu, Guangxi) ....2640 m
Underground river mapped by "China Caves 85".

8. **Maliu kang** (Nanxu, Guangxi) ....2550 m
Underground river mapped by "China Caves 85".

CYPRUS
Kipros/Kibris

The cretaceous limestones of the island (Mts. Kyrenea, Mts. Kapas, Mts. Pentadaktylos) have not yet been studied intensively. In 1965, British cavers from the Westminster S. G. descended to –134 m in **Pentadaktylos** (Kythrea).

INDIA
Bhārat

Thanks to the research of Daniel Gebauer and his valuable work *Caves of India and Nepal* (1983) we know much more about Indian caves than we used to. When India was a British colony, B. B. Osmaston and Robert B. Foote made some scientific investigations at the end of the 19th century, while E. A. Glennie followed in their footsteps beginning in 1930. Then in the 1970's two British expeditions (T. Waltham and J. Conway) investigated caves in the Himalayan mountains. It was starting in 1980 that D. Gebauer revealed the rich caving potential of the Deccan plateau in the state of Andhra Pradesh. Nonethe-less, it is the religious cave of **Amarnath** (Jammu and Cachemire), only 50 m long, which remains the most famous cave in India.

DEEP CAVES:

1. **Borra guhalu** .... –86 m
(Vizakhapatnam, Andhra Pradesh)
Alt 715 m. By Gebauer and A. Abele in 1982 (Gebauer, *op. cit.*, profile).

2. **Billam guha** .... –77 m
(Cuddapah, Andhra Pradesh)
By Narayana Reddy in 1984.

3. **Kuruva Bali guha** .... –76 m
(Anantapur, Andhra Pradesh)
Alt. 420 m. By Gebauer and Abele in 1984 (Abh. Karst-u. Höhlenkunde, *Kurnool*, 1984, profile).

4. **Lower Swift Hole** .... –74 m
(Dehra Dun, Uttar Pradesh)
Alt. 2600 m. Descended in 1945 by E.A. Glennie.

5. **Sough goffar** .... –61 m
(Arki, Himashal Pradesh)
Alt. 1580 m. In 1970 by the British Himalaya Spel. Exp.

6. **Upper Swift Hole** .... –54 m
(Dehra Dun, Uttar Pradesh)
Alt. 2610 m.

7. **Moila swallet** .... –50 m
(Chakrata, Uttar Pradesh)
Alt. 2480 m. Descended in 1890 by Osmaston and in 1945 by Glennie.

LONG CAVES:

1. **Belum guhalu** ....3225 m
(Kurnool, Andhra Pradesh)
Alt. 335 m. Explored in 1982-1984 by Gebauer and Abele (Gebauer, *op. cit.*, map; *Kurnool* 1984, map).

2. **Dobhakhol cave** ....1092 m
(Siju, Meghalaya)
Alt. 266 m. In 1922 by S. Kemp and B. Chopra.

3. **Borra guhalu** ....824 m
(Vizakhapatnam, Andhra Pradesh)
In 1982 by Gebauer and Abele (Gebauer, *op. cit.*, map).

4. **Yerra Zari Gabbi** ....684 m
(Kurnool, Andhra Pradesh)
Alt. 600 m. In 1981 and 1982 by Gebauer and Abele (Gebauer, *op. cit.*, map).

5. **Munagamanu gavi** ....440 m
(Kurnool, Andhra Pradesh)
Alt. 500 m. In 1983 and 1984 by Gebauer, Abele and C. Chabert (*Kurnool* 1984, map).

6. **Bellam cave** (Guntur, Andhra Pradesh) ...... 440 m
Alt. 213 m. In 1984 by Narayan Reddy.

7. **Gupteshwadam** (Rohtas, Bihar) ...... approx. 350 m
Alt. 140 m. In 1906 by L. O'Malley.

8. **Kuruva Bali guha** .... 318 m
(Kurnool, Andhra Pradesh)
In 1984 by Gebauer and Abele (*Kurnool* 1984, map).

9. **Billa sorgam** (Kurnool, Andhra Pradesh) ...... 300 m
(Gebauer, *op. cit.*, map).

INDONESIA

With its 13,677 islands spread out over some 5000 km, Indonesia is a country with a lot of promise for cave exploration in the future. Although thanks to the work of the Hungarian Dénes Balázs (published in *Karszt-és Barlangkututás*, 1963-1967, V) the karst areas of this country have been known for some time, it was only in 1982 that systematic exploration began with the creation of the Indonesian Federation of Speleology (or FINSPAC), President Robert Ko. Thanks to its work, numerous groups of foreign cavers have visited Indonesia, notably the British (Java, 1982-1984; Sumba, 1985, Irian Jaya, 1985), the Belgians (Java, 1982) and the French (Kalimantan, 1982-1983; Java, 1982; Sumba, 1985). It should be noted, however, that almost all the large Indonesian caves have been explored by locals hunting for birds nests, so that foreign cavers are more mappers than explorers!

Aside from the still poorly studied karsts of Irian Jaya, the depth potential in Indonesia appears to be limited. On the other hand, there is good potential for long caves. The principal karst areas (from west to east) are found at Sumatra (Barisan), Java (with its famous cone karsts of Gunung Sewu), Kalimantan (the Mangkalihat peninsula in particular), Sumba, Sulawesi (Maros), and Irian Jaya (Ajamaru, Fak-Fak, Kumawa).

DEEP CAVES:

1. **luweng Ombo** .... −230 m
(Gunung Sewu, Java Tengah)
Explored by the 1982 French expedition.

2. **gua Sallukan Kallang** .... (−184, +21) −205 m
(Maros, Sulawesi Selatan)
Assoc. Pyrénées de Spel. 1985 and 1986 (*Spelunca* 28, map).

3. **luweng Buhputih** (Gunung Sewu, Java) .... −200 m
Explored in 1982 by the British team led by T. Waltham (*Trans. B.C.R.A.*, 1983, 10(2) profile).

4. **luweng Puleireng** .... −196 m
(Gunung Sewu, Java)

5. **luweng Ngepoh** (Gunung Sewu, Java) .... −182 m
British expedition of 1982 (*Trans. B.C.R.A.*, 1983, 10(2) profile).

6. **luweng Soga** (Gunung Sewu, Java) .... −177 m

7. **luweng Ngiratan** (Gunung Sewu, Java) ..... −168 m
British expedition of 1982 (*Trans. B.C.R.A.*, 1983, 10(2) profile).

8. **gua Lebak Bareng** .... −166 m
(Gunung Sewu, Java)
Explored in 1982 by the British (*Trans. B.C.R.A.* 1983, 10 (2) profile).

9. **luweng Jero** (Gunung Sewu, Java) .... −163 m
Explored in 1982-1983 by the British (*Trans. B.C.R.A.* 1982 10 (2) profile to −151 m).

10. **luweng Grubug** (Gunung Sewu, Java) .... −161 m
Connected with luweng Jomblang. Explored by a French-Belgian team (*Java Karst 82*, report, profile).

11. **gua Si Wulung** (Cibodas, Java Barat) .... −160 m

12. **luweng Jaran** (Gunung Sewu, Java) .... −158 m

13. **gua Ngowe-Owe** (Gunung Sewu, Java) .... −156 m

14. **luweng Jowa** (Gunung Sewu, Java) .... −151 m
Explored in 1982 by the Belgian team led by Y. Quinif (*Spéléo-Flash*, 1982 (134) profile).

15. **gua Kapayau** .... +143 m
(Sangkulirang, Kalimantan Timur)

16. **luweng Tong Pocot** .... −142 m
(Gunung Sewu, Java)
Explored in 1982 by the British (*Trans. B.C.R.A.*, 1983, 10 (2) profile).

17. **luweng Setro** (Gunung Sewu, Java) .... −140 m
Explored in 1982 by the British (*Trans. B.C.R.A.*, 1983, 10 (2) profile).

18. **Yogoluk** (Wamena, Irian Jaya) .... −140 m

_____

28. **luweng Puniran** .... −100 m
(Gunung Sewu, Java)
Note: administratively, the Gunung Sewu covers two provinces, Yogyakarta (to the west), and Java Tengah (to the east).

LONG CAVES:

1. **gua Sallukan Kallang** .... 12,263 m
(Maros, Sulawesi Selatan)
Explored in 1985 and 1986 by the Assoc. Pyrénéenne de Spéléologie.

2. **luweng Jaran** .... 11,072 m
(Gunung Sewu, Java)
Explored in 1984 by a British-Australian team (*Caves & Caving*, 1984 (26) schem. plan).

3. **gua Sodong** ....4290 m
(Mudal, Gunung Sewu, Java)
Explored in 1982 by the British (*Trans. B.C.R.A.*, 1983, 10 (2) profile).

4. **lubang Dunia** ....4220 m
(Sangkulirang, Kalimantan Timur)
Explored in 1983 by the French "Borneo 83" expedition (map in report of expedition, 1983).

5. **gua Airhangat** ....approx. 4000 m
(Barisan, Padang, Sumatera Barat)
Swallet cave explored by the G.E.S.M. of Barcelona in 1977 (*Phénomènes Karstiques*, 1982, III, schem. map).

6. **Leang Assuloang** ....3500 m
(Maros, Sulawesi Selatan)
Explored in 1986 by Italian cavers.

7. **gua Bribin** ....3400 m
(Gunung Sewu, Java)
Explored in 1982 by the British (*Trans. B.C.R.A.*, 1983, 10 (2) profile).

8. **gua Jomblang** ....3325 m
(Gunung Sewu, Java)
Not to be confused with luweng Jomblang.

9. **Watu Karamba Kondameha/**
Lai baba Kondameha ....2542 m
(Lewa, Sumba)
Explored in 1985 by British expedition (Trans. BCRA 13 (1), map).

10. **gua Kapayau** ....2426 m
(Sangkulirang, Kalimantan Timur)
Explored in 1982 by the French (*Expedition Report*, 1982, map).

11. **luweng Ombo** ....2300 m
(Gunung Sewu, Java Tengah)
Explored in 1982 by the French (*Expedition Report*, 1982, map).

12. **luweng Grubug** (Gunung Sewu, Java) ....2290 m
(*Grottes & Gouffres*, 1983, (90) map).

13. **gua Malihau** ....2174 m
(Muara Napu, Kalimantan Selatan)
1982 French expedition (*Expedition Report*, 1982, map).

14. **luweng Pace** (Gunung Sewu, Java) ....2173 m
(*Cave Science*, 1985, 12(2) map).

15. **gua Gunung Talikur** ....2161 m
(Tapin, Kalimantan Selatan)
Explored in 1982 by the French (*Expedition Report*, 1982, map).

16. **Surupan Jumblengan/**
gua Mudalen Papringen ....2075 m
(Tulukan, Java)

Explored in 1986 by British Army Caving Assoc. (*Descent* 73, map).

17. **gua Sodong** ....2075 m
(Dadapayu, Gunung, Sewu, Java)
Explored in 1982 by the British (*Trans. B.C.R.A.*, 1983, 10 (2) profile).

18. **gua Kali Alan** (Tulakan, Java) ....1879 m
Explored in 1986 by British cavers.

19. **luweng Demplo** ....1538 m
(Gunung Sewu, Java)
(*Cave Science*, 1985, 12 (2) map).

20. **gua Batu** (Gunung Sewu, Java) ....1500 m
Explored in 1984 by British and Australians.
_____

31. **luweng Bedesan** ....1025 m
(Gunung Sewu, Java)

IRAN

There are many large limestone areas in Iran, but not all of them are karstified due to various geologic factors. The greatest speleological potential is in the Zagros mountain chain. This area has attracted foreign expeditions (primarily British, French, and Polish) between 1971 and 1977. The most active was the Frenchman Jean-Pierre Farcy (Centre Niçois d'Expéditions Spéléologiques), who worked from 1973 to 1975 in this area. Previous work dates back to 1896 (the Frenchman J. de Morgan) and 1947 (the Swede Knut Lindberg).

DEEP CAVES:

1. *ghar Parau* (Kermanshah) .... **–751 m**
This high altitude (3050 m) fossil insurgence is located on a middle Cretaceous age plateau just south of the Kuh-i-Parau, in the Zagros massif, not far from the Iraqi border and 25 km to the NE of Kermanshah. It was first explored in 1971 by the Speleological Reconnaissance Expedition to Iran led by John Middleton. They reached the depth of –732 m. In 1972, the Expedition to Ghar Parau pursued exploration but was stopped a few meters further on by a sump a –751 m. Since the resurgence is at an altitude of about 1350 m, there is still 1000 m of additional vertical potential possible if the sump could be passed!
Map: survey by J. Harper, M. Jenkins, P. Kaye, G. Edwards and P. Standing, furnished by J. Middleton.
Bibliography: *Transactions of Cave Research Group of G. B.*, 1973, 15 (1); Judson (D.) - *Ghar Parau*, London, 1973, 216 p.

2. *ghar Shah Bandu* ....approx. –275 m
(Kuh-i-Shahu, Kermanshah)
Explored in 1977 by the British Speleological Expedition.

3. **ghor-i-Cyrus** ....approx. –265 m
(Kuh-i-Shanu, Kermanshah)
Explored in 1975 by J. P. Farcy and G. Cappa (*Spelunca*, 1974 (4) profile).

4. **ghar-e-Morghan** .... –240 m
(Kuh-i-Parau, Kermanshah)
Explored in 1974 by a Polish team led by B. Koisar.

5. **ghar-e-Mariz** .... –132 m
(Kuh-i-Parau, Kermanshah)
Explored in 1974 by the Polish.

6. **ghar-e-Boland** .... –120 m
(Kuh-i-Parau, Kermanshah)
Explored in 1974 by the Polish.

7. **ghar Garun** .... –112 m
(Kuh-i-Garun, Nahavand)
Alt. 3210 m. Explored in 1973 by the A.S.B.T.P. of Nice (*Spelunca*, 1974 (4) profile).

8. **ghar Acker** .... –110 m
(Kuh-i-Parau, Kermanshah)
Explored in 1971 by the S.R.E. (*Yorkshire Ramblers' Club Journal*, 1973 (36) profile).

9. **ghar-i-Shahu** .... –110 m
(Kuh-i-Shahu, Kermanshah)
Explored in 1975 by the C.N.E.S.
It is reported that the British 1977 expedition descended two caves –180 m and –120 m in the Kuh-i-Shahu.

LONG CAVES:

1. **ghar Parau** ....1364 m

2. **ghar Shaban Kale** ....650 m
(Kermanshah)
In 1977 by the S. C. Lyon.

IRAQ
'Irak

Because of limited access, the caving potential of Iraq is almost untouched. The Kurdish limestone areas, in the north and east parts of the country (Dukan Dam, le Grand Zab, Dinok, Rawandiz) are known to be rich in caves and karst phenomena. Reports indicate caves estimated at 1000 m long (Fasaya) and 800 m long (Pigeon cave at Hadithah). The only mapped cave appears to be **Useiba** (570 m) by the Frenchman Kuster (*Sous Terre*, 1978-1979 (20) map). The most famous cave is **Shanigar**, known for its archaeological treasures.

ISRAEL
Yisra'el

Most of the mountains regions of Israel are composed of limestones, dolomites, and chalks. Most of the carbonate rock caves are in the northern and central Israel, at altitudes between 0 and 1200 m. They are generally old horizontal phreatic mazes formed at the top of the watertable. One also finds more recent pits of vadose origin that are not over 50 m deep and rarely connect to the horizontal passages below.

The arid zones of the country have few carbonate rock caves. Of special interest are the caves formed in salt on Mount Sedom: they are the longest and deepest salt caves known in the world. They are formed by water penetrating through vertical pits and travelling through horizontal passages down to base level (which in this case is the Dead Sea, 400 m below the level of the Mediterrenean Sea). These are young caves, not over 14,000 years old. They continue to form, but only once a year on average during the rare desert storms (annual precipitaion is only 50 cm). The Israeli cavers are represented by the Israel Cave Research Center. It has about a hundred members and publishes a bulletin entitled *Niqrot Zurim*.

Amos Frumkin.

DEEP CAVES:

1. **Hutat Jermak** (Galil) .... –157 m
(*Niqrot Zurim*, 10, profile).

2. **mearat Malham** (Mt. Sedom) .... –135 m
Explored in collaboration with Italians (*Speleologia*, 1983 (9) profile to –128 m).

3. **Colonel** (Mt. Sedom) .... –127 m

4. **Alma** (Galil) .... –108 m

5. **El Husfi** (Shomron) .... –100 m

6. **Zechuchit** (Mt. Sedom) .... –99 m

7. **Bor 8, Peqi'in** (Galil) .... –93 m
(*Niqrot Zurim*, 10, profile).

8. **Notsa** (Mt. Sedom) .... –90 m

9. **Sedom** (Mt. Sedom) .... –85 m

10. **Buah** (Mt. Sedom) .... –81 m

LONG CAVES:

1. **mearat Malham** (Mt. Sedom) ....5447 m
(*Speleologia*, 1983 (9) map showing 2300 m).

2. **Hariton** (Harei Yehuda) ....3450 m
(*Nahal Teqoa*, map).

3.. **Sedom** (Mt. Sedom) ....1799 m
(*Speleologia*, 1984 (11) map showing 1063 m).

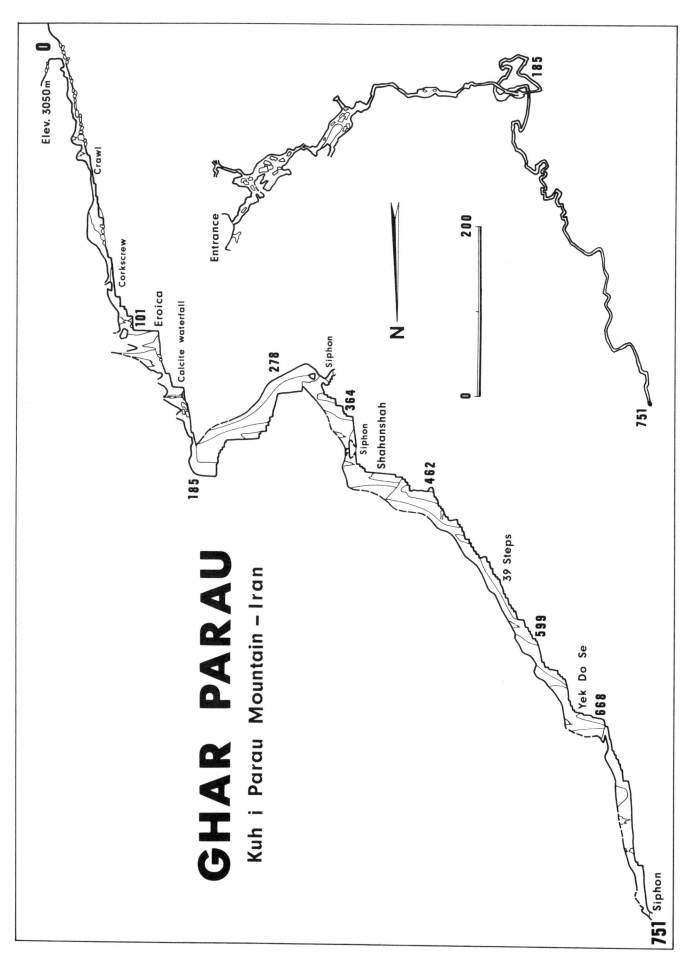

GHAR PARAU

Kuh i Parau Mountain – Iran

Elev. 3050 m

Crawl

Corkscrew

Eroica

Calcite waterfall

Entrance

Siphon

Siphon

Shahanshah

39 Steps

Yek Do Se

Siphon

N

200

0

0

101

185

278

364

462

599

668

751

185

751

4. **Colonel** (Mt. Sedom) .... 1448 m

5. **Zechuchit** (Mt. Sedom) .... 1135 m

6. **Falafel** (Mt. Sedom) .... 700 m

7. **Arak Na'asaneh** (Shomron) .... 621 m
    (*Niqrot Zurim*, 4, map).

8. **Peteq** (Mt. Sedom) .... 524 m

9. **Alma** (Galil) .... 496 m
    (*Niqrot Zurim*, 10, map).

10. **Qolnoa** (Mt. Sedom) .... 462 m

Note: All the caves on Mount Sedom (Har Sedom) are formed in salt.

JAPAN

Nippon/Hihon

Almost all the limestone in Japan dates from the Paleozoic and Mesozoic ages, although there is some Quaternary limestone on Okinawa. The major karst areas are those of Akka (Iwateken), Takine (Fukushima), Okutama (Tokyo), Maikomi-daira (Niigata), Hachiman (Gifu), Atetsu-dai (Okayama), Taishaku-dai (Hiroshima), Shikoku (Ehime and Kochi), Ayiyoshi-dai (Yamaguchi), Hirao-dai (Fukuoka), Kuma (Kumamoto), Tsukumi (Oita) and the islands of Okinawa.

The study of caves began in the 1930's. Hiroshi Yamauchi, a professor at the University of Ehime, was the pioneer of Japanese speleology. Along with biologists and paleontologists, he made constant cave exploration trips until the war. During the 1950's, activity was coordinated by the Japan Biospeleological Research Group and the Shikoku Caving Club. In 1959, H. Yamauchi founded the Japan Caving Association which began to make caving popular. Many university clubs were formed in the 1960's. Student groups made significant explorations at Ōmi-senri-dō (University of Kansai), at Akiyoshi-dai (University of Yamaguchi), at Shikoku and Okinawa (University of Ehime), and in many other places. They discovered the longest cave, Akka-dō in 1962 and the deepest, Byakuren-dō, in 1975. There are presently 38 significant caving clubs in Japan, and more than four national organizations: the Speleological Society of Japan, founded in 1975, which published an annual *Journal*; the Japan Caving Association, which publishes *Japan Caving* annually, the Association of Japanese Cavers formed in 1978 which publishes *Dōjin*; and the Yamaguchi Caving Club formed in 1962 that publishes an annual *Bulletin*.

Natsumi Kamiya.

DEEP CAVES:

1. *Byakuren-dō* .... **–422 m**
    (Ōmi-chō, Nishikubiki-gun, Niigata-ken)
    Byakuren-dō is a swallow cave located on the Maikomi plateau (Paleozoic limestone) at an altitude of 702 m, about 200 km NW of Tokyo, near Ōmi-Machi, in the municipality of Niigata. The 7 m by 4 m opening is in a 20 m deep doline. The first exploration was on July 24, 1972, by the Kansai University Exploration Club (K.U.E.C.): they dropped the entrance pit. Two years later, in two expeditions, they succeeded in reaching the terminal sump.
    From July 29 to August 20, 1976, the Nagato Caving Club (N.C.C.) mapped down to –422 m, for a length of 1060 m, stopping shortly before the sump which is estimated to be at –450 m. Deeper exploration was blocked on August 14 by a flood which resulted in a rescue. The deepest pit, the Elegant Lady Waterfall, is at the –142 m level and is 45 m deep. The resurgence, Fukugakuchi-dai-shōnyūdō, is 3 km to the north of Maikomi-daira, at an altitude of 200 m.
    Map: from the drawings of the N.C.C., sent by Natsumi Kamiya.
    Bibliography: *Grottes & Gouffres*, 1979 (73): 27; *Report of Yamaguchi Caving Club* N°15.

2. **Ōmi-senri-dō** .... –365 m
    (Ōmi-chō, Nishikubiki-gun, Niigata-ken)
    Alt. 382 m. Part of the same system as Byakuren-dō, explored from 1965 to 1968 by the K.U.E.C. (*Dōjin*, 1983, 4 (1-2) profile).

3. **Nunagawa-dō** .... –345 m
    (Ōmi-chō, Nishikubiki-gun, Niigata-ken)
    Alt. 673 m. Same system as the two above caves. Swallow cave explored from 1969 to 1970 by the K.U.E.C. It floods completely. (*Dōjin*, 1983, 4 (1-2) profile).

4. **Ginpo-dō** .... –333 m
    (Ōmi-chō, Nishikubiki-gun, Niigata-ken)
    Alt. 780 m. Explored in 1971 by the K.U.E.C. Has the deepest shaft in Japan (98 m deep) (*Dōjin*, 1984, 4 (1-2) profile).

5. **Gomagara-dō** .... –215 m
    (Tsukumi-shi, Ōita-ken)
    Alt 235 m. Explored in 1970 by the Ehime U.C.C.

6. **Irimi-anamido-no-ana** .... –204 m
    (Ōmine-chō, Mine-gun, Yamaguchi-ken)
    Alt. 236 m. Explored in 1976, 1981 by the C.C.S.U. Yamaguchi (*Darkness under the Earth*, in Japanese, 1983, profile).

7. **Abukuma-dō/Ōtakine-dō/Oni-ana** .... –183 m
    (Takine-chō, Tamura-gun, Fukushima-ken)

8. **Himesaka-kanachi-ana/**
    Hishi-ana/Ō-ana .... –164 m
    (Niimi-shi, Okayama-ken)

9. **Ryūō-dō** .... –160 m
    (Nomura-mura, Higashi-uwa-gun, Ehime-ken)
    Alt. 1132 m. In 1964 by the Waseda U.E.C., Ehime U.E.C.

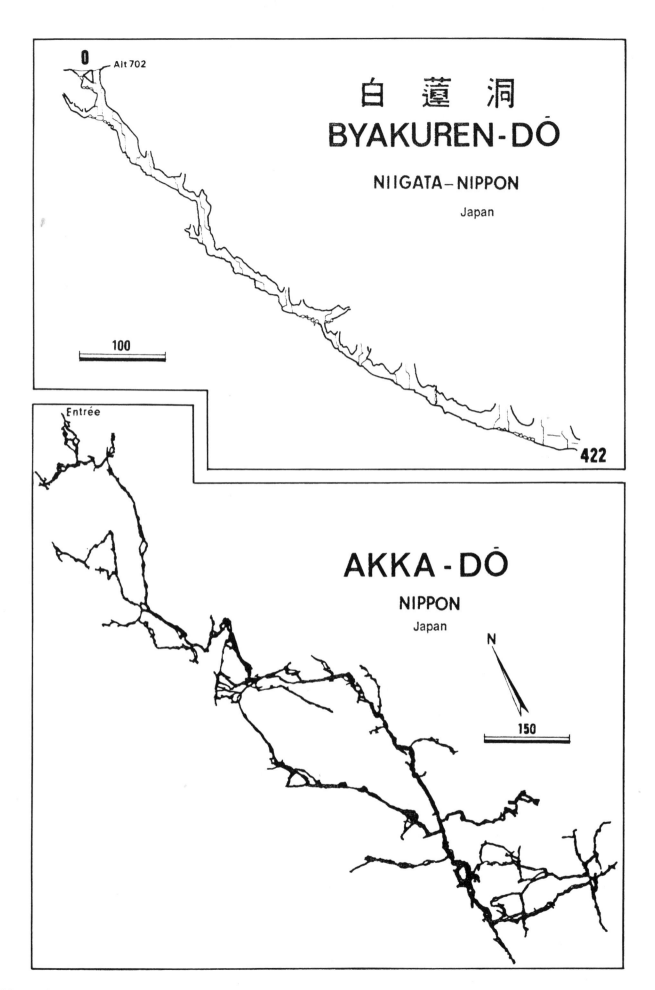

0　Alt 702

白 蓮 洞
BYAKUREN-DO

NIIGATA—NIPPON

Japan

100

422

Entrée

AKKA-DO

NIPPON

Japan

N

150

10. **Takaga-ana** .... –157 m
(Shūhō-chō, Mine-gun, Yamaguchi-ken)
Alt. 240 m. In 1963 by the C.C.S. Yamaguchi U.
(*Japan Caving*, 1976, 8 (1) profile to –128 m).

11. **Kumaishi-dō** .... –155 m
(Hachima-chō, Gujō-gun, Gifu-ken)
Alt. 650 m. Speleological Survey Group in 1973.

12. **Kyusen-dō** ....(–90, +60) 150 m
(Kuma-mura, Kuma-gun, Kumamoto-ken)
Alt. 200 m. Explored in 1975 by the Ehime U.E.C.

13. **Iwanaga-dai-P4-no-ana** .... –146 m
(Isa-chō, Mine-shi, Yamaguchi-ken)

14. **Gōtō-ana** .... –145 m
(Tōjō-chō, Hiba-gun, Hiroshima)

15. **Nishiyama-no-tate-ana** .... –145 m
(Shūho-chō, Mine-gun, Yamaguchi-ken)

16. **Ōgamine-dō** .... –140 m
(Izumi-mura, Yashiro-gun, Kumamoto-ken)

17. **Nippara-shōnyūdō** .... –135 m
(Okutama-chō, Okutama-gun, Tokyo)

18. **Akiyoshi-dō/Kaza-ana** .... +132 m
(Shūhō-chō, Mine-gun, Yamaguchi-ken)

19. **Ja-ga-ana** .... –130 m
(Shūhō-chō, Mine-gun, Yamaguchi-ken)

20. **Ryusen-dō** ....(–64, +65) 129 m
(Iwalzumi-chō, Shimohei-gun, Iwate-ken)

---

25. **Ōtaki-shōnyūdō** .... –110 m
(Hachiman-chō, Gujō-gun, Gifu-ken)

LONG CAVES:

1. *Akka-dō*..................................... **7650 m**
(Iwaizumi-chō, Shimohei-gun, Iwate-ken)
Akka-dō is located on the south bank of the Akka river, not far from Iwaisawa (from which comes its other name of Iwaisawa-no-ana), in the city of Iwaizumi, Shimohei district, Iwate province, northern Japan. It is formed in Paleozoic limestone, in the Akka formation. The entrance is at an altitude of 290 m. It was explored and mapped by the Japan Caving Association (J.C.A.) in August of 1959 and 1961 for a length of 7650 m. In August 1962 the J.C.A. found three new passages to bring the length to an estimated 8000 m.
   Map: from the drawings of the J.C.A. (1961-1962) send by N. Kamiya.
   Bibliography: J.C.A.- Report of speleological research of the caves in Iwate-ken, Government of Iwalzumi-cho, 1962.

2. **Kyūsen-dō** .... 4800 m
(Kuma-mura, Kuma-gun, Kumamoto-ken)
Explored from 1973 to 1983 by the Ehime U.E.C.

3. **Takaga-ana** ....4525 m
(Shūhō-chō, Mine-gun, Yamaguchi-ken)
Explored from 1973 to 1975 by the N.C.C. for a length of 4402 m (*Japan Caving*, 1976, 8 (1) map) then in 1979 by the Spel. Survey Group.

4. **Gyokusen-dō** ....3600 m
(Tamagusuku-mura, Okinawa-ken)
Opened as a show cave in 1972. Explored from 1963-1967 by H. Yamauchi and the Ehime U.E.C.

5. **Uchimagi-dō** ....3340 m
(Yamagata-mura, Kunohe-gun, Iwate-ken)
In 1978 by the J.C.C. and the E.C. Nihon Univ.

6. **Nippara-Mitsumata-dō** ....3320 m
(Okutama-chō, Nishitama-gun, Tokyo)
From 1977 to 1979 by the Tokyo S.S.G.

7. **Sora-ana No. 2** ....3200 m
(Sumita-chō, Kesen-gun, Iwate-ken)
Also called Rōkan-shin-dō.

8. **Maria-Iza** ....3200 m
(Ishigaki-shi, Okinawa-ken)

9. **Gonbōzone/Hongoya-no-ana** ....3060 m
(Niimi-shi, Okayama-ken)

10. **Shigawatari-dō** ....2882 m
(Iwaizumi-chō,Shimohei-gun, Iwate-ken)

11. **Kumaishi-dō**...................................2853 m
Hachiman-chō, Gujō-gun, Gifu-ken)

12. **Ōtakine-dō/Abukuma-dō/Oni-ana** ....2600 m
(Takine-chō, Tamura-gun, Fukushima-ken)

13. **Irishimuku-gama** ....2570 m
(Yomitan-mura, Okinawa-ken)

14. **Tateishi-shonyūdō** ....2370 m
(Kashima-chō, Soma-gun, Fukushima-ken)

15. **Ryuga-dō** ....2300 m
(Yamada-chō, Tosa-gun, Kochi-ken)

16. **Todoroki-dō** ....2196 m
(Ginowan-shi, Okinawa-ken)

17. **Ginsui-dō**.....................................2090 m
(China-chō, Ōshima-gun, Okinoerabu)

18. **Ginga-dō** ....2052 m
(Iseh-chō, Ōshima-gun, Kagoshima-ken)

19. **Akiyoshi-dō/Kaza-ana** ....2000 m
(Shuhō-chō, Mine-gun, Yamaguchi-ken)
Show cave first explored in 1907. Length estimated at 2650 m.

JORDAN
Urdunn

This country has many limestone landscapes, especially in the mountain chain along the Dead Sea. Unfortunately the limestone is generally in very thin layers and interbedded with other rocks, making for poor cave development. Few caves are known, but the country has not been systematically explored either. While travelling on business in 1980 and 1983 P. Courbon found the following small caves:

–Bir Adnâniyeh (Karak) .... –21 m
(*Grottes et Gouffres*, 1981 (81) profile).
–Dja'et Zagh (Irbid) .... –17 m

Several small caves rarely passing 30 m in length are known along the banks of the Jordan river, unusual in having their entrances well below sea level (between –160 m and 250 m altitude).

KOREA
(North and South)
Joseon and Daehan

Since 1977, information which has reached us on Korea has been sporadic at best, so the following list may well be out of date. The most interesting caves are lava tubes. Their systematic exploration began in 1966 with the formation of the Korean Speleological Society, which has been helped on occasion by Japanese cavers. A well-documented work by Byung-Hoon Lee describing 92 caves appeared in 1974 in *Annales de Spéléologie* (Vol. 29, N° 3). South Korea has a show cave formed in limestone that is 2200 m long: **Seongryu-gul** (Kyung-Sang).

Note: For convenience, the only cave we know of in North Korea has been added in with those of South Korea. Ri designates the municipality, Gun the county, and Do the province.

DEEP CAVE:

1. **Namgamduk-gul** .... –181 m

LONG CAVES:

1. **Manjung-gul** .... 13,268 m
(Geumnyeong Ri, Bugjeju Gun, Je Ju Do)
Also called Manjang-gul. Lava tube in several segments whose first explorion dates back to 1964 (Pu. Chonhyu). In 1977 Hong-Shi-hwan mapped 8994 m. In 1981, Koreans and Japanese reached 13,268 m (*Cascade Caver*, 1980, 19 (6) map with segments 4632 m and 613 m long).

2. **Bilremos-gul** .... 11,749 m
(Eo-eum Ri, Bugjeju Gun, Je Ju Do)
Or Bilemot-gul. Lava tube. Length of 8 km reached in

1971, length of 11,749 m reached in 1981. Mapped by the K.S.S. and Japanese cavers.

3. **Susan-gul** .... 4674 m
(Susan Ri, Bugjeju Gun, Je Ju Do)
Lava tube.

4. **Chodang-gul** .... 4000 m
(Hamaengbang Ri, Samcheog Gun, Gang Weon Do)

5. **Hwanseon-gul** .... 3920 m
(Daei Ri, Samcheog Gun, Gang Weon Do)

6. **Kosi-gul** .... 2980 m
(Handong-myeon Ri, Yeongweol Gun, Gang Weon Do)

7. **Seongryu-gul** .... 2200 m
(Noeum Ri, Uljin Gun, Kyeonsang-puk Do)

8. **Socheon-gul** .... 2186 m
(Hyeobjae Ri, Bugjeju Gun, Je Ju Do)
Lava tube, with another segment 404 m long (*Cascade Caver*, 1982, 21 (1-3) map).

9. **Micheon-gul** .... 1695 m
(Samdal Ri, Bugjeju Gun, Je Ju Do)
Lava tube.

10. **Dongryong-gul** .... approx. 1500 m
(Pyung An, North Korea)

LAOS
Lao

Laos is known to have a great richness and variety of caves, as indicated by the numerous underground river caves used by locals for transportation, and the dry maze caves used by the guerillas. The political situation has prevented systematic exploration of these karst features, however, so that our only real reference is the invaluable work of Paul Macey, "Cours d'eau souterrains du Cammon au Laos" in *Spelunca, Bull. et Mém. de la Soc. de Spél.*, N° 52, published in 1908, in which he mentions several large unexplored karst phenomena (for example the three swallets (swallow holes) of the Nam Kadinh).

LONG CAVES:

1. **Sé-Bang-Faï underground river** .... 4200 m
(Pou Pa Koue Ti Thame, Cammon)
Swallet cave explored by Macey in 1905 (Macey, *op. cit.*, map).

2. **Nam Hin-Boun
underground river** .... approx. 4000 m
(Pou Kong L'io, Ban-Nam-Thone, Cammon)
Swallet cave used by locals for transportation (Macey, *op. cit.*, map).

3. **Nam-Nhom resurgence** ....3027 m
(Ban-Tha-Thot, Maha-Xai, Thakhek)
Explored around 1950 by H. Cassen *et al.* (*Sciences et Voyages*, 1950 (59) map).

4. **Houei Thame Khi Heup**
underground river ....2000 m
(Pou Thame Khi Heup, Cammon)
Swallet explored in 1902 by Macey (Macey, *op. cit.*, map).

LEBANON

Lubnān

Lebanon is rich in karst phenomena on a grand scale and contains massive limestone deposits. Culminating at an altitude of 3083 m, the mountain chains run from north to south and cover almost the entire country, representing a tremendous speleological potential. Because the country has been torn by civil war since 1975, the activities of the Lebanon Caving Club, formed in 1952, have been slowed. This club has (often with the invited help of foreign cavers, especially French and British) made most of discoveries to date. Albert Anavy and especially Sami Karkabi have been the prime movers in the explorations of the large cave systems, assuring Lebanon a significant international rating. Today the L.C.C. is presided over by Hani Abdul-Nour and publishes a bulletin *Al Ouat'Ouate*.

DEEP CAVES:

1. *houet Faouar Dara* .... –622 m
(djebel Zaarour, Mteïn)
This is an insurgence cave at an altitude of 1598 m (with an upper entrance at 1616 m) in Jurassic limestone. The seasonal waters which it swallows originate in the djebel Zaarour. Located to the north of the village of Medjel Tarlich, access to the cave is by the Beirut-Tarlich-Zahlé road and the trail leading to aïn Dara, which comes within 500 m of the cave. A dye trace in 1965 established the resurgence as Faouar Antelias, on the Mediterranean coast (alt. 25 m), 1573 m below the swallet (or swallow hole). The cave was found in 1955 by Sami Karkabi and explored beginning in 1957 by the L.C.C. which reached the depth of –85 m that year. In 1959 the depth of –370 m was reached, then –512 m in 1961. The terminal sump at –622 m was reached in 1962. In 1968, with the Yorkshire Rambler's Club, horizontal extensions to the cave were discovered.
Map: sent by Sami Karkabi.
Bibliography: Middleton (J.) - Underground in the Lebanon, 1968 and 1969, *Yorkshire Rambler's Club Journal*, 1970, X (35): 315-335.

2. **houet Balouh Baatara** .... –240 m
(djebel Laqluq)
Or Cave of the Three Bridges (superimposed!) for which the 0 point is difficult to establish.

3. **houet Mechmiché** .... –240 m
Cave today filled in by sand!

4. **jouret el Abed** .... –230 m
(djebel Laqluq)

5. **houet el Badaouiyé** .... –202 m
Contains a vertical drop of 164 m.

6. **houet Balouh Balaa** .... –150 m
(djebel Laqluq)

7. **houet Tarchich** .... –147 m
This cave is being filled in by sand (only –65 m in 1973).

8. **grotte de Jeita** ....+141 m

9. **houet el Atoué** .... –120 m

10. **houet Alain** .... –115 m
Explored in 1977.

11. **ain el Kadah** .... –110 m

12. **houet Blaïtat** .... –106 m

13. **houet Sehaïlé** .... –104 m

LONG CAVES:

1. *grotte de Jeita*..............................**8330 m**
This is probably the most famous show cave of the Middle East. A powerful river flows through the cave to give birth to the nahr el Kelb (Dog River). From Beirut, one travels towards Zouk Mosbeh after the nahr el Kelb bridge. From there, a special road leads to the entrance. The history of its exploration began in 1837 when Thomson discovered the upper entrance. From September 23 to 30, 1873, engineers from the Water Office of Beirut, Maxwell and Bliss, followed the river upstream for about 1000 m and were stopped at the Hell rapids. These were conquered in 1924 by Dr. Lamarche who turned back 50 m farther on. In 1926 Dr. West, in three days, progressed an additional 400 m upstream. In 1927 Thompson, better equipped, got to a point 1600 m from the entrance. In 1946 four young Lebanese (including A. Anavy and L. Ghorra) began exploration. These same four founded the Lebanon C.C. in 1952. They reached 1800 m in 1946, 2200 m in 1949, 2800 m in 1951, 3600 m in 1952, 4400 m in 1953, and finally 6200 m in 1954 when the upstream sump was discovered. Later dives were unable to pass this sump. In 1958 the discovery of upper levels (opened to tourists in 1969, thanks to a tunnel) brought the length to 8000 m. In 1969 a survey using theodolites done by G. Dobroff and S. Karbaki took 200 sessions to map 6200 m of river passage and permitted the digging of a second artificial entrance.
Map: from the plan of S. Karkabi and G. Dobroff, 1972.
Bibliography: Karkabi (S.) *La grotte de Jiita*, Beirut, S.D., 44 p. Karkabi (S.) - *Jeita, architecture des temps*, Beirut, 1972.

2. **mgharet el Roueiss** ....5066 m
(Aaqoura, El Mnaïtra)
Maze cave with two entrances, explored since 1952,

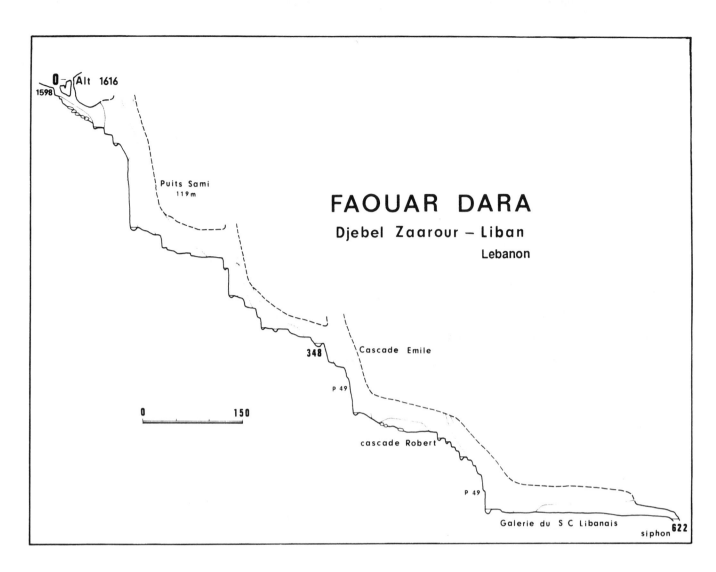

FAOUAR DARA

Djebel Zaarour – Liban

Lebanon

O Alt 1616
1598
Puits Sami
119 m

348
P 49
Cascade Emile

cascade Robert

P 49
Galerie du S C Libanais
siphon **622**

0 ——— 150

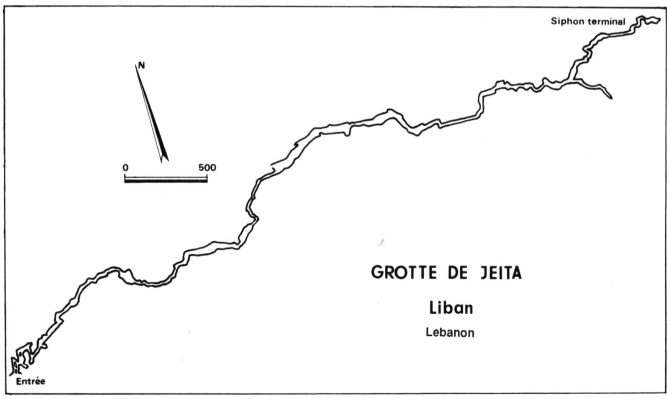

Siphon terminal

N

0 ——— 500

GROTTE DE JEITA

Liban

Lebanon

Entrée

partially mapped in 1974 by the L.C.C. and C. Chabert (*Grottes et Gouffres*, 1976 (61) map).

3. **mgharet Afqa** ....3600 m
   (Afqa, El Mnaïtra)
   Maze cave with impressive entrance explored in 1974.

4. **mgharet Nabeh el Chataoui** ....2550 m

5. **houet Faouar Dara** ....approx. 2500 m
   Explored in 1968 (*Y.R.C. Journal*, 1970, X (35) partial map).

6. **mgharet el Kassarat** (Antelias) ....2400 m
   Explored in 1969 (*Y.R.C. Journal*, 1970, X (35) map).

7. **mgharet Dahr el Ain** ....1500 m

8. **ain Lebné** ....approx. 1300 m
   Explored in 1975.

9. **mgharet el Rahoué** ....1100 m

MALAYSIA

Caves are abundant in all three Malaysian provinces, Malaya, Sarawak, and Sabah, but are generally clustered in small isolated outcrops of limestone. The first explorers in many caves were the local people, searching for valuable guano deposits or birds' nests, or using the caves as burial sites. The first scientists to become interested were British biospeleologists at the end of the 19th century and in the first quarter of the 20th century, led first by H. Ridley, then by H.C. Abraham and C. Dover from the British Association for the Advancement of Sciences. Maps of some of the more accesible caves were then produced by various Western scientists in the course of their work. These included geologists from the tin mine in Malaya, and archeologists in Niah (Sarawak). The extraordinary potential of the limestone deposits of Gunung Mulu National Park (in northern Sarawak, not far from Brunéi) was first noted in 1856 by Spencer St. John, but not pursued until the British expeditions of 1978, 1980-1981 and 1984. The expeditions were sponsored by the Royal Geographical Society and mapped over one hundred and fifty kilometers of underground passage in caves of huge dimensions, such as **gua Payau** and **lubang Nasib Bagus**, which contains (by far) the largest cave room in the world.

Tony Waltham.

DEEP CAVES:

1. **lubang Nasib Bagus** ....+423 m
   (Gunung Api, Mulu, Sarawak)
   (*Caves of Mulu'80*, 1981, profile). (See below).

2. **gua Air Jernih** ....+355 m
   (gunung Api, Mulu, Sarawak)
   (*Caves of Mulu*, 1978, profile: *Caves of Mulu '84*, 1984, partial profile).

3. **lubang Benarat** ....–340 m
   (gunung Benarat, Mulu, Sarawak)
   1980-1981, 1984 B.M.S.E.S. (*Caves of Mulu '84*, profile).

4. **lubang Hijau** ....(–245, +75) 320 m
   (Mulu Selatan, Sarawak)
   Explored in 1978 and 1980-1981 by the B.M.S.E.S. (*Caves of Mulu '80*, profile with –245 m).

5. **gua Harimau** ....–302 m
   (gunung Benarat, Mulu, Sarawak)
   1984, B.M.S.E.S. (*Caves of Mulu '84*, profile).

6. **Sendirian** ....–260 m
   (gunung Api, Mulu, Sarawak)
   1978, R.G.S. (*Caves of Mulu*, 1978, profile).

7. **gua Payau** ....+220 m
   (Mulu Selatan, Sarawak)
   1978, R.G.S. (*Caves of Mulu*, 1978, profile).

8. **lubang Sakai** ....168 m
   (gunung Benarat, Mulu, Sarawak)
   1980-1981, 1984, B.M.S.E.S. (*Caves of Mulu '80*, 1981, profile with 140 m).

9. **lubang Angin** ....+140 m
   (gunung Api, Mulu, Sarawak)
   1980-1981, B.M.S.E.S. (*Caves of Mulu '80*, profile).

10. **gua Ajaib** ....(–85, +45) 130 m
    (gunung Api, Mulu, Sarawak)
    1978, R.G.S. (*Caves of Mulu*, 1978, profile).

11. **lubang Pangkal Harimau** ....–128 m
    (gunung Benarat, Mulu, Sarawak)
    1978, R.G.S. (*Caves of Mulu*, 1978, profile).

12. **gua Teluk Cahaya Bulan Yang** ....–119 m
    (gunung Benarat, Mulu, Sarawak)
    1980-1981, B.M.S.E.S. (*Caves of Mulu '80*, 1981, profile).

13. **lubang Sarang Laba-Laba** ....–116 m
    (gunung Benarat, Mulu, Sarawak)
    1984, B.M.S.E.S. (*Caves of Mulu '84*, profile).

14. **lubang Lagan** (Mulu, Sarawak) .... –106 m
    1984, B.M.S.E.S. (*Caves of Mulu '84*, profile).

15. **lubang Lipah Sileh** ....–106 m
    (Mulu, Sarawak)
    1984, B.M.S.E.S. (*Caves of Mulu '84*, profile).

LONG CAVES:

1. *gua Air Jernih* ....**51,600 m**
   (gunung Api, Mulu, Sarawak)
   Gua Air Jernih is formed along the western edge of the gunung Api, near the sungai Melinau. To reach the cave from the base camp at Long Pala one takes a boat along this

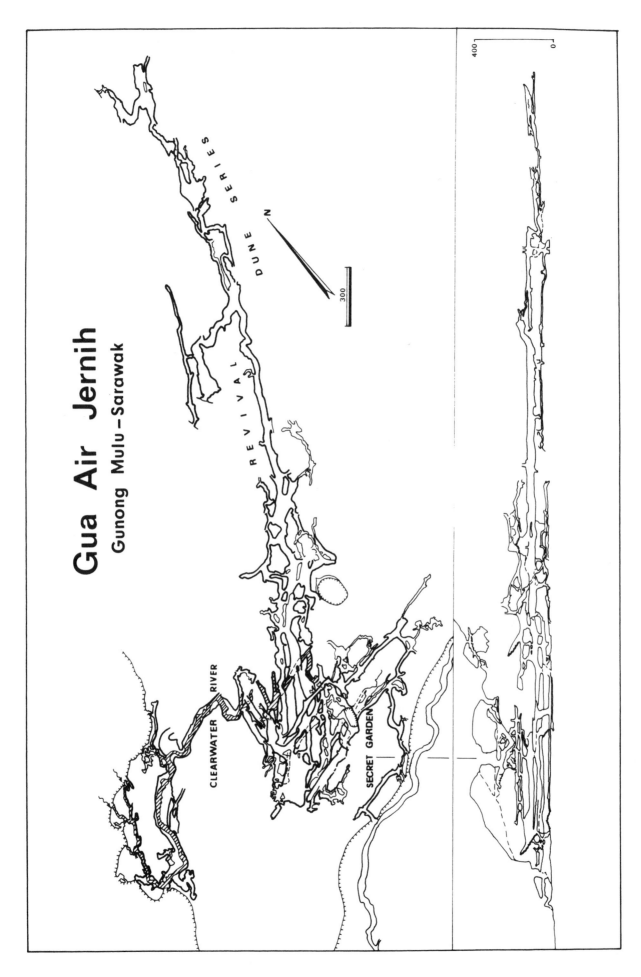

Gua Air Jernih
Gunong Mulu – Sarawak

DUNE SERIES

REVIVAL

CLEARWATER RIVER

SECRET GARDEN

N

300

400   0

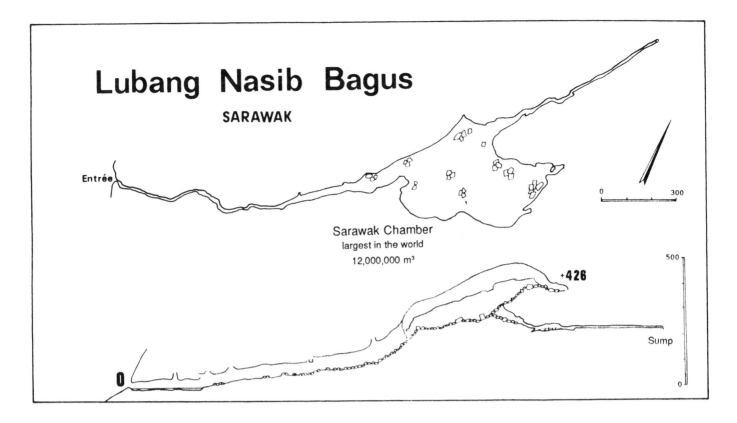

Lubang Nasib Bagus

SARAWAK

Entrée

Sarawak Chamber
largest in the world
12,000,000 m³

+426

0   300

Sump

500

0

0

river. The main entrance is at an altitude of 200 m at the SW end of the outcrop, only 30 m from the river, not far from one of the principal resurgences that drain the gunung Api. The exploration of this cave took place in three periods. In 1978 the Royal Geographical Society Mulu Expedition Speleological Team (D. Brook, P. Chapman, A. Eavis, M. Farnworth, B. Lyon, T. Waltham) mapped a total of 24,240 m. This promising start prompted A. Eavis and B. Lyon to undertake an expedition in common with the Forestry Department of Sarawak: the British-Malaysian Speleological Expedition to Sarawak (B.M.S.E.S.) had 22 British members and 8 Malaysians. They concentrated on the upper levels of the cave and by the end of 1980, 37,585 m had been mapped in the cave. The return expedition in 1984 increased the length to 51,600 m. The potential of this cave remains large and there are several possibilities for connections with nearby caves. Gunung Api is formed of Tertiary reef limestones of Eocene age.

Map: from the maps published in *Caves of Mulu '80* and *Caves of Mulu '84*.

Bibliography: *Caves of Mulu*, R.G.S., London, 1978, 44 p., *Caves of Mulu '80*, R.G.S., London, 1981, 52 p.; *Caves of Mulu '84*, 56 p.

2. **lubang Sarang Laba-Laba** .... 15,185 m
(gunung Benarat, Mulu, Sarawak)
Cave explored in 1984 by the B.M.S.E.S. (*Caves of Mulu '84*, map).

3. **gua Teluk Cahaya Bulan Yang** .... 9400 m
(gunung Benarat, Mulu, Sarawak)
Explored in 1980 by the B.M.S.E.S. (*Caves of Mulu '80*, map).

4. **lubang Limau** .... approx. 8500 m
(Subis, Sarawak)

5. **lubang Benarat** .... 8320 m
(gunung Benarat, Mulu, Sarawak)
Explored in 1978, 1980, and 1984 (*Caves of Mulu '84*, map).

6. **lubang Angin** .... 7510 m
(gunung Api, Mulu, Sarawak)
Explored in 1978 (2500 m), 1980 (6550 m) and 1984 (*Caves of Mulu '84*, map).

7. **lubang Tang Baan** .... 5800 m
(Kuching, Sarawak)

8. **gua Ajaib** .... 4770 m
(gunung Api, Mulu, Sarawak)
Explored in 1978, R.G.S. (*Caves of Mulu*, 1978, map).

9. **gua Jambusan** .... approx. 4000 m
(Bau-Serian, Kuching, Sarawak)

10. **lubang Labang** .... 3900 m
(Mulu, Sarawak)
Explored in 1984, B.M.S.E.S. (*Caves of Mulu '84*, map).

11. **gua Sungai Terikan Timur** .... 3840 m
(gunung Benarat, Mulu, Sarawak)
Explored in 1978, R.G.S. (*Caves of Mulu*, 1978, plan).

12. **gua Harimau** .... 3800 m
(gunung Benarat, Mulu, Sarawak)
Explored 1984, B.M.S.E.S. (*Caves of Mulu '84*, map).

13. **lubang Hijau** ....3445 m
(gunung Selatan, Mulu, Sarawak)
Explored in 1978 (2890 m) and 1980 (*Caves of Mulu*, 1978, map).

14. **lubang Lipah Sileh** ....3400 m
(Mulu, Sarawak)
Explored in 1984, B.M.S.E.S. (*Caves of Mulu '84*, map).

15. **gua Harimau Bintang** ....3375 m
(gunung Api, Mulu, Sarawak)
Explored in 1980, B.M.S.E.S. (*Caves of Mulu '80*, map).

16. **gua Niah** (Subis, Sarawak) ....3200 m
(*Malaysia Bull.*, 1964 (6) map).

17. *lubang Nasib Bagus* ....**2900 m**
(gunung Api, Mulu, Sarawak)
The exploration of this cave and the discovery of its huge chamber, Sarawak, was the high point of the British-Malaysian Speleological Expedition to Sarawak in 1980-1981. The search for resurgences along the southern edge of the gunung Api (which feeds the sungai Melinau Paku, which joins with the sungai Melinau at Long Pala) led to the discovery of lubang Nasib Bagus, "Good Luck Cave" by Hans Friederich and Danny Lawi on December 28, 1980. A strong wind blows from the entrance, which is 9 km as the crow flies from Long Pala.
The cave was explored in January 1981 and it was at 1300 m from the entrance that the main passage, after leaving the river, pops into the Sarawak Chamber, which has the colossal dimensions of 700 m long, up to 450 m wide, and an average height of 100 m. The top of the chamber is 423 m above the entrance to the cave. This chamber is formed along the contact of the limestone and an impermeable layer over which the underground river flows. The river has cut into the formation, allowing the huge collapse that formed the chamber.
Map: from the maps of the Mulu '80 expedition.
Bibliography: *Caves of Mulu '80*, R.G.S., London, 1981, pp.14-17.

18. **lubang Sakai** ....2645 m
(gunung Benerat Mulu, Sarawak)
Explored in 1980 (1120 m) and 1984, B.M.S.E.S. (*Caves of Mulu '84*, map).

19. **gua Sungai Terikan Barat** ....2270 m
(gunung Benarat, Mulu, Sarawak)
Explored in 1978, R.G.S. (*Caves of Mulu*, 1978, plan).

20. **gua Payau** ....2160 m
(gunung Selatan, Mulu, Sarawak)
Explored in 1978 (1760 m) and 1980, B.M.S.E.S. (*Caves of Mulu '80* map).

———————

31. **lubang Penyu** ....1155 m
(gunung Buda, Mulu, Sarawak)

NEPAL

Nepal has often evoked images of great caves in the minds of speleologists. A first expedition in 1970, British Karst Research Expedition to Himalaya, led by A. C. Waltham, was followed by others in 1974 and 1976. They all obtained disappointing results, confirming the findings of French geological reconaissance trips that the huge limestone outcrops are poorly karsitified in this country. A good summary of his own and previous work can be found in the work of Daniel Gebauer, *Caves of India & Nepal* (1983).

DEEP CAVES:

1. **Alope gupha**.. –63 m
(Kusma, Dhaulagiri)

2. **Pathale Chhango** .... –48 m
(Pokhara, Gandaki)

LONG CAVES:

1. **Pathale Chhango** .... 2959 m
(Chhorepatan, Pokhara, Gandaki)
Cave with 6 entrances formed in conglomerate (tourist site). Explored in 1970 by the B.K.R.E. Himalaya for 1479 m and extended in 1980 and 1982 by D. Gebauer (Gebauer, *op. cit.*, 1983, map).

2. **Chakra Tirtha gupha**....................approx. 1250 m
(Kathmandu)
The Paranamsa and Chakra Tirtha caves were connected to form this system in 1985 by the Czech expedition "Himalaya 85".

3. **Alope gupha** ....491 m
(Kuje Bisauna, Kusma, Dhaulagiri)

4. **"Western powerstation" gupha** ....293 m
(Phureskhola, Pokhara, Gandaki)
Alt. approx. 800 m. Explored in 1980 by D. Gebauer (Gebauer, *op. cit.*, plan).

5. **Mahendra gupha**..275 m
(Batlechhaur, Pokhara, Gandaki)
Alt. 900 m. In conglomerate. Explored by the British expedition of 1976, then D. Gebauer in 1980 (Gebauer, *op. cit.*, map).

6. **Chakhubha gupha** ....200 m
(Kathmandu)

7. **Gupteswary gupha** ....190 m
(Kusma, Dhaulagiri)
Alt. 760 m (*B.K.R.E. Himalaya report*, 1970, map).

8. **"Eastern powerstation" gupha** ....173 m
(Phusreskhola, Pokhara, Gandaki)
(Gebauer, *op. cit.*, map).

OMAN

'Uman

The limestone mountains of both northern and southern Oman contain a number of caves, and many more may exist which have not yet been explored. All of the explorations have been organized or sponsored by the Public Authority of Water Resources (PAWR).

Tony Waltham.

DEEP CAVES:

1. **Funnel Cave** .... –300 m
   (Selmeh Plateau, Jabal Bani, Jabir)
   Explored in 1987 by the Public Authority for Water Resources. Has a 177 m pit.

2. **Seventh Hole** .... –290 m
   (Selmeh Plateau, Jabal Bani, Jabir)
   Explored in 1987 by PAWR. Exploration is incomplete.

3. **Kahf Hoti** .... –262 m
   (Jabal al-Akhdar)
   Through system. (Trans. B.C.R.A., 1985, 12 (3) map).

4. **Arch Cave** .... –260 m
   (Selmeh Plateau, Jabal Bani, Jabir)
   Explored in 1987 by PAWR. Exploration is incomplete.

5. **Hufrah Misfah** .... –200 m
   (Jabal al-Akhdar)
   Exploration incomblete (*Trans. B.C.R.A.*, 1985, 12 (3) map).

6. **Majlis Al Jinn** .... –178 m
   (Selmeh Plateau, Jabal Bani, Jabir)
   Mapped by the PAWR. Has a large room (4 million m³ with three skylight entrances. The deepest entrance drop is 158 m.

LONG CAVES:

1. **Kahf Hoti** (Jabal al-Akhdar) ....4975 m
   Swallet Cave (See above).

PAKISTAN

The British efforts to find caves in 1976 in the Chitral district were fruitless. It is in the Nanga Parbat massif, at an altitude of 6645 m, that the highest known cave in the world is found: it was discovered on July 2, 1963 by the Meinzinger-Caldwell team on the south spine of Rakhoit Peak. It is 75 m long and formed in marble.

PHILIPPINES

Pilipinas

Philippino karst, although looked at by a few pioneers (P. Proust de la Gironière, around 1820, at Montalban, with H. Lindsay and E. Simon in 1890 at Libmanan, in southern Luzon) was poorly known until recently. In the 1970's, the Hungarian Dénes Balázs reconnoitered the karsts of the archipelego and explored Saint-Paul cave for 6 km (*Actes 6th Congr. Int. Spél.*, 1973, v.2). The first systematic investigations began in 1979-1980 (L. Daharveng, J. Orousset) at Sagada, followed beginning in 1982 by C. Mouret (with Y. Bousquet notably), then by Japanese and Italian cavers. Other karst areas (Libmanan, Biak-na-bato, Montalban, Kalinawan, Baguio, North Mindoro, Samar, Bohol, Cebu) were explored by French (1982-1984) and Japanese teams (Yamaguchi Caving Club, 1983-1985). To date a total of 35 to 40 km have been mapped.

The karst areas of the Philippines are numerous, but often not very extensive. Dénes Balázs (1973) and Louis Deharveng (1980) have made maps showing the different areas. The most extensive of these are at Samar, Cebu, southern Luzon and Mindanao. Most of the limestone is at a low elevation, and most of the caves are horizontal rather than vertical. Some areas, however, such as Mindoro and Mindanao, are difficult to reach, but offer a moderate vertical potential.

Claude Mouret.

DEEP CAVES:

1. **Latipan-Lokohong cave system** .... –163 m
   (Sagada, Mountain prov., Luzon)
   Alt. 1400 m approx. (*Mouret, Mém.* S.C. Paris, 1985, 12, profile to –145 m).

2. **Saint-Paul cave** ....+100 m
   (Mt. Saint-Paul, Bahile, Palawan prov., Luzon)

3. **Natividad cave** .... –71 m
   (Sagada, Mountain prov., Luzon)

4. **Kitungan Kampus pit** .... –59 m
   (Sagada, Mountain prov., Luzon)

5. **Agoyo-Ige cave system** .... –58 m
   (Sagada, Mountain prov., Luzon)

6. **Tataya En chasm** .... –54 m
   (Sagada, Mountain prov., Luzon)

7. **Kiosk cave** .... –48 m
   (Sagada, Mountain prov., Luzon)

LONG CAVES:

1. **Saint-Paul cave** ....8200 m
   (Mt Saint-Paul, Bahile, Palawan prov., Palawan)
   Cave with three entrances (resurgence, alt. 0 m; fossil passage, 23 m; Pennings cave, 100 m). Explored by D. Balázs, then by the Sydney Spel. Soc. in 1980 and 1981. S.S.S. map N° 651.

2. **Latipan-Lokohong cave system** ....3975 m
(Sagada, Mountain prov., Luzon)
   System with six entrances (Latipan, Sumaging 1 and 2, Lomyang, Lokohong and unnamed). Explored in 1980 by Deharveng and Orousset and in 1982-1984 by Mouret and Bousquet. (*Mém. S.C. Paris*, 1985 (12) plan showing 2280 m).

3. **Colaphitan cave** ....2856 m
(Tumanan, Libmanan, Carmarines Sur prov., Luzon)
   Explored in 1983 by Mouret, Y. Bousquet and J.-J. Matieu.

4. **Agoyo-Ige cave system** ....1816 m
(Sagada, Mountain prov., Luzon)
   Explored in 1983 by C. Mouret and S. Duflot.

5. **Tataya En chasm** ....1452 m
(Sagada, Mountain prov., Luzon)
   Explored in 1980 by Deharveng and in 1983 by C. Mouret and Y. Bousquet. (*Spelunca*, 1985 (20) map).

6. **Kalinawan cave** ....1368 m
(Tandang Kutyo, Tanay, Rizal prov., Luzon)
   Explored in 1983 by C. Mouret and J. -J. Matieu (*Spelunca*, 1984 (16) map).

7. **Irong-Ni-Kahuraw** ....1305 m
(Sohoton, Basey, Samar prov., Samar)
   Explored in 1983 by the Japanese Kamiya and Imamura (*Yamaguchi Caving Club Bull.*, N° 19, map).

8. **Balangagan cave** ....1280 m
(Sagada, Mountain prov., Luzon)
   (Deharveng, *Spéléologie aux Philippines*, 1980, plan).

9. **Natividad cave** ....1088 m
(Sagada, Mountain prov., Luzon)
   By C. Mouret and Y. Bousquet in 1982-1983.

10. **Sogong-Dokiw cave system** ....1021 m
(Sagada, Mountain prov., Luzon)
   Explored in 1980 (L. Deharveng) and 1984 (C. Mouret *et al.*).

SAUDI ARABIA

Al'Arabiyah as Sa'Udiyah

Mountains with crystalized rock, sandstones, and sand cover the greater portion of this country. The only karst regions are found around Riyadh and in the east of the country: the el Khardj region, the Hofuf, and in the stretches that separate this oasis from Kuwait. Cave potential in this desert climate would seem to be rather limited. Not open to tourists, Saudi Arabian caves have mostly been explored by geologists or those with contracts to work in the country, among them Americans B. L. Davis and Kochinski or the Frenchman, P. Courbon. The most spectacular cave is **dahl Hit**, 35 km S.W. of Riyadh. It is an enormous Vauclusian fountain at the bottom of which water appears among large breakdown blocks.

DEEP CAVES:

1. **dahl Hit** (Riyadh) .... –88 m
   By P. Courbon in 1983 (*Grottes et Gouffres*, 1985 (95) sketch map).

2. **dahl Abu Sukhayl** ....approx. –75 m
(Hofuf, Eastern Province)
   By Kochinski and B.L. Davis (*N.S.S. News*, 1983 (11) sketch map).

Awaiting further information:
   ain Dhila ....approx. –100 m
(Khardj, Eastern Province)

   ain Samha.................................approx. –100 m
(Khardj, Eastern Province)

LONG CAVES:

1. **dahl Hit** (Riyadh) ....approx. 250 m

2. **ghar an Nashab**..160 m
(Hofuf, Eastern Province)

3. **dahl Sabsab** ....approx. 150 m
(Haradh, Eastern Province)
   Explored in 1981 by B.L. Davis (*N.S.S. News*, 1983 (11) sketch map).

SRI LANKA

The Miocene limestone to the north and Precambrian rocks (in the center and south) found on this island seem to hold little speleological potential, and caving trips to Sri Lanka have been rare. Better known are caves of historical interest because of their use as habitations or for their paintings (for example at Sigiriya), as studied by Deraniyagala. Caves have been found by biologists from the University of Kyoto in 1960 and by Aellen and Strinati in 1970, and by the French caver M. Siffre in 1961. Siffre gave a length of 600 m to **Istripura** cave (in Pannala), while P. Strinati found **Istri-gal-Iena** cave (in Welimada) to also be 600 m long. Czech divers reached − 65 m in 1977 in a flooded pit near Jaffna (*Karszt ès Barlang*, 1978, I-II, profile).

SYRIA

Suriyah

The limestone areas of Syria have not yet motivated intensive study. In the 1970's, Sami Karkabi and the Lebanon Caving Club made some furtive visits, locating and exploring some relatively short (200 m typically) or shallow caves, such as the **Machtel el Helou** pit, –70 m deep. The **Fijeh** spring, which provides Damascus with water, was dived in 1971 by the Lebanon C. C. for a depth of 40 m and a length of 80 m. More recently (1984) the French society, Hydrokarst explored it for about 160 m, reaching a depth of 77 m.

THAILAND
Prathet Thai

Karst areas in Thailand are numerous and often quite extensive. They are generally of Permian age and form plateaus or small moutain chains, generally less than 1300 m high, except for the Doï Chiang Dao which rises to over 2200 m.

The first studies were those of Heinrich Kusch who visited and described various caves throughout the country. In 1981 L. Deharveng and A. Gouze explored and mapped several major caves. In 1985 an Australian team visited the Mae Hong Son region, an American team visited the Chiang-Dao-Fang region, and French cavers explored in the NW and southern sections of the country. French and Australian cavers made return trips in 1986 and 1987. The caving potential appears to be good, especially for horizontal caves. Many caves serve as Bhuddist sanctuaries and are decorated with giant Bhuddas, alters, and arrangements for comfortable meditation. It is usually possible to get permission to explore these caves. It should be noted that concentrations of carbon dioxide in caves in the NW can render exploration difficult or impossible.
Louis Deharveng.

LONG CAVES:

1. **tham Nam Mae Lana** (Mae Hong Son) .... 8390 m
   River cave explored by Australians (*Caves of northwest Thailand*, SRC, Sydney)

2. **tham Nam Lang** .... 8350 m
   (Mae Hong Son)
   Explored in 1984 and 1985 by Australian cavers (*Helictite* 23, 1).

3. **tham Luang** (Mae Saï) .... 6200 m
   Explored in 1986 and 1987 by A.P.S.

4. **tham Chiang Dao** .... 5170 m
   (Chiang Dao, Chiang Maï)
   Show cave mapped in 1980 by L. Deharveng and A. Gouze (Deharveng, Gouze, *Expédition en Thaïlande*, report, 1981, plan showing 4850 m) and in 1985 by the Association Pyrénéenne de Spéléologie.

5. **tham Pha Mon** .... 3989 m
   (Mae Hong Son)
   Explored in 1985 by the A.P.S.

6. **tham Nam Khlong Ngu** .... 3000 m
   (Kanchanaburi)
   River cave. (*Caves & Caving*, 38).

7. **tham Klaeb Yai** .... 2190 m
   (Chiang Dao)
   Explored in 1985, A.P.S.

8. **tham Boddhisat** (Sarabun) .... 2000 m

9. **tham Luang Nang Non** (Chiang Mai) .... 1600 m

10. **tham Kaeng-Khao** .... 1400 m
    (Mae Hong Son)

11. **tham Pha Thaï** .... 1180 m
    (Ngao, Lampang)

12. **tham Susa** (Mae Hong Son) .... 1160 m

13. **tham Poung Chang** .... 1150 m
    (Phangnga)
    Explored in 1985, A.P.S.

14. **tham Nim Lot** (Chumphon) .... 1150 m

15. **tham Tapan** (Phangnga) .... 1070 m

16. **tham Lot** (Mae Hong Son) .... 1030 m

17. **tham Huoy Khhun** (Mae Hong Son) .... 1000 m

18. **tham Thung Kwien** (Chiang Mai) .... 1000 m

19. **tham Pang Kham** (Mae Hong Son) .... 1000 m

TURKEY
Türkiye

Like Mexico and Papua New Guinea, but for different reasons, Turkey is a a country characterized by large karst phenomena. Twenty percent of its surface area of 767,000 km² is covered by carbonate rocks. The principal limestone band, the Taurus, is over 1000 km long and averages 100 km wide. In the west (Konya, Antalya) dye traces using up to 400 kg of fluorescein have established the existence of hydrologic systems 75 km long (Hormat Bürnü-Yedi Miyarlar or system of the Kembos), 56 km long (Homat Bürnü-Oluk Köprüsü), and 35 km long (Akpinar-Oymapinar, or Eynif system). Among the many poljes or karst valleys, that of Kembos is especially large at 17 km long and 2 to 3 km wide. The collapse doline Kayaağil çukuru has an opening 500 m by 180 m. Dumanlı, a resurgence with an average flow of 50 m³/sec (the source of the water is still unknown!) is now 120 m underwater due to the construction of the Oymapinar dam. Unfortunately, these huge karst areas have so far not produced any very large caves on the international scale.

In the north, the limestone outcrops between Zonguldak and Kastamonu have interesting caves that are regularly explored by the Turks. Since 1950, Turkey has been explored by biospeleological expeditions (H. Coiffait, P. Strinati), followed since 1964 by regular caving expeditions. In was in this year that the Turkish Speleological Society was formed. The work of its president, Temuçin Aygen, was particularly important in opening up the Western Taurus region. They were joined beginning in 1965 by the Spéléo-Club de Paris as well as the Club Martel de Nice (from 1969) and numerous British clubs (Chelsea, Sheffield, Nottingham). Turkey is a case where foreign collaboration with a local club (the T.S.S.) has worked very well. Today, cavers from the B.Ü.M.A.K. (Istanbul) have taken over and are finding many new caves. In his recent *Türkiye Mağaralari* (1984), Temuçin Aygen has collected information on the principal "Caves of Turkey".

DEEP CAVES:

1. **Düdencik** .... –330 m
   (Cevizli, Akseki, Antalya)
   Alt. 1040 m. Swallet cave explored in 1967 by the S. C. Paris and the T.S.S. (*Grottes et Gouffres*, 1967 (40) profile).

2. **Sakal Tutan düdeni** .... –303 m
   (Değirmenlik, Akseki, Antalya)
   Alt. 1650 m. Swallet cave explored in 1979 by the S.C.P and the Club Martel de Nice (*Grottes and Gouffres*, 1980 (75) profile).

3. **Sakal Tutan deliği** .... –302 m
   (Değirmenlik, Akseki, Antalya)
   Alt. 1650 m. Explored in 1979 (C.M.N., S.C.P.) and 1980 (A.S.P. Monaco, C.M.N.) (*Explorations spél. dans le Taurus Occid.*, 1976-1980, report, C.A.F., 1982, profile).

4. **Ilgarini** .... –250 m
   (Yamanlar, Ulus, Kastamonu)
   Explored in 1982 by the B.Ü.M.A.K. (*Delta*, Istanbul, 1983 (1) profile).

5. **Pınargözü** ....+248 m
   (Yenişarbademli, Isparta)
   Alt. 1550 m. In 1968-1969 (+138 m, Chelsea S.S.), 1970-1971 (C.A.F. Paris, Millau, Nice).

6. **Gölcük düdeni** .... –245 m
   (Seydişehir, Konya)
   Alt. 1565 m. Swallet cave, bottomed in 1976 by the U. Sheffield U.S.S. (*Trans. B.C.R.A.*, 1977, 4 (4) profile).

7. **Ürküten Oruğu 1** .... –243 m
   (Çimi, Akseki, Antalya)
   Alt 2300 m. 1979, B.Ü.M.A.K. and Imperial College (Britain) (Imperial College, *Expedition to Turkey*, report, 1979, profile).

8. **Koyungöbedi** .... – 235 m
   (Sadıklar, Antalya)
   Alt. 1080 m. Swallet cave explored in 1966 by the S.C. Orsay Faculté, T.S.S., Brit. Spel. Exped. to Turkey (*Grottes et Gouffres*, 1967 (40) profile showing –210 m). Extended in 1977 by the T.S.S.

9. **Karabayır düdeni** .... –225 m
   (Seydişehir, Konya)
   Alt. 1600 m. 1976, SUSS (*Trans. B.C.R.A.*, 1977, 4 (4) profile).

10. **Sorkun düdeni** .... –200 m
    (Yamanlar, Ulus, Kastamonu)
    1983, B.Ü.M.A.K. (*Delta*, 1983 (1) profile).

11. **Döngelyanı kuylucu** .... –195 m
    (Kazla, Ulus, Kastamonu)
    1985, B.Ü.M.A.K.

12. **Dünekdibi orugu** .... –192 m
    (Çimi, Akseki, Antalya)
    Alt. 2040 m. 1979, B.Ü.M.A.K. and Imperial College (I. C. *op. cit.*, 1979, profile).

13. **Büyük Obruk mağarası** .... –190 m
    (Çumra, Karasemir, Konya)
    1978, S.C. Causses and Lodève (S.C.C.-S.C.L., *Expéd. Spél. en Turquie d'Asie*, report, 1978, profile).

14. **Kapaklı kuylucu** .... –185 m
    (Kazla, Ulus, Kastamonu)
    1985. B.Ü.M.A.K.

15. **Tınaz Tepe düdeni** .... –171 m
    (Seydişehir, Konya)
    Alt. 1500 m. Swallet cave. Bottom reached in 1976 by the SUSS (*Trans. B.C.R.A.*, 1977, 4 (4) profile).

16. **Felengı mağarası** .... –170 m
    (Asmaköy, Konya)
    Alt. 1350 m. Mapped in 1973 by the Expéd. Spél. Française (*Grottes et Gouffres*, 1977 (65) profile).

17. **Kayaağıl çukuru** .... –160 m
    (Demirçal, Akseki, Antalya)
    Alt. 1200 m. Mapped in 1973 by the E.S.F. (*Grottes et Gouffres*, 1977 (65) profile).

18. **Ödemis subatanı** .... –160 m
    (Ödemiş, Izmir)
    1981, B.Ü.M.A.K.

19. **Tilkiler düdeni** ....(–66, +93) 159 m
    (Manavgat, Amtalya)

20. **Çayirönü düdeni** .... –155 m
    (Akseki, Antalya)
    Alt. 1050 m. Mapped 1966 by the S.C.O.F. and B.S.E.T. (*Grottes et Gouffres*, 1967 (40) profile; 1977 (65) profile to –110 m).

21. **Tefekli mağarası** .... –147 m
    (Avason, Manavgat, Antalya)
    Alt. 610 m. Bottom reached in 1976 by the S.C.P. and the C.M.N. (*Mém. S.C.P.*, 1976 (4) profile). Cave in conglomerate.

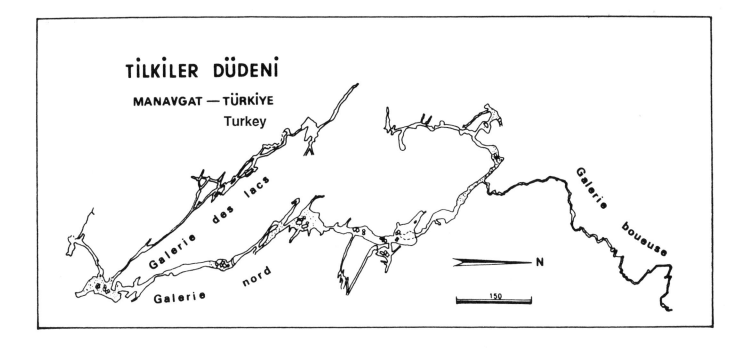

TİLKİLER DÜDENİ

MANAVGAT — TÜRKİYE
Turkey

Galerie des lacs

Galerie nord

Galerie

Galerie boueuse

N

150

LONG CAVES:

1. **Tilkiler düdeni** ....6600 m
(Manavıgat, Antalya)

Tilkiler düdeni is about 25 km north of Manavgat; by leaving the road shortly before the Oymapınar dam, one arrives at the cave by taking a trail which follows the Düden deresi, a stream which is dry in summer. The entrance is a tunnel at the foot of the first steep hills that lead to the village of Tilkiler. This tunnel is at an altitude of 128 m, at the contact of the Tefekli formation conglomerate and impermeable rocks. It was dug in 1974, during studies related to the construction of the Oymapinar dam, 4 km to the ESE. The tunnel intersects the cave 450 m from the entrance, where a strong air current can be felt.

The cave was explored in 1976 by the Spéléo-Club de Paris and the Club Martel de Nice, with the help of Temuçin Aygen and the T.S.S. A total of 2755 m were mapped for a vertical extent of 159 m (− 66, +93). In 1977 the length reached 4845 m and the "muddy passage" was opened up. This carries a strong air current and was explored in 1978 (length 5585 m) and 1979.

This cave is formed in Miocene conglomerates, except for the southern extremity, which is in Burdigalian limestone. It is part of the Eynif hydrologic system and functions as an overflow route for the Oymapinar resurgences (alt. 32 m, 10 m³/sec) in winter.

Map: from drawings of the S.C. Paris and C.M. Nice.

Bibliography: Chabert (C.) - Recherches sur les systèmes de Kembos et d'Eynif, *Mém. du S. C. Paris*, 1976 (4): 68 p.; *Grottes et Gouffres*, 1978 (67(; 1979 (71) and 1980 (75).

2. **Kızılelma mağarası** ....6250 m
(Karadon, Zonguldak)

Explored from 1976 to 1977 (*Trent Polytechnic Spel. Exped.*, report, 1978, map).

3. **Pınargözü** ....5275 m
(Yenişarbademli, Isparta)

Spring explored from 1965 to 1975 (S. C. Paris, C.A.F. Millau, C.M.N., 4675 m in 1971) (*Grottes et Gouffres*, 1972 (48) plan), the Red Rose, C.P.C. and U. Bristol S.S. (*R.R.C.P.C. Journal*, 1977 (7) partial plan). Cave wind measured at 155 km/hour!

4. **Dupnisa mağarası** ....3150 m
(Demirköy, Kırklareli)

1978-1979, B.Ü.M.A.K. (*B.Ü.M.A.K. Yayinları*, 1980 (3) map).

5. **Gökgöl mağarası** ....2970 m
(Zonguldak)

Explored from 1976 to 1978 (Trent Polytechnic, *op. cit.*, 1978, map).

6. **Atçi ini** ....2725 m
(Safranbolu, Zonguldak)

1976-1978 (Trent Polytechnic, *op. cit.*, 1978, map).

7. **Düdensuyu** ....1840 m
(Ürünlü, Akseki, Antalya)

1966-1967, S.C.P. (*Grottes et Gouffres*, 1967 (40) map).

8. **Güvercin Taşı deliği** ....1638 m
(Seydişenir, Konya)

1969, C.M.N. (*Grottes et Gouffres*, 1972 (48) map).

9. **Ayva ini** ....1400 m
(Ayva, Ulubat gölü, Bursa)

Swallet cave explored in 1970 by the Spanish "Turquia 1970" (*Mem. de la Exped. Espel.*, 1970, map). Length estimated at 5500 m.

10. **Tınaz Tepe düdeni** ....1380 m
(Seydişenir, Konya)

1968, S.C.P. (*Grottes et Gouffres*, 1968 (42) map

showing 750 m) and 1976, SUSS (*Trans. B.C.R.A.*, 1977, 4 (4) map).

11. **Körük ini** .... 1320 m
    (Çamlık, Gencek, Konya)
    1967, S.C.P. (*Grottes et Gouffres*, 1967 (40) map).

12. **Cumayanı mağarası** .... 1085 m
    (Zonguldak)
    1976-1978, S.C.P.-C.M.N., then Trent Polytechnic (*report*, 1978, map).

13. **Sakal Tutan düdeni** .... 1055 m
    (Değirmenlik, Akseki, Antalya)
    1979, S.C.P., C.M.N. (*Grottes et Gouffres*, 1980 (75) map).

VIETNAM

Viet-Nam Dan-Chu Cong-Hoa

The tropical karsts of Vietnam have been made famous by the Along Bay landscapes. The most developed karsts are in the north of the country, between Ha-giang and the Gulf of Tonkin, as well as to the west of Hanoi (Son La and Hoa Binh provinces). Karst studies have been undertaken by both Poles and Vietnamese, but the results are difficult to obtain. In the meantime, we will cite only two caves of the thirty explored by a biospeleological group led by Miss Colani in 1927: **Ban Than** (Muong Muôi, Thuan, Son La), 280 m long, and **Ban Sai** (ibid.), 220 m long. In addition, many caves were studied by prehistorians during the period of French colonialization.

V
EUROPE

ALBANIA
Shqïpëria

This is a closed country whose boundaries include the limestone ranges of Yugoslavia (Crna Gora) from the north and those of Greece (Pindhos) from the south. We know that the Hungarian Dénes Balázs was able to explore these ranges, and that a Western diplomat visited some caves in the south, such as **Mezhgoranet** (Tepelenë), about 2 km long (or possibly 8 km according to other sources) and **Korite** (Cerovada), about 3 km long.

AUSTRIA
Österreich

The classic period of speleological exploration in Austria began in 1879 with the formation of a caving club in Vienna. This organization, the first in the world to specialize in the study of caves and pits, also had a scientific orientation. Because of the political climate in Europe at the time, most of the work was done in the Moravian karst or in the classic karsts of Trieste and Yugoslavia. However, there were also a respectable number of explorations in the Alps between 1879 and 1894. A series of major explorations and discoveries in the limestones of the Alps uncovered the existence of great systems in the Dachstein (High Austria), in the Tennengebirge (Salzburg), and in Styrie, near Graz (the Lurgrotte). Around this time, the first speleological federation, the Verein für Höhlenkunde in Österreich, was formed and began putting out a newsletter.

After WWI, Austria acquired its present boundaries. The interest of cavers focused on the karst limestones of the Alps, which now occupied about 14% of the land area of the country. Scientific studies were coordinated by a national speleological institute, the Speläologisches Institut, which published an annual review, the *Speläologisches Jahrbuch*. Regular courses on speleology were offered at the University of Vienna beginning in 1924, and 1928 saw the passage of a cave protection law. Regional caving clubs became interested in vertical caving and joined together in a national club called the Hauptverband deutscher Höhlenforscher. As a consequence, the Geldloch, in Lower Austria, became the deepest cave in the world at −464 m. The systematic collection of data on all explored caves was undertaken. Many cave descriptions were published in the tri-annual bulletin the *Mitteilungen über Höhlen-und Karstforschung*.

The HdH federation brought together both sport and scientific cavers from several countries. All activities ceased during WWII, but began again immediately following the end of hostilities in 1945, thanks to the regional clubs of Graz, Linz, Salzburg, and Vienna. Major expeditions were mounted to explore in detail the systems of Lurhöhle, Geldloch, Dachstein-Mammuthöhle, and Tantalhöhle.

In 1949 caving clubs and show cave owners banded together to form the Verband Österreichischer Höhlenforscher (Federation of Austrian Speleologists). This federation has published a tri-annual bulletin *Die Höhle* since 1950. Up to 1974 they cooperated with the authorities to enforce the federal cave protection law. They also instruct the show cave guides (which must have a government certificate), and contribute to the documentation of caves and cave exploration. Until 1984, this information was archived in the Bureau for the Protection of Monuments (which had a special branch "Protection of Caves"). Since this date, a re-organization under the Insitut für Höhlenforschung, affiliated with the Naturhistorisches Museum of Vienna, has replaced the old department.

The period from 1961 to 1985 was characterized by an explosion in caving explorations and unexpected discoveries. Because most of the deep caves occur at high elevations, most of the expeditions are in the summer months. Because this is also the vacation period for the Austrians, many of the discoveries have been made by foreign teams, which has not made for very rigorous cataloguing of results. Nonetheless, the Verband Österreichischer Höhlenforscher and the Institut für Höhlenforschung of Vienna have attempted to prepare as accurate a list of long and deep caves as possible (as of 1986), choosing among the 8800 presently registered. A detailed document on all the caves longer than 2 km and deeper than 200 m was in preparation for the summer of that year: Theo Pfarr and Günter Stummer, *Die Iängsten und tiefsten Höhlen Österreichs*.

Hubert Trimmel.

DEEP CAVES:

1. **Schwersystem** .... **−1219 m**
(Tennengebirge, Salzburg) (1511/268)
The entrance to this cave is in the valley of Schwer, in the Tennengebirge range, at an altitude of 1843 m. One travels there from Salzburg passing through Golling, Abtenau, and Scheffau, then by a major foot trail passing the

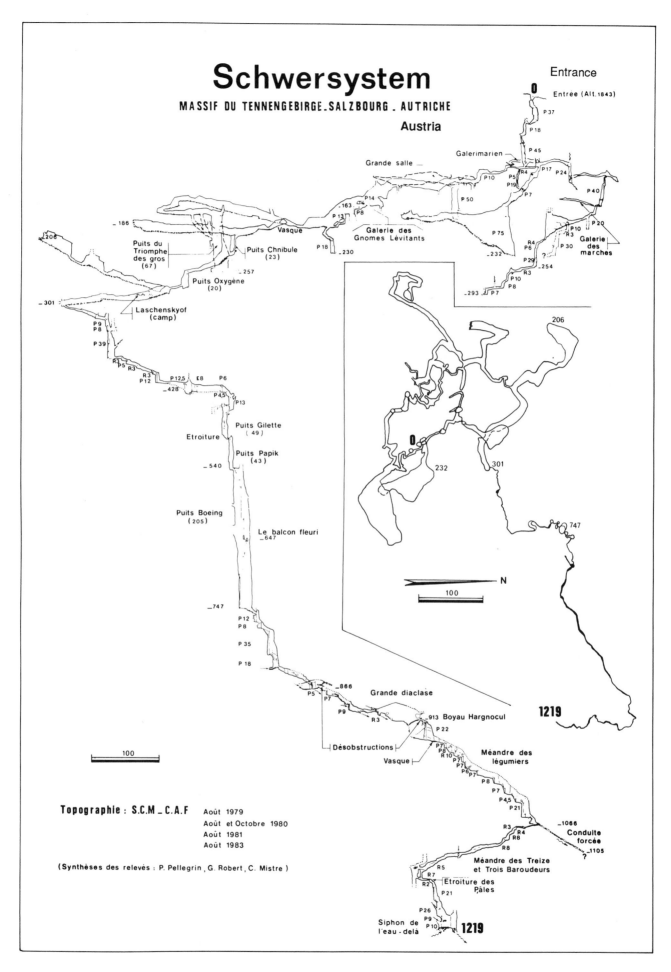

Schwersystem

MASSIF DU TENNENGEBIRGE - SALZBOURG - AUTRICHE

Austria

Entrance
Entrée (Alt. 1843)

Topographie : S.C.M _ C.A.F   Août 1979
Août et Octobre 1980
Août 1981
Août 1983

(Synthèses des relevés : P. Pellegrin , G. Robert , C. Mistre)

Rosberg hut and heading towards the Scheibling KG peak via the Schwer valley.

Location on the Hallein 1/50,000 map (N° 94) is 47° 32' 40.2" North and 13°16' 29.5" East. The presumed resurgence is the powerful Winnerfall, located some 1150 m lower at an altitude of 700 m.

The discovery of the cave was made by the Spéléo-Club de Marseille (Club Alpin Français) in August, 1979. They descended to –254 m in a branch that turned out to be a dead end at –293 m in 1980. Exploration resumed higher up, and a passage was found at –110 m that led to a new series of pits. The Marseillans stopped at the lip of a large 205 m pit at a depth of –540 m. In October of the same year, the S.C.M. attacked the pit but were stopped at –647 m for lack of rope.

The S.C.M., reinforced by individuals from Marseille, returned in 1981 and descended in stages to –1105 m, where a dangerous narrow phreatic tube stopped exploration. In 1982, this passage was proven to be completely blocked, but a way around was found at –1066 m. This was explored in 1983. A rift entirely filled with the turbulent river was reached at –1219 m.

Map: S. C. Marseille C.A.F., issue of *Spelunca Memoires*, 1983 (13).

Bibliography: Pellegrin (P)., Robert G. - Batman-höhle (sic.). Massif du Tennengebirge. Autriche, *Spelunca*, 1982 (6): 25-30.
S. C. Marseille - Expédition Autriche 1983, *Spéléopérations*, N° spécial, n. d., n.p.

2. *Dachstein-Mammuthöhle* ....(+423, –757) 1180 m
(Dachstein, Oberösterreich) (1547/9a-g)

This system has seven entrances: Mammuthöhle Alter Osteingang (1324 m), Westeingang (1392 m) and Neuer Osteingang (1368 m), Oedleingang (1334 m), Unterer Einstieg Wasserschacht (1259 m), Oberer Einstieg Wasserschact (1274 m), and Däumelkogelschacht (1815 m).

It is located in the Dachstein range, 3 km SE of Obertraun, 500 m west of the Schöhbergalpe hut, which is reached by a cable car from Obertraun. The show cave entrance is a fifteen minute walk from the hut.

The map is the Bad Ischl 1/50,000 (N° 96).

Explorations began around 1910. They were interrupted twice by the two world wars. After 1945, explorations were organized by the Landesvereins für Höhlenkunde in Vienna, who reached the vertical extent of 315 m in 1952 (–173, +142 m). In 1964, the connection of Mammuthöhle and Oedlhöhle gave a length of 16 km. This figure was increased to 17.6 km in 1966. In 1972, 25,280 m were reached, then 26,280 m in 1973, 26,783 m in 1974, and 27,746 m in 1975, with a vertical extent of 405 m (–287, +118). The length increased steadily, reaching 30,630 m in 1978, 33,457 m in 1979, and 35,825 m in 1980, with a vertical extent of 883 m.

With the connection to Däumelkogelschacht on the third of September, 1981, the vertical extent reached 1180 m (–757, +423), and the length was increased to 37,046 m. Additional surveys brought this to 38,185 m in 1984 and 38,529 m in 1985.

Map: from *Die Höhle*, suppl. N° 14, 1966 and profile from documents supplied by G. Stummer.

Bibliography: Stummer (G.) - *Atlas der Dachstein-Mammuthöhle 1 : 1000°*, Vienna, 1980, 100 p.

3. *Jubiläumsschacht* .... –1173 m
(Hoher Göll, Salzburg) (1336/70)

Located in the Hoher Göll range, west of the towns of Kuchl and Golling, the entrance is at an altitude of 2009 m, on the slopes east of the lapiaz of the Gruberhorn cirque (see Gruberhornhöhle).

1/50,000 map N° 94 (Hallein).

The cave was discovered on August 21, 1980 by the Hungarian Lázsló Lukács, member of an expedition by the Polish club P.T.T.K. Gdynia. The depth of about –175 m was reached in three trips between August 21 and 29.

The following year, the Polish Alpine Association of Warsaw organized an eleven member expedition directed by Christian Parma. They dedicated the months of August and September to the exploration of the cave. They negotiated the 201 m deep Amadeusschacht pit and installed a camp at –900 m. From there, by a series of wet pits, they reached the terminal sump at –1173 m.

Map: from L. Lakács *et al.*, *Spelunca*, 1984 (13).

Bibliography: Lukács (L.) - A Jubileum-barlang, *Karszt és Barlang*, 1980, II: 107-108.
Parma (C.) in *Spelunca*, 1984 (13):16-18.

4. *Schneeloch* ....(+132, –969) 1101 m
(Tennengebirge, Salzburg) (1511/7)

This cave, the first one in Austria to pass the 1000 m depth mark, is located on the slopes of Kuchelberg, in the Tennengebirge range, at an altitude of 1610 m. The entrance is marked on the topo map.

1/50,000 map N° 94 (Hallein).

Schneeloch is formed in the Dachsteinkalk limestones. The strike is to the east, with a dip of 30 to 35 degrees. The presumed resurgence is that of the Schwarzbach, 1 km to the north, at an altitude of 630 m, and the presumed overflow is the Winnerfallhöhle at 700 m.

The cave has been known for a long time. The large entrance was normally blocked inside by snow. On All-Saints day in 1975, the Landesvereins für Höhlenkunde in Salzburg noted that the entrance was open. They explored upwards to +70 m, while the Groupe Spéléo Alpin Belge explored downwards to –100 m. In September, 1977, the Polish of the Warszawa A.K.S. and the G.S.A.B. were stopped by a difficult and tight winding canyon at –330 m.

From August 19 to 28, 1978, the G.S.A.B. pushed the ascending branch up to over 132 m, and the lower one to –902 m, in spite of the squeezes. They returned in November to find a sump at a depth of –954 m. A final expedition by the G.S.A.B. was held in 1979, during which Georges Feller dove the sump to a depth of 15 m, bringing the total vertical extent to 1101 m.

The substantial cave winds suggest the likelihood of connections to higher entrances.

Map: from the G.S.A.B. (1978), furnished by J.-C. Hans.

Bibliography: Braun (J. -P.), Degrave (E.) - Le gouffre du Schneeloch (sic) (Autriche), *Spelunca*, 1979 (1): 11-14.

5. *Jägerbrunntrogsystem* .... –1078 m
(Hagengebirge, Salzburg) (1335/35)

The Jägerbrunntrogsystem gathers together the following caves: Petrefaktencanyon (alt. 2135 m and 2125 m), Zwillingsschacht (alt. 2069 m), Jägerbrunntroghöhle east (alt. 2135 m) and west (alt. 1907 m), Roithnerkar-schacht (alt. 1860 m), and Sulzenkarseishöhle (alt. 1831 m). It is developed in the Hagengebirge range, between Jägerbrunntrog (2248 m) on the west, and Raucheck (2215 m) on the east.

One gets to the cave from Blühnbachtal, climbing to Untere Sulzenkarl, then to Obere Sulzenkarl. The east entrance of Jägerbrunntroghöhle is about 50 m above pylon 154 of a power line, while the west entrance is half way between pylons 154 and 155.

1/50,000 map N° 124 (Saalfelden am Steinernen Meer).

The system is formed in the limestones and dolomites of the Dachsteinkalk. The resurgence is the Schwarze Torrenquelle (alt. 511 m).

The east entrance (Osteingang) of Jägerbrunntroghöhle has long been known by hunters. It was noted in 1946 by S. Felber. Roithnerkar-schacht was discovered in 1953 and Sulzenkareishöhle in 1959. Later, the Landesvereins für Höhlenkunde in Salzburg, L.H.S. discovered Petrefaktencanyon (1976).

Before their connection, these different caves were explored by the L.H.S, who descended to –280 m in Roithnerkar-schacht in 1954, and to –220 m in 1960 in the Jägerbrunntroghöhle after opening an ice-filled passage in 1959. Expeditions succeeded one another beginning in 1976: the L.H.S. reached –330 m in the Petrefaktencanyon, following which the Polish of Katowice (K.K.S.) reached –451 m in 1977-1978. The Austrians reached –330 m in Zwillingsschacht in 1978.

Connections began in 1980: Petrefaktencanyon, Zwillingsschacht, and Jägerbrunntroghöhle were connected by the K.K.S, bringing the total vertical extent to 647 m and the total length to 9452 m.

In 1981, the connection with Roithnerkar-schacht was made by the K.K.S., bringing the depth to –894 m and the length to 15,270 m. In 1982, the Polish just barely passed the –1000 m (1006 m) and the 20 km (20,080 m) marks. In 1983 the length grew significantly to 25,680 m, while the depth increased slightly to –1061 m. These figures were pushed to 28,026 m in length and –1078 m in depth in 1984.

Map: from the drawings of M. Cziepel, W. Klappacher, J. Kubiac et al., furnished by Günter Stummer.

Bibliography: .Klappacher (W.) et al. - *Salzburger Höhlenbuch*, Salzburg, 1979, Vol. 3.
Napierala (M.) - Exploration in the Jägerbrunntrogsystem and the search for the connection with Tantalhöhle, *Caves & Caving*, 1984 (25): 16-18.
Kardás (R.) - Polski alpinizm jaskiniowy 1983-1984, *Taternik*, 1984 (2): 79-82.

6. **Mäanderhöhle** or **Herbsthöhle** .... –1028 m
(Tennengebirge, Salzburg) (1511/272)
Has also been called Laufenhöhle. It is at an altitude of 1750 m, near Gr. Grießkogel, NE of Tennengebirge. 1/50,000 map N° 94 (Hallein). Formed in the Dachstein kalk, it is assumed that the cave is hydrologically connected with the Tricklfall and Dachserfall karst springs.

Exploration began in 1979, with the depth of –350 m being reached in 1983 by the Zagan and Brelsko Bcala caving clubs (Poland). They got down to about –725 m in 1984 by descending the Laufen shaft. They returned in 1986 to reach –800 m, –900 m and finally –1028 m by

following a series of muddy shafts. Strong winds suggest possible connections with higher entrances.

Map: from Palka, Lysien (1979) and Zyzanski (1983), furnished by G. Stummer showing –725 m.

Bibliography: *British Caver*, Vol. 102, (26) p 1 Summer 1987.
Salzburger Höhlenbuch, Salzburg, 1985, Vo. 4, pp. 451-452.

7. ***Lamprechtsofen*** ....(–10, +995) **1005 m**
(Leoganger Steinberge, Salzburg) (1324/1)
The entrance to this show cave, at an altitude of 664 m, is located at the side of the road from Lofer to Saalfelden, south of St-Martin-bel-Lofer. The sign shows where the entrance is.

1/50,000 map N° 92.

The watershed for Lamprechtsofen extends from 1500 to 2000 m in altitude, in the shadow of Nabelsbergkar, where two major caves, Wieserloch (alt. 2050 m, depth 730 m), and Rothöhle (alt. 2020 m, depth 280 m) are very close to connecting with Lamprechtsofen. All three caves are developed in the Dachsteinkalk dolomites. The dry season flow of the underground river in Lamprechtsofen is 10 liters per second.

The cave has been known for a long time. It was used as a hiding place for treasure in the first centuries of the modern era, and noted in 1503. The chronicles of Lofer indicate that the first caving exploration was the work of Ferchel in 1833. A caver from Salzburg named Posselt visited the cave in 1878. Feschl reconnoitered the first 800 m in 1882. Work to make it into a show cave began in 1905. During the following years, cavers such as Bock, Czörnig, Oedl, and others explored the off-trial portions. Even so, only 1200 m of passage was known by 1959.

The first systematic exploration of Lamprechtsofen by the L.H.S. began on February 2, 1964, when the group lowered the level of the Bock sump, which is 600 m from the entrance. This permitted penetration deeper into the mountain and the reconnaissance of 3600 m of passage. By 1965 the vertical extent had reached 292 m (–10, +282) and the length 6500 m. A passage was discovered in 1967 that bypassed the Bock sump.

Thanks to the achievement of the difficult climb up the Superklamm in 1969, the vertical extent made a leap to +511 m, with the length increasing to 9100 m. The cavers of the L.H.S. were stopped by impenetrable breakdown at 740 m above the entrance in 1973.

Even though the height of the cave remained unchanged, 1974 brought an increase in the length to 12,203 m, followed by 12,527 m in 1975. Discoveries of new passages permitted exploration up to +810 m in 1976. In early 1977, the Polish cavers from Kraków took over from the Austrians and progressed upwards to +850 m. The same teams progressed even higher in 1978, reaching a breakdown blockage at +952 m that was passed by other passages in 1979 to reach +995 m, giving Lamprechtsofen a final vertical extent of 1005 m and a length of 14,657 m. Only 100 m more in elevation is needed to reach the plateau above!

Map: from Kulbicki, Kleszynski, and Wiśniewski.
Bibliography: *Die Höhle*, 1964 (2); 1969 (4); 1970 (1). Klappacher W.), Knapczyk (H.) - *Salzburger Höhlenbuch*, Salzburg, 1977, Vol. 2: 57-97, article and bibliography.

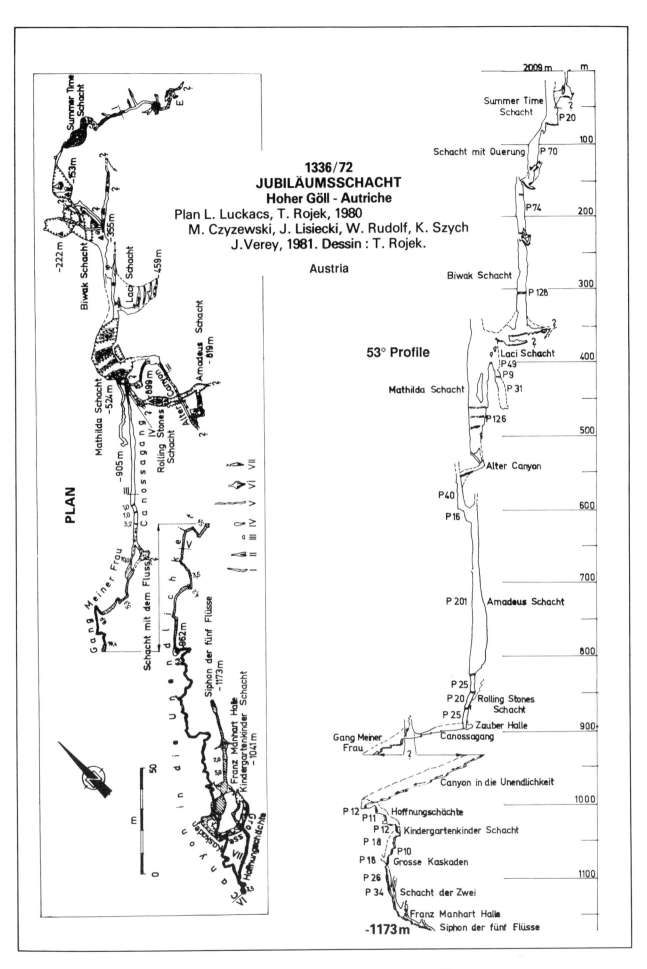

1336/72
JUBILÄUMSSCHACHT
Hoher Göll - Autriche
Plan L. Luckacs, T. Rojek, 1980
M. Czyzewski, J. Lisiecki, W. Rudolf, K. Szych
J.Verey, 1981. Dessin : T. Rojek.

Austria

53° Profile

PLAN

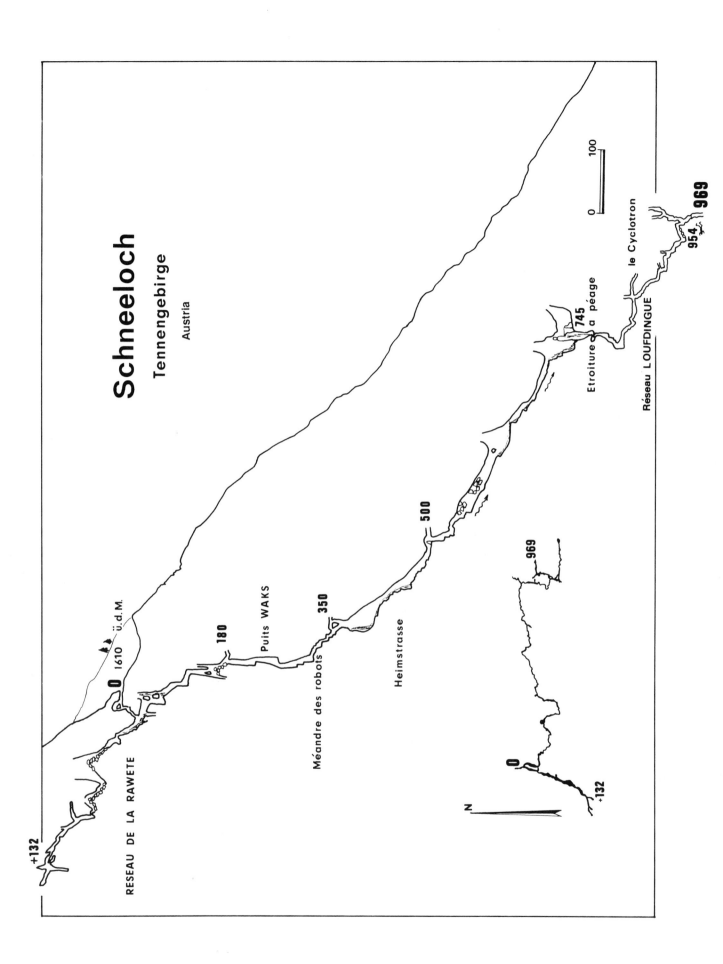

Schneeloch

Tennengebirge

Austria

RESEAU DE LA RAWETE

Puits WAKS

Méandre des robots

Heimstrasse

Etroiture à péage

Réseau LOUFDINGUE

le Cyclotron

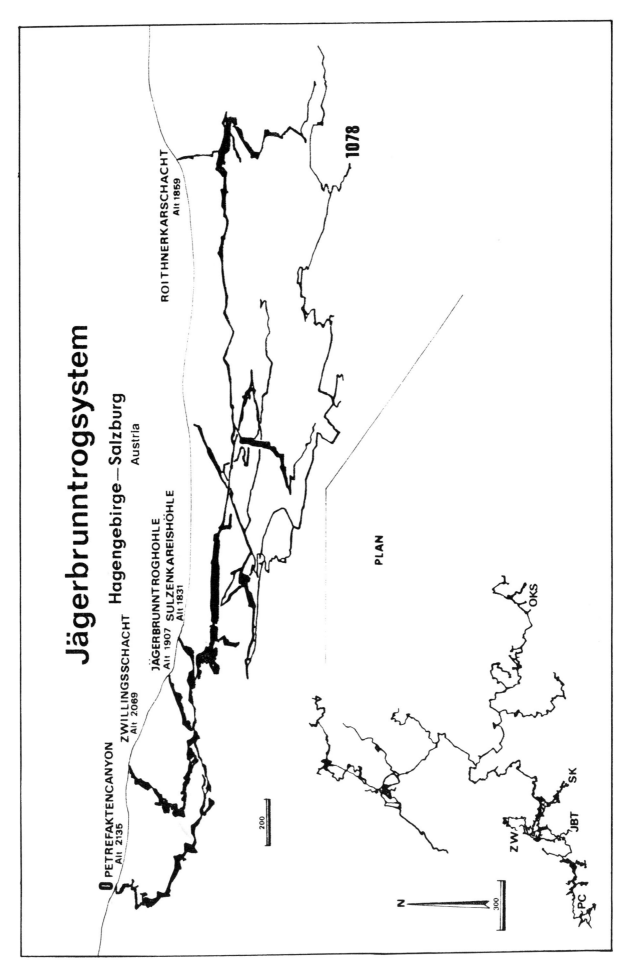

Jägerbrunntrogsystem

Hagengebirge — Salzburg
Austria

PETREFAKTENCANYON
Alt 2135

ZWILLINGSSCHACHT
Alt 2069

JÄGERBRUNNTROGHÖHLE
Alt 1907   SULZENKAREISHÖHLE
Alt 1831

ROITHNERKARSCHACHT
Alt 1859

1078

PLAN

200

300

N

ZW
JBT
SK
PC
OKS

(alt. 1860 m), and Sulzenkarseishöhle (alt. 1831 m). It is developed in the Hagengebirge range, between Jägerbrunntrog (2248 m) on the west, and Raucheck (2215 m) on the east.

One gets to the cave from Blühnbachtal, climbing to Untere Sulzenkarl, then to Obere Sulzenkarl. The east entrance of Jägerbrunntroghöhle is about 50 m above pylon 154 of a power line, while the west entrance is half way between pylons 154 and 155.

1/50,000 map N° 124 (Saalfelden am Steinernen Meer).

The system is formed in the limestones and dolomites of the Dachsteinkalk. The resurgence is the Schwarze Torrenquelle (alt. 511 m).

The east entrance (Osteingang) of Jägerbrunntroghöhle has long been known by hunters. It was noted in 1946 by S. Felber. Roithnerkar-schacht was discovered in 1953 and Sulzenkareishöhle in 1959. Later, the Landesvereins für Höhlenkunde in Salzburg, L.H.S. discovered Petrefaktencanyon (1976).

Before their connection, these different caves were explored by the L.H.S, who descended to –280 m in Roithnerkar-schacht in 1954, and to –220 m in 1960 in the Jägerbrunntroghöhle after opening an ice-filled passage in 1959. Expeditions succeeded one another beginning in 1976: the L.H.S. reached –330 m in the Petrefaktencanyon, following which the Polish of Katowice (K.K.S.) reached –451 m in 1977-1978. The Austrians reached –330 m in Zwillingsschacht in 1978.

Connections began in 1980: Petrefaktencanyon, Zwillingsschacht, and Jägerbrunntroghöhle were connected by the K.K.S, bringing the total vertical extent to 647 m and the total length to 9452 m.

In 1981, the connection with Roithnerkar-schacht was made by the K.K.S., bringing the depth to –894 m and the length to 15,270 m. In 1982, the Polish just barely passed the –1000 m (1006 m) and the 20 km (20,080 m) marks. In 1983 the length grew significantly to 25,680 m, while the depth increased slightly to –1061 m. These figures were pushed to 28,026 m in length and –1078 m in depth in 1984.

Map: from the drawings of M. Cziepel, W. Klappacher, J. Kubiac et al., furnished by Günter Stummer.

Bibliography: .Klappacher (W.) et al. - Salzburger Höhlenbuch, Salzburg, 1979, Vol. 3.
Napierala (M.) - Exploration in the Jägerbrunntrogsystem and the search for the connection with Tantalhöhle, Caves & Caving, 1984 (25): 16-18.
Kardás (R.) - Polski alpinizm jaskiniowy 1983-1984, Taternik, 1984 (2): 79-82.

6. *Mäanderhöhle* or *Herbsthöhle* .... –1020 m
(Tennengebirge, Salzburg) (1511/272)

Has also been called Laufenhöhle. It is at an altitude of 1750 m, near Gr. Grießkogel, NE of Tennengebirge. 1/50,000 map N° 94 (Hallein). Formed in the Dachstein kalk, it is assumed that the cave is hydrologically connected with the Tricklfall and Dachserfall karst springs.

Exploration began in 1979, with the depth of –350 m being reached in 1983 by the Zagan and Brelsko Bcala caving clubs (Poland). They got down to about –725 m in 1984 by descending the Laufen shaft. They returned in 1986 to reach –800 m, –900 m and finally –1028 m by

following a series of muddy shafts. Strong winds suggest possible connections with higher entrances.

Map: from Palka, Lysien (1979) and Zyzanski (1983), furnished by G. Stummer showing –725 m.

Bibliography: British Caver, Vol. 102, (26) p 1 Summer 1987.
Salzburger Höhlenbuch, Salzburg, 1985, Vo. 4, pp. 451-452.

7. *Lamprechtsofen* ....(–10, +995) 1005 m
(Leoganger Steinberge, Salzburg) (1324/1)

The entrance to this show cave, at an altitude of 664 m, is located at the side of the road from Lofer to Saalfelden, south of St-Martin-bel-Lofer. The sign shows where the entrance is.

1/50,000 map N° 92.

The watershed for Lamprechtsofen extends from 1500 to 2000 m in altitude, in the shadow of Nabelsbergkar, where two major caves, Wieserloch (alt. 2050 m and depth 730 m), and Rothöhle (alt. 2020 m, depth 280 m) are very close to connecting with Lamprechtsofen. All three caves are developed in the Dachsteinkalk dolomites. The dry season flow of the underground river in Lamprechtsofen is 10 liters per second.

The cave has been known for a long time. It was used as a hiding place for treasure in the first centuries of the modern era, and noted in 1503. The chronicles of Lofer indicate that the first caving exploration was the work of Ferchel in 1833. A caver from Salzburg named Posselt visited the cave in 1878. Feschl reconnoitered the first 800 m in 1882. Work to make it into a show cave began in 1905. During the following years, cavers such as Bock, Czörnig, Oedl, and others explored the off-trail portions. Even so, only 1200 m of passage was known by 1959.

The first systematic exploration of Lamprechtsofen by the L.H.S. began on February 2, 1964, when the group lowered the level of the Bock sump, which is 600 m from the entrance. This permitted penetration deeper into the mountain and the reconnaissance of 3600 m of passage. By 1965 the vertical extent had reached 292 m (–10, +282) and the length 6500 m. A passage was discovered in 1967 that bypassed the Bock sump.

Thanks to the achievement of the difficult climb up the Superklamm in 1969, the vertical extent made a leap to +511 m, with the length increasing to 9100 m. The cavers of the L.H.S. were stopped by impenetrable breakdown at 740 m above the entrance in 1973.

Even though the height of the cave remained unchanged, the year 1974 brought an increase in the length to 12,203 m, followed by 12,527 m in 1975. Discoveries of new passages permitted exploration up to +810 m in 1976. In early 1977, the Polish cavers from Kraków took over from the Austrians and progressed upwards to +850 m. The same teams progressed even higher in 1978, reaching a breakdown blockage at +952 m that was passed by other passages in 1979 to reach +995 m, giving Lamprechtsofen a final vertical extent of 1005 m and a length of 14,657 m. Only 100 m more in elevation is needed to reach the plateau above!

Map: from Kulbicki, Kleszynski, and Wiśniewski.

Bibliography: Die Höhle, 1964 (2); 1969 (4); 1970 (1).
Klappacher W.), Knapczyk (H.) - Salzburger Höhlenbuch, Salzburg, 1977, Vol. 2: 57-97, article and bibliography.

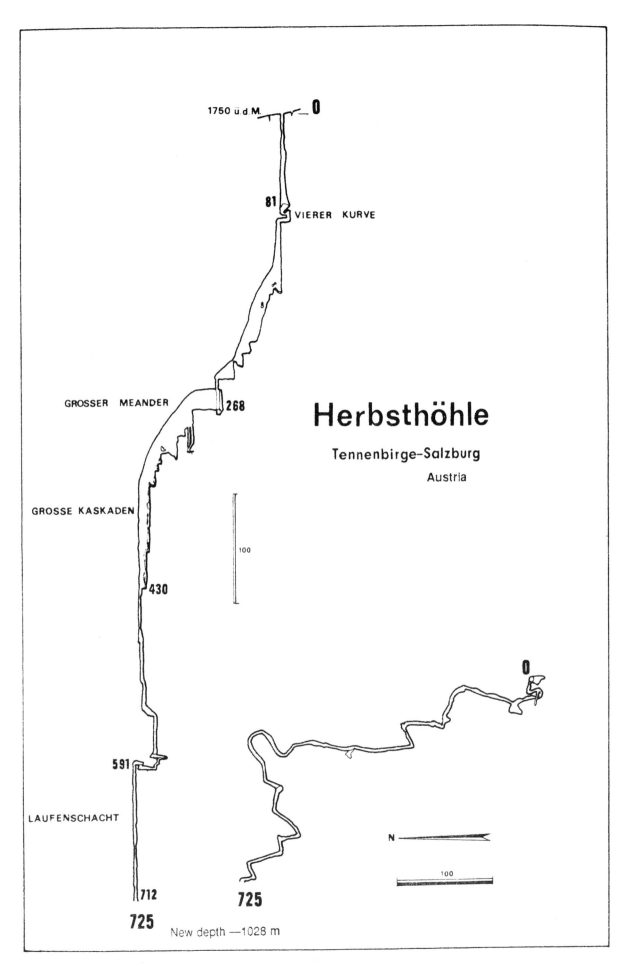

1750 ü.d.M.　　0

81　VIERER KURVE

GROSSER MEANDER　268

Herbsthöhle

Tennenbirge-Salzburg

Austria

GROSSE KASKADEN

100

430

0

591

LAUFENSCHACHT

N

100

712

725

725

New depth —1028 m

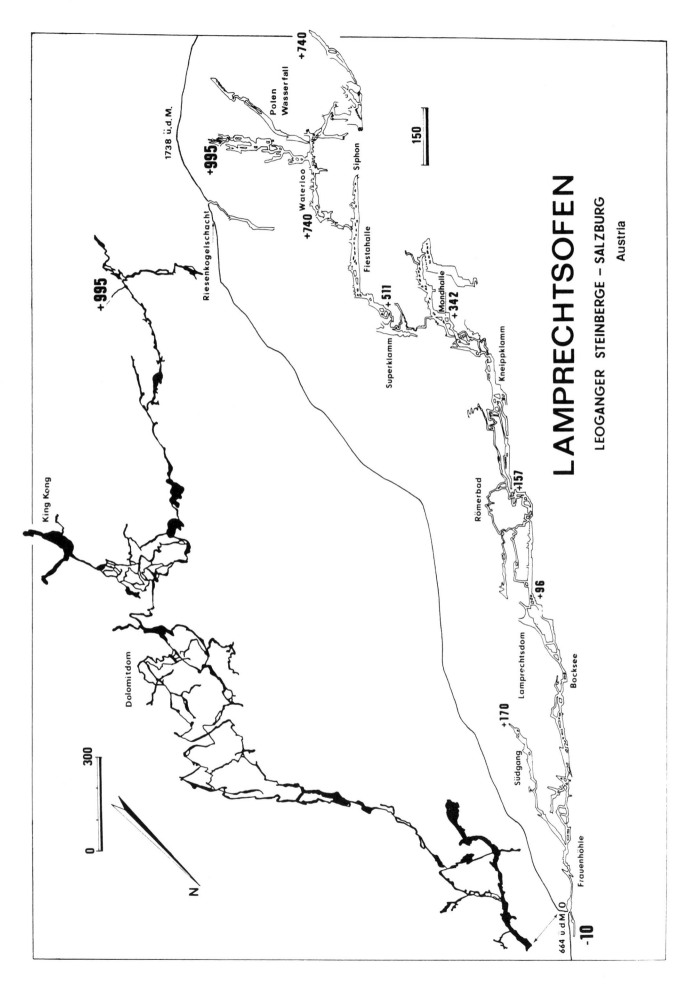

LAMPRECHTSOFEN

LEOGANGER STEINBERGE – SALZBURG
Austria

King Kong

Dolomitdom

Riesenkogelschacht

1738 ü.d.M.

+995

+995

Polen
Wasserfall

+740

+740 Waterloo

Siphon

Fiestahalle

Superklamm

+511

Mondhalle

+342

Kneippklamm

Römerbad

+157

+96

Lamprechtsdom

Bocksee

Südgang

+170

Frauenhöhle

664 ü.d.M.

-10

300

0

N

150

Klappacher (W.) - Le Lamprechtsofen (Autriche), *Spelunca*, 1978 (3): 103-104.

8. *Berger-Platteneck-System* .... −915 m
(Tennengebirge, Salzburg) (1511/164 a-g)

This system is located in the Bergeralpel, at the NW foot of Platteneck (1947 m), to the NW of Tennengebirge. Is is comprised of three sections: Plateneck eishöhle (alt. 1450-1615 m), with four entrance, Bergerhöhle (1511/163) (alt. 1050 and 1030 m), and Bierloch (1511/175) (alt. 1019 m).

Access is made starting from Pass Lueg (573 m), near Golling. One must travel to the north, towards Wirreck (1465 m), in the Bergeralpel, until one reaches an old wooden hut fitted out by cavers (alt. 1100 m). The main entrance (1450 m) is a bit further on, to the SE. 1/50,000 map N° 94 (Hallein).

The system is hydrologically related to the lower caves of Brunneckerhöhle (alt. 525 m), Petrefaktenhöhle at Pass Lueg, Infang-Wasserloch at Alm, and Winnerfallhöhle in the Schwarzbach-Quellegebiet. A connection is also possible with Jungebabaschacht (alt. 1855 m). The whole set of caves is developed in the Dachsteinkalk.

The lower entrance of Platteneckhöhle has long been known to hunters. It was not until July, 1966 that Salzburg cavers began exploration. The cavers discovered 1800 m of passages with a vertical extent of 110 m (−30, +80). Shortly thereafter, a connection was made with Eishöhle, bringing the vertical extent to about 180 m. Bergerhöhle was explored in parallel to a length of 2100 m.

In 1968 the depth reached 338 m and the length reached 4600 m. As a result of the Bergeralpel Austrian expedition, the Bergerhöhle reached a length of 7348 m.

Connection with Spitzbogenhöhle (1511/203) in 1970 pushed the length to 6200 m. In 1973 the 200 m deep Polish Pit was found and dropped, while 9700 m of passage was explored in Bergerhöhle.

1975-1976 were important years. First of all, cavers from Warsaw (Warszawa A.K.S.) reached −750 m in the Platteneck by following a narrow series of canyons, the Lehmberghalle (length 7500 m). The same year (1975), Bergerhöhle reached a length of 11,500 m.

The Platteneck-Bergerhöhle connection was made in September, 1976 by the W.A.K.S, making a system 21,000 m long and with a vertical extent of 879 m (−861, +18). A brief continuation to a sump at −877 m was found in 1977 by Belgians (the "Gours") and Polish (W.A.K.S.). Today the system has a vertical extent of 915 m, with 25,315 m of passage.

Map: from Kulbicki, Domoslavska, Klappacher, and Ausobsky, furnished by G. Stummer.

Bibliography: *Salzburger Höhlenbuch*, 1985, Vol. 4, p. 323.

9. *Feuertalsystem* .... −913 m
(Totes Gebirge, Oberösterreich) (1626/120)

Feuertalsystem has three entrances: Kacherlschacht (1940 m), "Vélo-Tracteur" (1774 m) and "Carcajau" (1736 m). It is located in the Schönberg range (Totes Gebirge), a one hour walk from Hochkogelhütte. Access is via Ebensee and Schwarzenbachstube to the hut at 1558 m. The cave is on the southern flank of Feuertal. 1/50,000 map N° 96 (Bad Ischl). The system seems to be related to Ahnenschacht and Trunkenboldschact caves. The resurgences

seem to be located to the north, at Gimbach and Schwarzenbachtal, at an altitude of 600 - 700 m.

The upper entrance (Kacherlschacht) was discovered in 1973 by the G. S. Alpin Belge. It was then rediscovered in July 1976 by a French group that rapidly reached about −700 m, having mapped down to −540 m. In 1977, the same group discovered the two lower entrances, connected with Kacherlschacht cave, and reached the depth of −913 m. The total length of the system is 12,014 m.

Map: from the notes of the explorers, furnished by D. Martinez.

Bibliography: XXX - Les karsts autrichiens à la façon provençale, *Spelunca*, 1976 (4): 159-162.

10. *Schnellzughöhle* .... − 898 m
(Totes Gebirge, Steiermark) (1623/115)

This system is located SW of Vorderer Schwarzmooskogel (1843 m), near to Loser, between 1467 and 1624 m altitude. Access is from Loserhütte.

1/50,000 map N° 96 (Bad Ischl).

It is developed in the Dachsteinkalk. Its low point is only 20 m above the level of the Altausseersee.

Schnellzughöhle has five entrances. The first exploration into one of them, Stellerweghöhle (1623/41) goes back to 1940 when German cavers descended to about −220 m.

Schnellzughöhle was explored beginning in 1980 by cavers from the University of Bristol Speleological Society and the Cambridge University Caving Club. They first reached −360 m, then −680 m in 1981 by connecting with Stellerweghöhle. They first reached −898 m in 1982, then made a voice connection with Schwa-schact (1623/143 and 144) to give a vertical extent of 972 m.

Map: from the Cambridge University Caving Club, 1982, furnished by G. Stummer.

Bibliography:.Proc. U.B.S.S., 1981, 16 (1): *Caves & Caving*, 1982 (15): *Cambridge Underground*, 1983.

11. *Hochlecken-Großhöhle* ....(−788, +102) 890 m
(Höllengebirge, Oberösterreich) (1567/29)

This cave is located NE of Hochleckenkogel (1691 m), above Aurachkar, at an altitude of 1520 m. Access is through the Grossalmstraße (Taferlklause), between Traun and Attersee, from which one climbs to the Hochlecken hut (1573 m). 1/25,000 map 66/3 Grosser Höllkogel sheet.

It is surmised that the resurgence for the cave is either below Traunsee or Attersee (alt. 440 m).

Hochlecken-Großhöhle has been known since at least 1923. Exploration began in 1963 when mountaineers from Vöcklabruck discovered 1700 m of basically horizontal passages. The L.H.S. found a passage not far from the entrance that led them to the top of the Stierwascher pit in November, 1972 (see section on "Great Pits").

This pit was the subject of several assaults by the L.H.S., who pushed the vertical extent to 552 m (−450, +102) in 1973, then 596 m (−494, +102) in 1974, with the length increasing to 3384 m. In August, 1975, French teams from Toulon, Cavaillon, and Vèdene renewed the attack on the pit which is wet below −200 m and were stopped at −554 m for lack of equipment. In September the French teams returned and got down to −701 m, at the top of a new wet pitch.

The French bottomed a series of wet drops to reach −753 m in February, 1977, increasing the length to 4300

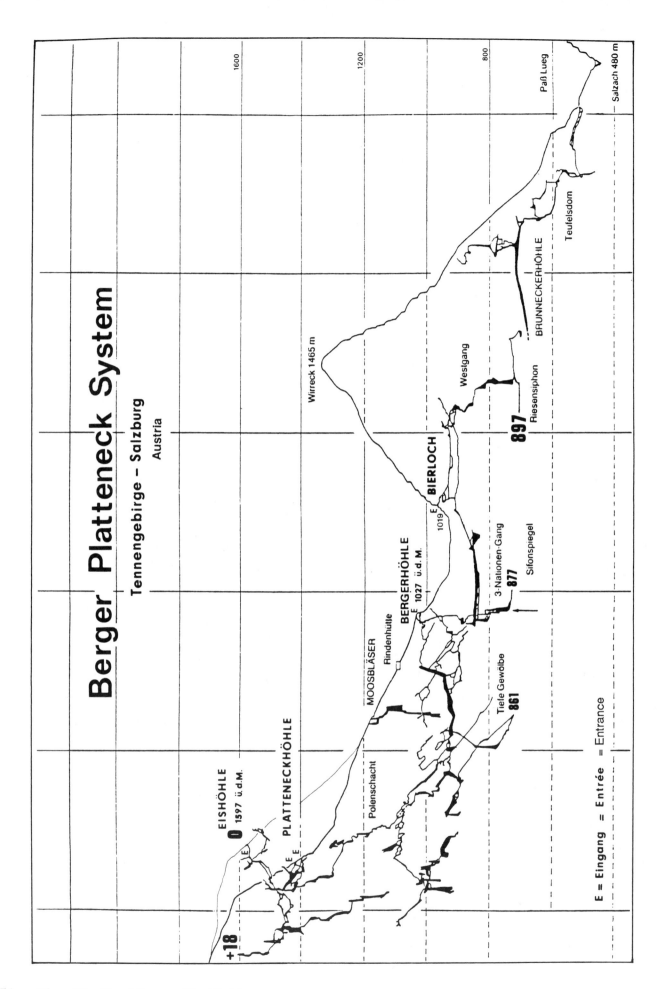

Berger Platteneck System

Tennengebirge – Salzburg
Austria

EISHÖHLE 1597 ü.d.M.

PLATTENECKHÖHLE

+18

MOOSBLÄSER

Rindenhütte

Polenschacht

BERGERHÖHLE 1027 ü.d.M.

BIERLOCH 1019

Wirreck 1465 m

3-Nationen-Gang 877

Sifonspiegel

Tiefe Gewölbe 861

Westgang

897

Riesensiphon

BRUNNECKERHÖHLE

Teufelsdom

Paß Lueg

Salzach 480 m

E = Eingang = Entrée = Entrance

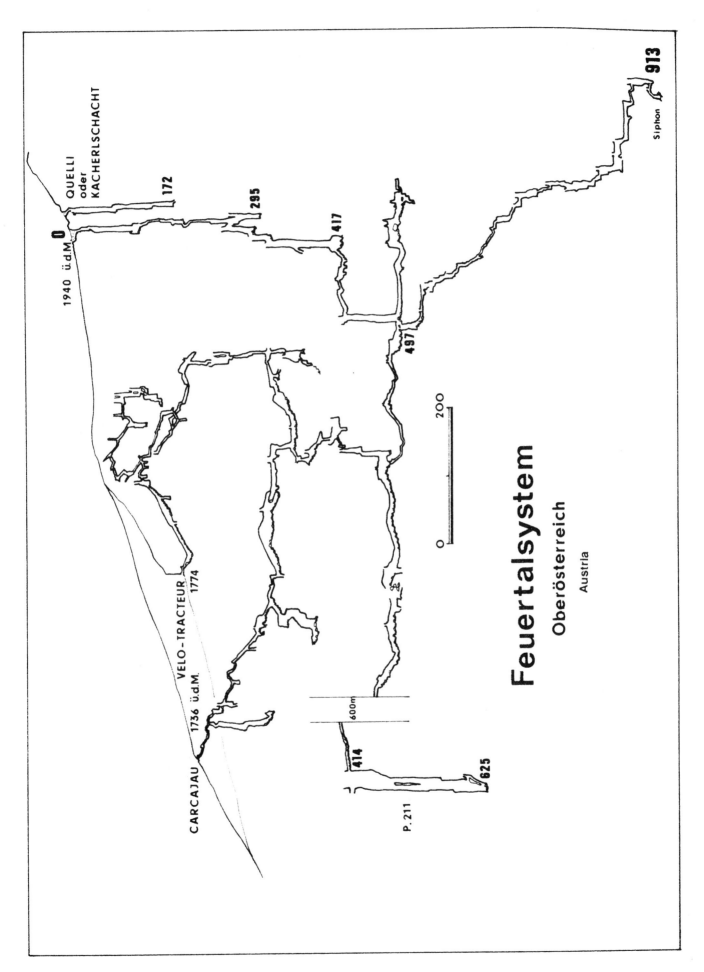

Feuertalsystem

Oberösterreich

Austria

913

Siphon

QUELLI oder KACHERLSCHACHT

1940 ü.d.M. 0

172

295

417

497

CARCAJAU 1736 ü.d.M.

VELO-TRACTEUR 1774 ü.d.M.

P.211

414

625

600m

200

0

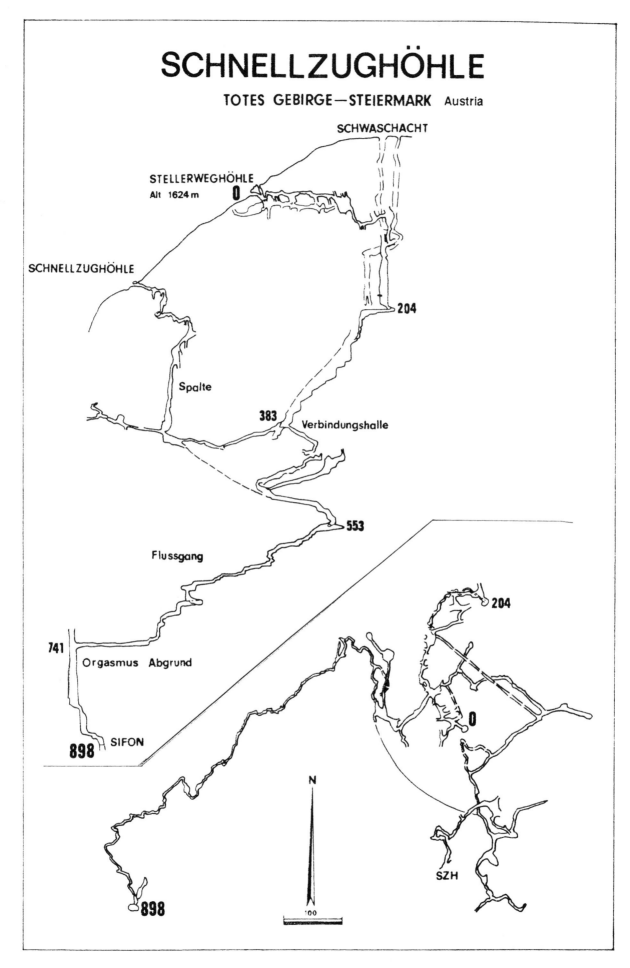

SCHNELLZUGHÖHLE

TOTES GEBIRGE—STEIERMARK   Austria

SCHWASCHACHT

STELLERWEGHÖHLE
Alt 1624 m   **0**

SCHNELLZUGHÖHLE

204

Spalte

383   Verbindungshalle

553

Flussgang

741
Orgasmus Abgrund

898 SIFON

898

204

0

SZH

N

¹⁰⁰

Hochlecken Grosshöhle

Oberösterreich   Austria

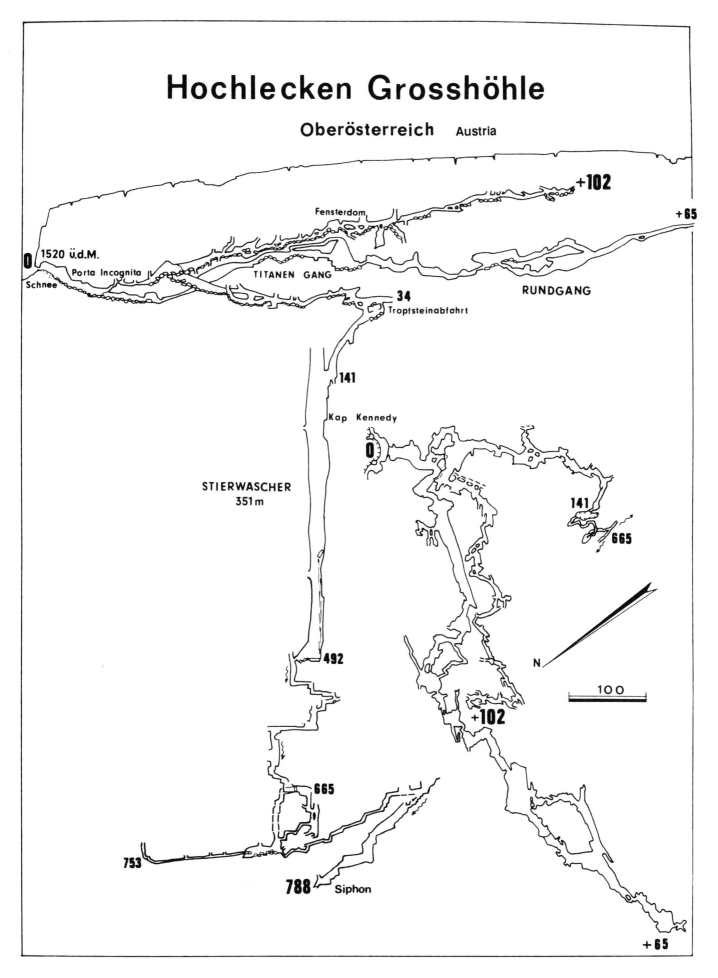

+102

+65

Fensterdom

0 1520 ü.d.M.

Porta Incognita

Schnee

TITANEN GANG

RUNDGANG

34

Tropfsteinabfahrt

141

Kap Kennedy

0

141

665

STIERWASCHER
351 m

N

492

100

+102

665

753

788 Siphon

+65

m. At the bottom they found an ascending passage, which they followed in 1979 to a stream which led to a sump at –788 m. The length then reached 5460 m.

Map: compendium of notes of L. H. Oberösterreich and the Provençaux of Vèdene, Toulon, and Cavaillon by P. Courbon.

Bibliography: Martinez (D.) - Hivernale au Hochlecken Grosshöhle, *Spelunca*, 1977 (4): 168-169.

12. Wildbaderhöhle .... –874 m
(Totes Gebirge, Steiermark) (1625/150)

The entrance to this cave is at an altitude of about 2100 m, in the SW Totes Gebirge. It is less than one km SE of Planka Mira (2188 m). 1/50,000 map N° 97 (Bad Mitterndorf). Dachsteinkalk.

The Wildbaderhöhle was discovered in July, 1977 by the L. H. in Steiermark, who explored it down to –130 m, stopping at the top of a 160 m deep shaft. They invited French cavers from the G. S. Doubs, G. S. Clerval, and C.A.F. du Haut-Doubs to pursue the exploration. Still in 1977, these cavers descended the 160 m pit and those following to be stopped by a squeeze at –470 m. They also explored a parallel branch to –360 m.

Three years later (in September, 1981), the tight spot at –470 m was forced by the S.A.C., who then descended to –644 m by a series of wet pits. In September 1982, a drier season permitted the S.A.C. to reach a fossil river passage, unfortunately blocked at –874 m. The length of the cave is 1644 m.

Map: from the notes of the G.S.D. and the S.A.C., in *Spelunca*, 1983 (11).

Bibliography: XXX - Planka Mira 1977. Expédition internationale en Autriche, *Bull. Assoc. Spél. Est*, 1978 (15): 9-24.

13. Trunkenboldschacht ....(–854, +5) 859 m
(Totes Gebirge, Oberösterreich) (1626/117)

This cave, located in the Totes Gebirge, is not far from Feuertalsystem and the Ahnenschacht. It is at an altitude of 1610 m on the crest of the cliffs that dominate the west side of Hintergrass. 1/50,000 map N° 96 (Bad Ischl).

The cave was discovered in July, 1976 by Fred Vergier. A French expedition, comprised of cavers from Toulon (A. C. Toulon, S. C. Toulon), Cavaillon (Darboun), and Vèdene (Spéléo Regaïe) explored the cave to an approximate depth of –490 m from July 12 to 23, 1976. A flood interrupted exploration by the original route, but another way discovered at the – 300 m level led to a large parallel pit 242 m deep. After many tries, this pit was conquered, but a second flood forced them to give up at the –590 m level at the top of a 40 m pit. The same teams returned in 1977 and continued the descent to sumps at the –854 m level.

Map: from the notes of D. Martinez and the exploration summary of P. Courbon.

Bibliography: XXX - Les karsts autrichiens à la façon provençale, *Spelunca*, 1976 (4): 159-162.

14. Gruberhornhöhle ....(–784, +70) 854 m
(Hoher Göll, Salzburg) (1336/29)

This two entrance cave (alt. 1870 and 1840 m) is located on the south flank of Gruberhorn (2230 m), west of Golling. 1/50,000 map N° 94 (Hallein).

The cave is part of the Schwarzbach hydrological system (alt. 580 m), resurging to the NE with a maximum flow of 20 cubic meters per second. It is formed in the Dachstein-Riffkalk limestones.

Even though the entrance had been spotted earlier, it was caver Walter Klappacher who discovered the cave on May 22, 1960. The L.H.S. began exploration, with 3500 m being mapped in 1962. As expeditions followed one another, the cave was discovered to be vertical as well as horizontal. The depth of –480 m was reached in 1963, for a length of 4200 m. In 1964, the L.H.S. reached –640 m, and –682 m in 1965, which gave a total vertical extent of 710 m (–682, +28). No depth was gained in 1966, but the length increased to 6200 m. In 1970, with the help of a Polish team, the L.H.S. got down to –784 m, for a vertical extent of 854 m (–784, +70) and a length of 6700 m.

Map: from the notes of the L.H.S. and *Speleologia* (Varsovie), 1971, VI (1-2).

Bibliography: Klappacher (W.), Knapczyk (H.) - *Salzburger Höhlenbuch*, 1979, 3, pp. 388-404.

15. Gipfelloch .... –840 m
(Tennengebirge, Salzburg) (1511/355)

Gipfelloch, in the Tennengebirge, is located on the SE flank of Wieselstein (2300 m). It is reached from the Leopold Happish-Haus hut. From there, a maintained trail climbs Wieselstein: one must leave it before the last slope and go a hundred meters to the SW, in the lapiaz, to discover, in a little depression, the entrance at an altitude of 2194 m. 1/50,000 map N° 94 (Hallein). There are two likely resurgences possible for the cave: Brunneckerhöhle (alt. 500 m), which is 4300 m away, and Winnerfallhöhle (alt. 700 m), which is 5 km away.

The opening was discovered on August 13, 1982, while the Foyer des Jeunes de Seyssins (Isère, France) were prospecting in the area, but the cave was not explored until the following year when the French cavers reached a depth of about –360 m (the mapping stopped at –293 m), leaving a large number of undescended pits, of which six were at a depth greater than –300 m.

The Gipfelloch was the main objective of the 1983 expedition by the Foyer des Jeunes de Seyssins: a camp was organized from July 28 to August 24 during which a sump was discovered on August 20 at –840 m by exploring the "meandre des Cantonniers".

The large number of undropped pits remaining give hope that at least the length, presently around 3 km, has good potential for increasing. In addition, the strong air currents suggest a connection with a higher entrance.

Map: from *Salzburger Höhlenbuch*, 1985, Vol. 4.

Bibliography: Parein (R.) - Le Gipfel-Loch, *Scialet*, 1984 (13): 121-127.

16. Burgunderschacht .... –827 m
(Totes Gebirge, Oberösterreich) (1625/20)

This cave is located on the SE flank of the Totes Gebirge range, in the immense lapiaz area of the Tauplitz zone. Its entrance is at an altitude of 1850 m, not far from Jungbauerkreuz. 1/50,000 map N° 97 (Bad Mitterndorf).

According to a water trace done in the 1950's using pollen, it seems that the Sagtümpfel spring (alt. 980 m) is the resurgence for Burgunderschacht. In 1951 an Austrian expedition directed by Hubert Trimmel discovered

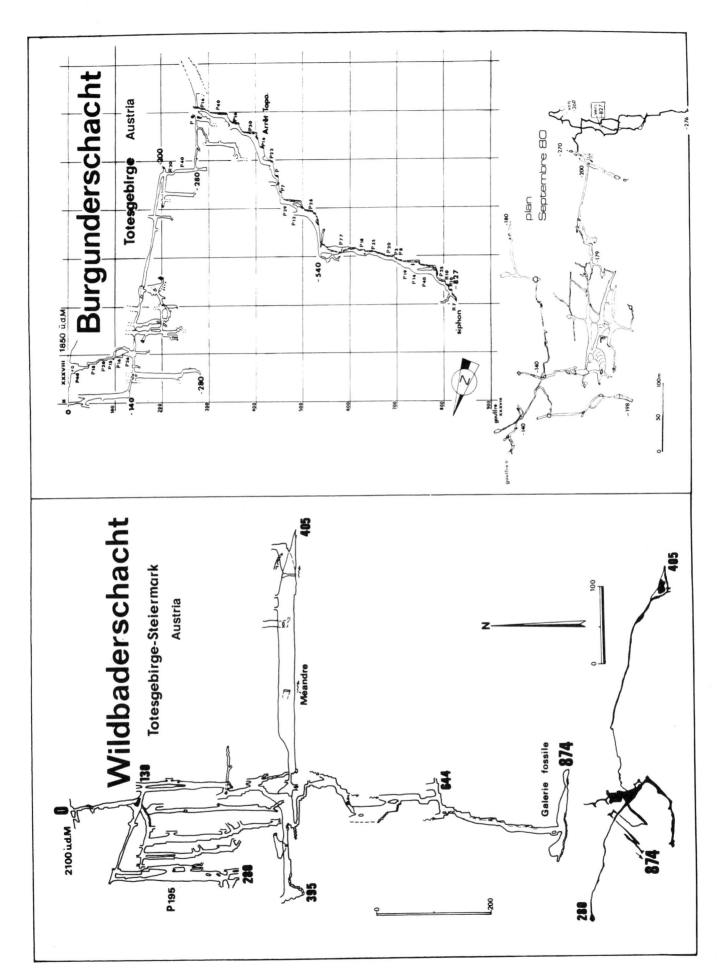

Burgunderschacht

Totesgebirge ● Austria

XXXVIII 1850 ü.d.M

plan
Septembre 80

Wildbaderschacht

**Totesgebirge-Steiermark
Austria**

2100 ü.d.M

Méandre

Galerie fossile

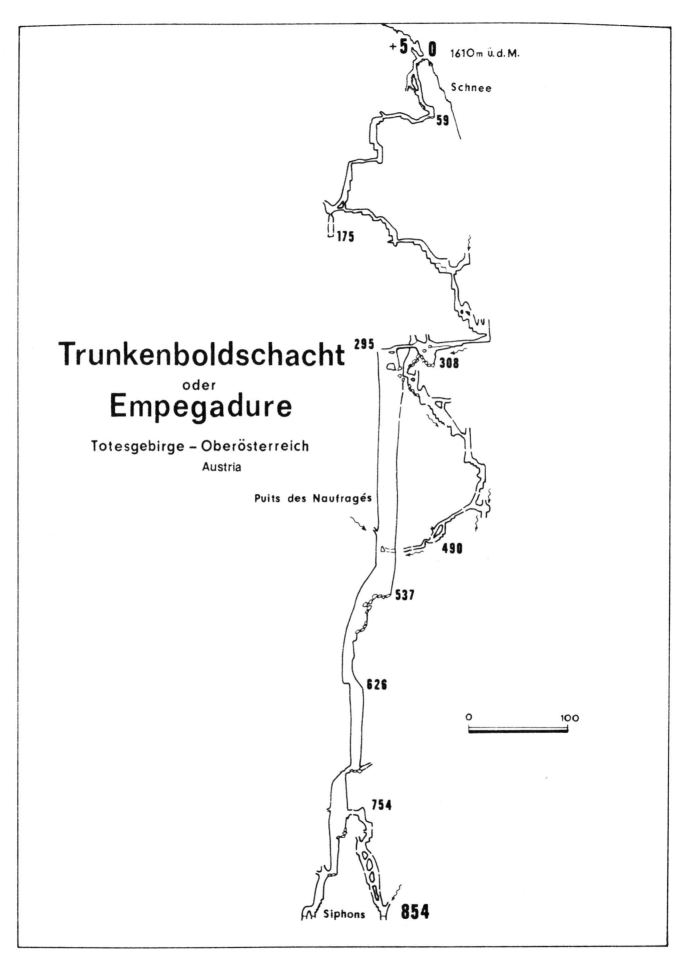

Trunkenboldschacht
oder
Empegadure

Totesgebirge – Oberösterreich
Austria

+5 0 1610m ü.d.M.

Schnee

59

175

295

308

VU

Puits des Naufragés

490

537

626

754

Siphons **854**

0   100

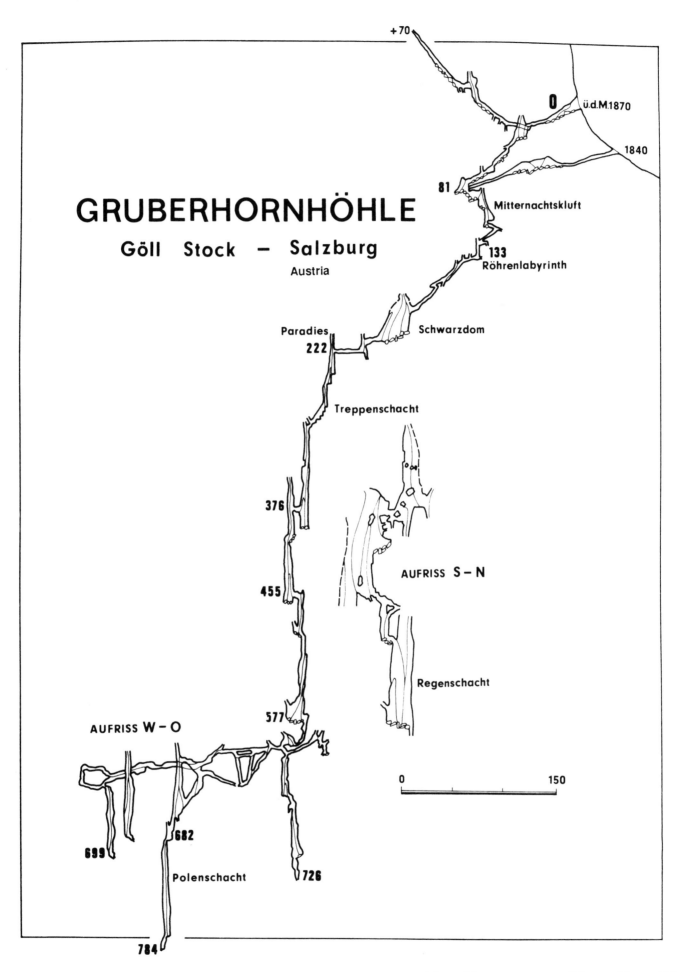

GRUBERHORNHÖHLE

Göll Stock – Salzburg
Austria

+70

0

ü.d.M.1870

1840

81

Mitternachtskluft

133

Röhrenlabyrinth

Paradies

222

Schwarzdom

Treppenschacht

376

455

AUFRISS S–N

577

Regenschacht

AUFRISS W–O

682

699

Polenschacht

726

784

0   150

GIPFELLOCH

Tennengebirge — Salzburg   Austria

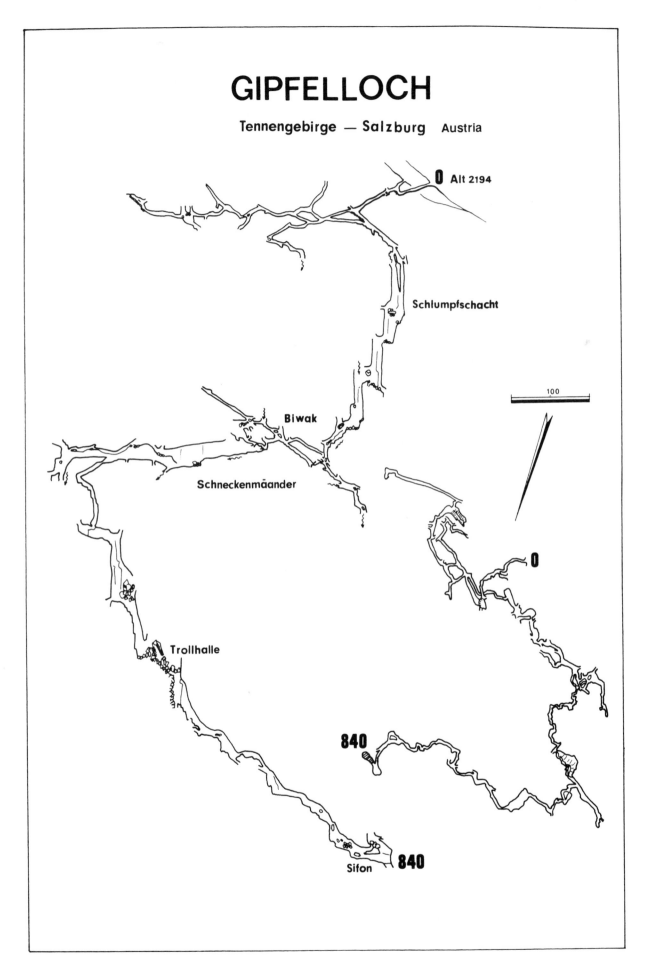

Alt 2194

Schlumpfschacht

100

Biwak

Schneckenmäander

0

Trollhalle

840

840

Sifon

the cave (which they numbered XXXVIII) and descended to –140 m. A quarter of a century passed before exploration was resumed. In August, 1975 an expedition bringing together cavers from Austria (Club de Bad Mitterndorf, led by Gunter Graf) and France (from G.S. Doubs, Clerval-Baume, S.C.V., and S.S.C.) became interested in the cave and discovered horizontal passages at the –140 m level which gradually dropped to –200 m over a length of 2500 m. In August, 1976, a second Franco-Austrian expedition continued exploration and reached –280 m. The French alone (G. S. Doubs, G.S. Clerval-Baume, Société des Amateurs de Cavernes de Rioz) reached –620 m, –373 m, and –310 m in different branches of the cave on an expedition from September 1 to 10.

It was only in 1980 that these same teams reached the terminal sump at –827 m, bringing the length to 3250 m.

Map: from the G.S.D. and the S.A.C., *Spelunca*, 1983 (9).

Bibliography: Perrin (D.) *et al.*, Le Burgunderschacht, massif des Totesgebirge, Autriche, *Spelunca*, 1983 (9): 22-24.

17. "Cabri-höhle" (Tennengebirge, Salzburg) .... –800 m

Located 200 m above Schwersystem (see this cave), it was discovered in 1984 by the C.A.F. Marseille, who descended to –345 m. Bad weather interrupted the 1985 explorations at –800 m.

18. Wieserloch .... –730 m
(Leoganger Steinberge, Salzburg) (1324/16)

This cave is connected hydrologically with Lamprechtsofen, with hopes being alive for a connection. It is located at an altitude of 2050 m, in the Leoganger Steinberge range, south of Weissbach. It is found on the slopes of Nebelsbergkar, near the east face of Rothorn (2405 m). 1/50,000 map N° 92. 16 is formed in the Dachsteinkalk dolomites and its resurgence is at an altitude of 600 m.

The entrance was discovered in the 1960's by a shepherd looking for a lost sheep. In 1975, he helped the L.H.S. to find the entrance, which permitted the Austrians to descend to –210 m the same year.

Reinforced by Polish cavers from Kraków, the L.H.S. reached an estimated –400 m in 1976, in spite of bad weather, wet pits, and waterfalls enlarged by the heavy rains. The Polish cavers set up a camp at –400 m in 1977 and were stopped at –580 m by a tight fissure. Thanks to the discovery of a parallel pit at – 520 m, they were able to reach –730 m in 1979, finding an underground stream and almost 3000 m of passages, but were unsuccessful in connecting with nearby Lamprechtsofen.

Map: adapted from that of J. Rozen, origin unknown.

Bibliography: Kleszynski (C.) - Caving in Austria. The search for Lamprechtsofen's upper entrance, *Caving International Magazine*, 1981 (12): 6-10.

19. Salzburger Vogelschacht .... –726 m
(Leoganger Steinberge, Salzburg) (1324/47)

Like Wieserloch, the Salzburger Vogelschacht is hydrologically connected with Lamprechtsofen. It is located in the Nebelsbergkar lapiaz, at an altitude of 2200 m. 47°28'45" N, 12°43'44" E. 1/50,000 map N° 92.

A stream with a maximum flow of 4 liters per second appears at – 53 m. Swollen by successive infeeders, it eventually reaches 30 liters/sec maximum. Its resurgence is at an altitude of 600 m (see Lamprechtsofen and Wieserloch).

The chronology of exploration is very brief: it was explored in a single season (August, 1984) by the Groupe Spéléologique des Vulcains (Lyon, France) who reached –726 m without interruption.

A cold breeze (3° C) blows through the cave. The huge boulders that one encounters (notably at the bottom of the 137 m pit which is remarkable for its volume) indicate a strong tectonic influence in its formation.

Map: from the G. S. Vulcains, issue of the *Echo des Vulcains*, 1984 (44).

Bibliography: Ohl (C.) - Le Vogelschacht, *Echo de Vulcains*, Lyon, 1984 (44): 3-5.

20. Raucherkarhöhle ....(–718, +7) 725 m
(Totes Gebirge, Oberösterreich) (1626/55)

This complex system (it has 17 entrances, the highest being at an altitude of 1630 m) is located in the western Totes Gebirge, near the edge of Oberösterreich. Access is via a 30 minute walk from the Ischl hut. One must start at Bad-Ischl, passing through the Rettenbach valley. 1/50,000 map N° 96 (Bad Ischl). 13°46'25" E; 47°42'10"N.

The direction of the underground drainage seems to be to the south, in the direction of the high valley of Rettenbach. The probable resurgence is the Nagelsteghöhle (alt.. 850 m).

The exploration of the cave began when the son of the guardian of the Ischl hut found a skull from a brown bear. The reconnaissance trip of the L. H. Linz in 1961 suggested the existence of a great system, and in 1964 the L.H.L discovered 4100 m of passage. In 1965 they brought the length to 6060 m and the depth to – 246 m.

An Austrian national expedition mounted the attack on Raucherkarhöhle in 1966. They reached a sump in the "Hinterland" at a depth of – 530 m, bringing the length to 10,374 m.

Following this the L.H.L. pursued regular explorations: in 1970 then length reached 17,728 m with the discovery of the "Dunklen Grund", but a sump at – 718 m stopped downward exploration.

On the other hand the length continued to grow: it reached 18,160 m in 1974. It reached 19,053 m in 1975 with the connection to Planer-Eishöhle, followed by 20,155 m in 1977 and 24,300 m in 1979.

The L.H.L. came close to 30 km in 1980: 29,214 m. A jump was made in 1981: Raucherkarhöhle reached 35,035 m, then 38,085 m in 1983. The 40 km mark (40,267 m) was passed in 1984. In 1985, Raucherkarhöhle was the longest cave in Austria with 44,111 m.

Map: from the notes of the L.H.Oberösterreich, furnished by E. Fritsch.

Bibliography: Fritsch (E.) in *Die Höhle*, 1966 (2); Jahrbuch des Österreichischen Alpenvereins, Innsbruck, 1967.
Trimmel (H.) - Österreich längste und tiefste Höhlen, *Die Höhle*, suppl. N° 14, 1966.

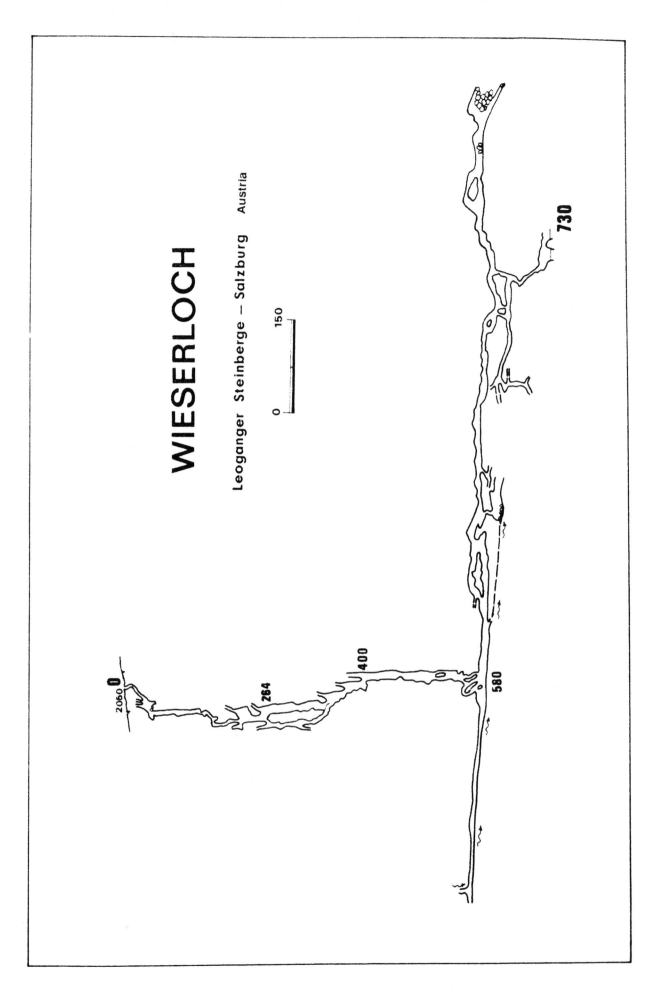

WIESERLOCH

Leoganger Steinberge — Salzburg   Austria

Salzburger Vogelschacht

Leoganger Steinberge   Austria

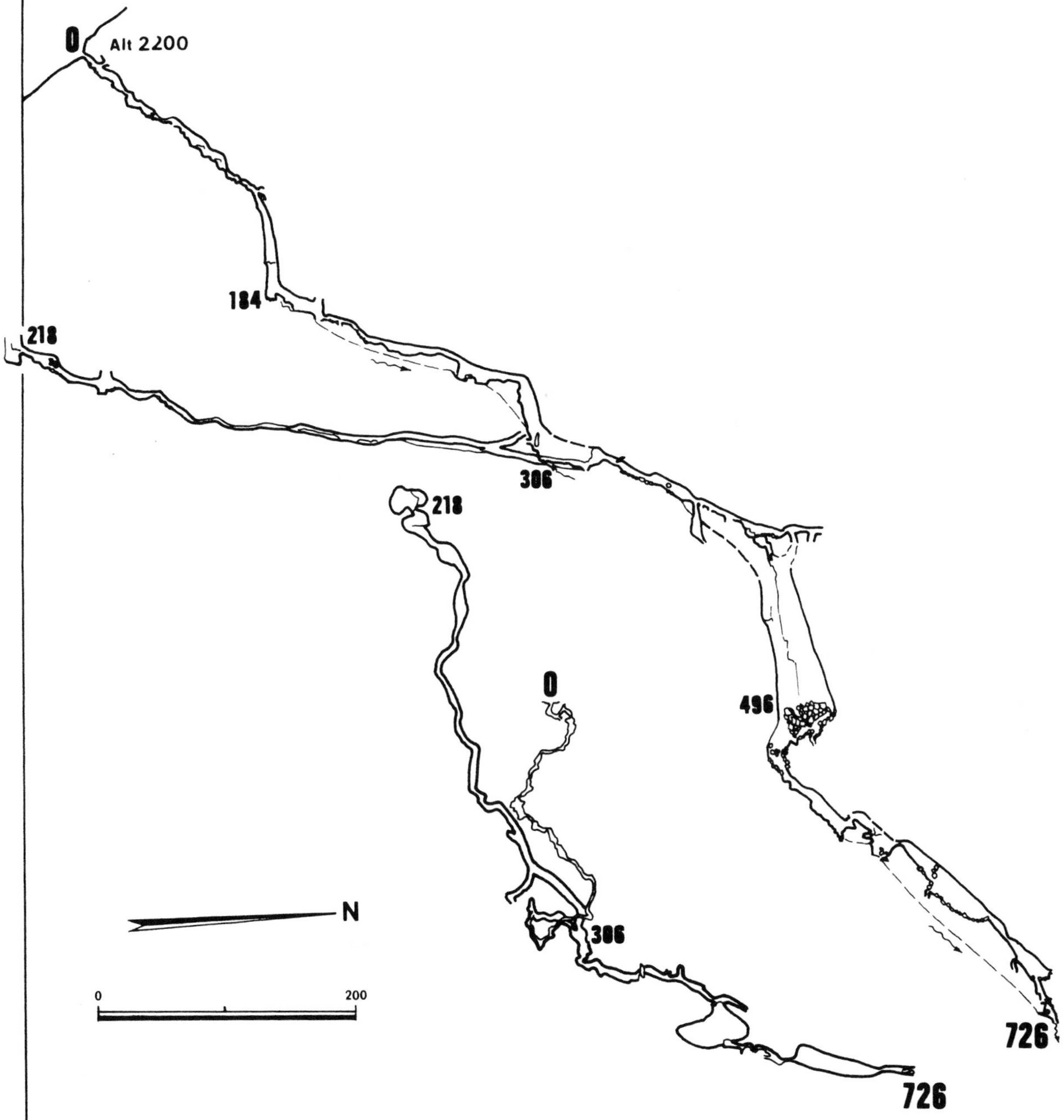

0 Alt 2.200

184

218

386

218

0

496

N

0   200

386

726

726

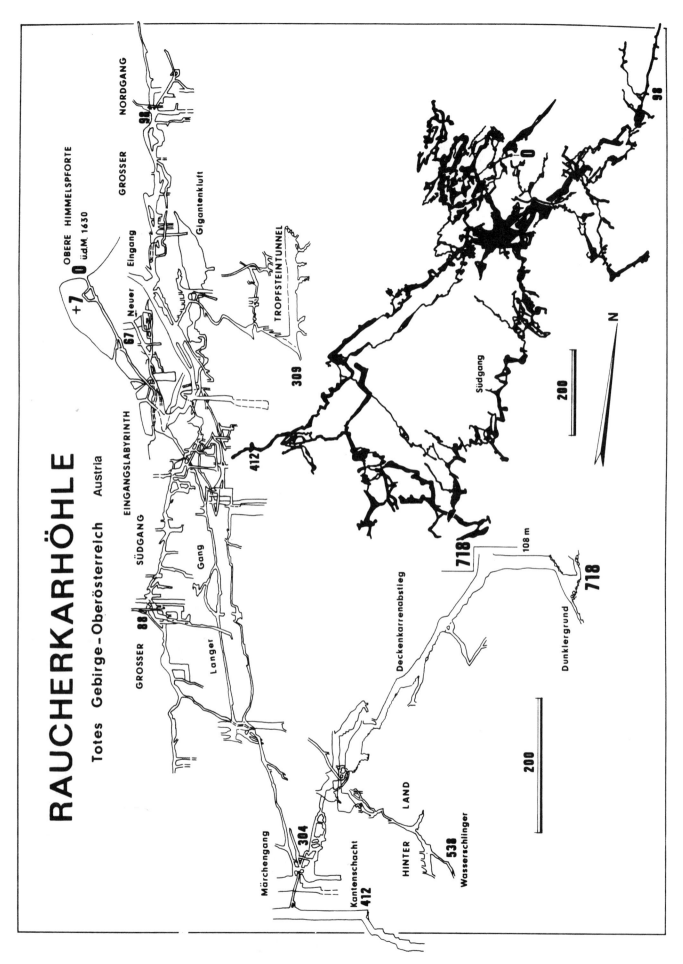

RAUCHERKARHÖHLE

Totes Gebirge - Oberösterreich   Austria

OBERE HIMMELSPFORTE
üdM. 1650

GROSSER   NORDGANG

98

+7
0

GROSSER

67   Neuer   Eingang

Gigantenkluft

309

TROPFSTEINTUNNEL

412

718

108 m

718

Dunklergrund

Südgang

N

200

EINGANGSLABYRINTH

SÜDGANG

Gang

Langer

GROSSER   88

Deckenkarrenabstieg

200

Märchengang

304

Kantenschacht
412

HINTER   LAND

538

Wasserschlinger

21. *schacht der Verlorenen* .... –724 m
(Tennengebirge, Salzburg) (1511/275)

The "Perdus Cave" is at an altitude of 2030 m in the lapiaz that covers the west flank of Wieselstein, in the Tennengebirge range (see also Gipfelloch). 1/50,000 map N° 94 (Hallein). This cave could be connected either with the Eisriesenwelt, via the Bretterschacht to the NE, or with Frauenofen (alt. 1635 m), although the latter is less likely.

The cave was discovered in September, 1980, during the last days of the camp by the Groupe Spéléologique de Seyssins (Isère, France). It was descended in two rapid incursions to a depth of about –250 m. In spite of bad weather, the G. S. de Seyssins reached –527 in pushes from August 17 to 30, 1981, but had to stop at the top of a 10 m waterfall for lack of rope. They returned from July 26 to August 18, 1982 to descend a series of pits and waterfalls and discovered that the water disappears into a gravel bed at –724 m, after passing through several narrow or almost sumping passages. About 3.7 km were mapped. An expedition the following year saw no increase in depth, but the length was increased to 4050 m.

Map: from *Atlantis*, 1983 (1).

Bibliography: Parein (R.) - Schacht de Verlorenen, *Scialet*, Isère, 1981 (10): 126-129, *ibid.*, 1982 (11): 120-130.

22. *Bretterschacht* .... –715 m
(Tennengebirge, Salzburg) (1511/141)

This cave is at an altitude of 1932 m in the west side of the Tennengebirge range, near to Happischhaus, and east of Windischkogel (2252 m). 1/50,000 map N° 94 (Hallein). Developed in the Dachsteinkalk. It reaches base level in the karst.

The first exploration was in 1920, when the L.H.S. reached — 56 m. They reached –150 m in 1951. Exploration in 1979 by the G. S. Alpin Belge reached a depth of –487 m and a length of 3156 m. In 1980 they reached –715 m for a length of 4510 m.

Map and Bibliography: *Salzburger Höhlenbuch*, Salzburg, 1985, Vol. 4.

23. Bärengasse-Windschacht .... –687 m
(Dachstein, Oberösterreich) (1543/69)

At an altitude of 1823 in the Dachstein range, near Wiesberghaus (1/50,00° map N° 96, Bad Ischl), the Bärengasse-Windschacht was discovered in 1978 by the Bristol Explorers Club, who reached –140 m in 1979 and –400 m in 1980. With the help of the N.C.C., they reached –490 m in 1983. The following year the B.E.C. got down to –622 m, finally reaching –687 m in 1985 (*Caves and Caving*, 1984 (24); 1985 (28)).

24. Hirlatzhöhle ....(–148, +464) 612 m
(Dachstein, Oberösterreich) (1546/7)
See below.

25. Ahnenschacht .... –607 m
(Totes Gebirge, Oberösterreich) (1626/50)

The "Cave of the Ancients" is located in the Schönberg range, not far from Salzburg (13°47'50"E, 47°43'N, alt. 1890 m. 1/50,000 map N° 96 (Bad Ischl)). Access is from Schwarzenbachstube, passing by the Hochkogel hut (see Feuertalsystem, which appears to be hydrologically connected).

The entrance was discovered in 1956 by F. Schernhuber and O. Kerschbaummayr. The L.H. Oberösterreich reached –336 m in 1958. The Bristol Explorers Club returned in 1967 to reach –395 m. The G. S. Alpin Belge "Les Gours" discovered over 5 km of galleries in 1972 and 1973, reaching a depth of –607 m (figure revised in 1977). (Courbon, *Atlas des Grands Gouffres du Monde*, 1979, pp 160, 171, profile).

26. Salzburgerschacht .... –606 m
(Untersberg, Salzburg) (1339/69)

This cave in on the Untersberg, north of Berchtesgaden, at an altitude of 1790 m. (1/50,000 map N° 93, Bad Reichenhall). First discovered in 1923. Gustav Abel explored it to –170 m in 1935. The G. S. Alpin Belge returned in February 1977 to drop to –430 m. The Austrians reached –550 m in March, then the Belgians reached –580 m in May of the same year. The two teams were then stopped at a sump at –606 m (*Salzburg Höhlenbuch*, 1975, 1; Courbon, *op. cit.*, 1979, p 160). The two teams pushed the length to 5500 m in 1979, then 6070 m in the following years. The cave has a second, lower, entrance (the Brunntalschacht entrance) 500 m to the north.

27. Kuchelberg Windhöhle .... –605 m
(Tennengebirge, Salzburg) (1511/207)

This cave formed in the Dachsteinkalk is at an altitude of 1610 m, and is located near Kuchelbergalm-Röth (1/50,000 map N° 94, Hallein). The entrance was discovered in 1966 by the L. H. Salzburg who explored it in 1976-1977. The G. S. Alpin Belge reached the bottom at –605 m in 1981 (*Salzburger Höhlenbuch*, Salzburg, 1985, Vol. 4, pp 386-388).

28. Frauenmauer-Langstein-Höhlensystem ... –595 m
(Hochschwab, Steiermark) (1742/1a-g)

This system has seven entrances: Langsteinschacht (alt. 1623 m), an entrance east of Frauenmauerhöhle (1589 m), Bärenlocheishöhle, (1520 m), an entrance west of Frauenmauerhöhle (1467 m), Langsteintropfsteinhöhle, Alter Eingang (1581 m), Bänenloch, Oberer Eingang (1540 m), and Schneeschacht (1501 m), which is accessible from Eisenerz-Trofeng by a marked trail above Gsoll-Alm (1/50,000 map N° 101, Eisenerz). The natural tunnel that connects the eastern and western entrances has long been known. The mapping began in 1928 (Bock, Ausobsky, and Gangl) for a length of 3 km. The Langstein-Tropfsteinhöhle, which was also long known, was connected with Frauenmauerhöhle in 1961 by the L. H. Steiermark. The system was deepened in 1962, and the low point (initially measured at –610 m but later corrected to –536) was reached in 1963. The vertical extent remained more or less unchanged, but over the years the length steadily grew, to 13,255 m in 1973, 16,280 m in 1977, and 20,215 m in 1984 (Courbon, *Atlas des Grands Gouffres du Monde*, 1979, pp 161, 173, profile to –536 m).

29. Knallsteinplattenschacht .... –584 m
(Tennengebirge, Salzburg) (1511/276)

This cave, at an altitude of 1885 m, is located east of Knallstein, north of the Tennengebirge (1/50,000 map N° 94, Hallein). In the Dachsteinkalk. It was explored in 1980

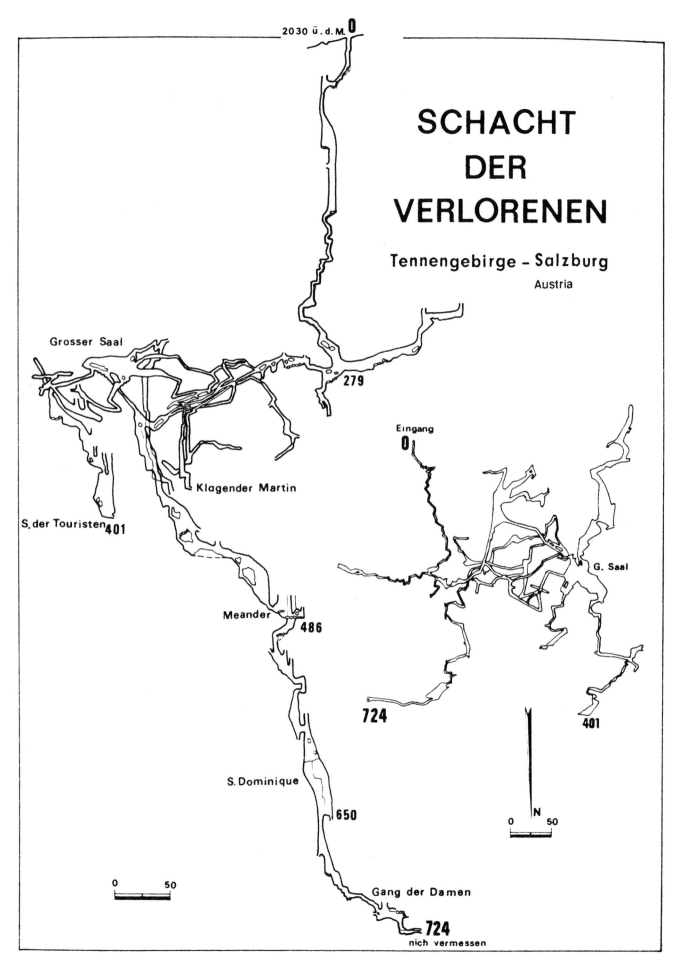

2030 ü.d.M. **0**

SCHACHT DER VERLORENEN

Tennengebirge – Salzburg

Austria

Grosser Saal

279

Eingang
0

Klagender Martin

G. Saal

S. der Touristen **401**

401

Meander **486**

724

N

0 ___ 50

S. Dominique

650

0 ___ 50

Gang der Damen

724
nich vermessen

by Bulgarian cavers. The profile was drawn by N. Gladniski (*Salzburger Höhlenbuch*, Salzburg, 1985, Vol. 4, pp 461-463).

30. **Kolkbläser-Monsterhöhle** .... (–492, +78) 570 m
(Steinernes Meer, Salzburg) (1331/141 and 1331/25)
System with two entrances (2120 and 2085 m) located near the Ingolstadterhaus, on the NW face of Schindelköpfe (1/50,000 map N° 124, Saalfelden). It is developed horizontally between 2150 and 1950 m altitude. The Pfingtschacht leads to the lower levels, explored in 1984 and 1985 (*Atlantis*, 1985 (1-2)).

31. **Ufo-Schacht** .... –565 m
(Totes Gebirge, Oberösterreich) (1626/122)
This cave, located west of the Totes Gebirge, in the Hintergras, at an altitude of 1520 m (1/50,000 map N° 96, Bad Ischl), was discovered in August 1975 by D. Motte. A strong breeze encouraged the G.S.A.B. to unplug the entrance in 1975 and 1976, then to explore to –201 m. This same group was stopped at –565 m in 1977 by an active and extremely tight meandering canyon. (*Spéalp, Belgium*, 1978 (2): 14-20, profile).

32. **Kitzsteinhornhöhle** .... –560 m
(Horhe Tauern, Salzburg) (2573/2)
Located at an altitude of 2450 m in the north part of the "Glocknergruppe" (1/50,000 map N° 153, Grossglockner). The cave was discovered by workers of the "Tauernkraftwerke" in 1980. It was explored by the L.H.S. as well as Polish cavers. (*Atlantis*, 1982 (2-3)).

33. **Zentrumshöhle** .... –557 m
(Hagengebirge, Salzburg) (1335/100)
Located in the center of the Hagengebirge plateau, in the Obere Lengtal, at an altitude of 1915 m. Access to this cave via Golling, passing by the Verbundshütte (1/50,000 map N° 93, Berchtesgaden). It was discovered in 1976 by Tilinghast and explored the same year by the L.H.S. and the G.S.A.B. to –438 m, stopping at a tight near-sump (*Salzburger Höhlenkunde*, 1973, 2, profile). It was only in 1981 that the G.S.A.B. was able to pass this obstacle and reach –557 m (*Spéléo-Flash* (136) profile).

34. **Mondhöhle** .... –546 m
(Hoher Göll, Salzburg) (1336/60)
The opening to this cave is at an altitude of 2005 m, below the summit of the Gruberhorn (see Gruberhornhöhle). 1/50,000 map N° 94 (Hallein). This cave is hydrologically related to the Gruberhornhöhle.

It was discovered on August 16, 1974 by the Polish S.P.T.T.K group from Czestochowa, who reached –376 m later that same month. The 223 m deep Euphorieschacht pit was explored using prussic knots. The Austrians and the Poles got down to –546 m in 1975 (Courbon, *op. cit.* 1979, pp 161, 172, profile, *Salzburger Höhlenbuch*, 1979, 3, pp 390, 417-419, profile).

35. **Im Zwisch-Höhle** .... –543 m
(Totes Gebirge, Steiermark) (1625/350)
This cave formed in the Dachsteinkalk is located between the Plankamira and the Hochweiß, north of Tauplitzalm (1/50,000 map N° 97, Bad Mitterndorf). It was

discovered in 1983 by French cavers from the Franche-Comté who descended to –543 m in 1984 (*Sac au Cul*, Soc. des Amateurs de Cavernes, 1985 (11), profile).

36. **Geldloch** .... (–434, +101) 535 m
(Ötscher, Niederösterreich)
This cave is at the foot of the Ötscher range (1894 m), by the side of the trail between Otschergräben and Spiel Büchler-Haus at Rauhen Kamim. It is at an altitude of 1460 m (1/50,000 map N° 72, Mariazell).

The cave has been known a long time: on August 24, 1591 it was explored by Reichart Strein on the orders of Emperor Rudolf II. Temperature measurements were made inside by J. N. Nagel on July 12, 1747. A survey was made in 1855 by A. Schmidl. Studies leading to the discovery of a 53 m pit were made by H. Crammer and R. Sieger in 1897 and by E. Berr, H. Hassinger, and A. Radio-Radils in 1900-1902. An expedition led by F. Mühlhofer between August 1 and August 10, 1923 reached the depth of –368 m, for a vertical extent of 464 m, which was a world record at the time.

The explorations of the Verband Österreichischer Höhlenforscher in 1953 brought the vertical extent to 524 m (–432, +92) and the length to 1800 m (Courbon, *op. cit.*, 1979, pp 161-162, 172, profile). The length grew steadily to 3810 m by 1980, 5150 m in 1984, and 6885 m in 1985. These extensions were discovered by the Landesverein für Höhlenkunde in Wien (Hartmann, *Die Höhlen Niederösterreichs*, 1985, Band 3, suppl *Die Höhle*, N° 30.).

37. **Fledermausschacht** .... –523 m
(Tonionalpe, Steiermark) (1762/1)
This cave (at an altitude of 1485 m) is located in Herrenboden, in the community of Gusswerk, not far from Marlazell, and SE of Tonionalpe (1/50,000 map N° 103, Kindberg). The descent of the entrance shaft (Ex-Tonionschacht) was made by cavers from Mariazell in 1926 (Beigel, F. Picheler), who continued their explorations in 1928 and 1929. On their last trip in 1936 they reached a depth of –527 m. The Eclaireurs de France du Clan de la Verna remapped the cave in 1953 and reached –517 m (later revised to –446 m). The Austrian cavers of the L. H. Steiermark and the L.H.S. found a continuation in 1977 and were stopped at a tight squeeze at –523 m. (*Die Höhle*, suppl. no 14, 1966, profile to –446 m, *Die Höhle*, 1977 (4): 110-114 (profile).

38. **Warnix** .... –507 m
(Untersberg, Salzburg) (1339/166)
Located near Klingeralm 1000 m from the German border, NE of Mitterberg, at an altitude of 1670 m (1/50,000 map N° 93, Bad-Reichenhall).

The cave was explored in 1976 by German cavers from Munich to a sump at –507 m (the mapping stopped at the –300 m level).

39. **Eislufthöhle** .... –506 m
(Totes Gebirge) (1623/106)
This cave has two entrances, both at about the 1650 m level. It is located near Loser, NE of Vorderer Schwarzmooskogel, and north of Stellerweghöhle (1/50,000 map N° 96, Bad Ischl).

English cavers from the Cambridge University Caving

Club discovered and explored the cave to a depth of –150 m in 1977, then to –350 m in 1978 (*Mitt. Landesverein für Höhlenkunde Steiermark*, 1979, 8 (2) profile). The C.U.C.C., together with a Polish team, reached –506 m in 1979 (*Speleo. Kraków*, 1980 (1/2) profile).

40. **Taubenloch** ....(–454, +52) 506 m
(Ötscher, Niederösterreich) (1816/14)
Located in the "Rauker Kamm" (1/50,000 map N° 72, Mariazell), the cave has been known since 1747, when Nagel made the first survey. It was explored by Austrian cavers in 1981, with the vertical range of 413 m (–392, +21) being reached, followed by 506 m (–454, +52) in 1983 (W. Hartmann, *Die Höhlen Niederösterreichs*, 1985, Vol. 3, suppl. *Die Höhle*, N° 30).

41. **Blitzwasserschacht** .... –505 m
(Leoganger Steinberge, Salzburg) (1324/43)
Cave located in the Nebelsbergkar (alt. 2012 m) (1/50,000 map N° 123, Zell am See). It was discovered and explored in 1979 to –310 m by the Polish (*Speleo, Kraków*, 1980 (1/2), profile), then to –505 m in 1981 by the Polish with the L.H.S. (*Atlantis*, 1982 (2-3) profile).

42. **Jungebabaschacht** .... –495 m
(Tennengebirge, Salzburg) (15111/258)
Alt. 1859 m (*Speleo Kraków*, 1980 (1/2) profile to –435 m).

43. **Edelweißhüttenschacht** .... –482 m
(Tennengebirge, Salzburg) (1511/52)
Alt 2340 m, 1980 (L.H.S., –193 m) and 1983 (G.S.A.B., –482 m) (*Salzburger Höhlenbuch*, 1985, band 4, pp 229-232, profile).

44. **Altes Murmeltier** .... –478 m
(Tennengebirge, Salzburg) (1511/302)
See section on "Great Pits".

45. **Lechnerweidhöhle** .... –470 m
(Dürrenstein, Niederösterreich) (1815/32)
Alt. 1374, 1384, and 1363 m. 1973, L. H. Wien.

46. **Höllenhöhle** .... –455 m
(Tennengebirge, Salzburg) (1511/274)
See section on the "Great Pits".

47. **Brunneckerhöhle** .... –445 m
(Tennengebirge, Salzburg) (1511/1)

48. **Tantalhöhle** .... –435 m
(Hagengebirge, Salzburg) (1335/30)
See below.

49. **Lou-Toti-Höhle** .... (–360, +70) 430 m
(Totes Gebirge, Oberösterreich) (1626/33)

50. **Kein-Problemschacht** .... (–342, +87) 429 m
(Totes Gebirge, Steiermark) (1626/2)

51. **Wolfhöhle** .... –426 m
(Totes Gebirge, Steiermark) (1623/145)
Alt. 1685 m. 1984, Cambridge U.C.C. (*Caves and Caving*, 1985 (29) profile).

52. **Eiskogelhöhle** .... –420 m
(Tennengebirge, Salzburg) (1511/101)

53. **Schachtsystem in der Knallsteinplatte** .... –420 m
(Tennengebirge, Salzburg) (1511/277)

54. **Internationalschacht** .... –410 m
(Tennengebirge, Salzburg) (1511/261)

55. **Eisriesenwelt** .... (–138, +269) 407 m
(Tennengebirge, Salzburg) (1511/24)
See below.

74. **Rothorn-Seehöhle** .... –300 m
(Leoganger Steinberge, Salzburg) (1324/48)

LONG CAVES:

1. **Hirlatzhöhle** .... **57,000 m**
(Dachstein Oberösterreich) (1546/7)
Hirlatzhöhle has two entrances (866 and 899 m), and is located in the Hirlatzwand, which is the North side of the Vorderer Hirlatz, dominating the Echerntal valley, 2 km NNW of the village of Hallstatt.
1/50,000 map N° 96 (Bad Ischl). Dachsteinkalk.
The cave had only 4200 m known in 1957, with an additional 2 or 3 km being explored in the 1960's. The L. H. Linz pushed the length to 8051 m in 1974, with progress to only 8478 m being achieved by 1981. After this date there was a veritable explosion of discoveries as a result of the efforts of the members of the Verein für Höhlenkunde Halstatt-Obertraun. Hirlatzhöhle had reached 25 km by 1984, 42,700 m in 1985, and 44,600 m by March, 1986.
In parallel with the growth in length, the vertical extent grew to over 600 m (–148, +464) by May 1985, while it was only 332 m back in 1972.

2. *Raucherkarhöhle* .... **48,033 m**
(Totes Gebirge, Oberösterreich)
See above.

3. *Eisriesenwelt* .... **42,000 m**
(Tennengebirge, Salzburg) (1511/24)
This famous show cave is located 4 km north of Werfen, or 34 km SSE of Salzburg. A cable car climbs to the Friedrich-Oedl-Haus hut from Wimmerhütte. A steep trail (10 minutes) leads to the entrance, which is at an altitude of 1664 m (1/50,000 map N° 94, Hallein). The first explorations were made by A. von Posselt-Czorich in 1879. They were continued in 1912, 1913, and 1914 (Alexander von Mork, Angermayer, and Rihi), then interrupted by the war. After the fighting, exploration was done by cavers from Salzburg and Vienna (W. von Czoerng-Czernhausen, the brothers Friedrich and Robert Oedl, Gustav Abel, and Hermann Gruber notably).
By the time WWII came around, most of the cave had been explored, with a vertical extent of 407 m (–138, +269). A new survey of the cave is presently underway.
The cave contains 30,000 square meters of ice and 1 km of ice-floored passages.
Map: from *Die Höhle*, suppl. N° 14, 1966.
Bibliography: Angermayer (E.) - *Kleiner Führer durch die Eisriesenwelt*, Salzburg, 1971, 36 p.

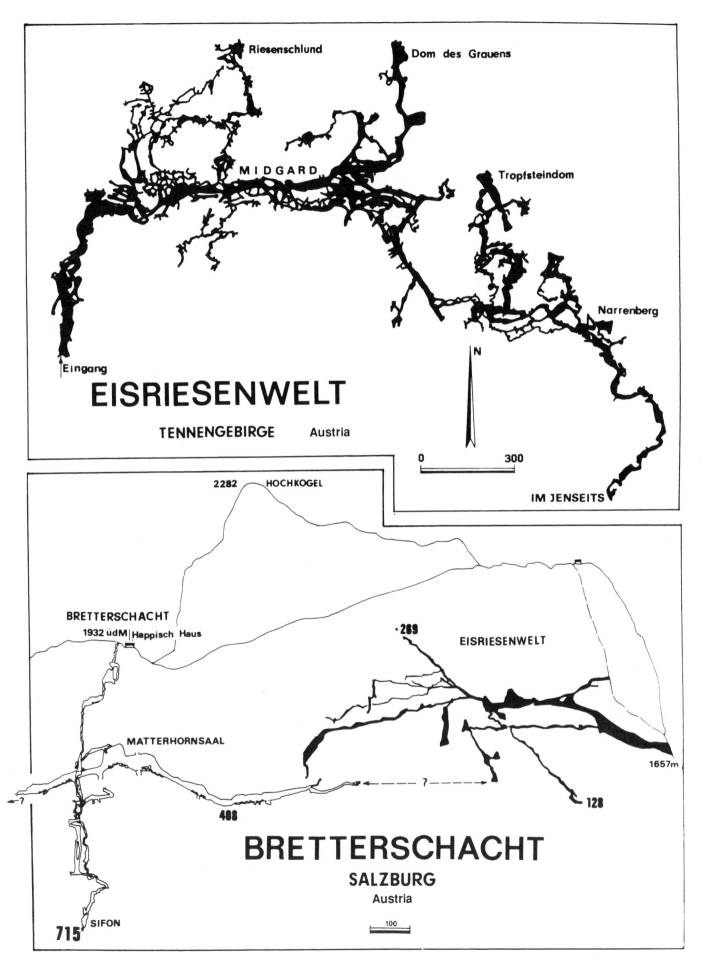

EISRIESENWELT

TENNENGEBIRGE   Austria

BRETTERSCHACHT

SALZBURG

Austria

4. **Dachstein-Mammuthöhle** ....38,529 m
(Dachstein, Oberösterreich) (1547/9)
See above.

5. *Tantalhöhle* ....**30,850 m**
(Hagengebirge, Salzburg) (1335/30)
Located in the Hagengebirge range, north of the Blühnbach, east of Tenneck, the cave is located at an altitude of 1710 m, 400 m south of Spiralenlang (2223 m). It is developed under the Tantalkölpfe plateau in the dolomitic limestones of the Dachsteinkalk. 1/50,000 map N° 125 (Bischofshofen).

The entrance was discovered in 1947 by A. Koppenwallner. Exploration by the L.H.S began right away, with the depth of –400 m being reached in 1948. The depth only slowly increased (–420 m in 1949, –435 m presently), but the length increased substantially. It stayed around the 16 km mark in the 1960's, then went to 28,700 m in 1972 and 30,200 m in 1973, today reaching 30,850 m.

Map: from Ager ánd Klappacher (L.H.S.), issue of *NSS News*, 1973 (31): 210.

Bibliography: Klappacher (W.), Knapczyk (H.) - *Salzburger Höhlenbuch*, Salzburg, 1979, Vol. 3, pp. 165-199.
Ager (H.), Klappacher (W.), Wimmer (A.) - 25 Jahre Tantalhöhle (Hagengebirge, Salzburg), *Die Höhle*, 1973 (2).

6. **Jägerbrunntrogsystem** ....25,315 m
(Hagengebirge, Salzburg) (1511/162)
See above.

7. **Frauenmauer-Langstein-Höhlensystem** .20,215 m
(Hochschwab, Steiermark) (1742/1)
See above.

8. **Lamprechtsofen** ....14,657 m
(Leoganger Steinberge, Salzburg)
See above.

9. **Feuertalsystem** ....12,014 m
(Totes Gebirge, Oberösterreich)
See above.

10. **Kolkbläser-Monsterhöhle** ....11,189 m
(Steinernes Meer, Salzburg) (1331/141)

11. **Gamslöcher-Kolowratsystem** ....10,175 m
(Untersberg, Salzburg) (1339/1)

12. **Hüttstatthöhle** ....8140 m
(Totes Gebirge, Steiermark)

13. **Salzgrabenhöhle** ....7797 m
(Steinernes Meer, Salzburg) (1331/29)

14. **Karrenschacht** ....7277 m
(Totes Gebirge, Oberösterreich) (1625/49)

15. **Geldloch** ....6870 m
(Otscher Niederösterreich) (1816/6)

16. **Gruberhornhöhle** ....6700 m
(Hoher Göll, Salzburg) (1336/29)
See above.

17. **Eiskogelhöhlensystem** ....6500 m
(Tennengebirge, Salzburg) (1511/101 and 160)

18. **Almberg-Eis-und-Tropfsteinhöhle** ....6293 m
(Totes Gebirge, Steiermark) (1624/18)

19. **Schwersystem** ....6101 m
(Tennengebirge, Salzburg (1511/268)
See above.

20. **Frauenofen** ....6076 m
(Tennengebirge, Salzburg) (1511/18)

21. **Salzburgerschacht** ....6070 m
(Untersberg, Salzburg) (1339/69)

22. **Langsteineishöhle** ....6051 m
(Hochschwab, Steiermark) (1744/1)

23. **Windlöcher-Klingertalschacht** ....6000 m
(Untersberg, Salzburg) (1339/31)

24. **Lurgrotte** ....5975 m
(Grazer Bergland, Steiermark) (2836/1)

25. **Hochlecken-Großhöhle** ....5460 m
(Höllengebirge, Oberösterreich) (1567/29)
See above.

26. **Elmhöhlensystem** ....5300 m
(Totes Gebirge, Steiermark (1624/38)

27. **Lechnerweidhöhle** ....5252 m
(Dürrenstein, Niederösterreich) (1815/32)

28. **Ahnenschacht** ....approx. 5000 m
(Totes Gebirge, Oberösterreich) (1626/50)
See above.

29. **Schönberghöhle** ....4830 m
(Dachstein, Oberösterreich) (1547/70)

30. **Bretterschacht** ....4510 m
(Tennengebirge, Salzburg) (1511/141)
See above.

31. **Brunneckerhöhle** ....4430 m
(Tennengebirge, Salzburg) (1511/1)

22. **Drachenhöhle** ....4127 m
(Grazer Bergland, Steiermark) (2839/1)

33. **Hermannshöhle** ....4112 m
(Bucklige Welt, Niederösterreich) (2871/7)

34. **Taubenloch** ....4053 m
(Otscher, Niederösterreich) (1816/14)

35. **schacht der Verlorenen** ....4050 m
(Tennengebirge, Salzburg)
See above.

————

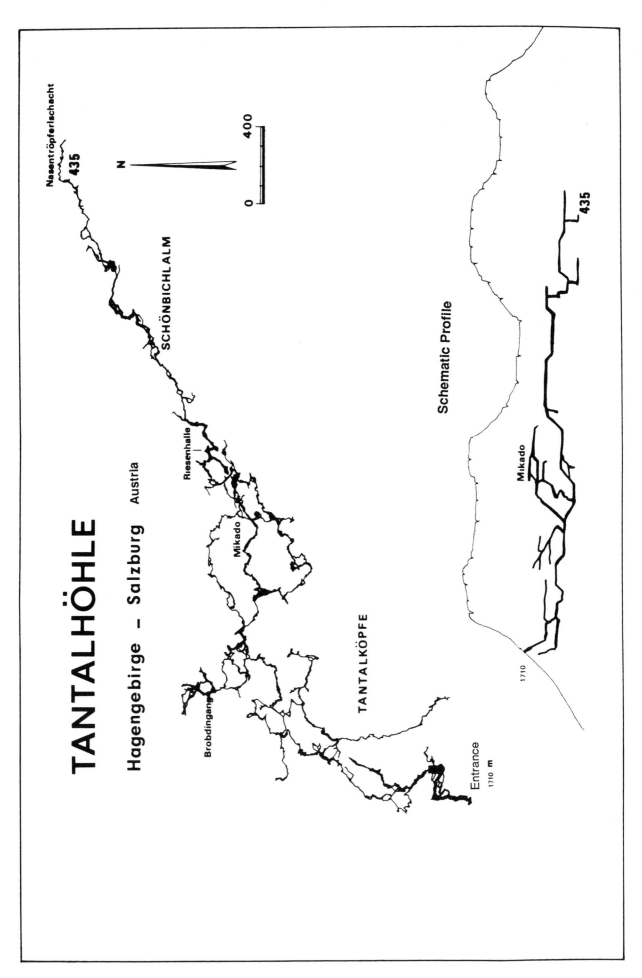

TANTALHÖHLE

Hagengebirge – Salzburg   Austria

Nasentröpferlschacht
435

SCHÖNBICHL ALM

Riesenhalle

Mikado

Brobdingang

TANTALKÖPFE

Entrance
1710 m

400

N

Schematic Profile

435

Mikado

1710

43. **Westwandhöhle** ....3000 m
(Totes Gebirge) (1625/351)

BELGIUM

Caves in Belgium have been visited for a long time. The abbé de Feller began the exploration of the famous Han-sur-Lesse cave in 1771, which was pursued in 1818 by Robiano de Boesbeck. Quicks and Quetelet made a precise survey in 1823 (1138 m not counting the side passages), in preparation for its opening as a show cave on May 15, 1828.

At the end of the 19th century, Van den Broeck became interested in the limestones of Belgium: in 1910 he published (in collaboration with Martel and Rahir) *Les Cavernes et les rivières souterraines de la Belgique*, a massive 1842 page work in two volumes that has served as the foundation for further study.

Probably the best-known explorer between the two world wars was Max Cosyns, but he did most of his work in the French Pyrenees.

Félix Anciaux wrote *Cavernes* in 1950 (ed. Bellvue, Dinant, 315 p.), an inventory of Belgian caves. Following this, cavers began to organize themselves. Research was coordinated by structured clubs with ties to scientific groups. Three major publications have appeared: the *Atlas des grottes de Belgique* by Paul Vandersleyen, the cave inventories of Doemen (1968) and Gevaerts (1970), and the bibliographies of De Block and Fontaine in 1968, 1974, 1976, 1981, and 1983.

The expansion of exploration has given rise to associations of various clubs: by 1980 there were no less than four national "federations". On the recommendation of the Physical Education and Sports Administration, they were joined together on January 1, 1986 to form the single Union Belge de Spéléologie.

Belgian speleology remains very active at home as well as abroad, as witnessed by recent expeditions to Algeria, Indonesia, Mexico and Zaïre.

Guy De Block.

DEEP CAVES:

1. **trou Bernard** (Maillen, Namur) .... –140 m
Alt. 230 m. Dug open and explored in 1949 by B. Margos to –120 m. Dived another 20 m by G. De Block in 1969 (P. Vandersleyen, *Atlas des Grottes de Belgique*, 1, profile).

2. **trou Weron** (Maillen, Namur) .... –110 m
In 1941 by Cosyns, Cardeyn, and Lecomte (P. Vandersleyen, *op. cit.*, 1, profile).

3. **trou-qui-Fume** (Furfooz, Namur) .... –87 m

4. **grotte Persévérance** (Sprimont, Liège) .... –84 m

5. **trou de l'Eglise** (Mont-sur-Meuse, Namur) ... –82 m
(P. Vandersleyen, *op. cit.*, 1, profile).

6. **chantoire de la Chawresse** (Tilff, Liège) .... –81 m
(*Actes Journées de la Spéléologie*, 1978, profile).

7. **grotte du Père Noël** (Wavreille, Namur) .... –80 m
In 1964 by the C.Y.R.E.S. (P. Vandersleyen, *op. cit.*, 3, profile).

8. **trou Picot** (Wavreille, Namur) .... –70 m
In 1953 by the S.C.U.C. Louvain (P. Vandersleyen, *op. cit.*, 3 profile).

9. **puits aux Lampes** (Jemelle, Namur) .... –70 m
In 1953 by the S. S. Namur (P. Vandersleyen, *op. cit.*, 2, profile).

10. **trou des Manants** (Tilff, Liège) .... –70 m

LONG CAVES:

1. **grotte de Han-sur-Lesse** .... 5720 m
(Han-sur-Lesse, Namur)
Resurgence of the Lesse river, which sinks into the gouffre de Belvaux (drève des Etançons). Long known, it was explored in the 19th and 20th centuries as well as being open to tourists. The dives of M. Jasinski and R. Sténuit in 1959 and the discoveries of 1962 and 1966 brought the length to 5720 m (P. Vandersleyen, *op. cit.*, 3, plan).

2. **grotte de Hotton** ....approx. 3500 m
(Hotton, Luxembourg)
Map showing 1914 m in P. Vandersleyen, *op. cit.*, 2.

3. **grotte de Remouchamps** .... approx. 2800 m
(Sougné-Remouchamps, Liège)
Show cave explored in the 19th and 20th centuries (P. Vandersleyen, *op. cit.*, 2, plan).

4. **drève des Etançons** ....2560 m
(Han-sur-Lesse, Namur)
Insurgence cave explored from 1972 to 1976 (*Speleologia Belgica*, 1975 (3) plan with 1650 m).

5. **chantoire de la Chawresse** ....approx. 2000 m
(Tilff, Liège)
Connected on May 20, 1984 with la grotte Veronika (*Spelunca*, 1985 (17) schematic map).

6. **grotte du Père Noël** ....approx. 1800 m
(Wavreille, Namur)
(P. Vandersleyen, *op. cit.*, 3, map).

7. **galerie des Sources** ....approx. 1800 m
(Hulsonniaux, Namur)
(*Clair-Obscur*, 1980 (26) map showing 1480 m).

8. **grotte de Rochefort** (Rochefort, Namur) ..... 1450 m
Show cave (P. Vandersleyen, *op. cit.*, 2, map).

9. **trou d'Haquin** (Lustin, Namur) .... 1360 m
(P. Vandersleyen, *op. cit.*, 3, map).

10. **trou des Crevés** (Wavreille, Namur)............. 1350 m
(P. Vandersleyen, *op. cit.*, 2, map).

11. **grotte de Tilff** (Tilff, Liège) .... 1200 m
(P. Vandersleyen, *op. cit.*, 3, map).

12. **grotte de Ramioul** (Ramioul, Liège) approx. 1200 m

13. **résurgence Lucienne** (Lustin, Namur) .... 1160 m

14. **grotte de la Vilaine Source** .... 1059 m
(Arbre, Namur)
(*Subterra*, 1976, (68) map).

15. **grotte de l'Adugeoir** (Pétigny, Namur) .... 1050 m
(P. Vandersleyen, *op. cit.*, 1, map).

16. **grotte Nys** (Aishe-Heyd, Luxembourg) .... 1035 m
(P. Vandersleyen, *op. cit.*, 1, map).

BULGARIA
Bâlgarija

In spite of its large variety of karst phenomena, Bulgaria is not well known as a caving area. The karst is essentially confined to the north, in the Stara Planina (in the mid-level plateaus), notably around Vraca and Trojan, and in the south, in the Pirin mountain (Pre-Cambrian marbles) and the Rodhopie mountain, south of Plovdiv. Caves began attracting interest beginning in the 19th century (biological work of E. Mercki in 1878, archaeological studies of H. Skorpil in 1882 and 1884). The karst and caves were described by H. and K. Skorpil in 1900, R. Popov from 1904 to 1940, and J. Radev in 1915. The first caving club was formed in 1929. The various clubs have been organized under the Federajica po Pechterno Delo since 1973, under the auspisce of the Bulgarian Tourist Union. Since the Bulgarians rarely communicate with Western cavers, the following lists should be considered to be only approximate.

DEEP CAVES:

1. **Barkite 14** (Vračanska) .... –415 m

2. **Rajčova dupka** .... (–372 + 10) 382 m
(Černi Os'm, Loveč)

3. **Tizoin** (Gubeš, Sofia) .... (–376, + 2) 378 m

4. **Jamata na Kipilova** .... –350 m
(Kotel, Slivenski okr.)

5. **Beljar** .... –273 m
(Gorno Ozirovo, Bezirk Mihajlograd)

6. **Ledenika** (Kotel, Slivenski okr.) .... –242 m

7. **Malkata jama** (Černi Os'm, Loveč) .... –232 m

8. **Maglivata** (Kotel, Slivenski okr.) .... –214 m

9. **Kacite** (Zimovica, Sofia) .... –205 m

10. **Kasana** (Bansko, Blagoevgrad) .... –193 m

11. **Barkite 18** (Vračanska) .... –186 m

12. **Pukoja** (Pavolče, Vraca) .... –178 m

13. **Drangaleškata dupka** .... –165 m
(Mugla, Smoljan, Rhodopi)

14. **Uža't na Imanjarite** (Kotel, Slivenski okr.) ... –158 m

15. **Borova dupka** (Černi Os'm, Loveč) .... –156 m

LONG CAVES:

1. **Duxlata** (Bosnek, Perniski, Vitoša) .... 15,128 m

2. **Orlova čuka** (Pepelina, Rusenski) .... 13,155 m

3. **Jagodinska** (Smoljensk) .... approx. 9000 m

4. **Temnata dupka** .... approx. 7000 m
(Lakatnik, Sofijski okr.)

5. **Imamova dupka** .... 6450 m
(Jagodina, Smoljanski, Rhodopi)

6. **Baltovitki** .... approx. 4000 m

7. **Bamborova** .... 3500 m

8. **Vodnata peščera** (Cerovo, Sofijski okr.) ...... 3264 m

9. **Prikazna** (Kotel, Slivenski okr.) .... 3100 m

10. **Morovica** (Gložene, Loveč) .... 3020 m

11. **And'ka** (Drjanovo, Gabrovski okr.) .... 3000 m

12. **Ponora** (Ciren, Vracanski okr.) .... 2948 m

13. **Boninska peščera** .... 2753 m
(Krušuna, Loveški okr.)

14. **Kacite** (Zimovica, Sofia) .... 2500 m

15. **Magura** (Rabiša, Bezirk Vidin) .... approx. 2500 m

CZECHOSLOVAKIA
Československo

There are two principal mountain chains in Czechoslovakia: the western Carpathians and the Czech range. The most extensive karst regions are found in the western Carpathians, and are from the middle Trias (about 2000 km²). These are dark grey limestones from the Guttenstein (karst areas of Lower Tatras, of Velká Fatra, and Belanské Tatry) and the light grey limestones of the Wetterstein (karst areas of the Slovak Paradis, and southern Slovakia). On the other hand, the Czech range has limestone islands from the Devonian and the Silurian of small dimensions (300 km²). Four principal morphological types can be distinguished: the plateau type (southern Slovakia, Slovak Paradis, Moravia), the slotted type (Lower Tatras, Belanská Tatry, Czech karst to the SW of Prague), the covered type (Sumiac, Poniky) and the cryptokarst, cut in crystaline limestones (Ochtiná, Chynov near Tábor).

The history of speleology in Czechoslovakia began in the 18th century (Jaraj Buchholz, Nagel...) and took off in the 20th century thanks to the efforts of Karel Absolon (1877-1945). To date, over 2,000 caves have been catalogued.

Karst and caves in Slovakia are explored by the Slovak Karst Museum and the Nature Conservancy of Liptovský Mikuláš and the Slovak Speleological Society, while those of Bohemia and Moravia are studied by the Czech Speleological Society which is based in Prague.

Dr Anton Droppa
Slovak Speleological Society.

DEEP CAVES:

1. **Starý Hrad priepast'** .... –424 m
(Nízke Tatry, Liptovský Mikuláš, Slovakia)
Or jaskyňa Čierna. Alt. 1488 m. Discovered in 1967, the cave was explored in 1970 to – 152 m, then – 277 m in 1979, – 322 m in 1980, – 343 m in 1981 (cave dive) and – 424 m in 1982-1984 (*Speleofórum*, 1985, profile).

2. **jaskyňa Záskočie-Na Predných** .... –284 m
(Nízke Tatry, Liptovský Mikuláš, Slovakia)
Explored in 1958, 1962, 1973, 1974 (*Lidé + Zeme*), 1973 (9) profile of 263 m (– 146, +117).

3. **jaskyňa Mŕtvych Netopierov** ..... (–223, +20) 243 m
(Nízke Tatry, Slovakia)
In 1981-1983 (*Speleofórum*, 1985, profile).

4. **Tristárska** (Slovakia) .... –200 m

5. **systém Amatérskej a Punkevnej jeskyně** .. –192 m
(Moravský Kras, Ostrov, Moravia)

6. **propast Čertova Diera** .... –186 m
(Rožňava, Slovakia)
Alt. 790 m. (F. Skřivánek, J. Rubín, *Caves in Czechoslovakia*, 1973, profile to – 125 m).

7. **Stratenská jaskyňa** (Rožňava, Slovakia) ..... –184 m
Alt. 1003 m. Bottomed in 1980.

8. **Rudické propadání** .... –180 m
(Moravský kras, Rudice, Moravia)

9. **Brázda priepast'** (Rožňava, Slovakia) .... –180 m
Alt. 600 m. In 1953 and 1964 (Skřivánek/Rubín, *op. cit.*, 1973, profile to – 205 m).

10 **Hranická propast** .... –179 m
(Hůrka, Hranice, Moravia)
Including 110 m submerged (dive of 1981). Plumbed to –244 m.

11. **Belanská jaskyňa** .... –161 m
(Belianské Tatry, Spišska Bela, Slovakia)

12. **jaskyňa Slobody-Pustá** .... –153 m
(Nízke Tatry, Liptovský Mikuláš, Slovakia)

13. **priepast' Malá Žomboj** .... –142 m
(Rožňava, Slovakia)
A single shaft 142 m deep (Skřivánek/Rubín, *op. cit.*, 1973, profile).

14. **Bystrianska priepast'** .... –140 m
(Nízke Tatry, Bystra, Slovakia)
Alt. 676 and 536 m (Skřivánek/Rubín, *op. cit.*, 1973, profile).

15. **Meiselův závrt** .... –137 m
(Moravský kras, Moravia)

16. **Hedvábná jeskyně** .... –137 m
(Moravský kras, Moravia)

17. **Ponorna priepast'** .... –132 m
(Slovenský kras, Slovakia)

18. **priepast' Velká Buková Bikfa** .... –131 m
(Rožňava, Slovakia)

19. **Ohnište priepast'** .... –129 m
(Nízke Tatry, Liptovský Mikuláš, Slovakia)
Alt. 1537 m (Skřivánek/Rubín, *op. cit.*, 1973, profile to – 125 m).

20. **Diviačia priepast'** (Rožňava, Slovakia) .... –127 m
Alt. 590 m (Skřivánek/Rubín, *op. cit.*, 1973, profile to – 118 m).

21. **Dámsky závrt** .... –125 m
(Moravský kras, Sloupsko, Moravia)
(*Československy Kras*, 1972 (24) profile).

22. **Veterná priepast'** (Rožňava, Slovakia) .... –122 m
Alt. 720 m (Skřivánek/Rubín, *op. cit.*, 1973, profile).

23. **Harbešska jeskyně** .... –120 m
(Vilémovice, Moravia)
(Skřivánek/Rubín, *op. cit.*, 1973, profile).

24. **jaskyňa Arnoldka** .... –113 m
(Česky kras, Bohemia)

LONG CAVES:

1. *systém Amatérskej a Punkevní jeskyně*
(Moravský kras, Ostrov, Moravia) ....32,500 m
This system has six entrances including the famous Macocha Abyss (opening 76 m across and 174 m deep). One of the entrances, Punkevní jeskyně opened to tourists in 1914 and the resurgence of the system, was connected artificially to the Macocha pit. The other entrances are Amatérská jeskyně (or Nové Punkevní jeskyně), Holštejnká 13 C jeskyně, Spirálka jeskyně and Nová Rasovna jeskyně.

The Moravian karst in which the Amatérská system has formed in is found to the NE of Brno. The entrance to the show cave is located in the Pustý žleb ravine, on the road from Blansko to Sloup, 6 km ENE of Blansko.

Punkevní jeskyně is the principal resurgence for the Moravian karst, which is drained by the the underground Punkva river. The water collects at the bottom of Macocha

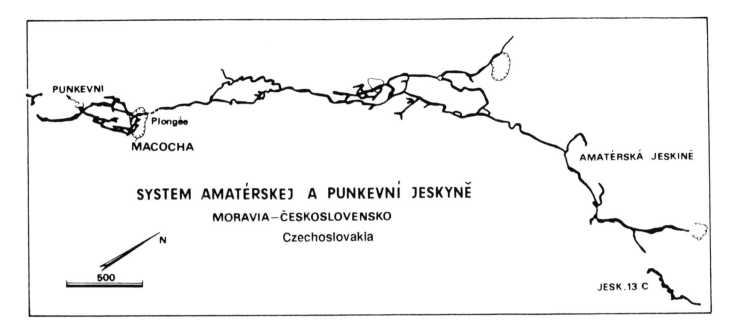

SYSTEM AMATÉRSKEJ A PUNKEVNÍ JESKYNĚ

MORAVIA-ČESKOSLOVENSKO

Czechoslovakia

N

500

JESK.13 C

and emerges into the Pustý žleb valley. The water mostly comes from the northern edge of the karst and from swallet caves near Sloup. The limestone is of Devonian age.

Exploration began with the descent of Macocha propast (alt. 490 m, − 138 m) in 1723 by Lazar Schoper. The descent of the mathamatician Nagel in 1748 is more generally known.

The exploration of the Moravian karst was the great work of professor Karel Absolon who led four expeditions from 1901 to 1909, and was especially interested in finding the underground course of the Punkva. In 1910 Macocha propast and Punkevní jeskyně were artificially linked. Karel Absolon worked again in the area in 1933 (pumping and dives in the Punkva), in 1935-1936 (pumping in the Nová Rasovna jeskyně), and 1939 (Hedvábná jeskyně).

In the 1960's Holštejnka jeskyně was explored by the Plánivy Group de Brmo (P.G.B.). In 1969 the passing of the entrance sump of Cigánsky závrt (later Amatérská jeskyně, alt. 510 m) by the P.G.B. gave access to the long sought-after underground system. About 1300 m were explored that year. In August 1970 a sudden storm led to the death of two cavers caught by the rising water. Thanks to various cave dives, the length reached 8,000 m in 1972.

On April 15, 1975 Amatérská jeskyně was connected to Macocha Abyss (the depth of which had been increased to 163 m by diving) by the passing of a 430 m long sump: the total vertical extent reached 192 m. In 1979 the length reached 17,202 m.

It was other connections which increased the length of the system, first with the Holstejnká jeskyně on December 17, 1983, then with Spirálka jeskyně on January 15, 1984, and finally with Nová Rasovna jeskyně in 1985 to bring the length to 32,500 m.

Map: from the document sent by O. Stelcl.

Bibliography: "Largest cave system of the Czech Socialist Republic in the Moravský kras", *Studia Geographica*, Brno, 1973.
Kučera (B.), Hromas (J.), Skřivánek (F.) - Jeskyně a propasti v Československu, *Praha*, 1981.

2. **Domica jaskyňa-Baradla barlang** .... 25,000 m
   (Rožňava, Slovakia/Hungary)
   See Hungary. The Czech section is 5,080 m long.

3. **Stratenská jaskyňa** (Rožňava, Slovakia) ... 16,700 m
   Discovered in 1972, explored in 1972-1974 (9500 m), 1980 (15,100 m) and 1983.

4. **Rudické propadáni** (Rudice, Moravia) ...... 12,300 m
   System explored in 1976, 1984-1985 (*Speleofórum*, 85, plan).

5. **jaskyňa Slobody-Pustá** .... 11,385 m
   (Liptovský Mikuláš, Slovakia)
   Part of the Demänova system. Artificial connection with jaskyňa Mieru. Explored in 1921, 1923, 1927, 1930, 1950-1951, 1982-1983.

6. **jaskyňa Mieru-Ladová** .... 8500 m
   (Nízke Tatry, Liptovský Mikuláš, Slovakia)
   Part of the Demänova system (artificial connection). The Mieru-Ladová connection was made in 1952 (Droppa, *Slovenské Jaskyne*, 1973, map with 6690 m).

7. **Sloupsko-Šošůvské jeskyně** .... 6500 m
   (Moravský kras, Sloup, Moravia)
   4000 m by K. Absolon in 1912.

8. **jaskyňa Záskočie-Na Predných** .... 5034 m
   (Liptovský Mikuláš, Slovakia)

9. **Starý Hrad priepast'** .... 4509 m
   (Liptovský Mikuláš, Slovakia)

10. **Javoříčské jeskyně** (Javoříčsko, Moravia) ... 3800 m

11. **jaskyňa Mŕtvych Netopierov** .... 2850 m
    (Nízke Tatry, Slovakia)

12. **Čachtická jaskyňa** (Slovakia) .... 2823 m

13. **Jasovská jaskyňa** (Košice, Slovakia) .... 2184 m

14. **Bobačka jaskyňa** (Muránský kras) .... 2134 m

15. **Liskovská jaskyňa** .... 2100 m
    (Ružomberok, Slovakia)

16. **Koněpruské jeskyně** .... 2050 m
    (Česky kras, Koněprusy, Bohemia)

17. **Vypustek jeskyně** (Moravia) .... 2000 m

EAST GERMANY
Deutsche Demokratische Republik

The caves of the German Democratic Republic are located in the west of the country, in the Harz. The longest are developed in Permian age gypsum (N° 1, 3 and 4) and Devonian age limestones. Most of them were discovered in the 19th century (the Bielshöhlen-system was opened to the public in 1788) and most of the exploration was done before the second world war. No cave is known to be deeper than 100 m.

LONG CAVES:

1. **Wimmelburger Schlotten** (Südharz) .... 2400 m

2. **Hermannshöhle** (Rübeland, Harz) .... 1750 m

3. **Heimkehlen system** .... 1710 m
   (Uftrungen, Südharz)

4. **Barbarossahöhle** (Kyffhäuser) .... 1000 m

5. **Bielshöhlen-system** (Rübeland, Harz) .... 1000 m

FINLAND
Suomi

Finland has a few small granite caves. The largest are **Kasbergsgrottan** (Ströma Kasberget, Åland), and **Torholagrottan** (Lohja), neither one more than 20 m long.

FRANCE

The poet René Char once wrote "France-des-cavernes", or "France, the country of caves". And the writings of the poets are always correct! With 45% of its surface area covered by calcareous rocks, the richness of speleology is great in this country. One finds cave in every 'departement' (i. e. county) of France. While the karsts on the plains and lower elevations have not been intensively studied, those in the mountain ranges are richer and have been minutely combed for over a century, so that today the tasks of inventorying and cataloguing the huge number of caves found are of vital importance.

French speleological tradition is very old, with hunters, peasants, and villagers having been interested in caves for a very long time, and the more scientifically oriented taking over in the seventeenth century. Caves with prehistoric usage, often with prehistoric art as well, are especially abundant in the Périgord region, and are summarized in two indispensable reference books: *Préhistoire de l'art occidental*, by A. Leroi-Gourhan (Mazenod, ed., 1965, 484 p.) and *L'Art des Cavernes* (coll. Impr. Nationale, 1984, 673 p.).

Springs, natural fountains, and underground rivers were of special interest to scientists in the sixteenth and seventeenth centuries (B. Palisay, 1547, 1563; J. de Clugny, 1666; P. Perrault, 1674). Beginning in the eighteenth century, the number of publications becomes hard to count. Famous examples are the survey of the grotte de Miremont (Dordogne) for a length of 4229 m in 1765, the exploration by Marsollier of the grotte des Demoiselles (Hérault) in 1780, and the descent by the abbé Carnus in the tindoul de la Vayssière (Aveyron) in 1785. Modern speleology truly began on June 28, 1888, when Edouard-Alfred Martel (1859-1938) made the first through trip in the Bramabiau system (Gard). The scientific, sporting, and cultural elements of modern twentieth century French speleology basically all came from Martel.

Martel has many well-known disciples (L. Armand, L. de Launay, G. Gaupillat, A. Viré, O. Decombaz, F. Mazauric...), among whom one must specially cite the names of Eugène Fournier (1871-1941), who studied the Franche-Comté karsts, and René Jeannel (with Racovitza), who established the foundations of modern French biospeleology.

The war of 1914-1918 basically put an end to the sporting career of Martel. The 1930's saw the rise of a new and vigorous generation of explorers: Robert de Joly (1887-1968) and Louis Balsan did a lot of work in the Causses in the Massif Central, Norbert Casteret was allured by the caves of the Pyrenées, Guy de Lavaur (1903-1986) launched the concept of cave diving in 1946, Pierre Chevalier, who with Trombe, revolutionized exploration techniques (he led the classic explorations in the Dent de Crolles in the Chartreuse, Isère), Félix Trombe (1906-1985) who combined science and sport in the caves of the Haute-Garonne as Martel and Chevalier had done, and Bernard Gèze who significantly advanced the science of biospeleology.

After the second world war, there began an explosion in the number of explorations of deep or difficult caves, building on the technical advances of Chevalier and Bruno Dressler. The birth of numerous clubs began to limit the importance of individual "stars" and confirmed the great richness of karst regions as Martel had presaged: many expeditions were guided by the 600 pages in his book *France Ignorée*. Today (in 1985) France has 162 caves over 300 m deep and 182 longer than 3 km.

It is also worthwhile to note the extraordinary progress made in cave diving beginning in the 1960's, spearheaded by such individuals as Michel Letrone, Jean-Louis Vernette, Bertrand Léger, and Francis Le Guen. J. Hasenmayer descended to -200 m in the fontaine de Vaucluse and O. Isler penetrated over 3200 m in the doux de Coly!

Martel founded the Société Spéléologique in 1895, hoping for international participation. Robert de Joly reactivated it in 1930 under the name of the Spéléo-Club de France, which became the Société Spéléologique de France in 1936, joining in 1963 with the Comité National de Spéléologie to form the Féderation Française de Spéléologie, an organization with over 6000 members in 450 clubs, organized into numerous active commissions. They published *Les Grandes Cavités Françaises* by C. Chabert in 1981, an inventory in which over 2000 caves are described.

DEEP CAVES:

1.   *réseau Jean Bernard* ....(–1494, +41) 1535 m
(Samoëns, Haute-Savoie)

This system has eight entrances: the B 21 (2210 m), B 22 (2190 m), B 19 (2150 m), V 11 (1970 m), V 5 (1870 m), V 4 bis (1860 m), V 6 (1850 m), and the V 4 (1840 m). Located in the vallon des Chambres, not far from the Fóllis hut, on the mountain of the same name. The upper entrances are north of the lac des Chambres. 1/50,000 map Samoëns x 944.44; y 132.48 (B 19); x 943.41; y 132.15 (V6). The cave gathers water from the Couarra and Avoudrues ranges and is developed in Urgonian limestones near a contact.

The G. S. des Vulcains at Lyon discovered the system and has led in its exploration. They began in 1964, but wet and difficult passages made for slow progress, so that by 1968 the depth of –450 m had been reached in the V 4. The sump at –623 m was reached in November 1969 with the help of other clubs. The exploration of a side passage in 1973 led to the greater depth of 663 m. The discovery of the V 5 in 1974 added 30 m to the system, now –693 m deep. The depth of –934 m was reached in 1975 in a room filled with breakdown.

In parallel, the B 19 (discovered in 1968) was being explored. The dynamiting of constrictions at –100 m and –170 m (in 1971), and a dig at –230 m (in 1973) led them to –450 m. On November 1, 1975, the connection with the V 4-5-6 was made, making a –1208 m deep system. A passage was found in the terminal breakdown in January 1976, allowing access to a sump at 1298 m.

The connection with the B 21 on July 5, 1979 made this the deepest cave in the world at –1358 m. The G. S. Vulcain then concentrated on the terminal sump. Successive cave dives in 1980, 1981, and 1982 increased the depth to –1402, –1455, and –1494 m (divers J.-L. Fantoli and P. Penez). At this depth a fourth, sand-filled sump has been declared undive-able.

With the connection to the B 22, an additional 41 m in vertical extent were gained in November 1983. The exact progression of the system's length as a function of time has not yet been researched, but the present total length is 17,900 m.

Map: from the work of the G. S. Vulcain, sent by C. Rigaldie.

Bibliography: Courbon, *Atlas des Grands Gouffres du Monde*, 1979, pp. 83-84, 95.

Maire (R.), Rigaldie (C.) - *Spéléo-sportive dans les Alpes de Haute-Savoie*, 1984, pp. 51-68.

2.   *réseau de la Pierre Saint-Martin* .... –1342 m
(Arette, Pyrénées Atlantiques)

This prestigious cave has an almost mythical fame due to several factors: its 320 m entrance pit located in a once remote area in Spain, just across the French border (see section on "The Great Pits"), the much publicized death of Marcel Loubens in 1952, the recovery of his body in 1954, the discovery of enormous passages and a huge room (La Verna, 270 by 230 by 180 m), to which we might add the attempt by the E.D.F. to capture the underground river for a possible hydro-electric project, and finally the allure associated with having been for many years the deepest known cave in the world.

The system has several entrance pits, one of them being in Spain. In chronological order they are 1) sima de la Piedra de San Martin (Isaba, Navarra, alt. 1717 m), discovered in August 1950 by G. Lépineux, B. Occhialini, and M. Cosyns; 2) Basaburuko Iezia (alt. 1882 m), discovered in August 1965 by the G. S. Hautes-Pyrénées (G.S.H.P.) and the C.C.F. Montpellier and connected on August 24, 1966 by the Basaburuko Lezentzat; 3) the gouffres M 3 and M 13 (alt. 1984 m), discovered in 1970 and connected on August 10, 1975; 4) the gouffre du Beffroi (alt. 2037 m) discovered in August 1975 and connected on August 23, 1975; 5) the gouffre du Pourtet (alt. 2058 m), connected in August 1982. There is also the artificial entrance, the tunnel built by the E.D.F. between 1956 and 1960 into the Salle Verna (alt. 1052 m, Sainte-Engrâce).

These entrances are located on the slopes of the pic d'Arlas and around the col de la Pierre-Saint-Martin, which is accessible via the Arette ski center (from France), or by the Isaba and Venta de Arraco road (from Spain). They are located in a wild and impressive lapiaz area that is extremely rich in caves and pits.

1/50,000 map Larrau. x 349.88; y 78.21 (Beffroi). x 348.72; y 78.32 (Basaburuko). x 346.60; y 78.90 (Piedra de San Martin).

The resurgence is the Bentia, in Sainte-Engrâce, alt. 445 m, which is also the resurgence for the Arres Planères and Arphidia caves (see below). These are also major caves, and show the vast potential in the area.

Conscious of this fact, the cavers exploring the Pierre Saint-Martin decided in 1966 to form the Association de Recherches Spéléologiques Internationales à la Pierre Saint-Martin (A.R.S.I.P.), which has coordinated explorations and scientific work since that date. Many clubs have worked under its auspices, such as the S. C. Poitevin, G.S.H.P., S. C. Paris, I.P.V. Pamplona, and others.

The cave potential of the area was first noted by Martel in 1908. In 1951, following the discovery of the "historic" entrance, Lépineux and Occhialini descended the great 320 m shaft using a winch and explored to about –450 m (on an expedition organized by Max Cosyns). The death of Loubens in 1952 did not slow explorations too much, for in 1953 the group La Verna de Lyon (M. Letrone, G. Ballandraux..), with Casteret and Lépineux, discovered the La Verna room at a depth of –689 m and 2,611 m from the entrance. A Franco-Spanish team explored the upstream section in 1954 as far as the tunnel du Vent (the Wind Tunnel).

In 1960, while the E.D.F. remapped the downstream section, the Spanish continued to explore upstream as far as the diaclase Hidalga. August 8, 1961 is an important date: entering by the newly finished E.D.F. tunnel (soon to become the classic entrance), Felix Ruiz de Arcaute, J. San Martin, and A. Aratibel climbed the side of the Verna room

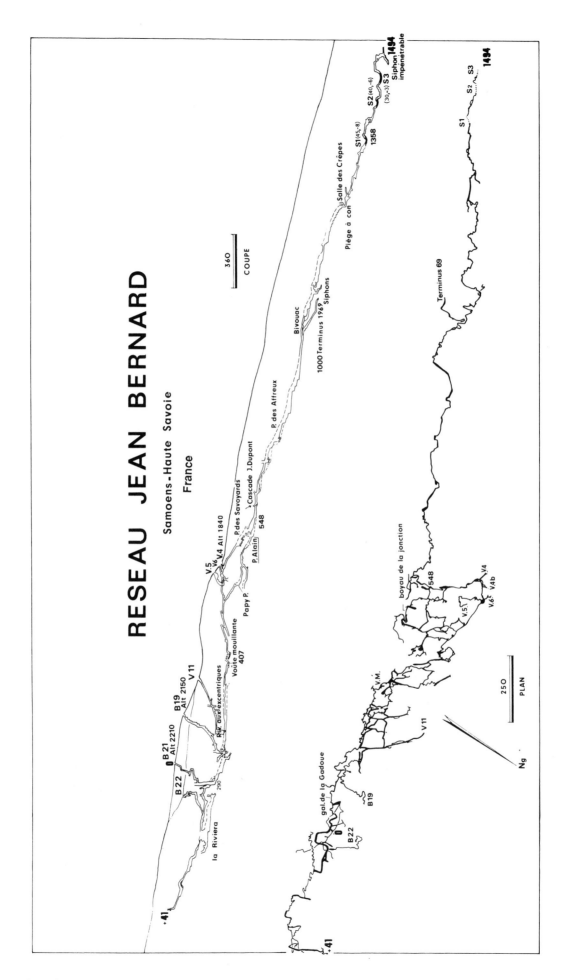

RESEAU JEAN BERNARD

Samoens - Haute Savoie

France

COUPE

PLAN

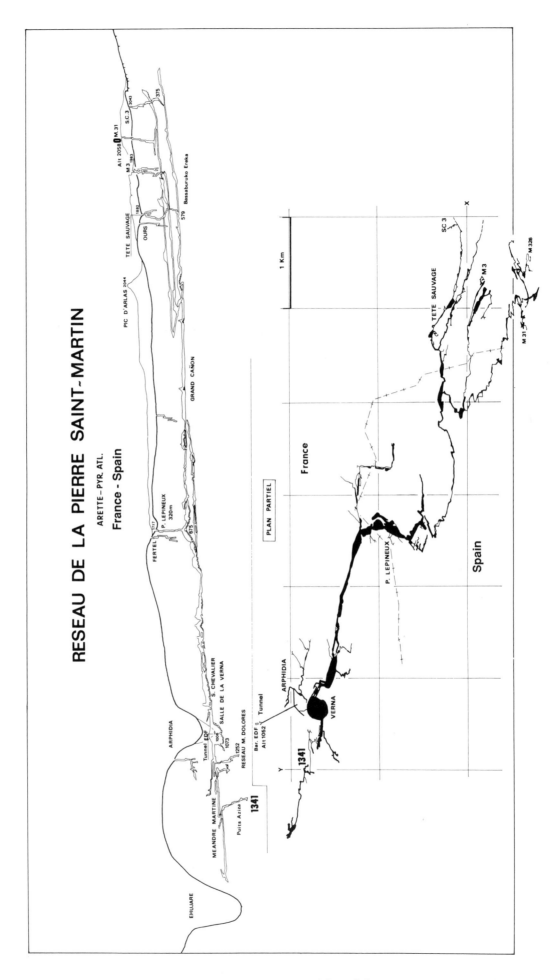

RESEAU DE LA PIERRE SAINT-MARTIN

ARETTE – PYR. ATL.

France - Spain

for 95 m and found the way on in the galerie Aranzadi, permitting a team led by J. Jolfre to descend to –767 m the same year, then –861 m in 1962. In another descending branch, the S. C. Rouen reached about –890 m in 1963, then –940 m in 1964, while the S. C. Paris pushed upstream. The S.C.R. reached what is currently the low point of the system at –1006 m in 1965 m. The S.C.P. made progress upstream in 1965-1966, with the Basaburuko Iezia being explored and connected on August 24, 1966 in the meantime. The system had a vertical extent of 1171 m at this point and a length of 13,050 m. Nine years later, with the length having reached 21,880 m in 1974, the connection with the gouffres M3-M13 gave a temporary vertical extent of 1273 m, which was increased to 1321 m on August 23 of the same year with the connection to the gouffre du Beffroi, for a length of 25,500 m. The length continued to increase, to 31,780 m in 1976, 34,900 m in 1977, 38,445 m in 1979, and 39,960 m in 1980.

The connection in 1982 with the gouffre du Pourtet added only 21 m to the vertical extent, but increased the length to 46,228 m. The explorations of the last few years have been in diverse sections of the cave and brought the length to 51,200 m, then to 52,000 m.

Map: synthesis of M. Douat (1979) and of S. Puisais (1986), furnished by the authors.

Bibliography: *Bull. de l'A.R.S.I.P.*, 1966-1980, N° 1 to 15. Institución Principe de Viana - *Larra, sima de San Martin*, Pamplona, 1964, 223 p.

3. *gouffre Berger* ....(–1191, +50) 1241 m
(Engins, Isère)

The gouffre Berger is a system with four entrances: the gouffre des Rhododendrons (1510 m), the gouffre Berger or scialet d'Engins N° 2 (1460 m), the puits Marry (1440 m), and the gouffre des Elfes (1425 m). It is developed in the northern apron of the Vercors plateau dominating the town of Grenoble. The entrances are in the wooded lapiaz in the Sornin basin.

1/50,000 map Grenoble. x 856.68; y 329.48 (gouffre Berger).

The resurgence is at the cuves de Sassenage (alt. 297 m), 4200 m to the ESE, having traversed Urgonian and Hauterivian strata.

The cave owes much of its fame to the fact that it was the first in the world to pass the 1000 m depth mark. Its entrance was found on May 24, 1953 by Jo Berger, of Grenoble, and exploration was immediately undertaken by the Spéléologues Grenoblois du C.A.F. (S.G.C.A.F.); the depth of –370 m was reached on November 9, 1953. The descent continued in 1954, stopping on September 25 at –903 m at the top of a large waterfall.

In 1955 they were stopped on July 29 at –985 m at the Hurricane pit. The S.G.C.A.F. decided to organize an international expedition and brought in cavers from several countries: on July 11, 1956 the famous sump at –1122 m was discovered, being 5890 m from the entrance.

The British attacked the sump in 1963, and in August Ken Pearce dove it and was stopped by another sump at –1135 m. He dove this second sump in 1967. The S. C. Seine (B. Léger) carried on in 1968 and came to a third sump at –1141 m, bringing the length to 10,805 m in 1969. This "terminal" sump was dived for 7 m in 1978 by F. Poggia.

On February 15, 1981, F. Poggia passed the sump in the gouffre des Rhododendrons and brought the depth to 1191 m, and on July 11, 1982, Patrick Penez, helped by the C.D.S. du Vaucluse, passed the four sumps in the bottom of the gouffre Berger and dove the fifth one to a depth of 50 m, bringing the depth of the system to 1241 m.

The work of local clubs from 1974 to 1981 (including the connection with the gouffre des Elfes in 1976) doubled the length of the cave to 20,665 m.

Map: from the synthesis of G. Bohec, sent by B. Faure.

Bibliography: Cadoux (j.) *et al.* - *Opération -1000*. Arthaud ed., 1955, 250 p.

Marry (G.) *Gouffre Berger, premier -1000*, from the author, 1977, 120 p. Lismonde (B.) *et al.* -*Grottes et Scialets du Vercors*, C.D.S. Isère, t. 2 1979.

4. *gouffre Mirolda* ....(–936, +275) 1211 m
(Samoëns, Haute-Savoie)

The gouffre Mirolda, also called the C.D.11, is located at an altitude of 1880 m, in the massif du Criou. It is found in the ridge to the east of the chalets of Criou (1671 m) toward the Tête du Fer à Cheval. The entrance is closed by a wooden trapdoor. 1/50,000 map Samoëns. x 942.70; y 130.67.

It is developed in Urgonian and Barremian limestones. The resurgence is not known.

The entrance was discovered in 1973 during a summer camp of the G. S. Aven. A tight canyon stopped the first explorers at –127 m. The S. C. Lyon found a way on in 1980, an upper passage at the top of a pit, and discovered a stream at –183 m. During the fall and winter the S.C.L., with the help of the G. S. Cavernicole and the S. C. Chablais, reached a terminal sump at –936 m.

In parallel the same clubs, reinforced by the S.S.S. de Genève, explored the fossil passages and, not far from the entrance, climbed an upstream passage to +110 m. A sump at this level was passed once and 2000 m of passages were explored, but not mapped! The vertical extent of the cave is certainly greater than the 1046 m mapped, while the length reached 9 km in 1982. The depth was increased to over 1211 m in 1987.

Map: from *Spelunca Mémoires* N° 13, and from D. Colliard.

Bibliography: S. C. Chablais and S.S.S. Genève - Le gouffre Mirolda 1980-1981 –950 m. massif du Crious, *Spelunca Mémoires*, 1983 (13); 118-120.

Maire (R.) and Rigaldie (C.) -*Spéléo-sportive dans les Alpes de Haute Savoie*, Edisud. 1984, pp 69-75.

5. *gouffre de Bracas de Thurugne 6* ....–1166 m
(Arette, Pyrénées-Atlantiques)

This cave is also known by the names of gouffre de la Bordure de Tourgne N° 6 and rivière de Soudet. It is located in the massif de la Pierre Saint-Martin (1/50,000 map Larrau. x 347.73; y 80.37) and opens at an altitude of 1618 m. Its pits give access to the Soudet river.

It was explored in 1970 by the S. C. Rouen who reached about –212 m (*Bull. A.R.S.I.P.*, 1970-1971 (5-6) profile), then in 1974 by the Club Léo-Lagrange de Pau. The C.L.A.C. du Blanc returned to the exploration and found the underground river which they followed in large passage to –687 m in 1985. They returned in 1986 with S. C. Blois, SGCAF and S. C. Belgique, reaching a sump at

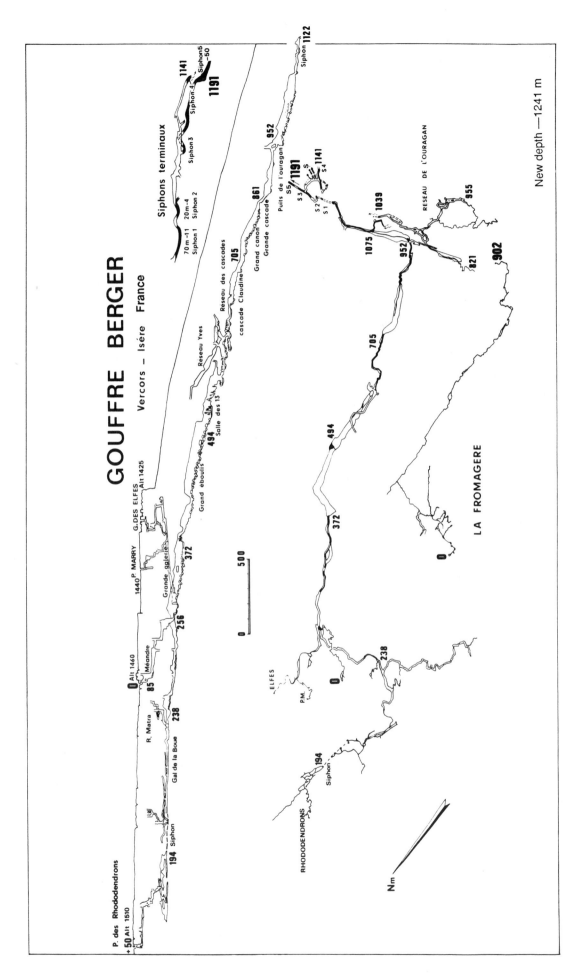

GOUFFRE BERGER

Vercors – Isére   France

Siphons terminaux

New depth —1241 m

LA FROMAGERE

RESEAU DE L'OURAGAN

RHODODENDRONS

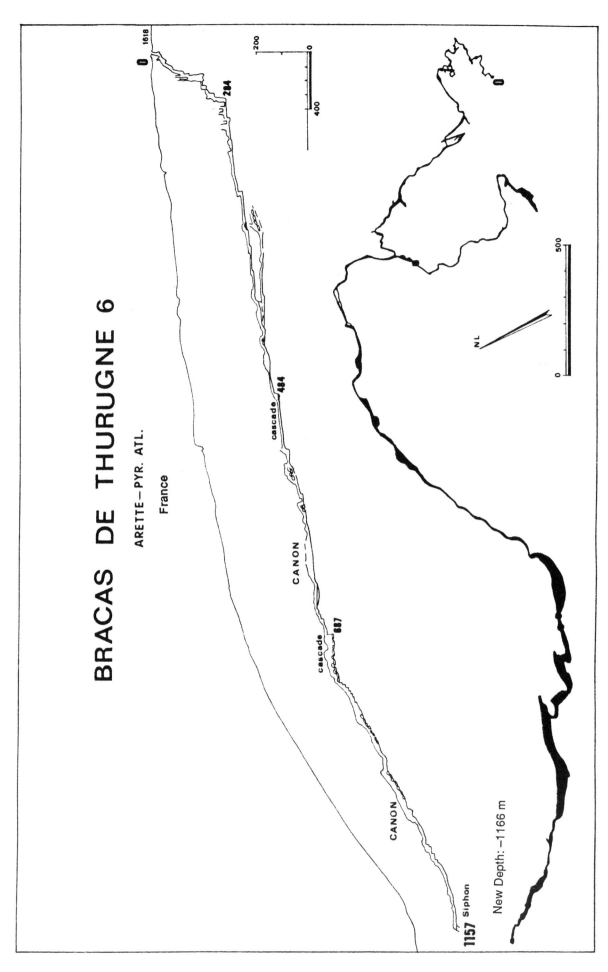

BRACAS DE THURUGNE 6

ARETTE – PYR. ATL.

France

CANON

CANON

cascade

cascade

New Depth: –1166 m

1618

0

284

484

687

1157 Siphon

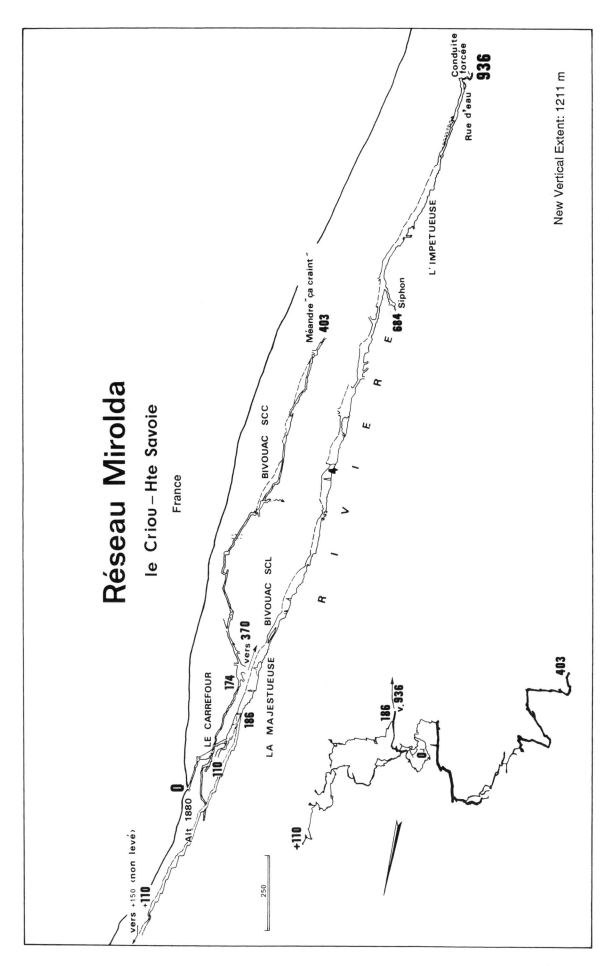

Réseau Mirolda

le Criou – Hte Savoie

France

vers +150 ⟨non levé⟩

+110

Alt 1880

0

LE CARREFOUR

174

vers 370

186

110

LA MAJESTUEUSE

BIVOUAC SCL

BIVOUAC SCC

Méandre "ça craint"

403

R I V I È R E

684 Siphon

L'IMPÉTUEUSE

Rue d'eau

Conduite forcée

936

New Vertical Extent: 1211 m

250

+110

186

v.936

0

403

–1152 m. The depth was increased to –1166 m in 1987.
Map: sent by P. Courbon.

6. *système de la Coumo d'Hyouernèdo* .... –1004 m
(Herran/Arbas, Haute-Garonne)

The massif d'Arbas, 100 km south of Toulouse, with its unique resurgence, the goueil di Her, contains a labyrinth of underground passages, to which 34 entrances give access. The limestone layers are in the Jurassic-Cretaceous series: the wooded lapiaz and rocks in which are found most of the entrances belong to the lower Cretaceous (Urgonian).

Although the goueil di Her is located near the village of Arbas, to the south, to get to the pit entrances, which are on the north side of the coume d'Hyouernèdo, one must pass by the hamlet of Labaderque (commune de Herran). At the foot of the Pène Blanque rock, visible from Arbas, one finds the cave of the same name.

1/50,000 map Aspet. x 479.76; y 75.08 (puits de l'If). 479.85; 75.08 (gouffre Raymonde). 479.97; 74.98 (trou Mile). 480.76; 75.28 (gouffre Pierre). 480.87; 76.08 (Pount dech Erbaou). 480.91; 76.62 (grotte de Pèneblanque) and 482.63; 76.69 (goueil di Her).

The history of this labyrinth cave is itself labyrinthine. Although a somewhat dry method, we will first list all the entrances, along with the connection dates, separating them into two groups.

The gouffre de la Coquille (alt. 1452 m) was discovered in 1971 and linked with the gouffre Raymonde on July 20, 1977 (M.J.C. Aubagne). The puits de l'If (1397 m) was discovered in 1959 (N. Casteret) and connected to the gouffre Raymonde on October 9, 1959 (N. and R. Casteret, J. Delfeil, E. Bugat). The puits des Sapins (1383 m) was discovered in 1959 (N. Casteret) and connected to the gouffre Raymonde on August 13, 1959 (N. Casteret, Scouts d'Aix-en-Provence, G. S. Provence). The gouffre Raymonde (1360 m) was discovered in 1957 (N. Casteret) and explored by N. Casteret, the G.S.P. and the Scouts d'Aix to a depth of 438 m. The trou Mile (1325 m) was connected to the gouffre Raymonde on July 26 1964 (Joifre, Lafranque, Nave). The trou du Vent (1276 m) was discovered in 1956 (Casteret) and connected to the gouffre Pierre on August 8, 1960. An ascending branch led to the trou Mile, the entrance to which was dug open from the inside on July 27, 1963 (Jolfre *et al.*). The gouffres Duplessis 1 and 2 (1257 m) were discovered in 1958 (G.S.P.) and connected to the gouffre Raymonde on December 9, 1973 (G. S. Pyrénées). The gouffre Pierre (1180 m) was discovered in 1956 (P. Gicquel), and connected to the trou du Vent. The gouffre Barnache (1149 m) was discovered in 1964 (by P. Gicquel) and connected to the gouffre Pierre on September 10, 1970 (G. S. Py., G.S.P.). The gouffre du Pont de Gerbaut (1080 and 1055 m), noted in 1873, was connected in the winter of 1964 to the trou du Vent (Jolfre, Laffranque *et al.*). The grotte de Pèneblanque (930 m) was the site of archeological work in 1874 by Filhol, Jeanbernat, and Timbal-Lagrave, and was connected to the Pont de Gerbaut on July 9, 1971 (G. S. Py.). The goueil di Her (486 m) was connected on August 28, 1979 to Pèneblanque (G. S. Py. *et al.*).

The second group of entrances is made up of the gouffre de la Henne-Morte or clot des Ourtigas (1339 m), connected to the gouffre Odon on August 21 1978 (G. S. Py.), the sarrat dech Méné (1328 m), discovered in 1940 (N. Casteret), connected to the Henne-Morte on August 3, 1956 (G.S.P., Scouts d'Aix), the gouffre Odon (1284 m), connected to the Henne-Morte on August 21, 1978 and to the Pont de Gerbaut on September 2, 1978 (G. S. Py.), a connection which was important because it linked the two groups of entrances.

Among the other entrances are the puits de la Couquette (1400 m), the puits des Champignons (1384 m), the puits Francis (1383 m), the puits du Bouvreuil (1383 m), the grottes des Ours Bruns 1 and 2 (970 m), the puits des Cendrillons (1360 m), the puits superieur de sarrat dech Méné (1353 m), the gouffre Michèle (1234 m), the gouffre Vincent (1192 m), the grotte des Commingeois (940 m), and the grotte du Grand Bourusse (790 m).

1908 may be considered as the year when exploration of the system began. Martel descended to –60 m in the grotte de Pèneblanque and stopped at the rockfall in the Pount dech Erbaou, a cave in which R. de Joly reached –65 m in 1931, then –100 m in 1936 with N. Casteret.

From 1940 to 1947 it was the Henne-Morte which received attention; Marcel Loubens and Josette Ségouffin, reinforced by Casteret, reached –180 m there in 1941; –290 m in 1942, and –358 m (original survey) in 1943. In 1947 a large expedition of the S. C. Paris permitted Casteret and Loubens to discover the sump at 358 m (then measured at –446 m).

In 1952 the S. C. Paris became interested in the Pèneblanque again, where they reached the sump at –355 m in 1963 (for a length of 5213 m), while the sump in the goueil di Her, first passed when dry in 1943, was passed in 1956 by Dufour, who died in the sump in 1957.

Beginning in 1956 and up to 1964, on the advice of Casteret, the G. S. Provence and the Scouts de Aix organized annual expeditions that gradually connected together to form an ever increasingly complex system.

Some of the highlights: in 1958, –515 m was reached in the gouffre Pierre and its connection with the trou du Vent in 1960 gave –610 m, then –650 m in 1962. The connection Mile-Vent-Pierre in 1963 formed a system 668 m deep.

In 1964, under the guidance of Jacques Jolfre, the Pont de Gerbaut was explored to –450 m and the connection with the trou du Vent gave a system 703 m deep (May, 1964), then 770 m (July 1964) when the Mile-Vent-Pierre-Pont system was connected to the If-Sapins-Francis-Raymonde complex.

Beginning in 1969, with already 17 km of length, the G. S. des Pyrénées, under the direction of Maurice Duchène, redoubled its efforts, with the crowning point being the connection in 1971 of the Pount dech Erbaou with the Pèneblanque (–830 m and 23 km). Efforts were then concentrated on the connection with the goueil di Her, in which the G.E.P.S. de Marseille (J.-L. Vernette) had passed the second sump and been stopped by a third in 1968. From 1972 to 1977 the length increased to 32,580 m.

1978 and 1979 were great years in the Massif d'Arbas. In 1978 the connection of the Raymonde/Pèneblanque and Henne-Morte systems made the length leap to 51,013 m. In 1979, Pèneblanque was connected to the goueil di Her by diving, giving a depth of –1004 m and a length of 59,500 m, a figure which increased again when the grotte du Grand Bourusse (length 3960 m in 1982) was integrated into the system.

COUMO D'HYOUERNEDO

Massif d'Arbas – Hte Garonne

France

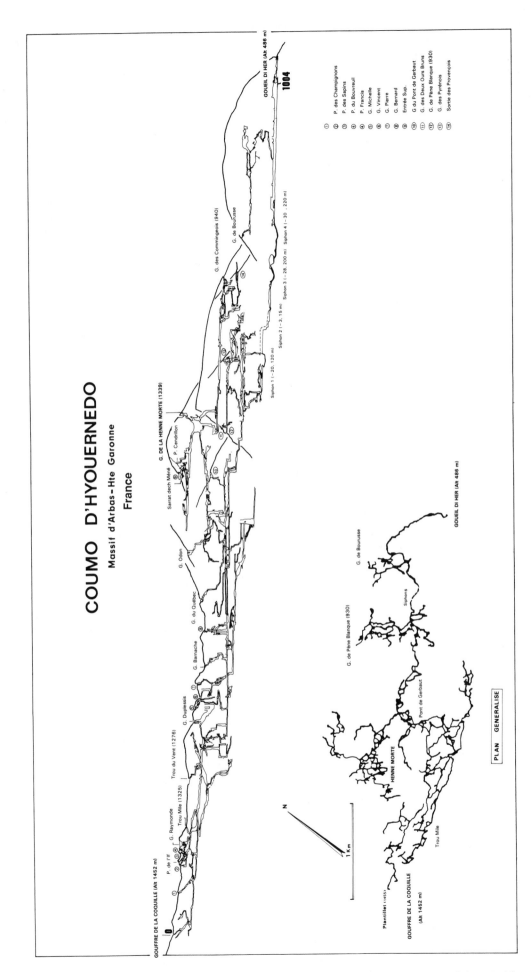

GOUFFRE DE LA COQUILLE (Alt 1452 m)

P. de l'If
G. Raymonde
Trou du Vent (1276)
Trou Mile (1325)

G. DE LA HENNE MORTE (1339)

Sarrat dech Mèné
P. Cendrillon

G. Odon

G. du Québec

G. Barnache

G. Duplessis

G. des Commingeois (940)

G. de Bourusse

GOUEIL DI HER (Alt 486 m)

1004

Siphon 1 (– 20, 120 m)
Siphon 2 (– 3, 15 m)
Siphon 3 (– 28, 200 m)
Siphon 4 (– 30 . 220 m)

① P. des Champignons
② P. des Sapins
③ P. du Bouvreuil
④ P. Francis
⑤ G. Michelle
⑥ G. Vincent
⑦ G. Pierre
⑧ G. Bernard
⑨ Entrée Sup.
⑩ G du Pont de Gerbaut
⑪ G. des Deux Ours Bruns
⑫ G. de Pène Blanque (930)
⑬ G. des Pyrénois
⑭ Sortie des Provençois

N

1 Km

Plantillet (1455)

GOUFFRE DE LA COQUILLE
(Alt 1452 m)

G. de Pène Blanque (930)

G. de Bourusse

Siphons

GOUEIL DI HER (Alt 486 m)

HENNE MORTE

Pont de Gerbaut

Trou Mile

PLAN GENERALISE

From 1982 to 1986 the various clubs exploring the system have pushed to the length to 82,500 m and the number of entrances to 34.

Map: from the sketches of the G. S. Pyrénées and the S. C. Comminges, synthesis of Boyer, Drillat, and Segura, sent by M. Duchêne.

Bibliography: Duchêne (M.) and Drillat (P.-A.) - *La Coumo d'Hyouernèdo.* G. S. Pyrénées ed., Toulouse, 1982, 346 p.

7. *réseau des Aiguilles* ....(+22, –958) 980 m
(Agnières-en-Dévoluy/La Cluse, Hautes-Alpes)

This system has two entrances: the chourum du Rama (La Cluse, alt. 2271 m), and the chourum des Aiguilles (alt. 1995 m), located in the massif du Dévoluy, on the north side of the vallon des Aiguilles, above a knoll dominating the Sainte-Aure waterfall.

1/50,000 map Saint-Bonnet, x 876.21; y 269.80.

A dye trace on May 31 1974, at –160 m, showed the connection of the system to the Gillardes spring (alt. 867 m), at Pellafol, Isère.

A shepherd showed the chourum des Aiguilles to J. Tourres in 1965 . The S. C. Alpin de Gap undertook the exploration right away and was stopped at –97 m at a squeeze. The depth of –430 m was reached in 1966, followed by –530 m in 1967 and –620 m in 1968.

The terminal sump at –682 m was reached in 1969 by an expedition with the S.C.A. Gap, the Abîme-Club de Toulon, the G. S. Catamarans de Sochaux, and the S. C. Salernes.

The A.C.T., helped by the S. C. Savoie, returned to the exploration of the chourum in 1972, discovering at − 205 m an ascending gallery in which they accomplished the exploit of gaining over 500 m in elevation, including climbing 10 and 40 m tall pits. In this way the chourum de Rama was discovered from inside. The A.C.T. and the S.C.S. climbed to +298 m relative to the Aiguilles entrance. In 1973, the same teams explored new passages above the Minotier pit, but none of them allowed bypassing the sump at –958 m. As a result of these discoveries, the length of the system grew to 6100 m. In 1974 the A.C.T. tried without success to pass the sump discovered the year before.

Map: from the drawings of P. Courbon, the S. C.A. Gap, the Catamarans (to –682 m) and of P. Paris (from –205 to +298 m).

Bibliography: Spéléo-Club Alpin de Gap - Le chourum des Aiguilles, *Spelunca*, 1970: 211-214. Paris (P.) - Le chourum des Aiguilles, *Spelunca*, 1974; 18-19.

8. *gouffre du Cambou de Liard* .... –926 m
(Accous, Pyrénées-Atlantiques).

Located near the Touya de Liet (see below), 1500 m to the SSW of the col d'Iseye, in the cirque de Liard, about 500 m to the ENE of the pic Ronglet.

1/50,000 map Laruns-Somport . x 366.53; y 75.415.

Formed in the steeply dipping (60° N) Santonian limestones, it has the same resurgence as the Touya de Liet: the fontaine des Fées (alt. 460 m), 7000 m away (dye trace of J.-P. Besson).

It has two entrances: the gouffre du Petit Coin (alt. 1975 m) and Cambou de Liard N° 1 (1957 m). The latter was discovered in August 1970 by the S.S.P.P.O. (Pau), who reached –400 m the same year. In 1971, helped by the S.G.C.A.F., they reached –768 m. In 1972 the same groups, with the assistance of cavers from the Ardèche, reached the terminal sump, 908 m below the entrance. The connection in August 1977 by the S.G.C.A.F. to the gouffre du Petit Coin, 18 m higher, increased the vertical extent to 926 m. The length is about 2500 m.

Map: from the drawings of the S.S.P.P.O. and the S.G.C.A.F., sent by J.-P. Besson.

Bibliography: Lismonde (B.) - Le gouffre du Cambou de Liard, *Scialet*, 1972 (1); 81-85.
Lismonde (B.) - Gouffre du Petit Coin, *Scialet*, 1977 (6) ; 131-138.

9. *gouffre de la Fromagère* (Engins, Isère) .. –902 m

One finds this cave on the maintained trail which leads to the gouffre Berger (see above), halfway between the latter and the plateau de la Molière, in the Vercors, at an altitude of 1540 m. It belongs to the same hydrologic system as the gouffre Berger.

1/50,000 map Grenoble. x 856.50; y 328.48.

A shepherd showed the entrance to cavers from Villard-de-Lans in 1937: it was descended to –32 m. The

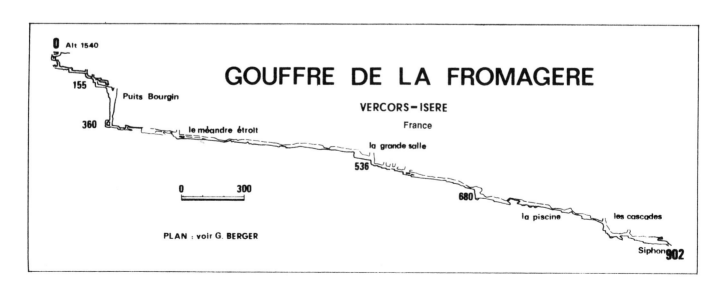

GOUFFRE DE LA FROMAGERE

VERCORS – ISERE

France

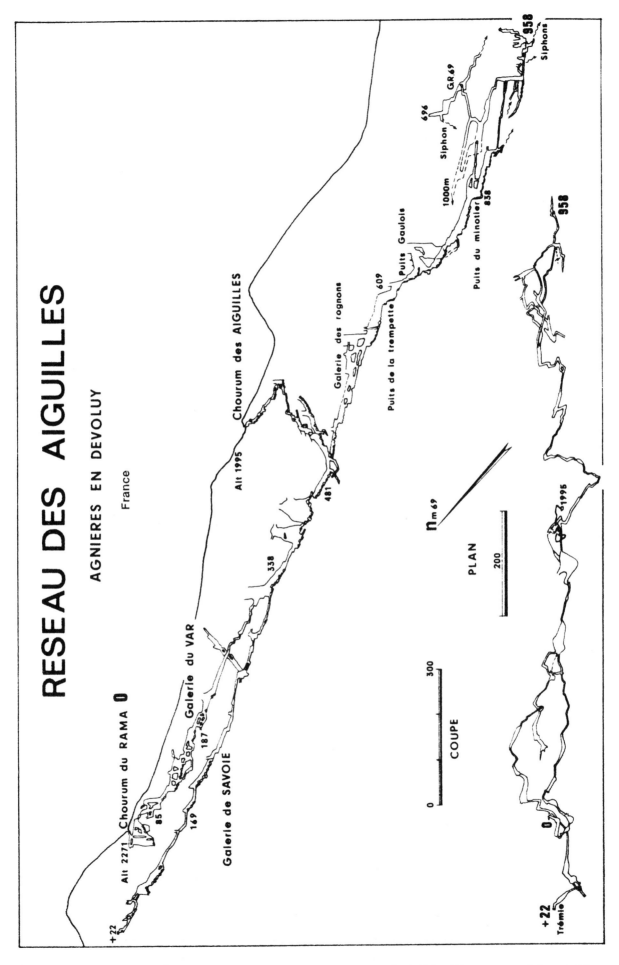

RESEAU DES AIGUILLES

AGNIERES EN DEVOLUY

France

Chourum du RAMA

Alt 2271 Chourum du RAMA

Galerie du VAR

Galerie de SAVOIE

Chourum des AIGUILLES

Alt 1995

Galerie des rognons

Puits de la trempette

Puits Gaulois

Puits du minotier

Siphon

GR69

Siphons

COUPE

PLAN

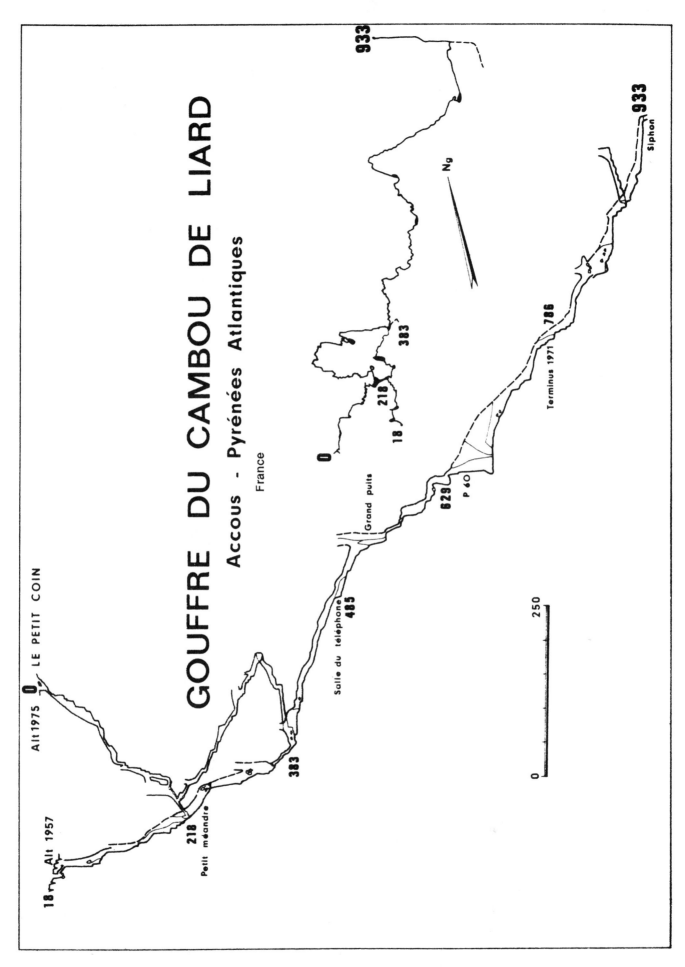

GOUFFRE DU CAMBOU DE LIARD

Accous - Pyrénées Atlantiques

France

S.G.C.A.F. tried without success to enlarge the tight fissure at 32 m in 1960. Cavers from Cannes and Ardèche succeeded in 1967, at the price of a spectacular amount of passage modification. This allowed them to reach –165 m in 1968, then –370 m in 1969. With the G. S. Montagne de Grenoble, they were stopped by a sump at –380 m in 1970.

Explorations in 1971 and 1972 added nothing to the depth. It was not until 1979 that the S.G.C.A.F. followed an air current and found a way on. In October and November they descended without much difficulties to –902 m. The S.G.C.A.F. was able to increase the length from 4067 m to 4900 m in 1980. This exploration was saddened by the death of two cavers in December 1979.

Map: from the drawings of the S.G.C.A.F., sent by B. Faure.

Bibliography: Lismonde (B.) et al. - *Grottes et Scialets du Vercors*, C.D.S. Isère, 1979, t. 2. Faure (B.) and Lismonde (B.) - *Le gouffre de la Fromagère*, *Scialet*, Grenoble, 1979 (8): 15-28.

10. *gouffre Touya de Liet* .... –894 m
(Accous, Pyrénées-Atlantiques)

Among the major caves in the lapiaz of the col d'Iseye is found the Touya de Liet, 700 m NE of the pic Permayou, at an altitude of 2058 m. Access is by a long walk from Accous or Laruns up to the cirque de Liet. The gouffre has four other entrances at 2045, 2032, 1990, and 1880 m altitude.

1/50,000 map Laruns-Somport . x 367.72; y 75.17.

The cave is formed in the Santonian and Campanian limestones. The resurgence is at the fontaine des Fées (Accous, alt. 460 m), in the Aspe valley.

The entrance was discovered in 1973 during the expedition organized by the Société Spéléologique et Préhistorique des Pyrénées Occidentales, regrouping including the C.D.S. de l'Ardèche and the S.G.C.A.F. The cave was pushed at that time to about –500 m.

In 1974 an expedition composed of the same groups reached the bottom of the cave (at –894 m) where there is a huge 302 m deep pit (see "The Great Pits"). The lower entrance, the gouffre de la Porte Etroite, descended in 1975, was connected the same year to the main cave. The following year another entrance, the gouffre de Liet N° 210 (alt. 1990 m), was also integrated into the Touya de Liet system by local cavers, bringing the length to 3750 m.

Map: collection of drawings by the explorers sent by J.-P. Besson and M. Chiron.

Bibliography: Poggia (F.) - *Le gouffre André Touya*. *Scialet*, Grenoble, 1974 (3); 55-61 and 1975 (4); 136-138, 141. Lismonde (B.) - *La Porte Etroite*. *Scialet*, 1975 (4): 133-135.

11. *tanne aux Cochons* .... –825 m
(Aillon-le-Jeune, Savoie)

This system has four entrances: the tanne itself (alt. 1495 m), the tanne du Névé (1468 m), connected in 1964, the Porte Cochère (1470 m), and the tanne Froide (1275 m), connected in 1983. It is located in the massif du Margériaz, not far from the Baban trail which services the location known as the "place à Baban" (cabin).

1/50,000 map Chambéry. x 890.01: y 77.30 (Cochons). x 890.975: y 77.710 (Froide).

The system is developed in Urgonian limestones. It traverses the Hauterivian (impermeable). The resurgence is the exsurgence du Pissieu (dye trace of August 14, 1960), 4 km from the terminal sumps (alt. 680 m).

The first explorer of the tanne aux Cochons was Pierre Chevalier, who descended to –106 m in 1953. In 1957 and 1958 the S. C. Lutèce and S. C. Paris returned to reach –365 m. In 1971 the S. C. Savoie became interested in the cave, bringing its length from 4000 m to 5389 m. In 1979 they found a way on that made a big jump in the dimensions of the cave, bringing it to –646 m deep and 7508 m long. On June 27, 1981 the S.C.S. (J.-L. Fantoli) dived the sump and reached –671 m.

The tanne Froide was known to –40 m in 1946. In June 1983, three individuals were able to find the way on and, in the same month, to connect with the main drain of the mountain; they reached –584 m, having traversed 4100 m of new passages. A pit parallel to the Baston pit permitted the connection with the tanne aux Cochons.

At –584 m (–804 m) there is a first sump which was dived on October 22, 1983 by J.-L. Fantoli. He got through after 10 m, then penetrated the second (30 m) and the third (45 m) to reach a fourth sump in which he penetrated 235 m distance and 21 m depth, bringing the depth of the cave to –825 m. The length in 1984 was 15,468 m.

Map: from the drawings of the S. C. Savoie, sent by P. Lesaulnier.

Bibliography: Spéléo-Club de Savoie - Massif du Margériaz, *Grottes de Savoie*, 1973 (3): 17-24. Lesaulnier (P.) - Tanne Froide. Margériaz, –605, *Spelunca*, 1984 (13): 40-42.

12. *puts dets Tachous* .... –804 m
(Saint-Pé-de-Bigorre, Hautes-Pyrénées)

The puts is located at an altitude of 1320 m, on the Arou mountain, in the massif du primont Pyrénéen, 10 km to the west of Lourdes. Access is from Saint-Pé-de-Bigorre or Asson, with 3 hours of hiking required.

1/50,000 map Lourdes. x 394.12: y 89.411.

The cave is developed in the Jurassic age dolomitic limestones (Kimmeridgian and Callovo-Oxfordian). The resurgence is unknown.

The G. S. des Hautes Pyrénées de Tarbes discovered the entrance in May 1983. Winter interrupted progress at –511 m, with a lot of work having been done in the meantime to enlarge a 40 m long squeeze at the –365 m level. These narrow passages, coupled with a very damp pit, slowed exploration in 1984. A sump was reached at the –637 m level. The latter was bypassed by a fossil passage at –585 m which returned to the main stream (Isarce river) which was followed to –725 m, when, again, winter put a stop to the explorations. In 1985 the main stream was followed for an additional two kilometers to a sump at –804 m.

This cave has many wet pits (from –365 to –600 m) and narrow passages which make exploration difficult. It is 3960 m long.

Map: from the drawings of the G.S.H.P., sent by M. Douat.

13. *réseau Ded* .... –780 m
(Saint-Pierre-de-Chartreuse, Isère)

This system has five entrances: the puits Kriska (alt. 1558 m), the puits de l'Escalade (1608 m), the puits de

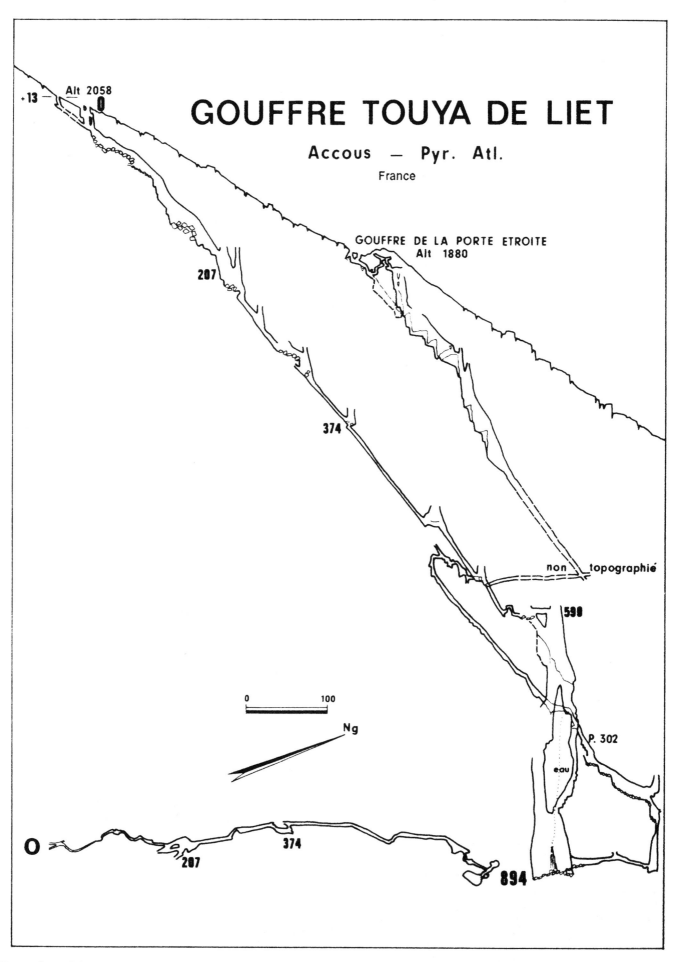

GOUFFRE TOUYA DE LIET

Accous — Pyr. Atl.

France

Alt 2058

+13

GOUFFRE DE LA PORTE ETROITE
Alt 1880

207

374

non topographié

590

0   100

Ng

P. 302

eau

O

207

374

894

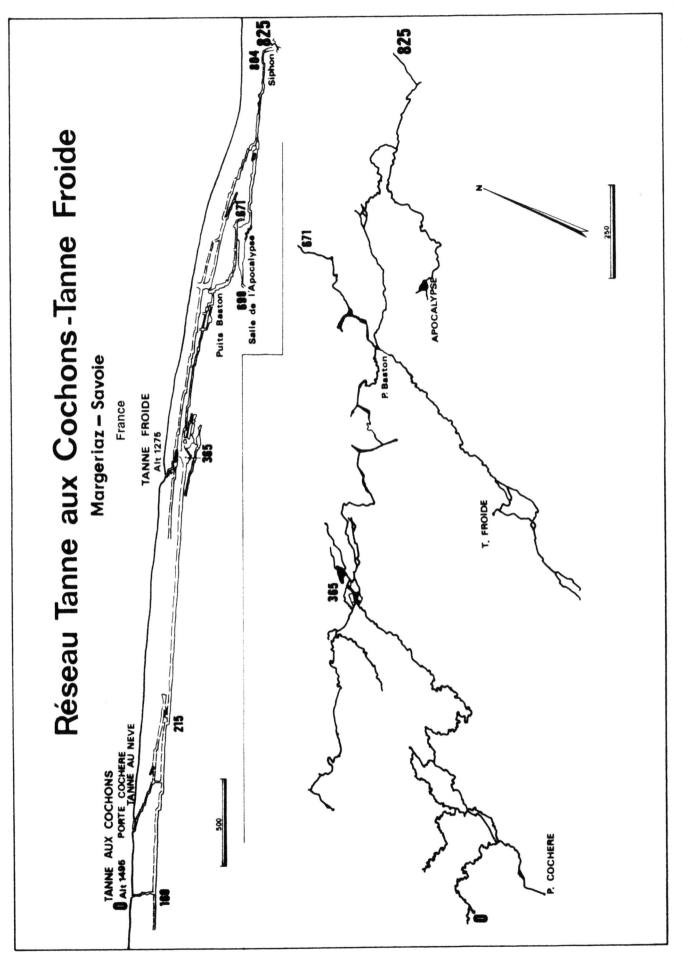

Réseau Tanne aux Cochons-Tanne Froide

Margeriaz — Savoie

France

TANNE FROIDE
Alt 1275

TANNE AUX COCHONS
Alt 1496

PORTE COCHERE
TANNE AU NEVE

160

215

365

671

698
Salle de l'Apocalypse

804

825

Siphon

Puits Baston

671

825

APOCALYPSE

P. Baston

365

T. FROIDE

P. COCHERE

0

250

500

N

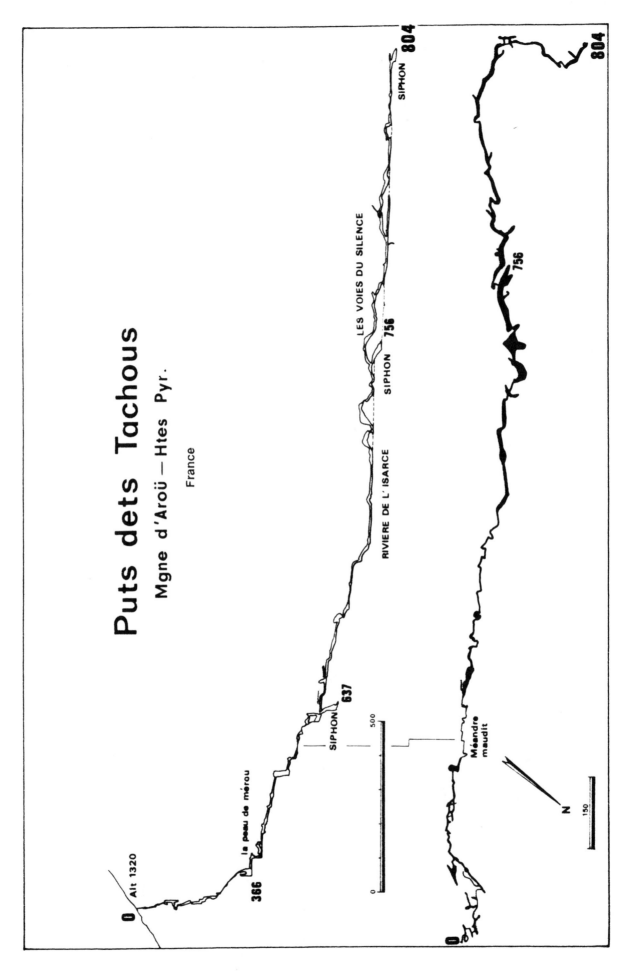

Puts dets Tachous

Mgne d'Aroü — Htes Pyr.

France

LES VOIES DU SILENCE

RIVIERE DE L'ISARCE

SIPHON 756

SIPHON 804

756

804

804

SIPHON 637

la peau de mérou

Alt 1320

0

366

500

Méandre maudit

N

150

0

0

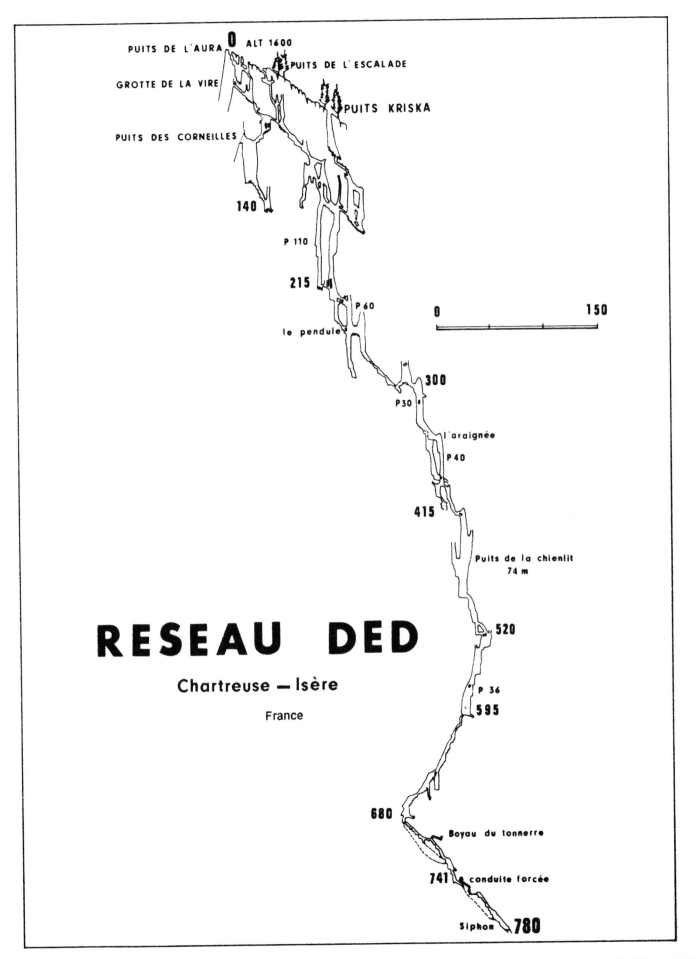

PUITS DE L'AURA **0** ALT 1600

PUITS DE L'ESCALADE

GROTTE DE LA VIRE

PUITS KRISKA

PUITS DES CORNEILLES

140

P 110

215

P 60

le pendule

0   150

300

P 30

l'araignée

P 40

415

Puits de la chienlit
74 m

520

RESEAU DED

Chartreuse — Isère

France

P 36

595

680

Boyau du tonnerre

741 conduite forcée

Siphon **780**

l'Aura, the grotte de la Vire, and the puits des Corneilles. It is located in the massif de la Chartreuse, about 20 minutes walk from the chalet du Charmant Som, on a trail (at the end of the Charmant Som road) heading towards the east face and descending to the Collet.

1/50,000 map Domène. x 869.14; y 341.83 (Kriska).

The probable resurgence is the source de la Porte de l'Enclos (alt. 780 m), 3 km to the NE. The whole system is in Urgonian limestone.

The puits "Criska" was discovered in 1962 by the S. C. Cartusien, thanks to a dog trained in finding pits! With the S.G.C.A.F., they removed rocks and dirt at − 27 m, following an air current, and were able to reach − 230 m. On September 4, 1966, J. -C. Dobrilla and A. Marbach found the way on, a very tight passage with good air movement. It took several uses of explosives (May 1968) by the G. S. de Fontaine-La Tronche in order to open this passage to cavers. It only remained to descend the pits that followed. A sump at − 535 m was bypassed and on July 15, 1968, the explorers were stopped at − 640 m.

The connections with the other entrances were made in 1969 (grotte de la Vire in May, puits de l'aura on November 11), adding 60 m of vertical extent. At the bottom of the cave, the G. S. F.L.T. descended to − 740 m in a tight rift which should only be entered in good weather. Thanks to very dry weather on September 27, 1970, they were able to follow the passage safely to a sump at − 780 m.

Map: furnished by J.-C. Dobrilla and A. Marbach.

Bibliography: Lismonde (B.) and Drouin (P.) - *Chartreuse Souteraine*, Grenoble, 1985, pp. 101-106.

14. *réseau des Arres Planères* .... −774 m
(Arette, Pyrénées-Atlantiques)

This cave is located in the massif de la Pierre Saint-Martin, and has five entrances, the AP7 at 1709 m altitude, the GL 4 (or Lonné-Peyret) and GL 4 bis at 1652 m, the T 102 at 1635 m, and the GL 80 at 1652 m. The entrances are located about 1000 m from the ski area of Arette and 500 m to the right of the road which leads to the col de la Pierre Saint-Martin. The cave is part of the same hydrological system as the Pierre Saint-Martin, to which it is strictly parallel. The resurgence is Bentia (alt. 446 m). The grotte d'Arphidia (see below) is very close to the bottom of the cave, but a connection has not yet been made.

1/50,000 map Larrau. x 347.15; y 79.82 (GL 4).

The exploration of the system began in the GL 4 (or Lonné-Peyret). The small entrance was found in 1968 by the S. C. Rouen cavers, who were stopped by a constriction at –10 m. Rediscovered in August, 1970 by two Belgians (who reached –120 m), it was again rediscovered the same month by the G. S. de Fontaine-La Tronche, who descended a series of pits to pop into a large river gallery at –360 m. On November 10, the bottom was reached at –717 m, with the upstream passage being explored for 1900 m (making the total length about 6 km).

In 1971 the G. S. de F.L.T. and the Centre Routier Spéléo (Bruxelles) explored further upstream (adding about 2 km), but enthusiasm was cooled by the death of Felix Ruiz de Arcaute who was killed when he became stuck in a waterfall.

The connection with an upper entrance (the AP 7) in 1977 by the S. C. Paris increased the depth to − 769 m and

the length to about 10 km. Efforts were concentrated on the upstream passage in 1978 and 1979. A third entrance was added in 1980, and a fourth in 1983, bringing the length of the system to an estimated 15 km.

Map: from the documents of M. Douat and the Bulletin ARSIP 9-11.

Bibliography: Marbach (A.) - Le gouffre Lonné-Peyret, *Spelunca*, 1972 (4): 103-105. *Bull. A.R.S.I.P.*, 1970-1971 (5-6) and 1977-1980 (12-15): 33-36.

15. *réseau de la Tête des Verds* .... −768 m
(Magland, Haute-Savoie)

This system has four entrances: the gouffre des Marmottes (with two openings, alt. 2078 m), the gouffre de la Tête des Verds (alt. 2070 m), and the gouffre du Petit Loir (alt. 2065 m). It is located in the désert de Platé and is situated to the SW of a large grassy butte (2128 m) from which its gets its name. Access is from Flaine by taking the trail which climbs under the cable car.

1/50,000 map Cluses. x 938.93; y 119.78 (Tête des Verds).

The system traverses the Senonian and the Albian to mainly develop in the Urgonian. The presumed resurgence is the source de Magland, at an altitude of 502 m.

The entrance to the gouffre de la Tête des Verds was long known to the shepherds (it was shown to E. Chaix in 1895). It was first explored in 1969 by M. Félix and G. Morel to − 140 m. In 1981 the G.E. du Karst Haut-Alpin found a way on at − 115 m, but were stopped at − 170 m by a constriction. A connection was made with the gouffre des Marmottes, explored in 1970 by N. Porret, M. Félix, and C. Moret (− 140 m) and in 1977 by the S. C. Lyon (− 185 m).

The gouffre du Petit Loir (whose entrance was dynamited open) was discovered on July 13, 1981, by the G.E.K.H.A., who overcame a 217 m deep pit, descended to − 268 m, then, in the autumn and in December of 1981, reached the flooded zone at − 750 m. During the winters of '82-'83 and '83-'84 the upstream river passage was explored. Connected to the gouffre des Marmottes in the summer of 1983, the vertical extent of the system increased to − 768 m and the length to about 7500 m.

Map: from the exploration notes (profile) and the drawings of the G.E.K.H.A. (plan), sent by G. Gros.

Bibliography: Maire (R.); Rigaldie (C.) - *Spéléo sportive dans les Alpes de Haute-Savoie*. 1984, pp. 107-111.

16. *gouffre des Bourrugues N° 3* .... −745 m
(Arette, Pyrénées-Atlantiques)

Located in the massif de la Pierre Saint-Martin, at an altitude of 1620 m, it is developed between the gouffres du Couey Lotge and the réseau des Arres Planères, near the Arette ski area. (1/50,000 map Larrau. x 349.02; y 80.05). The resurgence is still not known, but appears to be different that those of the Arres Planères or the Couey Lotge.

Discovered in 1972, it was explored the same year by the G. S. de Fontaine-La Tronche who reached − 305 m and reconnoitered about 3 km of passages. In 1973 a first 25 m long sump was passed by R. Jean, who stopped at a second sump. To the explorers' surprise, a recent rockfall impeded access to the upstream river passage which had been explored in 1972! During 1977-1978, various teams of the A.R.S.I.P. discovered passages increasing the length to 4050 m. The second sump (40 m long) was passed by

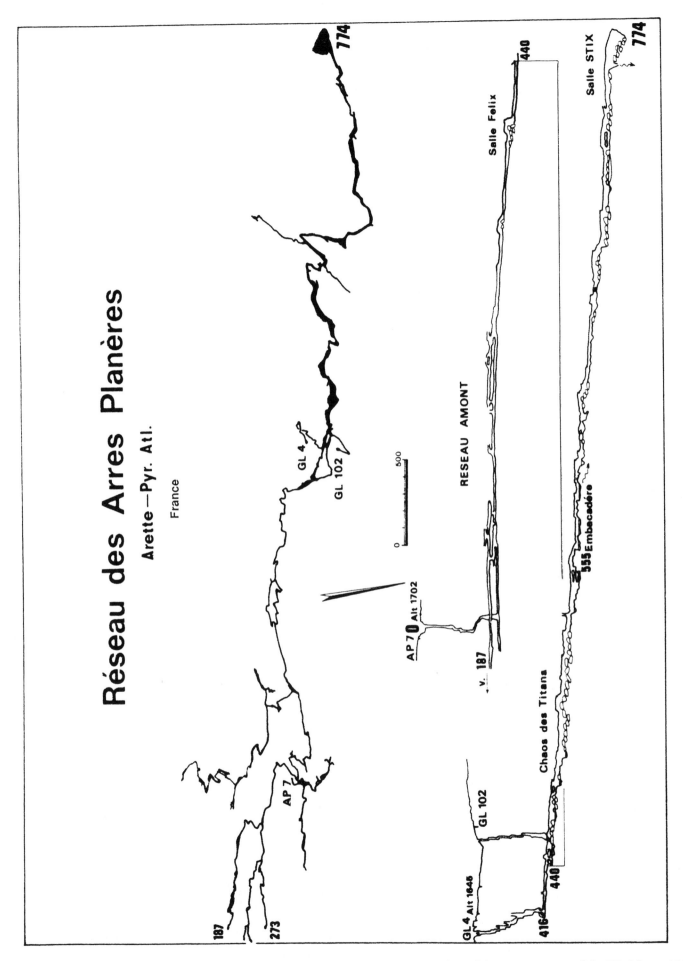

Réseau des Arres Planères

Arette—Pyr. Atl.

France

GL 4

GL 102

187

273

AP 7

774

AP 7 **0** Alt 1702

v. 187

RESEAU AMONT

Salle Felix

440

Salle STIX

774

555 Embacadère

Chaos des Titans

GL 102

GL 4 Alt 1846

440

416

0   500

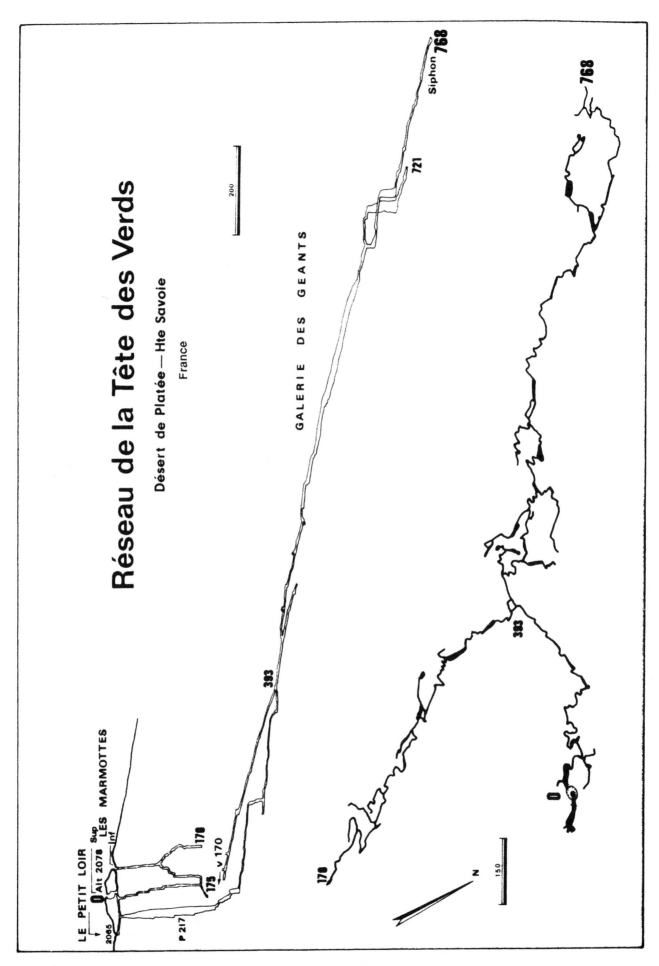

Réseau de la Tête des Verds

Désert de Platée — Hte Savoie

France

GALERIE DES GEANTS

LE PETIT LOIR

LES MARMOTTES

Siphon 768

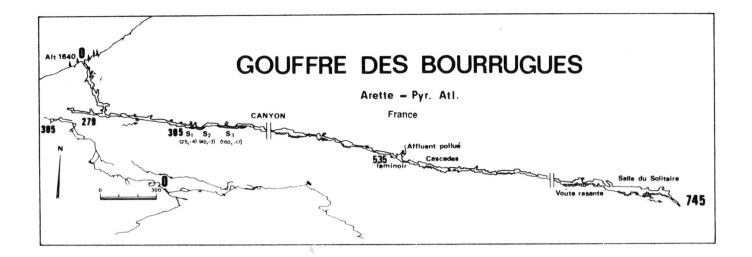

GOUFFRE DES BOURRUGUES

Arette – Pyr. Atl.

France

the C.D.S. des Deux-Sèvres, giving a vertical extent of 320 m. They reconnoitered the third sump for a distance of 60 m.

December 29, 1982: Frédéric Poggia reequipped the cave to do a solo exploration. This was his third attempt. On December 31, two members of the S. C. d'Aubagne accompanied him to the flooded zone. F. Poggia passed the first two sumps, then a third, to discover a river. He explored 1200 m of passage and stopped at –395 m at the top of an 8 m waterfall.

In late February 1983 he returned, again alone, to continue exploration. He passed his previous stopping point and passed a polluted side stream at –535 m and a large dome at –635 m. He passed a collapsed area where the river disappeared then, at –745 m, came to a room 5530 m from the entrance and from which he could find no way on!

Map: from the profile of F. Poggia in *Spelunca*, 1983 (10).

Bibliography: Linger (G.) - Le gouffre B 3, *Scialet*, Grenoble, 1973 (2): 81-85.
Poggia (F.) - Le gouffre de Bourrugues, *Spelunca*, 1983 (10): 21-24.

17.   **gouffre de la Ménère** .... –735 m
(Saint-Pé-de-Bigorre, Hautes-Pyrénées)
Located at an altitude of 1318 m, 600 m to the NE of the puts dets Tachous (see above), at the summit of the Ménère cliffs.   1/50,000 map Lourdes. x 394.315; y 89.876.

The resurgence is unknown. It was discovered in May 1983 and explored by the G. S. des Hautes-Pyrénées de Tarbes to –274 m. The following year the bottom of the last pit was dug open and the depth of –300 m reached. In 1985, the G.S.H.P reached the bottom of the pits at –484 m and explored a narrow canyon. The pits are wet and the deepest is 142 m deep.   In 1986 more pitches were descended to a streamway which sumps at –735 m (*Spelunca* 1986 (26), plan, profile).

18.   **puits Francis** .... –723 m
(Saint-Pierre-d'Entremont, Isère)
Also called the réseau de Bovinant (it has four entrances, alt. 1565, 1565, 1600, and 1605 m), it is located in the massif de la Chartreuse, not far from the Grand Som. Access is by a trail leading from the convent of the Grande Chartreuse that climbs to the col de Bovinant: it is located a little before the col, 20 m from the trail.

1/50,000 map Montmélian. x 872.384; y 348.865.

A dye trace by B. Talour showed the resurgence to be the source de Noirfond (alt. 600 m). Urgonian limestones.

The puits was discovered on July 24, 1966, by J.-C. Dobrilla and F. Lugiez (G. S. de F.L.T.) who descended to –100 m. In 1967, the G.S. F.L.T., with the S. C. Seine had a summer camp during which they dug open some passage and reached the terminal sump at –688 m on August 23.

In 1975, the Association Spéléologique Alpine undertook the exploration of a small cave 35 m higher that had been discovered in 1968 by the S. C. Villeurbanne, and succeeded in a connection with the puits Francis. Complementary explorations from 1975 to the present by S. Aviotte and J. -C. Dobrilla have increased the length of the system to 6515 m (from 3150 m in 1975).

Map: from the drawings of the G. S. Fontaine-La Tronche and J. -C. Dobrilla, furnished by the authors.

Bibliography: Lismonde (B.), Drouin (P.) - *Chartreuse Souterraine*, 1985, pp. 144-148.

19.   *Antre des Damnés* .... –723 m
(Corrençon-en-Vercors, Isère)
This cave is located in the clot de la Fure (massif du Vercors) at an altitude of 1760 m.   From Corrençon, go towards the cabin of the serre du Play, but before arriving there take the trail of the Ranc de l'Abbé to the Grand Pot cabin.   The cave is located on the left, about 500 m from the cabin, at the foot of a 5 m rock outcrop.

Map 3236 Ouest. x 851.10; y 303.00.

The antre des Damnés appears to be part of the drainage basin feeding the goule Blanche (alt. 832 m). It is formed in Urgonian and Barremian limestones.

It was discovered in early July 1982 by P.-A. Sibué, but exploration was impeded by breakdown through which an air current blew.   Three digging sessions removed the obstacles and permitted the S. C. Fontanil to reach the depth of –360 m on October 11, having conquered a 205 m deep pit.   In December, the explorers were stopped at –570 m at a dead end, but shortly before it a canyon was found that led the explorers in 1983 to two sumps at –720 m. The

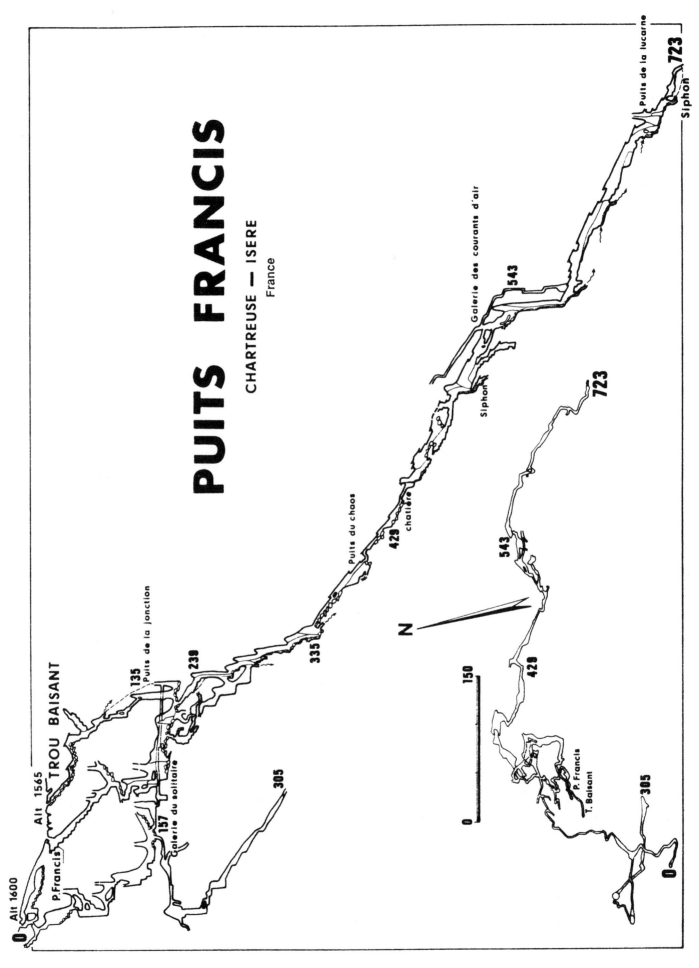

PUITS FRANCIS

CHARTREUSE — ISERE

France

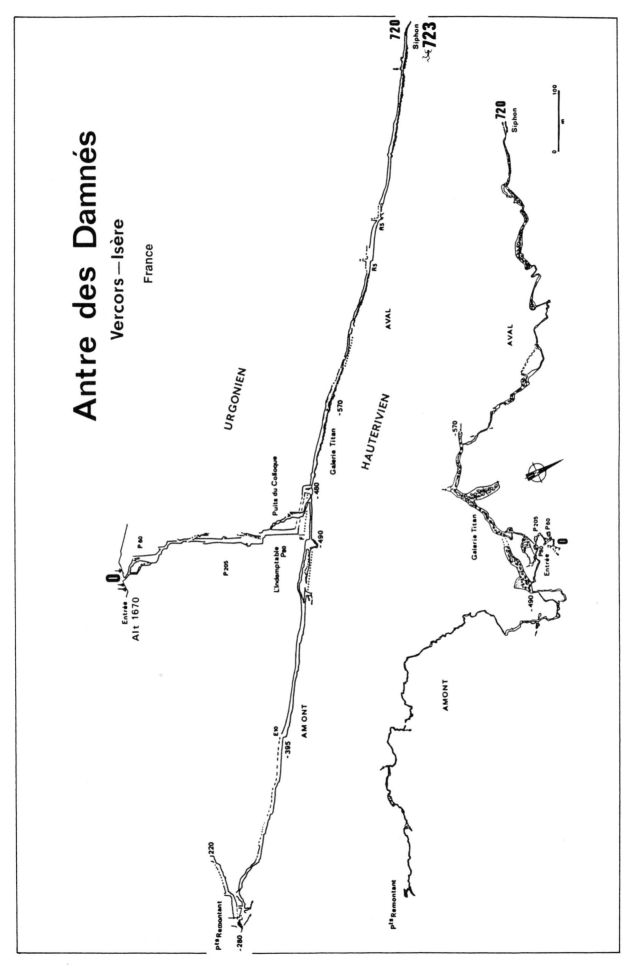

Antre des Damnés

Vercors—Isère

France

sumps were dived by F. Poggia (–3 m) but proved to be impenetrable.

Map: from the documents sent by J.-J. Delannov.

Bibliography: Sibué (G.), Pomot (C.) - L'antre des Damnés, *Scialet*, 1982 (11): 32-35 and C. Pomot, *ibid.*, 1984 (13): 36-37.

20. *gouffre du Mont-Béas* ....(–707, +7) 714 m
(Le Port, Ariège)

This system has six entrances: A 8 (alt. 1624 m), A 5 (1624 m), A 3 (1595 m), A 9 (1595 m), A 1 (1585 m), and the gouffre des Feuilles Mortes. It is located 500 m south of the Mont Béas (Pyrénées ariégeoises), on the wooded slopes that lead down towards the Lhers pond.

1/50,000 map Aulus - les-Bains. x 520.42; y 56.56 (A 5).

Formed in Cretaceous breccia, the resurgences are the springs of the Neuf Fontaines, 1 km NW of Aulus-les-Bains (alt. 720 m).

The A 5 entrance (–40 m) was discovered in 1946 by the abbé Glory, but exploration did not begin until 1967, when the Cordée Spéléologique du Languedoc reached a depth of –360 m. With the collaboration of local as well as Spanish cavers, the C.S.L. reached the bottom of the cave at –726 m, readjusted to –687 m in 1975 by the G. S. Massat who brought the length to 3490 m in 1976.

In 1983 J. Michel passed the terminal sump, then a second sump, adding 20 m to the vertical extent. In the same year, the S. C. Haut-Sabarthez connected the gouffre des Feuilles Mortes to the system.

Map: from the drawings of 1975 from the G. S. Massat and the A.R.S. de l'Oule.

Bibliography: Glory (A.) - Le bassin hydrologique de l'étang de Lhers, *Annales Spél.*, 1947: 63-66. Claria (J. -P.) and Petit (J. -P.) - Expédition étang de Lhers 1968..., *Spelunca*, 1970: 12-21. *L'Excentrique*, Bull. de la C.S.L., 1968 (13). *Bull. A.R.S. Oule et G. S. Massat*, 1976.

21. *gouffre de la Consolation* .... –711 m
(Accous, Pyrénées-Atlantiques)

Located between the gouffre du Cambou de Liard and the Touya de Liet (see above) in the lapiaz of the col d'Isère. The presumed resurgence is the fontaine des Fées (alt. 460 m), while the altitude of the entrance is 2125 m! It is formed along the contact of the Turonian and schisty Cenomanian limestones.

1/50,000 map Laruns-Sómport. x 366.92; y 75.15.

The entrance was discovered in August 1977 by M. Chiron of the S.G. C.A.F., who, on the following days, explored down to –405 m, stopping in a large room. In August 1978 the same club explored three parallel branches, the deepest of which stopped at about –580 m (the mapping went only to –520 m). The S.G. C.A.F. pursued other parallel branches in 1978. One of these led down to –711 m, where a narrow passage let only the breeze pass. The length is 3448 m.

Map: drawings of M. Chiron and of the S.G. C.A.F., sent by B. Faure and M. Chiron.

Bibliography: *Scialet*, Grenoble, 1977 (6): 143-147; 1978 (7): 82-92; 1979 (8): 94-108.

22. *grotte D'Arphidia* ....(+149, –561) 710 m
(Sainte-Engrâce, Pyrénées-Atlantiques)

The cave, located in the massif de la Pierre Saint-Martin, does not have a natural entrance. It was discovered in 1956 during the digging of the EDF tunnel (alt. 1052 m) (see above, réseau de la Pierre Saint-Martin). To gain access, one climbs up the ravin d'Arphidia from Sainte-Engrâce. Even though intertwined with both the réseau de la Pierre Saint-Martin and the Arres Planères, no connection with either one has yet been found.

1/50,000 map Larrau. x 344.51; y 80.35.

The upstream passage was explored in 1958 and 1962-1964 (+75 m) by C. Queffélec and his team, which included J. Choppy and Loriaux; in 1973 the height of 149 m was gained by chimneys and climbs (G.S.H.P. and G. S. Charentes).

Downstream the Queffélec team was stopped at –135 m. In 1966, exploration was undertaken by the A.R.S.I.P. (S. C. Rouen) who descended to –180 m, then –360 m in 1968, having overcome the "chaos du Baron". Between 1972 and 1976 the mapped length increased from 1192 m to 6647 m. In 1980, fossil passages (Arphidia III) were discovered and explored in 1981 (length 11,634 m). One branch went lower than the previous bottom (–493 m) and popped into a flooded pit. This was dived in 1983 to a depth of –68 m, giving the cave a vertical extent of 710 m, while the length reached 16,500 m. A voice connection with the salle de La Verna in the PSM was made from the Arphidia III area. In 1986 the G.S.H.P and the S. C. Gascogne discovered Arphidia IV to the north, giving a total length of over 18 km. In 1987 the length was over 21 km.

Map: from the drawings of the A.R.S.I.P. (S. C. Paris, C.R.S. Bruxelles, E. S. Pyrénées-Charente, G.S.H.P.).

Bibliography: Courbon, *Atlas de grands gouffres du monde*, 1979, pp. 89, 108.
Douat (M.), Sautereau de Chaffe (J.) in *Bull. A.R.S.I.P.*, 1974-1976 (9-11): 10-21.

23. *réseau Bel-Espoir-Diau* .... –701 m
(Dingy-Saint-Clair/Thorens-Glières, Haute-Savoie)

The system is comprised of the tanne du Tordu (alt. 1660 m), the tanne des Météores (1623 m), the tanne du Bel Espoir (1575 m), and the grotte de la Diau (962 m). It is formed under the massif du Parmelan. The grotte de la Diau has an impressive entrance in the back of a cirque at the foot of the plateau, 1700 m NE of Thorens-Glières. Access is by la Verrerie and a trail on the right bank of the torrent de la Filière. The three upper entrances are on the SE border of the plateau. One gets to them using the Avernioz trail at the chalet de l'Anglette.

1/50,000 map Annecy-Bonneville. x 904.35; y 113.14 (Diau). 905.80; 141.11 (Bel Espoir). 903.90; 113.29 (Tordu). 904.04; 112.785 (Météores). Urgonian limestones.

The grotte de la Diau has always been known. R. de Joly checked out the entrance in 1932. Pierre Chevalier made his first visit on August 14, 1937. From 1938 to 1940, Bocquet and his friends explored the entrance area. From 1942 to 1943 Chevalier and F. Petzl did the same. In 1949 and 1950, Chevalier and the S. C. Lyon reached a sump at +162 m, 2650 m from the entrance (cave length 5200 m).

It was not until 1975 that the S.G. C.A.F. returned to the exploration: they climbed upstream in a side stream

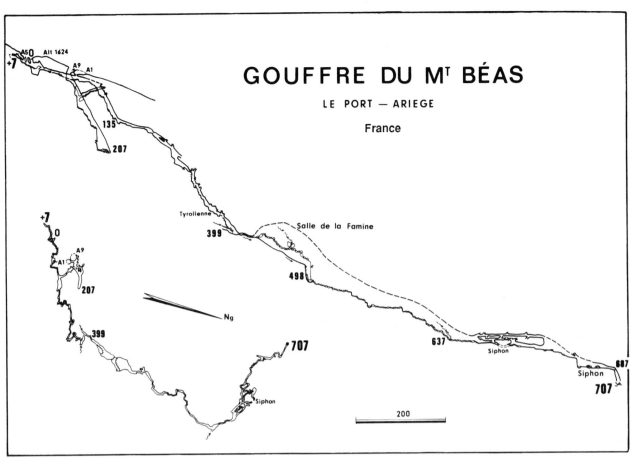

GOUFFRE DU Mᵀ BÉAS

LE PORT — ARIEGE

France

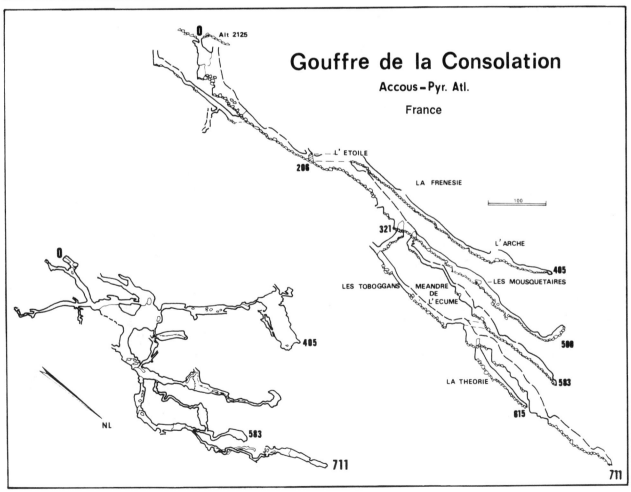

Gouffre de la Consolation

Accous — Pyr. Atl.

France

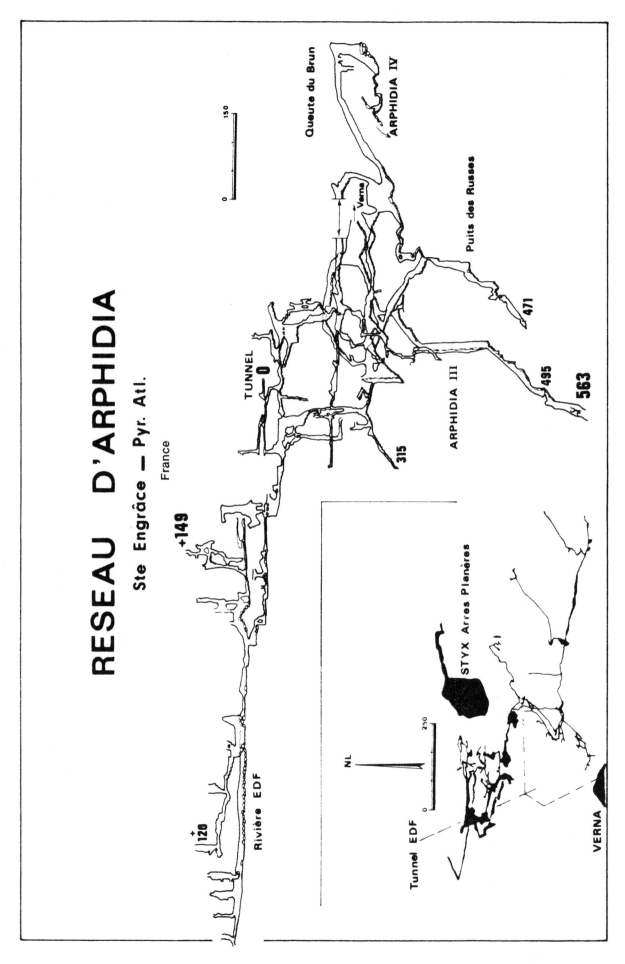

RESEAU D'ARPHIDIA

Ste Engrâce — Pyr. Atl.

France

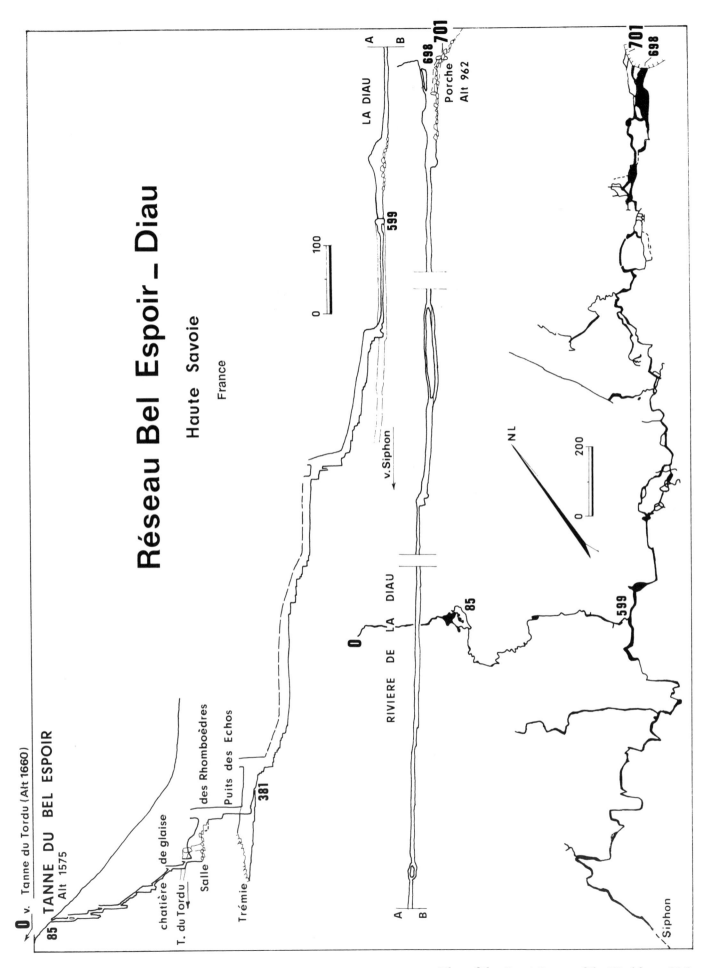

Réseau Bel Espoir_Diau

Haute Savoie

France

v. Tanne du Tordu (Alt 1660)

TANNE DU BEL ESPOIR
Alt 1575

85

chatière de glaise

T. du Tordu

Salle des Rhomboèdres

Puits des Echos

Trémie

381

0

RIVIERE DE LA DIAU

85

599

A
B

LA DIAU

A
B

599

v.Siphon

698

701

Porche
Alt 962

100

0

NL

200

0

701

698

Siphon

and reached +330 m for 7100 m of length. F. Poggia, alone, passed the sump that had stopped Chevalier and explored 3 km beyond!

On the plateau, B. Talour discovered the tanne du Bel Espoir (August 1975): it was explored in 1976 by the S.G. C.A.F. They connected on June 13, 1976 to the Diau, giving a depth of –613 m and an estimated length of 11 km. The tanne du Tordu, explored to –114 m in 1974 by the S. C. Annecy, was re-explored in 1978 by the S.G. C.A.F., who made the connection with the Diau on September 16, 1978, for a mapped length of 9600 m and a depth of –698 m. In 1979 F. Poggia, first with Dobrilla, then alone, dived sumps N° 2 and 3. While on the plateau, G. Masson and M. Bianco, after dropping 400 m in the pits of the tanne des Météores (discovered in August 1976 by G. Masson), connected to the Diau upstream of sump N° 1, at a depth of –559 m. The surveyed length reached 14,490 m.

Map: from the drawings of the S.G. C.A.F.

Bibliography: *Scialet*, Grenoble, 1975 (4); 1976 (5); 1978 (7); 1979 (8).
Talour (B.) - La traversée tanne du Bel Espoir-grotte de la Diau, *Spelunca*, 1976 (4): 146-148.

24. **gouffre de la Benoîte** .... –700 m
or **creux de la Benoîte** (Arith, Savoie)
Located in the Massif de Préporlain. Explored in 1986 by S. C. de Savoie (*Spelunca* 1986 (24), plan).

25. *grotte de Gournier* (Choranche, Isère) ...... **+680 m**
This resurgence cave in the Vercors is located near the Coufin show cave. It is located at the foot of the lovely cliffs which overlook the cirque de Choranche, at an altitude of 580 m. It is formed along the Urgonian and Hauterivian contact.

(1/50,000 map Romans. x 840.78; y 313.04).

The entrance has been long known, with O. Decombax reconnoitering the entrance in 1899. The Méduse was climbed in 1947 by the S. C. Pau (Jean Deudon) who explored the fossil trunk with André Bourgin in 1947 and 1949 for about 2500 m. In 1952 Pierre Chevalier and the S.C.A. de Lyon reached the Grande Muraille at +190 m. In 1964, the S. C. de la Seine, with the Marbach brothers, discovered the affluent des Parisiens and reached +260 m in 1965: then +270 m in 1966 with the G. S. de F.L.T., bringing the length to 6225 m, then 8309 m in 1969. The explorations were pursued by the S. C. Lyon in 1973; they dove the terminal sump and reached +460 m in 1975 for a length of 11,400 m, but in 1976 the death of three cavers caught by a flood interrupted their explorations. They were restarted by the S. C. Dijon in 1981 and 1982; they climbed the terminal waterfall of 1975, explored up-river, climbed several pitches and were stopped, +680 m above and 8325 m from the entrance at a 7 m pit filled with sand where the stream becomes impossible to follow. The length was then 15,125 m. Some hope of finding additional passage is given by the many side streams that feed the main river.

Map: from the drawings of S. C. Seine, G.S.F.L.T., S. C. Duchère, S. C. Lyon, S. C. Dijon, sent by P. Degouve.

Bibliography: Lismonde (B.) *et al.* - *Grottes et scialets du Vercors*, 1979, t. 2, pp. 148-153.

26. **gouffre de Génieux** .... –675 m
(Saint-Pierre-de-Chartreuse, Isère)

This cave is located in the forêt de Génieux, between the chalet du col de la Charmette and the col de la Petite Vache, in the massif de la Chartreuse, at an altitude of 1410 m. The probable resurgence is the grotte de la Passerelle (alt. 650 m)

1/50,000 map Grenoble. x 866.59 ; y 342.49. Urgonian limestones.

The cave was discovered in May, 1972, by J.-C. Dobrilla. It was explored to –340 m by the G. S. de F.L.T., then to –675 m by G. Marbach, J.-C. Dobrilla and S. Aviotte the same year (Lismonde, Drouin, *Chartreuse Souterraine*, 1985, pp. 148-151, profile).

27. **gouffre du Caladaïre** .... –668 m
(Montsalier, Alpes-de-Haute-Provence)
This cave belongs to the hydrologic system of the fontaine de Vaucluse, located 40 km away (alt. 105 m) (dye trace of 1966). It is located at an altitude of 885 m between the ferme de l'Obœf and the ruins of Montsalier-le-Vieux, to the east of the plateau de Vaucluse (1/50,000 map Sault. x 861.63; y 198.01).

Its first exploration (Marty and Servel d'Apt) goes back to 1944. In 1946, the S.S. d'Avignon with the Eclaireurs de France descended to –220 m, then –313 m in 1947 and –470 m in 1948. In 1949, a constriction stopped the explorers at –487 m. In 1964, the S.S.A. found at way on (–518 m) which permitted them to reach –640 m in 1965 and –668 m in 1966 (Courbon, *Atlas des gouffres de Provence et des Alpes de lumière*, 1980, profile).

28. **réseau de Krakoukas** .... –658 m
(Accous, Pyrénées-Atlantiques)
One of its entrances, the hosse de las Garças (alt. 1780 m), is located slightly below the gouffre Touya de Liet (see above), 1200 m NE of the pic Permayou. The other entrances are the gouffre des Gendarmes (1820 m), the gouffre des Jumeaux (1830 m), and the gouffre du Rateau (1830 m). The resurgence is the fontaine des Fées (alt. 460 m).

1/50,000 map Laruns-Somport . x 367.96; y 75.45; z 1780 m.

Long known to shepherds, the hosse was first explored by the S.S.P.P.O. in 1969 (–120 m). In 1974, the S.G. C.A.F. took over and quickly reached the terminal sump at –632 m. The other entrances were connected in July, 1976 by the same club. The length of the system is 3207 m (*Scialet*, 1976 (5), profile, plan).

29. **réseau de la Pointe de Sans Bet** .... –656 m
(Sixt, Haute-Savoie)
Made up of the tanne Cassina (1875 m), the entrée des Artistes (1775 m), the tanne Inaccessible (1775 m), the tanne des Béliers (1845 m), and the tanne du Bourdon (1750 m), located in the massif du Haut Griffe (Upper Jurassic limestones). It can be found on the west slopes of the Pointe de Sans Bet. 1/50,000 map Samoëns-Pas de Morgins. x 946.03; y 130.68; z 1875 m. 945.32; 130.56; 1750 m.

The tanne Cassina was discovered (1976) and explored by the G. S. de F.L.T. in 1976-1977 (–505 m). The connection with the entrée des Artistes (1977) and the tanne Inaccessible (1977) gave the system a vertical extent of 515 m, then 656 m with the connection to the tanne du

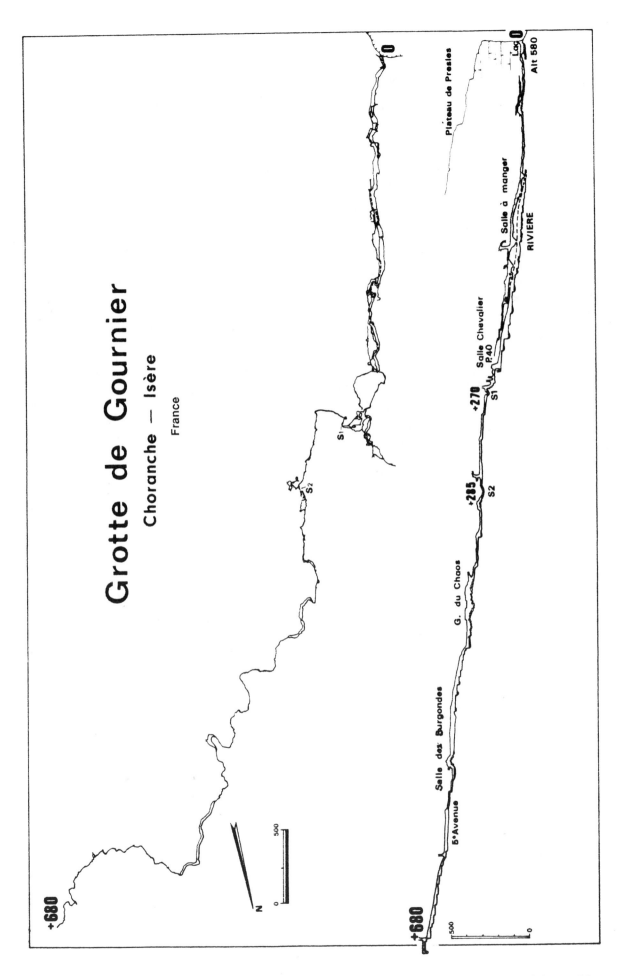

Grotte de Gournier

Choranche — Isère

France

+680

Plateau de Presles

+680

Salle des Burgondes

5°Avenue

G. du Chaos

+285

+270

S2

S1

Salle Chevalier

P.40

Salle à manger

RIVIÈRE

Log

Alt 580

S¹

S²

N

500

500

0

Bourdon (*Scialet*, 1976 (5), 1977 (6) profile; 1981 (10) profile).

30. aven du Vallon des Soupirs .... –640 m
(Saint-Christol, Vaucluse)

This cave at an altitude of 955 m is located on the plateau d'Albion, 2500 m to the WNW of Saint-Christol, and is hydrologically connected to the fontaine de Vaucluse (alt. 105 m) 1/50,000 map Sault. x 850.55; y 198.10.

The entrance was opened up in 1933 by F. Autran to –30 m (squeeze). This was enlarged in 1978 by the G. S. d'Albion who with the C.A.F. de Briançon reached –269 m. In 1979, cavers from Cavaillon and Toulouse "took over" the cave, passing a sump at –408 m and following a river trunk to another sump at –603 m. In 1981 F. Poggia passed this sump and came to another at –640 m that was judged impenetrable (Courbon, *op. cit.*, 1980, profile).

31. gouffre du Couey Lotge .... –637 m
(Arette, Pyrénées-Atlantiques)

Its entrance is at an altitude of 1478 m, in the lapiaz of the Braca de Guillers (massif de la Pierre Saint-Martin), near the road going to the Arette ski lifts. 1/50,000 map Larrau. x 349.21; y 81.16. The presumed resurgence is the main spring of the cave d'Issaux at Osse-en-Aspe (alt. 680 m).

The cave was discovered in 1957 by C. Queffélec. The C.D.S. Deux-Sèvres explored it in 1974 to –249 m). Between August 7 and 14, 1975 the same team reached the terminal sump at –625 m (length 2450 m, then 3800 m in 1976). A dive by P. Jolivet in 1985 deepened the cave by 12 m while its length reached 8200 m (*Bull. A.R.S.I.P.*, 1974-1976 (9-11) profile).

32. réseau de L'Alpe .... –614 m
(Saint-Vincient-de-Mercuze-Sainte-Marie-du-Mont/Chapareillan, Isère and Saint-Pierre-d'Entremont, Savoie)

This dendritic system has 35 entrances, of which the most famous or significant are the grotte du Biolet (alt. 1745 m, Savoie), the golet du Pompier (1737 m, St-Vincent...), and the gouffre de Source Vieille (1688 m). The upper entrance is the golet du Broyage (1782 m) and the lower entrance is the gouffre de la Combe des Arches (1550 m, Chapareillan).

It is located in the massif de l'Alpette, in Chartreuse. Access is from the Varvats, above Saint-Même, by climbing the trail on the west face, or from La Plagne by climbing to the col de l'Alpette.

Two dye traces (Biolet and insurgences of the chalet de l'Alpe) have shown that the resurgence is the source de Cernon, alt. 1160 m, at Chapareillan. The system is formed mostly in the Urgonian and in part in the lower Barremian.

1/50,000 map Montmélian. x 878.97; y 352.33 (Biolet). 879.78; 353.303 (Pompier). 880.385; 353.585 (Source Vieille). 878.99; 353.15; 1750 m (grotte aux Ours). 879.513; 353.411 (Broyage). 880.92; 354.585 (Combe des Arches).

Explorations began in 1936 when Pierre Chevalier discovered the grotte du Biolet, in which he reached –338 m in 1948. It was not until 1961 that explorations were undertaken by the S. C. Savoie who did most of the exploration and made most of the connections. In 1964, they reached –432 m in the Biolet, then –482 (–470, +12)

in 1965; the length was then 11,115 m. On September 16, 1971, thanks to the conquering of a 205 m deep pit, the golet du Tambourin was connected to the Biolet. This first important connection brought the length to 12,470 m. In 1975 the S.C.S. deepened the system; 493 m in the Biolet and –520 m in the golet du Pompier. On November 12, 1977, the Biolet-Ours connection gave the system a vertical extent of 527 m and a length of 21,698 m.

In 1981 the gouffre Brutus (alt. 1585 m) gave access to the main stream passage downstream of the grotte aux Ours. Connected to the Combe des Arches in 1982, it made a system of 7855 m. The year 1983 was important because it saw the connection (by the Clan des Tritons de Lyon) of the grotte du Biolet with the golet du Pompier, giving the system a length of 39,152 m.

In 1984, connections with the entrances at the head of the system (gouffre du Broyage) deepened the system to –602 m. Entering by the golet de la Source Vieille, connected in 1983 to the Biolet-Pompier system, the S.C.S. dove the terminal sump and established the connection with the Brutus-Combe des Arches system on August 15, 1984. These two great systems connected together form a huge system of 51,777 m in 1985. In 1986 three sumps were passed by J. -L. Fantoli to reach the head of a pit at –614 m.

Map: from the synthesis of B. Lismonde in *Chartreuse souterraine*, 1985, sent by B. Faure and general plan sent by J.-L. Fantoli.

Bibliography: Lismonde (B.), Drouin (P.) - *Chartreuse souterraine*, 1985, pp. 49-64. Fantoli (J. -L.) - Le golet de Source Vieille..., *Spelunca*, 1985 (20): 16-24.

33. trou Souffleur .... –610 m
(Saint-Cristol d'Ablion, Vaucluse)
Explored in August, 1986 by G. S. Bagnole Maroucle.

34. système de la Dent de Crolles .... –603 m
(Saint-Pierre-de-Chartreuse/Saint-Pancrasse, Isère)

This extraordinary labyrinth, made famous by the saga of Pierre Chevalier, is developed in a spur, the Dent de Crolles, which lies to the SE of the massif de la Chartreuse. It has seven entrances: the trou du Glaz (1697 m) and the grotte du Guiers Mort (1332 m) on the west face; the grotte d'Arguille (or A. Bouchacourt) (1685 m), the grotte Chevalier (1670 m) and the grotte des Montagnards (1765 m) on the east face, the P 40 (1935 m), and the gouffre Thérèse (1925 m) on the plateau. Thus, it traverses from east to west and from top to bottom the entire limestone mass (Hauterivian and Urgonian).

1/50,000 map Domène. x 875.75; y 340.64 (P 40); 875.51; 341.15 (Glaz). 875.85; 339.96 (Chevalier). 875.99; 342.15 (Guiers Mort).

The history of its exploration is one of the highlights in the history of speleology, and is principally due to Pierre Chevalier and his team from the S. C. Alpin de Lyon. From 1935 to 1947 they established a very large system with a vertical extent of 603 m and a length of 16,898 m, not neglecting the fact that Martel explored 350 m of galleries in the trou du Glaz in 1899, and R. de Joly descended to –199 m in 1933. Chevalier's goal was to connect the trou du Glaz to the resurgence, the Guiers Mort, a feat which was accomplished on August 9, 1941 (–365, +41; length 9164 m), then to the plateau, climbing to +70 m in 1943, +147

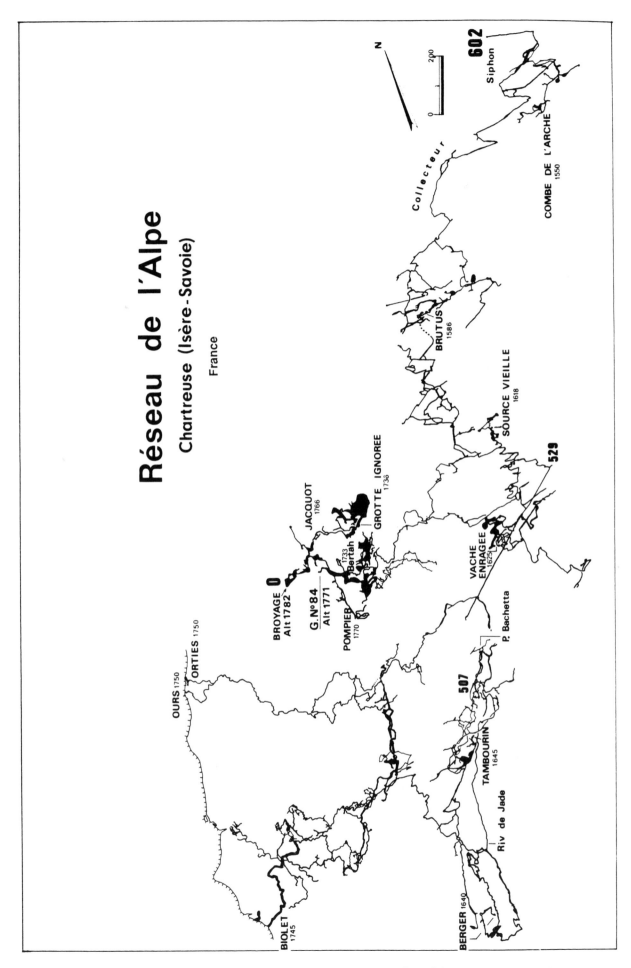

Réseau de l'Alpe

Chartreuse (Isère-Savoie)

France

602
Siphon
COMBE DE L'ARCHE
1550

Collecteur

BRUTUS
1586

SOURCE VIEILLE
1618

529

VACHE ENRAGEE
1625

JACQUOT
1766

GROTTE IGNOREE
1736

1733
Bertah

BROYAGE
Alt 1782

0

G. N°84
Alt 1771

POMPIER
1770

OURS 1750

ORTIES 1750

P. Bachetta

507

TAMBOURIN
1645

Riv de Jade

BERGER 1640

BIOLET
1745

N

200

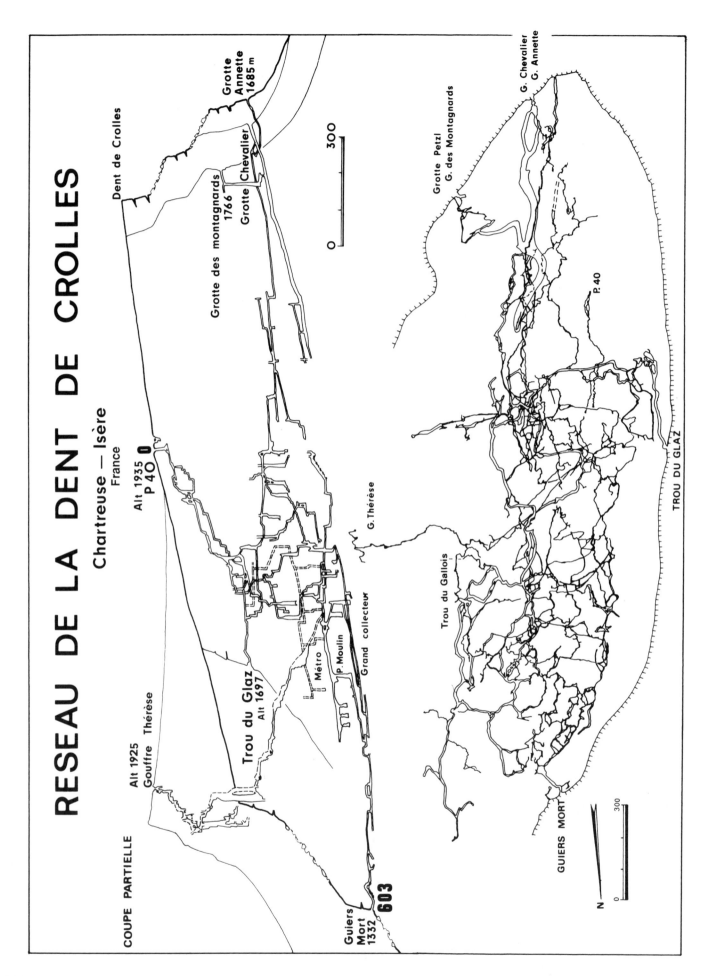

RESEAU DE LA DENT DE CROLLES

Chartreuse — Isère
France

COUPE PARTIELLE

Dent de Crolles

Grotte Annette 1685 m

Grotte Chevalier

Grotte des montagnards 1766

Alt 1935 Gouffre Thérèse

Alt 1935 P 40

P 40

Trou du Glaz Alt 1697

Métro

P. Moulin

Grand collecteur

G. Thérèse

Guiers Mort 1332

603

Grotte Petzl G. des Montagnards

G. Chevalier G. Annette

P. 40

Trou du Gallois

TROU DU GLAZ

GUIERS MORT

N

300

0

300

m in 1944, +184 m in 1945, and finally, on May 4 1947, after a dig, the trou du Glaz was connected to the P 40 to produce a total depth of –603 m.

In 1960 the clan des Tritons, under the leadership of Michel Letrone, returned to the exploration of the cave, with the goal of doubling the length surveyed by Chevalier. They succeeded in 1974 when they reached 32,400 m, the same year when a third wave of explorers came on the scene, led by Jo Groseil of the Furets Jaunes de Seyssins. This group brought the length of the system to over 40 km (40,460 m) in 1983, a year of intense exploration.

On September 9, 1984 the long awaited connection with the grotte Chevalier was finally realized (F.J.S.), giving a total length of 45,500 m, which was increased in 1985 to 53,800 m.

Map: from the synthesis of G. Bohec in *Chartreuse souterraine*, sent by B. Faure.

Bibliography: Chevalier (P.) - *Escalades souterraines* (*Underground Climbers*), Susse ed., 1948, 190 p. (Available in English from Cave Books). Lismonde (B.), Drouin (P.) - *Chartreuse soutterraine*, 1985, pp. 108-121.

35. scialet de la Combe de Fer .... –580 m
(Corrençon-en-Vercors, Isère)

Located at an altitude of 1555 m in the massif du Vercors, 2 km to the west of the petite Moucherolle. The resurgence is the goule Blanche (alt. 832 m) at Villard-de-Lans, from a dye trace in 1967. 1/50,000 map La Chapelle-en-Vercors. x 851.80; y 304.67. Urgonian limestone.

Martel explored the main gallery to –86 m. In 1937, A. Bourgin and the S. C. Paris stopped at –183 m at a constriction which was forced in 1963 by the A. S. Vercors (–367 m). In 1965 the A.S.V. reached –425 m and came to the terminal sump at –580 m in 1966. The projected length was 3400 m in 1975 (Lismonde *et al.*, *Grottes et scialets du Vercors*, 1979, t. 2, pp. 90-95, profile).

36. scialet du Clot de la Fure .... –580 m
(Corrençon-en-Vercors, Isère)

Located a half hour from the cabane du serre du Play, near Corrençon, in the Vercors, at an altitude of 1910 m. It apparently belongs to the drainage basin of the goule Blanche (alt. 832 m). 1/50,000 map La Chapelle-en-Vercors. x 851.86; y 302.98. Urgonian limestone, then lower Barremian.

Discovered on July 13, 1980 by the Montélimar A.S.C., who descended to –335 m. In 1981, the C.D.S. de la Drôme found the way on (following an air current), but came to a dead end at –580 m in 1982. (*Spelunca Mémoires*, 1983 (13), profile).

37. réseau de la Combe des Foges .... –577 m
(Samoëns, Haute-Savoie)

The system has three entrances, the puits du Solitaire (alt. 2122 m), the gouffre du Ver (2173 m), and the gouffre du Double S (2158 m). It is located in the combe des Foges, north of the désert de Platé. Access is from Flaine and the col Pelouse. 1/50,000 map Cluses. x 941.13; y 121.19; z 2173 m.

The resurgence is the source du Déchargeux (alt. 1370 m). The system crosses the Priabonian limestones and sandstones, the Senonian limestones, the Albian sandstone, and the Urgonian limestone.

In 1974 R. Maire discovered the puits du Solitaire which the S. C. Lyon explored to –350 m. In 1975 the S.S.S. de Genève discovered the gouffre du Ver and established the connection on July 27, while the S.C.L. brought the depth to –481 m, then –522 m in January, 1976. The discovery (1978) and the connection (August 29, 1979) of the Double S by the G.E.K.H.A. permitted the deepening of the system in 1980 to –577 m. (R. Maire, Rigaldie, *op. cit.*, 1984, pp. 133-142, profiles).

38. aven Jean Nouveau .... –573 m
(Sault-en-Vaucluse, Vaucluse)

Opening at an altitude of 831 m, near Saint-Jean-de-Durfort. 1/50,000 map Carpentras. x 844.50; y 196.10.

This is a cave with a 163 m entrance pit that was conquered by Martel on August 31, 1892, making it famous. Its resurgence is the fontaine de Vaucluse (alt. 105 m).

In 1933 R. de Joly deepened the cave to –186 m. Beginning in 1954 local clubs resumed the exploration and, after performing a dig, reached –337 m in 1957. Under the A.V.E.N. several annual explorations took place; –384 m (1961), –392 m (1967), –407 m (1964), –501 m (1965), –525 m (1966), and –573 m (terminal sump reached in 1969) (Courbon, *op. cit.*, 1980, profile).

39. les Cinq scialets .... –570 m
(Corrençon-en-Vercors, Isère)

Located a few hundred meters from the scialet de la Combe de Fer, in the locality of the same name. Discovered in the 1970's by the Association Spéléo Vercors, their explorations were resumed in 1985-1986 by the S. C. Villard-de-Lans. From the –150 m level, various clubs of the C.D.S. de la Drôme joined with the S.C.V.L. and descended to –570 m.

40. gouffre A 3 .... –560 m
(Samoëns, Haute-Savoie)

This cave is located at an altitude of 1969 m, in the massif de Bossetan, NE of Samoëns, and was explored by the S. C. Annemasse who reached successively –140 m, –180 m, –223 m in 1983 (*Spelunca*, 1985 (18), profile), and –560 m in 1985.

41. Behia Iezia .... –559 m
(Saint-Michel/Esterençuby, Pyrénées-Atlantiques)

The deepest cave in the massif de l'Urkulu, it is located a few meters from a forestry road, alt. 922 m, on the north side of the col d'Iropile, 900 m from the Spanish border. The resurgence is the source de la Nive (alt. 337 m). 1/50,000 map Saint-Jean-Pied-de-Port. x 310.95; y 90.50. Upper Cretaceous.

The entrance pit was sounded by Martel in 1908. The exploration did not begin until 1970. In 1972, the A. S. Charente reached about –200 m and –456 m in 1973, reinforced by cavers from le Havre and Grenoble. In 1978 and 1979 the Ziloko Gizonak de Bayonne et des Palois extended the cave and reached the terminal sump at –559 m. The surveyed length in 1983 was 9,450 m. (*Carst.* 1979 (2) partial profile; 1983 (4).

42. scialet Moussu .... –536 m
(Corrençon-en-Vercors, Isère)

It is located at the foot of the grande Moucherolle,

massif du Vercors, to the west of the hut of the combe de l'Ours, at an altitude of 1665 m. It is probably in the goule Blanche drainage basin (alt. 832 m). 1/50,000 map Vif. x 853.32; y 306.48. Urgonian then Hauterivian.

Discovered on June 10, 1957, by the Clan de la Verna who were stopped at –144 m by the squeeze. Having changed their name to the Clan des Tritons, they resumed exploration in 1965 (–411 m) and reached the bottom in 1966 (Lismonde *et al. Grottes et scialets du Vercors*, 1979, t. 2, pp. 192, 194-195, profile).

43. **grotte inférieure de Bury** (Izeron, Isère) ... –520 m
Located on the plateau de Presles (Vercors), it opens at an altitude of 1258 m, not far from the hamlet of Fas. It is linked with the grotte de Pré Martin (Choranche, alt. 230 m. Dye trace of 1910). 1/50,000 map Romans. x 843.48; y 318.94. Urgonian.

It was explored around 1936-1938 by the S. C. Paris (–70 m), then in 1954 and 1955 by the Cyclopes. In 1966 the G. S. de F L.T. were stopped 3013 m from the entrance at a depth of –354 m at a sump which they passed with the S.G. C.A.F. in 1972 (–385 m); they then reached about –405 m in 1973.

The dives of F. Poggia, helped by the Belgians, brought the explored cave past the fifth sump in 1984, for a depth of –520 m and a length of 4910 m (Lismonde *et al.*, *op. cit.*, 1979, t. 2, plan).

44. **grotte de la Luire** ....(–451, +63) 514 m
(Saint-Agnan-en-Vercors, Drôme)
This long-known show cave (alt. 880 m) is located on the west side of the plateau of the grand Veymont, between Rousset-en-Vercors and the hamlet of Brittière. The cave is a giant chimney; sometimes, in the winter, the Vernaison river climbs 450 m to resurge at the sources d'Arbois (Chatelus, Isère, alt. 406 m), making a sudden drain on the Luire. 1/50,000 map La Chapelle-en-Vercors. x 844.39; y 292.52.

Exploration began in 1896 (E. Mellier) and 1898 (O. Decombaz). The entrance pits were conquered in 1936 by the S. C. Paris (–185 m) who reached –218 m in 1945. Beginning in 1952, exploration was resumed by the G. S. Valentinois who reached –348 m in 1952, +63 m in 1961 for a length of 7,260 m, –393 m in 1976 for a length of 11,695 m, and 451 m in 1978 for a length of 11,851 m (Lismonde *et al.*, *op. cit.*, 1978, t. 1, plan).

45. **chourum de la Combe des Buissons** .... –511 m
(Agnières-en-Dévoluy, Hautes-Alpes)
This cave is located above the hamlet of la Chaup, in the massif du Dévoluy, not far from the chourum du Chaudron, at an altitude of 1745 m. The opening was cleared in 1984 by A. Pailler, giving access to a series of descending galleries cut by small pits. In 1984 a group of cavers from the C.D.S. 83 and the S. C. Voconces reached a sump at –450 m. In 1985 a 10 m climb gave access to another system where a series of digs in a very narrow canyon permitted them to reach –511 m at the bottom of a pit where further digging is planned.

46. **réseau du Pinet** ....(+22, –485) 507 m
(Saint-Pierre-d'Entremont, Savoie/Saint-Vincent-de-Mercuze-Sainte-Marie-du-Mont, Isère)

Located in the massif de l'Alpe (Chartreuse) and having thirteen entrances (between altitudes of 1770 and 1645 m), nine of them in Savoie, and four in Isère, of which the best known is the gouffre du Brouillard (alt. 1645 m). Nine entrances are found in the falaise du Pinet; the Brouillard is 900 m to the SW of the habert de l'Alpette. The system is a tributary of the réseau de l'Alpe. 1/50,000 map Montmélian. x 879.81; y 354.33. Urgonian limestone.

The gouffre du Brouillard was discovered and explored in 1964 (110 m), 1972 (–360 m) and 1973 (–360, +18) by the S. C. Savoie. In 1981 this club returned to exploration and connected, from June 19 to July 8, the nine Pinet entrances to the four Brouillard entrances (Lismonde, Drouin, *op. cit.*, 1985, partial profiles).

47. **gouffre de Mauvernay** .... –507 m
(Saint-Pierre-de-Chartreuse, Isère)
The cave is located in the massif de la Chartreuse above a bend in the west cliff near the col de Mauvernay, in the grand Som, at an altitude of 1815 m. Formed in the Urgonian, its probable resurgence is the source de Noirfond. 1/50,000 map Montmélian. x 871.77; y 347.43.

The entrance was discovered by J. -C. Dobrilla and Ch. Pomot in 1977, who explored to –250 m. In 1978 S. Aviotte found the way on and the cave was descended to –474 m (stopping at a sump), after first passing a sump at –433 m. Recent extensions have given the cave a vertical extent of 507 m (Lismonde, Drouin, *op. cit.*, 1985, profile to –474 m).

48. **gouffre des Myrlades** (Chapareillan, Isère) –505 m
In the Chartreuse area, at an altitude of 1790 m, reached from the Fontaine Neuve. The likely resurgence is the source des Eparres (alt. 950 m). 1/50,000 map Montmélian. x 880.8; y 57.42. Bedoulian and Barremian limestones.

The cave was discovered and explored to –70 m in 1962 by the S. C. des Ardennes. In 1978 the S. C. Savoie found a way on above a 35 m pit and explored the cave to –505 m the same year. The length is 5689 m (Lismonde, Drouin, *op. cit.*, 1985, pp. 199-202, plan, profile).

49. *gouffre d'Aphanize* .... **–504 m**
(Mendive, Pyrénées-Atlantiques)
See "The Great Pits". This cave is located in the Albian and its great pit traverses the Aptian (Urgonian). The resurgence is unknown. 1/50,000 map Tardets-Sorholus. x 323.38; y 94.90; z 1040 m.

(*Spelunca*, 1973 (2): 48-49, profile).

50. **gouffre des Trois Dents** ....(–495, +5) 500 m
(Eaux-Bonnes, Pyrénées-Atlantiques)
Located at 2450 m altitude in the massif de Ger, not far from Pène Medaa, its resurgence is the springs of Ley (alt. 1300 m) and of Iscoo (930 m). 1/50,000 map Laruns.

Discovered and explored in 1964 by the S. C. Périgueux to –410 m. In 1965 a squeeze stopped exploration at –425 m. The discovery of a parallel branch at –101 m permitted the S. C. Px to bypass the old terminus and to reach a sump at –495 m in 1968 (Courbon, *op. cit.*, 1979, profile).

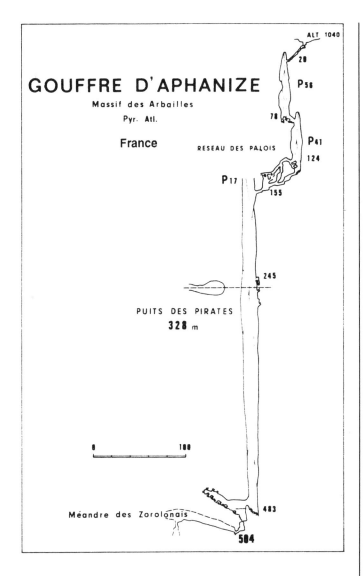

GOUFFRE D'APHANIZE

Massif des Arbailles

Pyr. Atl.

France

RESEAU DES PALOIS

ALT 1040

20

P₃₈

78

P₄₁

124

P₁₇

155

245

PUITS DES PIRATES

328 m

0   100

Méandre des Zorolonais

483

504

A considerable amount of literature has been devoted to this cave. Exploration using diving bells began on March 27, 1878: Ottonelli reached –23 m underwater (–44 m in the cave). On September 25, 1938 Negri went slightly deeper than Ottonelli.

On August 27, 1946 the Groupe de Recherches Sous-Marines (Cousteau, Taillez, Dumas) reached –46 m (–67 m in the cave).

In 1955 the Cousteau team made another attempt and got down to –74 m (–95 m). In 1967 the Office Français de Recherches Sous-Marines led by Cousteau passed the old record to reach –90 m (–111 m) while a mechanical device sounded the sump to –106 m (–127 m).

September 21, 1981, saw a theatrical event: the German diver Jochen Hasenmayer descended alone without any link to the surface, and at night, to a depth of –145 m. On October 11, 1981 Claude Touloumdjian, helped from the surface, reached –153 m. But on September 9, 1983 Hasenmayer, under the same conditions as before, dove again and reached –200 m (–221 m total).

From 1981 to 1985 the S.S. de Fontaine-de-Vaucluse has undertaken expeditions to sound the cave by a machine guided from the surface which has so far measured a depth of − 308 m!

Map: from the documents sent by M. Fradin of the S.S.F.V.

Bibliography: J. Guigue and J. Girard - *La Fontaine de Vaucluse*, Avignon, 1949, 158 p.

51. **gouffre Pentothal** .... –500 m
(La Brigue, Alpes-Maritimes)

Located in the enclave de La Brigue, massif du Marguareis, near a hut, at an altitude of 2125 m. The resurgence is the spring of the Pis del Pesio (Italy, alt. 1340 m). 1/50,000 map Viève. x 1025.7; y 221.29. Eocene, Cretaceous, Jurassic.

It was discovered and explored in 1981 by the G. S. CAI Piemontese (*Spéléologie*, Nice, 1981 (114) profile).

162. **gouffre du Petit Lapiaz**
du Soum de Lèche N° 18 .... –300 m
(massif de la Pierre Saint-Martin, Arette, Pyrénées-Atlantiques)

Fontaine de Vaucluse.................................... **–221 m**
(Fontaine-de-Vaucluse, Vaucluse)

This very famous spring resurges 26 km to the east of Avignon, at an altitude of 105 m. 1/50,000 map Cavaillon.

The official zero point for the water is 21 m below the top of the overflow dam. The perennial resurgences are at altitudes of 82 and 78 m. The average flow is 29 m³/sec (a maximum of 170 m³/sec in January, 1936 and a minimum of 4.5 m³/sec in December, 1884).

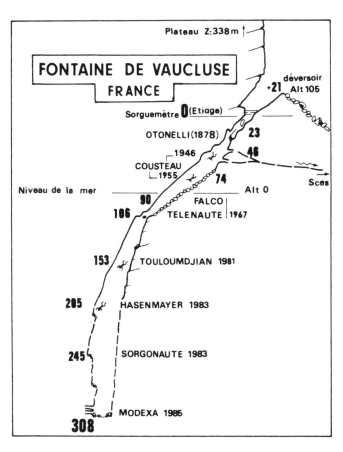

Plateau Z: 338 m

FONTAINE DE VAUCLUSE
FRANCE

déversoir
+21 Alt 105

Sorguemètre 0 (Etiage)

OTONELLI (1878)   23

1946   46

COUSTEAU
1955   74   Sces

Niveau de la mer   Alt 0

90   FALCO

106   TELENAUTE 1967

153   TOULOUMDJIAN 1981

205   HASENMAYER 1983

245   SORGONAUTE 1983

MODEXA 1985

308

LONG CAVES:

1. **système de la Coumo d'Hyouernedo** .... 90,000 m
(Herran/Arbas, Haute-Garonne)
See above.

2. **réseau de l'Alpe** (Isère/Savoie) .... 53,679 m
See above.

3. **système de la Dent de Crolles** .... 53,200 m
(Saint-Pierre-de-Chartreuse, Isère)
See above.

4. **réseau de la Pierre Saint-Martin** .... 52,000 m
(France/Spain)
See above.

5. *système Souterrain du Verneau* .... **32,100 m**
(Nans-sous-Sainte-Anne/Déservillers, Doubs)

The system contains the following entrances: gouffre de Jérusalem (alt. 758 m), gouffre des Biefs Boussets (765 m), gouffre de la Baume des Crêtes (790 m), gouffre de la Vieille Folle (685 m), grotte Baudin (Nans-sous-Sainte-Anne, 485 m) and the source du Vernau (Nans-sous-Sainte-Anne, 420 m).

It is located in Franche-Comté and developed under the Déservillers plateau, to resurge at Nans-sous-Sainte-Anne, at the source du Verneau, which feeds the Lison river. The latter is also fed by the gouffre de la Vieille Folle and the gouffre des Biefs Boussets. The system is formed in upper Jurassic limestones. 1/25,000 map Quingey. x 884.30; y 228.50 (Baume des Crêtes) and Salins-les Bains 3/4. x 879.08; y 226.34 (Verneau).

The history of exploration can be divided into two equally important periods: that of Eugène Fournier (1903-1912) who discovered all the entrances to the system, and that of the Société Hétéromorphe des Amateurs de Gouffres (Besançon) who under the leadership of Yves Aucant, beginning in 1969, explored essentially the entire system, thanks to the diving of sumps that stopped their predecessors.

On May 21, 1972, the S.H.A.G. connected the Biefs Boussets to the Vieille Folle. Then, on February 15, 1975, these two were connected to the source du Verneau which they dove in 1972 to reach the main river passage. The system was then 18 km long, then 20,100 m long at the end of 1975.

On July 14, 1976, Jérusalem and Biefs Boussets were connected (to give 25,220 m) and on October 1, 1977 it was the turn of the gouffre de la Baume des Crêtes to join the system (giving 27,750 m). In 1980 the length reached 28,150 m and in 1982, 28,225 m. On April 29, 1984, the G. S. Doubs made the connection with the grotte Baudin, bringing the length to 32,100 m.

Map: from the drawings of the S.H.A.G. (Y. Aucant, J.-P. Urlacher), sent by Alain Couturaud.

Bibliography: Aucant (Y.) *et al* - *Le Verneau souterrain*, S.H.A.G., ed., Besançon, 1985, 169 p.

6. **grotte de Saint-Marcel
d'Ardèche** .... approx. 27,000 m
(Bidon, Ardèche)

The cave is located in the lower Ardèche valley and was discovered in 1838 by a hunter. Martel, Gaupillat, Armand, Deloly were the first explorers, mapping 2260 m on August 21, 1892. Then R. de Joly and the S. C. de France in 1931, 1933, and 1947 (about 5,500 m). In 1960 the Equipe Spéléo de Bruxelles added 1250 m and in 1964 reached 10 km with R. Courbis. The G. S. Forez returned to the exploration and brought the length to 11,050 m (1965), 13,091 m (1966), then 14,417 m (1967). The G. S. Saint-Marcel took over in 1972; in 1974 the cave was 19,630 m long, then 21,300 m in 1976, and 24,757 m in 1977 (new survey). (*Scialet*, 1977 (6) plan). In 1982 the dives of the Spéléo-Ragaïe added about 2500 m.

7. **grotte des Ambouilla** .... approx. 25,000 m
or **"réseau A.-Lachambre"**
(Ria-Sirach-Urbanya/Corneilla-de-Conflent, Pyrénées-Orientales)

Length as given by Patrick Cabrol (April, 1986). Cave explored and mapped by the Conflent S. C. with the collaboration of Belgian and Barcelonian cavers. It was discovered thanks to a dig in the 1980's by André Lachambre. Mapped for 15,541 m in 1983-1984 by the G.E.S. Barcelona and the C.S.C.

8. **système de Foussoubie** .... 23,266 m
(Vagnas/Salavas, Ardèche)

Hydrological system made up of the goule de Foussoubie (alt. 197 m), the aven Cordier, and the upper and lower évents de Foussoubie (alt. 88 m), which resurge in the right bank of the gorges de l'Ardèche. Exploration began in 1892 (Armand, Gaupillat). In 1934 (R. de Joly), 1949 (S. C. Montpellier), 1958 (P. Renault *et al.*), the length advanced to only 1302 m. From 1959 to 1967, year of the connection by diving of the insurgence to the resurgence, the S. C. Lutèce, with the S.C.U.C. Louvain and the S. S. Namur increased the length from 6925 m to 17,886 m. In 1972 the Groupe de Recherches Biologiques returned to the exploration and mapping and reached 19,130 m in 1976, 20,380 m in 1978, and 23,134 m in 1981.

9. **système des Vers Luisants
et du Vertige** .... approx. 23,000 m
(Aviernoz/Dingy-Saint-Clair/Thorens-Glières, Haute-Savoie)

Located in the massif du Parmelan, it contains the gouffre des Vers Luisants (1525 m), glacière d'Aviernoz (1572 m), tanne au Vieux (1679 m), trou Noir (1682 m), Voie Lactée (1692 m), PA 130, trou du Lapin, grotte du Mirador, the gouffre de la Merveilleuse, gouffre du Ramoneur (1610 m), gouffre du Téléphone and the grotte supérieure de Bunant (1330 m). This 369 m deep hydrological system was explored beginning in 1977 by the G. S. des Troglodytes d'Annecy. The connection with the grotte de Bunant was made by diving in 1978 (S.S.S. Genève). The connection by diving in August, 1983 (Groupe Lémanien de Plongée Souterraine, S. C. Annecy) of the Vers Luisants system (8490 m long in 1982) and the Vertige system (7913 m long in 1982) made a new system with a length of about 23 km.

10. **grotte d'Arphidia** .... 21,000 m
(Sainte-Engrâce, Pyrénées-Atlantiques)
See above.

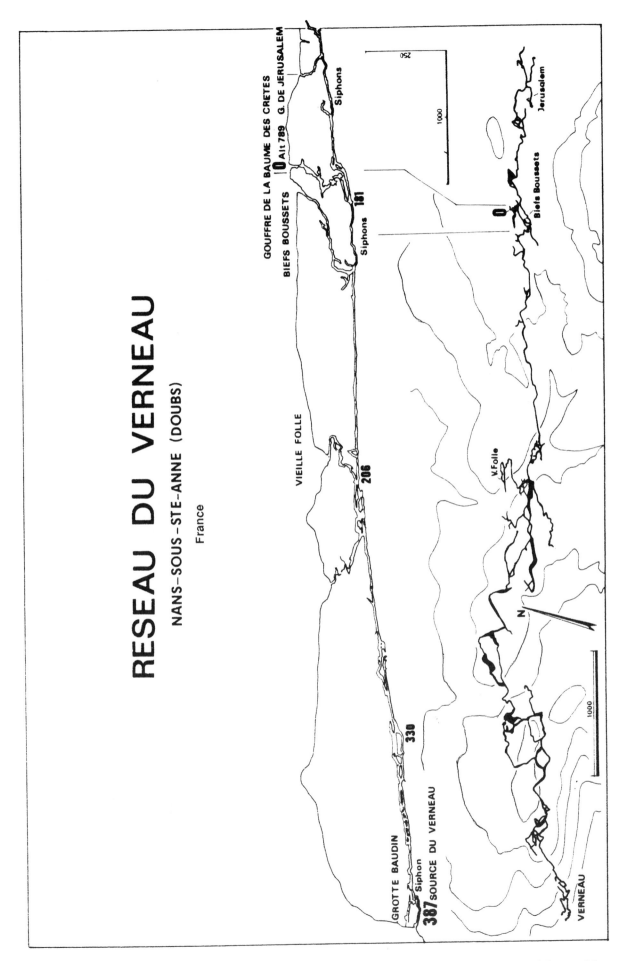

RESEAU DU VERNEAU

NANS–SOUS–STE–ANNE (DOUBS)

France

GOUFFRE DE LA BAUME DES CRETES

G. DE JERUSALEM

Alt 789

BIEFS BOUSSETS

Siphons

181

Siphons

VIEILLE FOLLE

206

330

GROTTE BAUDIN

Siphon

387 SOURCE DU VERNEAU

Jerusalem

Biefs Boussets

0

V.Folle

N

VERNEAU

250

1000

1000

11. **gouffre Berger** (Engins, Isère) ....20,665 m
See above.

12. **trou qui souffle** ....19,876 m
(Vercors, Méaudre, Isère)
Exploration of this cave began in 1937 (S. C. Paris), and was pursued in 1940, 1942 (A. Bourgin), 1953 and 1954. In 1962, 1964, 1965, and 1966 about 8 km were explored over a vertical extent of 280 m (Cyclopes). Since 1974, exploration has been discontinuous. Diving of sumps (G. S. de F.L.T.) and discoveries beyond the sumps (S.G. C.A.F.) brought the length from 11,965 m in 1981 (*Scialet*, 1981 (10) plan) to 19,876 m in 1983 for a vertical extent of 373 m (-313, +60).

13. **trou du Garde** (Les Déserts, Savoie) ....19,782 m
A cave in the massif de Révard-Féclaz (alt. 1362 m). Explored by the S. C. Savoie starting in 1971. In 1974, 1680 m were mapped and in 1976, 17,892 m. The explorations of 1977, 1978, and 1981 gave a total length of 19,782 m.

14. **grotte de Neuvon** ....18,650 m
(Plombières-les-Dijon, Côte-d'Or)
The entrance was opened by a flood on October 2, 1965, and the S. C. Dijon, after passing the entrance sump in 1975, explored 5200 m in 1976, 6300 m in 1977, 14,030 m in 1978, 18,220 m in 1983, and 18,650 m in 1985 (*Sous le Plancher*, 1977-1979, XVI, plan showing 14,200 m).

15. **réseau de Francheville** ....18,500 m
(Francheville, Côte-d'or)
System made by the connection of the gouffre de la Combe aux Prêtres with the creux du Soucy on September 22, 1984, by the S. C. Dijon. The latter was explored by Martel and Drioton (1904), then by the S.C.D. (1967-1978) for 4,205 m. The Combe aux Prêtres, discovered in 1969 (by quarrying) was explored by the S.C.D. for 2500 m in 1971, 3300 m in 1972, 4500 m in 1974, 6275 m in 1975, 7900 m in 1976, 8200 m in 1977, 10,280 m in 1978, and 12,525 m in 1979.

16. **réseau Jean Bernard** ....17,900 m
(Samoëns, Haute-Savoie)
See above.

17. **gouffre de Padirac** ....17,427 m
(causse de Gramat, Padirac, Lot)
This famous show cave was explored in 1889 and 1890 by Martel for 1900 m, in 1898 by Viré and Armand, and in 1899 by Martel (2275 m). Exploration was continued by G. de Lavaur in 1937, 1938 (with R. de Joly), 1948 and 1951 (with the S. C. Paris, 5500 m).
In 1962 the terminal sump was reached by the S.C.P. (10,700 m). From 1970 to 1976 local clubs brought the length to 17,247 m. The terminal sump was dived by the S. C. Dijon in 1983.

17a. **réseau des Arres Planères** ....approx. 17,000 m
(Arette, Pyrénées Atlantiques)
See above.

18. **borne aux Cassots** ....15,630 m
(Névy-sur-Seille, Jura)

The entrance was dug open in 1965-1966 by the G. S. Jurassien who explored the cave from 1967 to 1977. Survey of Frachon and Aucant in *Bull. A. S. Est*, 1980 (16).

19. **tanne aux Cochons** ....15,468 m
(Aillon-le-Jeune, Savoie)
See above.

20. **grotte de Gournier** (Choranche, Isère) ....15,125 m
See above.

21. **réseau de Coufin-Chevaline** ....15,063 m
(Vercors, Choranche, Isère)
The two caves, reconnoitered by Décombaz in 1897 and 1898, explored by various clubs in 1943, 1949, 1952, and 1954, were connected in September, 1966, by the S.G. C.A.F. and the G. S. de F.L.T. (length about 7,800 m). The G. S. Valentinois then took over and reached 10,064 m in 1972 and 15,063 m in 1979 (Lismonde *et al.*, *op. cit.*, 1979, t. 2, plan).

22. **système de la Diau** ....14,940 m
(Dingy-Saint-Clair/Thorens-Glières, Haute-Savoie)
See above.

23. **Io Gaugnas** (Cabrespine, Aude) ....14,737 m
Located on the hydrological system of Clamous-résurgence du Pestril (Lastours), this cave was explored in 1935 by the S. C. Montagne Noire et Espinouze. In 1969, they reached 3,850 m. Starting in 1972, the explorations of the S. C. Aude brought the length to 14,737 m in 1978. The total explored length is estimated at 17 km.

24. **creux de la Litorne** ....14,249 m
(massif de Banges, Arith, Savoie)
It has three entrances and was explored by the S. C. Savoie in 1980, 1981 (–328 m, 6284 m), 1984 (13,098 m) and 1985 (14,249 m).

25. **rivière souterraine des Vitarelles** ....14,200 m
(causse de Gramat, Gramat, Lot)
Explored in 1948 (military teams) and from 1956 to 1958 by the G. S. Gramat (5500 m) who returned in 1969 to reach 13,447 m in 1975 and 14,200 m in 1979.

26. **grotte de la Luire** ....13,600 m
(Saint-Agnan-en-Vercors, Drôme)
See above.

27. **système de Sauvas-Cocalière** ....11,668 m
(Saint-André-de-Cruzières/Saint-Paul-le-Jeune, Ardèche)
System with four entrances explored starting in the 19th century (Malbos, then Gaupillat in 1892). In 1937 R. de Joly (4200 m) and from 1953 to 1964, the S.S.P. du Gard-Ardèche did the exploration. On August 2, 1978 the Groupe Rhodanien de Plongées Souterraines connected the aven de la Cocalière to the goule de Sauvas. In 1980 the U. S. St-Ambroix mapped 9130 m. Extensions discovered in 1982 by J. Chauvet gave a length of 11,668 m.

28. **résurgence du Rupt-du-Puits** ....10,256 m
(Barrois, Beurey-sur-Saulx, Meuse)

Explored from 1966 to 1972, after passing the entrance sumps (445 m long).

29. **grotte de Prérouge** ....10,135 m
(massif de Banges, Arith, Savoie)
The cave was explored by P. Chevalier in 1937, Mercier in 1940 (length 2685 m), Vertut in 1950 (3710 m) and the S. C. Savoie from 1972 (4855 m) to 1983 (10,135 m).

30. **système de Bramabiau** ....10,100 m
(Saint-Sauveur-des-Pourcils, Gard)
River cave first explored by Martel and his team on June 28, 1888. Explored by F. Mazauric from 1890 to 1892 (6350 m). Beginning in 1982, the S. C. des Causses remapped the cave, bringing the length to 10,100 m in 1984.

31. **baume Layrou** ....9863 m
(Causse Bégon, Trèves, Gard)
(*Bull. S. C. Causses*, 1981 (4), plan).

32. **grotte de la Cigalère** ....9634 m
(Haut-Lez, Sentein, Ariège)
(*Spelunca*, 1975, (4), plan showing 9300 m).

33. **gouffre de Pourpevelle** (Soye, Doubs) ....9576 m

34. **Behia Iezia** ....9450 m
(Saint-Michel/Estérençuby, Pyrénées-Altantiques)
See above.

35. **réseau de trou qui fume** ....9380 m
(La Rochette, Charente)

36. **aven de la Leicasse** ....9300 m
(causse du Larzac, Saint-Maurice-Navacelles, Hérault)

37. **cuves de Sassenage** ....9200 m
(Vercors, Sassenage, Isère)

38. **gouffre du Couey Lotge** ....9100 m
(Arette, Pyrénées-Atlantiques)
See above.

39. **gouffre des Bourrugues N° 3** ....9010 m
(Arette, Pyrénées-Atlantiques)
See above.

40. **gouffre Mirolda** ....approx. 9000 m
(Samoëns, Haute-Savoie)
See above.

41. **réseau du Pinet** (Isère/Savoie) ....8614 m
See above.

42. **aven de Puech Nègre N° 2** ....approx. 8500 m
(causse Noir, Millau, Aveyron)

43. **réseau du Brudour-Toboggan** ....8313 m
(Vercors, Bouvante, Drôme)
(Lismonde *et al.*, *op. cit.*, 1978, t. 1, plan.)

44. **réseau de la Couze** ....8300 m
(Noailles/Chasteaux, Corrèze)
(*Spelunca*, 1982 (6), schem. plan).

45. **trou du Vent** (Bouzic, Dordogne) ...approx. 8000 m

46. **réseau de Niaux-Lombrives** ....approx. 8000 m
(Niaux/Ussat, Ariège)

47. **réseau de la Doria** ....7522 m
(massif de Révard-Féclaz, Saint-Jean-d'Arvay, les Déserts, Savoie)

48. **réseau de la Tête des Verds** ....7500 m
(Magland, Haute-Savoie)
See above.

49. **aven de la Combe Rajeau** ....7434 m
(Saint-Laurent-sous-Coiron, Ardèche)

50. **grotte de Lauzinas** ....approx. 7400 m
(Montagne Noire, Saint-Pons, Hérault)

51. **grotte du massif des Fanges** ....7400 m
(Caudiès-de-Fenouillèdes, Pyrénées-Orientales)
(*Spelunca*, 1985 (18), schem. plan.).

52. **grotte du Goutal** ....7300 m
(causse de Sauveterre, Les Vignes, Lozère)

53. **grotte de Chauveroche** (Ornans, Doubs) ...7220 m
(*Bull. A. S. Est*, 1974 (11), plans).

54. **grotte de Villars** (Villars, Dordogne) ....7000 m
(*Spéléo-Dordogne*, 1974 (50-51), plan).

55. **grotte de Miremont** ....approx. 7000 m
(Rouffignac, Dordogne)
(*Spéléo-Dordogne*, 1977 (63-64), plan).

56. **grotte d'Artenac-Ponteratz** ....approx. 7000 m
(Montagne Noire, Saint-pons, Hérault)

56a. **Bracas de Thurugne 6** ....6700 m
(Arette, Pyrénées-Atlantiques)
See above. Also known as rivière de Soudet or gouffre de la Bordure de Tourugne N° 6.

GREECE
Ellás

Sixty-five percent of Greece is formed of calcareous rocks, ranging from the Jurassic to the Quaternry. The best known areas for foreign cavers are the Pindhos Mountains in Ipeiros and the Levka ori in Crete. Greece contains a very large number of caves (over 7000 were counted in 1980), often of small dimensions, but very often having an archeological, historical, or mythological interest: for example caves were the first habitations of Paleolithic and Neolithic man, they were cult centers, sanctuaries, monasteries, hermitages, and more. The range of cultural uses to which they have been put is immense.

Modern speleology began under the influence of foreign scientists and inquisitive amateurs. In 1673 the marquis de Nointel explored Antiparos cave and in 1841 the German Findler explored Kataphki cave at Kythnos. In 1891 Martel and N.A. Sidéridès explored the katavothre of Taka on the central plateau of the Peloponnèse, a region in which, from 1892 to 1910, the same Sidéridès systematically explored the katavothres. A caving group was formed in 1936 under the wing of the Hellenic Alpine Club. In 1950 Jean Petrocheilou founded the Speleological Society of Greece which publishes a bulletin, *Deltion.* In *Caves of Greece* (1984, 160 p.), Anna Petrocheilou presents all the show caves of her country.

DEEP CAVES:

1. *Epos* (Pindhos oros, Ioannina, Ipeiros) ...... **–451 m**
Epos is located near the western border of the Astraka plateau, about 2500 m south of Provatina, about 800 m from the majestic Vicos gorges, or about 4500 m as the crow flies from the village of Vicos to the WNW. The town of Ioannina is 30 km to the south. The altitude of the cave is 1645 m, and it is formed in Eocene limestone.

This great cave was discovered by Peter Livesey while ridgewalking on the plateau in 1968. With the British team he brought in, the depth of –244 m was reached and the bottom reached in 1969 at a lake at –442 m. In 1979 a trip by the South Wales Caving Club increased the depth to –451 m.

Map: profile from W. Porobski and P. Ruska, from *Speleo,* Kraków, 1980.

Bibliography: Wigley (T.) - Caves of the Astraka Plateau, North-Central Greece, *The Canadian Caver,* 1970 (2): 17-24.

2. *Provatina* .... **–407 m**
(Pinhos oros, Ioannina, Ipeiros)
This great pit is located on the northern border of the same plateau as Epos, Astraka, directly overlooking the small village of Papigon from which access is gained to the plateau.

First known by shepherds, it was reported in 1965 by the Cambridge University Caving Club. The vertical entrance pitch (158 m) was dropped to –156 m in 1966 by Jim

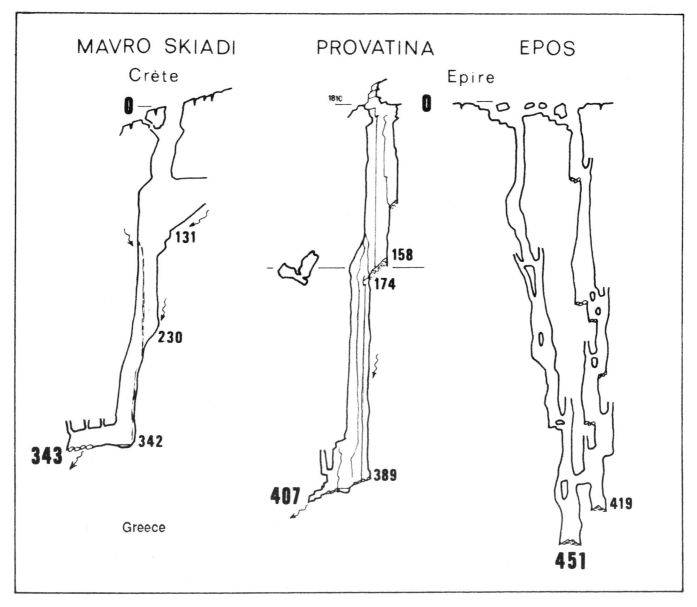

Eyre who then ran out of ladders. In 1967 a new British expedition reached –174 m on a steeply sloped snowy ledge from which they measured the depth of the second vertical drop: –214 m. In 1968 a British military expedition of the Royal Army Medical Corps used manual winches to reach the bottom of the pit which was later surveyed to be –405 m deep (see "The Great Pits").

Map: drawing of Ph. Rouiller and R. Wenger (1979), sent by Ph. Rouiller.

Bibliography: Wenger (R.) - Provatina, *Caving International Magazine*, 1980 (8): 28-31.

3. **tripa Ligeri** .... –386 m
(Pindhos oros, Ioannina, Ipeiros)
Alt. 1930 m. In 1979 by Sheffield U.S.S. (*S.U.S.S. Journal*, 1980, 3 (1) profile).

4. **Mavro Skiadi** .... –343 m
(Levka Ori, Melidoni, Crete)
See "The Great Pits".

5. **Propantes** .... –315 m
(Paleochorion, Leônidion, Arkadia)
Bottomed in 1974 by the S. S. Greece and French cavers, measured to be –418 m deep.

6. **tripas tis Nifis** or **Pandremeni Gynaïka** ... –299 m
(Pindhos, Ioannina, Ipeiros)
Explored in 1975 (Sydney S. S., U. Bristol S. S., U.N.S.W.S.S.), 1976 (British, –294 m) and 1977 (Americans) (*Trans. B.C.R.A.*, 1978, 5 (1) profile).

7. **Tzani spilios** or **Chonos** .... –280 m
(Omalos, Khania, Crete)
Depth reached in 1975 by the S.S.G., S. C. Seine.

8. **"Lo 50"** (Levka ori, Crete) .... –267 m
Explored in 1981, S. S. Plantaurel (*Bull. G.E.S.F.*, 1982 (2), profile).

9. **"gouffre des Vires"** .... –230 m
(Pindhos oros, Ioannina, Ipeiros)
Explored in 1979 by the S. S. Plantaurel (*Bull. G.E.S.F.*, 1982 (2), profile).

10. **Tspelovon spiara** or **Rokkoros** .... –228 m
(Pindhos oros, Ioannina, Ipeiros)
Explored in 1970 by the Westminster S. G.

11. **Stouros** .... –210 m
(*Echo des Vulcains*, 1981 (41), profile).

12. **barathro Skorpiôn** .... –208 m
(Leônidiou, Kynourias, Arkadia)
(*Deltion*, 1974 (6-7), profile).

13. **Defteroki Asma** (Lefka ori, Kriti) .... –207 m

14. **Neraïdotripa** .... approx. –200 m
(Portes, Patras, Achia)

LONG CAVES:

1. **Vlyhada** or **Glyphada** .... 3400 m
(Diros, Lakonie, Peloponnisos)
Show cave on the coast, explored in the 1960's by the Petrochilos and the S.S.G. (*Deltion*, 1974, (6-7) plan).

2. **Tzani spilios** .... approx. 2900 m
(Omolos, Khania, Crete)

3. **Maras** (Pigès, Drama) .... 2650 m
Spring explored in 1981 by the E.R.I.S.

4. **Ambartsiki** .... approx. 2000 m
(Alistrati, Sérrai, Makedonia)

5. **Limnospilaion tôn Limnôn** .... 1950 m
(Kastria de Kalavrita, Peloponnisos)
In 1967 by the S.S.G.

6. **Kyklops Polifimos** .... 1800 m
(Maronia, Komotini, Thrace)
In 1969 by the S.S.G.

7. **Drakolaki** (Levka ori, Anopolis, Crete) .... 1750 m
(*Caves & Caving*, 1982 (15), plan).

8. **Perama** (Perama, Ioannina, Ipeiros) .... 1700 m
In 1963 by the S.S.G.

9. **Kokkines Petres** .... 1500 m
(Petralona, Chalkidiki, Makedonia)
In 1964 by the S.S.G. (*Actes IVth Int. Congr. Spel.*, 1965, 4-5, plan).

10. **Megali spilia** (Levadhia, Voiotia) ..... approx. 1000 m

HUNGARY
Magyarország

There are about a dozen different karst areas in Hungary, of which the most significant, Bükk and Aggtelek, are located in the north of the country. The Bakony mountains to the north of Lake Balaton also contain large caves. The Pilis-Budai and Gerecse mountains near Budapest are famous for their geothermal caves.

Hungarians have a long speleological tradition (it goes back to 1037!); the first cave descriptions appeared in the 18th century. In 1848, speleologists banded together under the auspices of the Hungarian Geological Society, and in 1910 under the leadership of Lajos Lóczy formed a commission that became independent in 1913. A national society was formed in 1926, the Magyar Karszt és Barlangkutató Társulat. The fathers of modern speleology are Ottokár Kadić and Endre Dudich, followed by László Jakucs, Dénes Balázs, Attila Kósa. Thanks to the Karst Research Institute of Jósvafö, Hungarian scientific speleology is among the most dynamic in the world (the Hungarians were pioneers in speleotherapy). Read *Karszt és Barlang*, 1977 special issue, 76 p.

DEEP CAVES:

1. **Vecsembükki-zsomboly** .... –245 m
(Aggteleki-karszt)
Bottomed in 1970 (*Karszt és Barlang*, 1970, 1, profile).

2. **István-Lápai-barlang** (Bükk) .... –240 m

3. **Alba Regia barlang** (Bakony) .... –210 m
Alt. 453 m (Kordos, *Magyarország Barlangjai*, 1984, profile).

4. **Létrás-Tetői-barlang** (Bükk) .... –166 m
(*Karszt és barlang*, 1962, 2, profile).

5. **Borókási-Viznyelőbarlang N° 4** ..... approx. –160 m
(Bükk)

6. **Fekete-barlang** (Bükk) .... –153 m

7. **Pénz-pataki-viznyelőbarlang** (Bükk) .... –146 m

8. **Meteor-barlang** (Aggteleki-karszt) .... –132 m

9. **Középbérci-Bányász-barlang** (Bükk) .... –130 m

10. **Keselő-Hegyi-barlang** (Gerecse) .... –120 m

11. **Jubileum zsomboly** (Bakony) .... –120 m

12. **Hajnóczy-barlang** (Bükk) .... –118 m

13. **Baradla barlang-Domica jaskyňa** .... –116 m
(Aggteleki -karszt, Jósvafö)

14. **Jávor-Kuti-viznyelőbarlang** (Bükk) .... –112 m

15. **Kis-Kőháti-zsomboly** (Bükk) .... –110 m

16. **Szeleta-zsomboly** (Bükk) .... –110 m

17. **Borókási N° 2-viznyelőbarlang** ..... approx. –110 m
(Bükk)

18. **Mátyás-Hegyi-barlang** (Budai-hg) .... –106 m

19. **Csőszpuszta-barlang N° 1-12** (Bakony) ..... –103 m

LONG CAVES:

1. **Baradla Barlang-Domica Jaskyňa** .... 25,000 m
(Aggteleki-karszt, Jósfavö)
The Baradla-Domica system is developed north of Aggtelek and west of Jósfavó. It has seven entrances, including Aggtelekibarlang, Felszabadulás-ági-kijarat, Vörös-tói-bejárat, Jósvaföibejárat, and Domica jaskyňa, the later being in Czechoslovakia!

The cave is visited by tourists and has been known a long time. The first explorations were in the 18th century: in 1794 Joseph Sartory explored 1800 m and made the first map. Seven years later Keresztély Raisz carried the length to 3000 m. In 1825 Imre Vass estimated he had explored 8300 m (he published a second map of Baradla in 1831). In 1856 Schmidl added 300 m to bring the length to 8666 m. In 1926 Hubert Kessler and Jószef Sandrik made the connection with Domica, which was discovered the same year by Jan Majskó. The length passed 20 km in 1938 but has increased only slowly since: 22 km in 1960 and 25 km in 1975 by the Vörös Meteor Club. Seven km of passage extend underneath Czeckoslovakia.

Map: from *Cavernes*, Neuchâtel, 1974, 2.
Bibliography: Jakucs (L.) - *Aggtelek*, 1975, 163 p.

2. **Béke-barlang** (Aggteleki-karszt, Jósvafö) .... 8743 m
Cave explored since 1952.

3. **Mátyás-Hegyi-barlang** (Budai-hg) .... 4200 m

4. **Ferenc-Hegyi-barlang** (Budai-hg) .... 4000 m
(*Karszt és Barlang*, 1977, plan).

5. **Pál-Völgyi-barlang** (Budai-hg) .... 3200 m
Discovered on June 23, 1904 (*Karszt és Barlang*, 1980, 2, plan).

6. **István-Lápai-barlang** (Bükk) .... approx. 2940 m

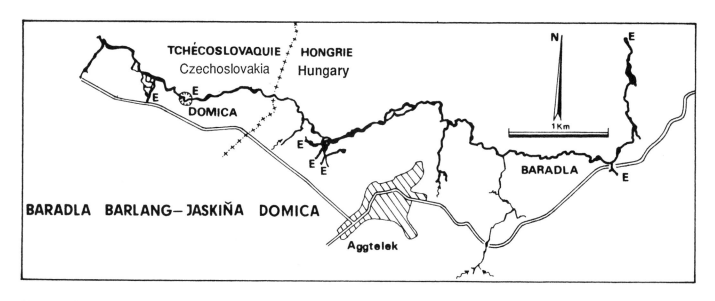

BARADLA BARLANG– JASKIŇA DOMICA

7. **Szabadság-barlang** (Aggteleki-karszt) .... 2717 m
Discovered in 1954.

8. **Alba Regia barlang** (Bakony) .... 2500 m

9. **Cserszegtomaji-kútbarlang** .... approx. 2000 m
(*Karszt és Barlang*, 1982, 1, plan).

10. **Létrás-Vizes-barlang** (Bükk) .... approx. 2000 m
(Kordos, *Magyarország barlangjai*, 1984, plan).

11. **Solymári-Ördög-Lyuk** (Budai-hg) .. approx. 2000 m

12. **Szemlő-Hegyi-barlang** (Budai-hg) .... 1962 m
Discovered on September 20, 1930 (*Karszt és Barlang*, 1965, 1, plan).

13. **Létrás-Tetői-barlang** (Bükk) .... 1160 m

14. **Abaligeti-barlang** (Mecsek) .... 1166 m
Discovered in 1768.

15. **Borókási-viznyelőbarlang N° 4** ..... approx. 1000 m
(Bükk)

16. **Fekete-barlang** (Bükk) .... approx. 1000 m

17. **Jávor-Kuti-Viznyelőbarlang** .... approx. 1000 m
(Bükk)

18. **Tapolcai-Tavas-barlang** .... approx. 1000 m
(Tapolca, Balaton)

19. **Vass-Imre-barlang** .... approx. 1000 m
(Aggteleki-karszt)

ICELAND
Island

The list of caves in Iceland we have assembled is certainly incomplete. This is a volcanic island with numerous glaciers around 2000 m altitude. Caves are found both under the glaciers and in lava tubes which have been explored for a long time (Olafsson in 1905, the Barcelonians in 1967 and British of the Shepton Mallet C. C. in 1970 and 1972.)

LONG CAVES:

1. **Kverkjöll** (Vatnajökull) .... 2850 m
Glacier cave with three entrances (*Spelunca*, 1985 (17) plan) explored in 1984 by the Swiss who reached the depth of − 525 m.

2. **Surtshellir** (Kalmanstunga) .... 1580 m
In 1905 by Olafsson and 1970/1972 by the Shepton Mallet C. C.

3. **Stephanshellir** (Kalmanstunga) .... 1520 m
A segment of Surtshellir.

4. **Vidgelmir** (Flotstunga) .... 1460 m

5. **Raufarhorshellir** (Hjalli) .... 1350 m
In 1970/1972 by the Shepton Mollet C.C.

IRELAND
Eire

This section does not include the northern part of the island (Northern Ireland), which is included under United Kingdom. The majority of the large caves (which generally develop horizontally and near the surface) are found in the county of Clare in the west of the country. E.-A. Martel, with his book *Irlande et Cavernes Anglaises* (1897) was the first to write extensively about the caves of the island. English clubs, particularly the Bristol Speleological Society, have subsequently done most of the exploration.

DEEP CAVES:

1. **Carrowmore Caverns** .... − 142 m
(Seighmairebaun-Pollnagolum)
(County Sligo)

2. **Polliska Pot** (County Sligo) .... − 126 m

3. **Poulnagollum-Poulelva**
Cave System (County Clare) .... approx. − 100 m

LONG CAVES:

1. **Poulnagollum-Poulelva**
Cave System (County Clare) .... 12,100 m

2. **Doolin-St. Catherine's**
Cave System (County Clare) .... 10,500 m
Three entrances. Explored by the U.B.S.S. (1953-1956).

3. **Cullaun 5** (County Clare) .... 5950 m

4. **Coolagh River Cave** (County Clare) .... 5280 m
(*Proc. U.B.S.S.*, 1975, 14 (1) map with 4,100 m).

5. **Crag Cave** (County Kerry) .... 3810 m

6. **Cullaun 3** (County Clare) .... approx. 3500 m

7. **Cullaun 2** (County Clare) .... 3330 m

8. **Poll Cahercloggaun West-One** .... 3170 m
(County Clare)

9. **Fergus River Cave** (County Clare) .... 2760 m

10. **Cullaun 1** (County Clare) .... 2610 m

11. **Mitchelstown Cave** (County Tipperary) ..... 2540 m
Explored for about 2000 m by Martel in 1895 (Martel, *op. cit.*, plan).

12. **Cloyne Cave** (County Cork) .... 2400 m

13. **Gragan West Cave** (County Clare) .... 2250 m

14. **Pollahuna** (County Cavan) .... approx. 2100 m

15. **Pollballiny** (County Clare) .... 2060 m

ITALY

Italia

Italy, with a surface area of 301,000 km², presents surface and underground karst features in all of its regions. The geologic ages of interest range from the Paleozoic to the Quaternary. The main karst zones are generally found in the Mesozoic limestones. Nonetheless, great caves are found in the Cambrian limestones in the SE region of Sardinia (grotta di S. Giovanni Dumusnovas, length 3068 m) and in the Miocene gypsum in Emilie (complesso Spipola-Acquafredda, length 7 km).

The variety of structural and climatic conditions is too great to survey every aspect, but we can present a schematic summary of some of the more important regions, as follows:

The karst areas with the greatest density of extensive caves are the denuded high altitude karsts (Marguareis, Canin, Alpi Apuane, Grigne, etc.), the medium elevation mountain karsts in the northern woods (Prealpi Lombarde, Lessini, Cansiglio-Cavallo, etc.) and in the central ranges (Monte-Cucco, Matese, and Alburni mountains), and finally the low altitude Mediterranean karsts (Trieste area, western Ligurie, Fiume-Vento, western Sardinia).

There are presently over 21,000 caves known in Italy, with the richest areas being the Friuli-Venezia Giulia (4502 caves), Veneto (2576), Lombardia (2500), Sardinia (1620), Pugglia (1600 approx.), and Liguria (1207). Geological processes have favored a predominance of vertical systems (95 caves are over 300 m deep), but there is no lack of horizontal caves (44 caves are over 3 km long).

Though the major caving exploits occurred in the last twenty years, Italian speleology has nonetheless a long tradition of research which began in the second half of the 19th century [1].

In spite of the recent caving explosion, there are still areas in the center-western alps and in southern Italy that are rarely visited by cavers. Exploration of the great systems in the heavily karstified areas is also far from being finished.

There are today more than 180 active caving clubs, distributed with 50% of them in the North, 25% in the Center, 9% in the South, and 16% in the islands. The Società Speleologica Italiana (S.S.I.) is the organization which represents Italian speleology internationally. It was founded in 1950 and has about 1100 members in 140 groups, or about 78% of the organized national movement.

The Club Alpino Italiano (C.A.I.), which has 65 member clubs (of which 50 belong to the S.S.I.) is mostly concerned with rescue (Soccorso Speleologico, created in 1967).

There are local federations in several Italian regions. In some cases (Umbria, Toscana, Piemonte, Veneto, Friuli-Venezia Giulia, Abruzzo, etc.) they work closely with local authorities.

So far there have been 14 national congresses, which presently occur every four years. In addition, in several areas, the principal sporting and scientific results are reported at periodic conferences (Convegno).

Among the gatherings with an international flavor in the last few years there have been the Convegno Internazionale sul carso di alta montagna (Imperia, 1982), the Simposio Int. sull'utilizzazione delle aree carsiche (Bari, 1982), the IV Simposio Int. di Vulcanospeleologia (Catania, 1983) and the Simposio Int. sul carsismo nelle evaporti (Bologna, 1985).

The national review *Speleologia*, the official publication of the S.S.I., has been published twice a year since 1979. For strictly scientific articles the Istituto Italiano di Speleologia de Bologna and the S.S.I. annually put out *Le Grotte D'Italia*. The S.S.I. also publishes the *International Journal of Speleology* of the U.I.S. The C.N.S.A.-Sezione Speleologica del C.A.I. puts out the annual *Bollettino*, edited in Trieste.

Among the major publications edited by the local clubs, the *Atti e Memorie* and *Progressione* (Commissione Grotte, E. Boegan, Trieste), *Sottoterra* (G. S. Bolognese, C.A.I.), *Grotte* (G. S. Piemontese C.A.I.-UGET Torino), *Bollettino del G. S. Imperiese C.A.I.*, *Mondo Sotterraneo* (Circolo Speleologico e Idrologico Friulano, Udine), *Speleologia Sarda* (G. S. Pio XI, Cagliari), and *Notizario del Circolo Speleologico Romano* are of particular interest.

Gilberto Calandri and Luigi Ramella.

(1) the Commission Grotte "Eugenio Beogan", Società Alpina delle Giulie, section of the C.A.I. de Trieste is the oldest speleological group in the world; it has been active since 1833!

DEEP CAVES:

1. ***complesso Fighiera-Farolfi-
   Antro Del Corchia*** .... **–1215 m**
   (Lucca, Toscana)

   This system has seven entrances: buca d'Eolo (Alt. 1100 m), buca del Serpente (930 m), ingressi alti (1268 and 1301 m), buca del Cacciatore or abisso Fighiera (1640 m), abisso Farolfi (1240 m), and abisso "Black & White" (1420 m).

   The historic entrance, buca d'Eolo, is an man-made tunnel in a marble quarry, on the slopes of the Monte Corchia, above the village of Levigliani (commune de Stazzema), in the Alpes Apuanes.

   1/25,000 map Monte Altissimo, 96 II SO. 2°09'23"; 44°01'44" (upper entrances: 2°09'29"; 44°02'04").

   The cave is developed in the upper Trias dolomites, lower Jurassic metamorphic limestones, and upper and middle Lias stratified limestones. A dye trace by the G. S. Florentino showed the resurgence to be near the village of Cardoso (alt. 350 m).

   Quarry workers discovered the cave (buca d'Eolo) around 1840. The first explorer was Francesco Simi who stopped at –65 m in 1841. Systematic exploration began in 1933 when the G. S. CAI Fiorentino reached –228 m, then –480 m in 1934 (the original surveys claiming –541 to –580 m), where a wet drop marked the end of progress for some time.

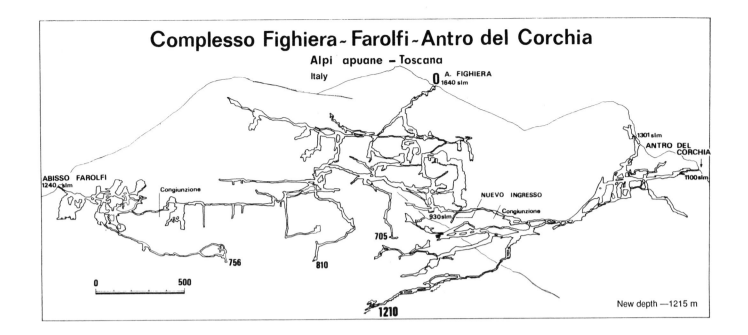

Complesso Fighiera ~ Farolfi ~ Antro del Corchia

Alpi apuane ~ Toscana
Italy

A. FIGHIERA
1640 slm

ABISSO FAROLFI
1240 slm

Congiunzione

NUEVO INGRESSO

Congiunzione

930 slm

1301 slm

ANTRO DEL CORCHIA

1100 slm

705

756

810

1210

0   500

New depth —1215 m

In 1956 the Adriatic Society of Science of Trieste resumed exploration and reached –560 m, then –580 m in 1958. In 1960 the G. S. CAI Bolognese and the S. C. Milano reached –668 m, which they claimed at the time to be –805 m (corrected in 1965 by the G.S.F. and the S. C. Roma).

In 1968 British cavers from the Derbyshire Caving Club and the G.S.F. discovered new ways on in a series of passages and pits on several levels. In 1971 the G.S.F. forced a constriction which led to new passages and a new entrance; the buca del Serpente. In 1977 the G.S.B, with the G.S.F. and the G. S. Pipistrelli de Fiesole deepened the cave to –871 m and climbed upwards in the cave to +79 m and discovered the third and fourth entrances.

In the meantime the buca del Cacciatore (or abisso Fighiera), already explored to –30 m in 1930 by the G. S. Fiorentino, was reexplored by the G. S. CAI Piemontese. On March 20, 1976 G. Baldracco opened the constriction at –30 m, permitting the G.S.P. au Centre Méditerranéen de Spéléologie, the U. S. Bolognese, and the G. S. Faentino to reach –574 m (G. Badino, G. Baldracco, L. Béranger, A. Gobetti). In 1977 the G.S.P., the G.M.S. and the G. S. Faenza reached –630 m and the G.S.P. alone went to –780 m. In 1980 various groups explored the abisso Farolfi to –350 m, then connected it to the buca del Cacciatore in 1981.

Finally, in 1983 these two systems were connected with the antro di Corchia, making a system 1210 m deep. The same year an upper entrance, the abisso "Black & White" was added to the six others.

The year by year length figures are not known, but the total presently stands at 45,000 m.

Map: from *Speleologia*, 1983 (9), sent by Luigi Ramella.

Bibliography: Badino (G.) - Abisso Claude Fighiera, *Speleologia*, 1980 (3): 2-6; Fighiera e Corchia, *ibid.*, 1983 (9): 9-12.

3. *spluga della Preta* (Verona, Veneto) .... **–985 m**
The spluga della Preta is located on the grassy slopes of the Corno d'Aquilio in the Monti Lessini, at an altitude of 1475 m, commune de S. Anna di Alfaedo, in the Venetian Prealps. Access is from Sant' Anna di Alfaedo by the dirt road from the Corno d'Aquilio.

1/25,000 map Monte Baldo, 35 ll SE. 1°30'06"; 4°40'43". The cave is formed in limestones of the lower Cretaceous, then the Jurassic (Malm, Dogger, and Lias).

It has been known a long time and was first written about in 1825. Its 131 m deep entrance pit was dropped in 1925 by J. Battisti (S. U. CAI Verona), who was tied to a rope and pulled up using horses! In 1926, with Cabianco, he reached –295 m (originally quoted as –520 m, then –340 m). In 1927 the S. U. CAI Verona descended to –376 m and stopped at a little lake. (The depth was quoted at the time as –637 m, making it the deepest cave in the world). The figure was adjusted in 1954 by cavers from Trieste from the geospeleological section of the S.A.S. who were attempting to find a way on.

The explorations were resumed by the G.E.S. Falchi Verona who only descended an additional 10 m in 1958 (to –390 m, quoted as –476 m at the time), then another 10 m in 1959, through a series of constrictions.

Helped by the Trieste cavers of the Commissione Grotte Boegan, the G.E.S.F.V. reached –460 m in 1960. In 1962 they organized a large expedition with 80 cavers from various clubs: the depth reached was –578 m (announced as –836 m).

In 1963 the G. S. Piemontese, G. S. Bolognese and the G. S. Città di Faenza reached the provisional bottom of the cave at –878 m and remapped the cave to get an accurate depth.

It was not until 1981 that the G. S. CAI de Verona and de Vittorio Veneto found a way on, to reach –985 m. The length is 2900 m.

Map: from the drawings of the G. S. Verona, G. S. Piemontese, G.S.B., G. S. Modena, and G. S. Faentino, furnished by L. Ramella.

Bibliography: G. S. CAI Verona - Preta: il risveglio di un colosso, *Speleologia*, 1982 (8): 6-7. Also see *Sottoterra*, 1963 (5).

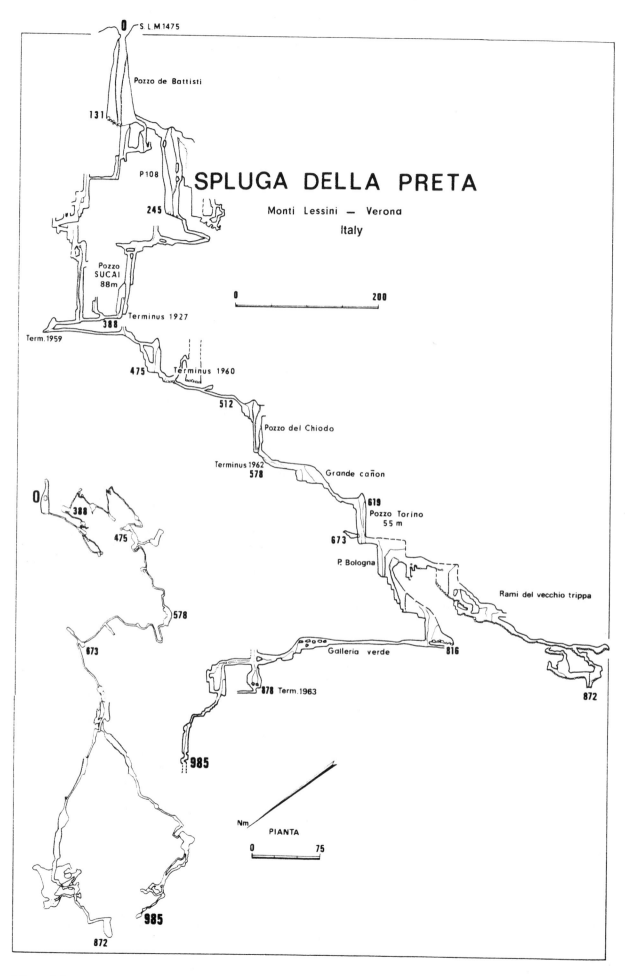

SPLUGA DELLA PRETA

Monti Lessini — Verona

Italy

S.L.M 1475

Pozzo de Battisti

131

P 108

245

Pozzo
SUCAI
88m

Terminus 1927

388

Term. 1959

475   Terminus 1960

512

Pozzo del Chiodo

Terminus 1962
578   Grande cañon

619
Pozzo Torino
55 m

673

P. Bologna

Rami del vecchio trippa

Galleria verde   816

878 Term. 1963

985

0

388

475

578

673

985

872

872

0   200

Nm

PIANTA

0   75

4. *complesso Sotterraneo del col delle Erbe* .... –935 m
or ***Michele Gortani-Davanzo*** (Udine, Friuli)

The system has seven entrances: abisso Michele Gortani or A 10 (alt. 1900 m), P 4 (1928 m), A 12 or abisso Gianni Venturi (1860 m), U 2 (1874 m), abisso I del Col delle Erbe or abisso Enrico Davanzo (1920 and 1910 m) and meandro del Plucia (1721 m).

It is located near the Col delle Erbe, on the Monte Canin, in the Julian Alps, not far from the Yugoslavian border, in Chiusaforte. It is located in the heart of a very rich karst area which can only be reached by a cable car from the Sella Nevea to the Gilberti hut, then by the Col delle Erbe trail which goes by the Sella Canin.

1/25,000 map Monte Canin, 14 ll SE. 0°59'42"; 46"22'24" (1900 m) and 0°59'27"; 46°22'39" (1920 m).

The resurgence is the fontanon di Goriuda (alt. 868 m). Dachstein limestones, upper Noriano and Rethiano.

The abisso Michele Gortani was discovered in 1965 by the Commissione Grotte Boegan de Trieste who alone explored essentially the entire system. In 1965 they descended to –240 m, then –342 m in 1966 and –675 m in 1967 to a sumped lake which stopped exploration. The discovery of another branch permitted –763 m to be reached in 1968 and –866 m in the summer of 1969, when a new sumped lake stopped exploration. In December the level of the lake dropped enough to permit the explorers to continue to a terminal sump at –892 m. This winter trip cost the life of three cavers on the way back home: Marino Vianello, Enrico Davanzo, and Paolo Picciola were carried away by an avalanche.

In 1970 the discovery of an upper entrance added 28 m to the vertical extent: 920 m. The abisso A 12 was discovered by the C. G. Boegan in 1972 and explored to –200 m. In 1973, with the U. S. Bolognese and the G. S. Faenza, they got to –305 m and made the connection with the Gortani at –446 m in July, 1975.

The abisso U 2 was discovered in 1973 by the U.S.B., the G.S.F. and the S. C. Forli. In 1975 the C. G. Boegan got down to – 242 m and made the connection.

The exploration of the abisso Enrico Davanzo was almost simultaneous with that of the Gortani. Discovered in 1965 by the C. G. Boegan, it was explored to –172 m. Exploration was resumed in 1970 and brought to –404 m in a fossil branch and –402 m in the active branch, then –446 m in the dry and –420 m in the wet passage in 1971. The stream passage was followed to –735 m in 1972. In August, 1973 the gain in vertical extent was only 2 m (sumped lake). New ways on were found in 1982 and 1983, permitting the C. G. Boegan to connect the Davanzo to the Gortani in 1984. In 1985 Trieste cavers dived the terminal sump in the abisso Gortani bringing the depth to –935 m.

The estimated length for the entire system is 13 km.

Map: from the C. G. Boegan, the U. S. Bolognese, the S. C. Forli, and the G. S. Faentino, furnished by the C. G. Boegan and Suzanna Martinuzzi.

Bibliography: Casale (A.), Vaia (F.) - Relazioni fra schema deformativo a cavità carsiche nell'abisso M. Gortani, *Atti e Memorie della C.G.E. Boegan*, 1971, XI: 67-94; Badino (G.), Bonelli (R.) - *op. cit.*, 1984.

5. *grotta di Monte Cucco* ....(–803, +119) 922 m
(Perugia, Umbria)

The main entrance to the cave is located on the slopes of Monte Cucco at an altitude of 1390 m, at Sigillo, in Costacciaro, in the Appennins d'Ombrie. From Sigillo, one takes the Val di Ranco road to its end, then a mule trail which leads to the cave. A second entrance to the cave was dug open in 1974 and is at an altitude of 1509 m.

1/25,000 map Costacciaro, 116 ll SO. 0°17'47"; 43°22'16".

The resurgence is the major spring at Scirca (alt. 528 m), 2500 m from the entrance. Lower Lias limestones.

The cave was noted by F. Gabrielli in 1759 (in his work of 1775, *Descrizione delle Grutta di M. Cucco. Storia dell' Agro Pasarese*, Bologna), but there are inscriptions inside dating from the 15th century. Jean-Baptiste Miliani made the first exploration in 1889, exploring and mapping the horizontal passages and rooms that make up the first part of the cave.

In August, 1967 the G. S. CAI Perugia found a descending lead from the southern end of the the Marguerite room (140 x 30 x 70 m), reaching a depth of –380 m in December, 1967. In 1968 the same group reached –700 m for a length of about 4000 m. In Spring 1969, with the help of the G. S. Piemontese, the C. G. Boegan de Trieste and two Belgians, they arrived at the terminal sump at –803 m, giving a vertical extent of 829 m. In the following years the length increased and went from about 12,000 m in 1972 to 13,235 m in 1973, the year in which an ascending branch leading from the Marguerite room was followed to +38 m, for a vertical extent of 841 m. In the summer of 1974, with the G. S. Sigillo, the old cave sections were remapped and an ascending gallery was found leading up from the "galleria dei Laghetti" (–12 m) to a dome pit, the "pozzo del Nibbio", which was climbed for 63 m to reach the upper entrance and give a vertical extent of 922 m.

In 1975 the length was increased to 15,023 m, then 20,867 m in 1980/1981, 24,390 m in 1983 (the year in which the G. S. Marchigiano descended to – 740 in the "galleria dei Barbari") and to 26,135 m in 1984.

Map: from the drawings of the G. S. Perugia, put together by Francesco Salvatori, furnished by F. Salvatori.

Bibliography: Iuretig (L.) - La grotta di M. Cucco, *Atti Congr. Naz.*, Genova 1972, I: 95-101. Badino (G.), *Gli abissi italiani. Guida ai grandi mondi sotteranei*, Bologna, ed. Zanichelli, 1984, 126 p.

6. ***complesso di Piaggia Bella*** .... –911 m
(Cuneo, Piemonte)

The Piaggia Bella system presently has eleven entrances which include the abisso S 2 or abisso Carciofo (alt. 2357 m), chiesa di Bac or abisso Caracas (2297 m, two openings), gola del Visconte (2272 m), abisso "des Pensées" or abisso Jean Noir (2197 m), grotta dell' Indiano or abisso Velchan (2190 m), buco delle Radio (2175 m), grotta di Piaggia Bella or voragine del (Colle del) Pas (2157 m), abisso Solaï (2038 m), abisso Raymond Gaché (2513 m) and grotte des Mastrelle (1921 m).

It is located in the Marguareis range which spans the Franco-Italian border (commune de Briga Alta, Alpi Maritime). The easiest access is from the col des Seigneurs which one reaches by a strategic road coming from the col de Tende and from Limone. From there, an hour of walking

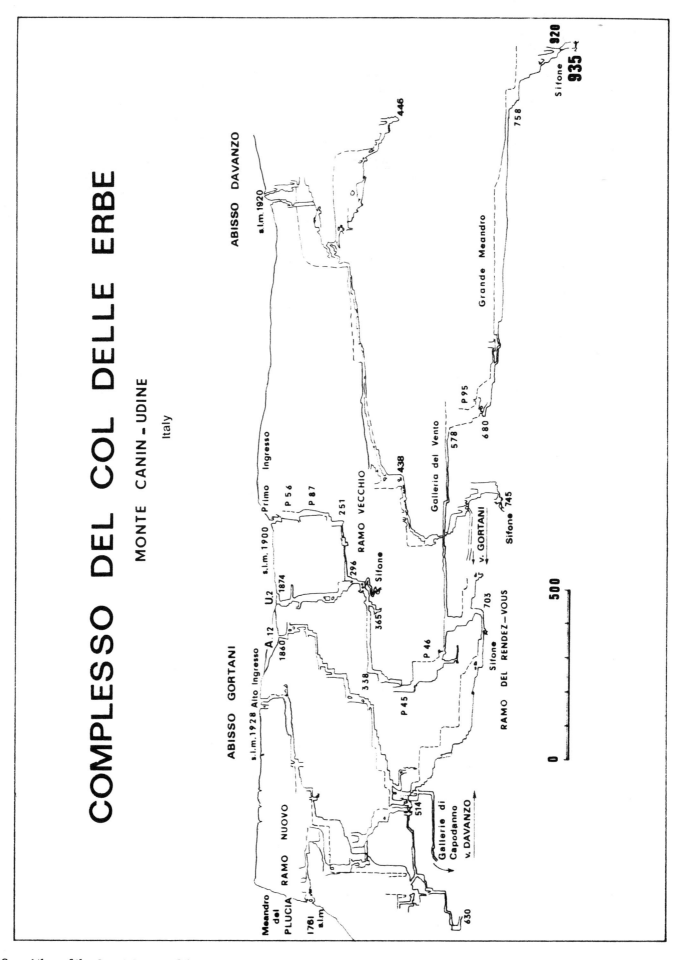

COMPLESSO DEL COL DELLE ERBE

MONTE CANIN – UDINE

Italy

ABISSO DAVANZO

s.l.m. 1920

446

Grande Meandro

758

Sifone 920

Sifone 935

Primo Ingresso

P 56

P 87

251

RAMO VECCHIO

296 · Sifone

438

Galleria del Vento

578

680

P 95

s.l.m. 1900

U.2

1874

365

V. GORTANI

Sifone 745

ABISSO GORTANI

s.l.m. 1928 Alto Ingresso

A.12

1860

338

P 46

703

Sifone

RAMO DEL RENDEZ—VOUS

P 45

514

Gallerie di
Capodanno
v. DAVANZO

500

Meandro del
PLUCIA RAMO NUOVO

1761 s.l.m.

630

0

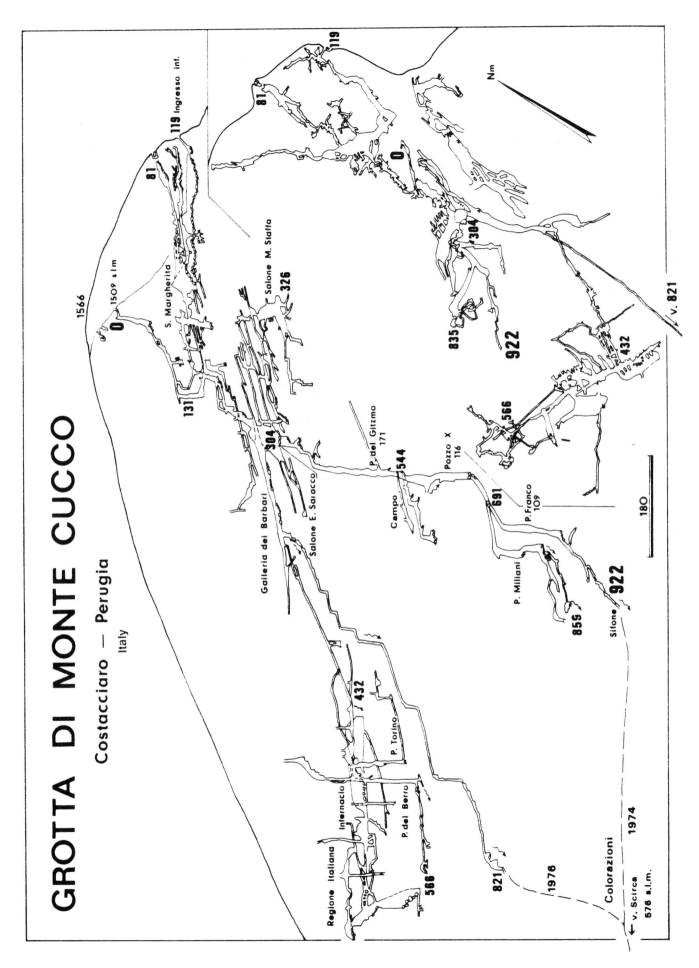

GROTTA DI MONTE CUCCO

Costacciaro — Perugia
Italy

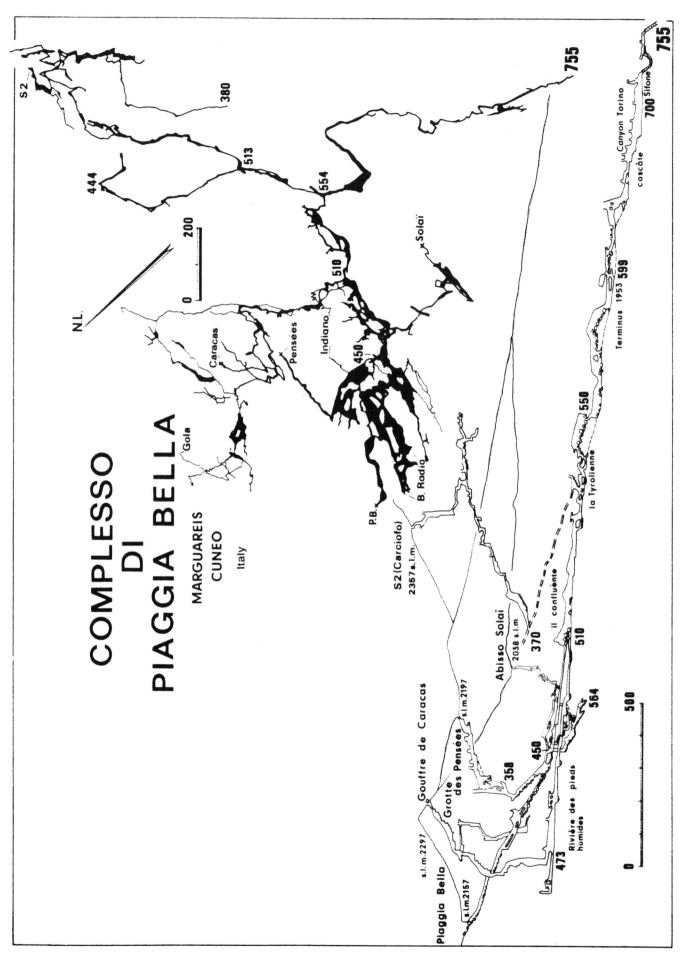

COMPLESSO
DI
PIAGGIA BELLA

MARGUAREIS
CUNEO

Italy

along the side of the col, to the east, is necessary to arrive in the cirque which dominates the crest of Pian Ballaur and where the grotta di Piaggia Bella is found, not far from the hut recently built at Saracco-Volante.

1/50,000 map Viève; x 1029.18; y 2221.30 (France) or 1/25,000 map Monte Mongioie, 91 I SO and Viozene, 91 II NO.

The system is formed in the Briançon limestone series (Jurassic and middle Trias). A dye trace has shown the resurgence to be the résurgence de la Fus (alt. 1180 m) in the vallée du Negrone.

The entrance to the grotta di Piaggia Bella, into which flows a small stream, was known to shepherds who spent the summers in the area. It was noted in 1944 by Capello who explored to –166 m. Systematic exploration began in 1952 with the Expéditions Spéléologiques Françaises (bringing together notably the Spéléo-Club de Paris, the Club Martel de Nice and the G. S. C.A.F. de Millau). In this year the depth of –354 m was reached. In 1953 the depth of –399 m was reached, where a sump blocked the way. The abisso Jean Noir was discovered and descended to –100 m.

In 1954 the E.S.F. explored both the abisso Jean Noir (–125 m) and the abisso Caracas (–60 m), known to Capello and rediscovered by the Venezuelan Eugenio de Bellard Pietri.

In 1955 the E.S.F. and the G.G. Debeljak de Trieste arrived at –177 m in the abisso Jean Noir. The connection with Piaggia Bella was made in August 1956 at –311 m, making this system 443 m deep.

In 1956 Caracas was pushed by the E.S.F. to –115 m, then to –336 m in 1957 with the help of the G. S. CAI Alpi Maritime de Cuneo. In 1958, at –413 m, the connection was made with Piaggia Bella. During the same period, the G. S. Piemontese bypassed the sump and other obstacles that had stopped the French in the bottom of Piaggia Bella and discovered the large "canyon Torino", stopping at a terminal sump at –500 m, giving a vertical extent of 640 for the system (according to a resurvey by the Centre Méditerranéen de Spéléologie (Nice) in 1974). The length reached 5,800 m.

The abisso Solaï, discovered in 1971 by the C.M.S. and pushed by them to –240 m, was connected to the system in 1975 by the G.S.P. The length, thanks to the explorations by the C.M.S. during the preceding years, was increased to 13,700 m.

It was the G.S.P. that in 1977 added two new entrances: the buco delle Radio and the grotta dell' Indiano. In 1978-1979 they discovered extensions ("gallerie Belladonna", "Khyber Pass") that brought the length to 15,800 m.

The terminal sump was dived in 1980 by Fred Vergier, increasing the vertical extent by 56 m to 696 m.

An upper entrance, the abisso S 2 (discovered in 1981 by the G. S. Imperiese) was connected by them to the Piaggia Bella on June 26, 1982, bringing the depth to –755 m and the length to about 21,000 m. Finally, in 1983, an eighth entrance, the gola del Visconte was added to the others. In 1984 the discoveries of the G.S.P. and the G.S.I. brought the length to over 25 km.

The abisso Raymond Gaché was discovered in 1954 by the Expéditions Spéléologiques Françaises who reached –314 m. In 1955, the young Lucio Mersi was killed in the 127 m deep pit, shortly before the E.S.F. reached –375 m. It was in 1961 that the G. S. Piemontese succeeded in passing the old terminus and got to –390 m. In 1962, with the G. S. Bolognese, they reached the low point of the cave at –558 m. In 1983, the G.S.P. explored the abisso Essebue (alt. 2525 m) which connected to the Gaché at –450 m. The Gaché was connected to Piaggia Bella in 1986, making a system –911 m deep.

Map: synthesis of the drawings of the E.S.F., the G.S.P., the C.M.S., and the G. S. Imperiese sent by L. Ramella, by P. Courbon. (Does not include abisso Raymona Gaché).

Bibliography: Dematteis (G.) - Il sistema carsico sotterraneo Piaggia Bella-Fascetta (A. Liguri), *Rass. Spel. Italiana*, 1966, 18 (3/4): 87-121.
Calandri (G.) - Osservazioni geomorfologiche e idrologiche sull'abisso S 2 ed il settore Arpetti-Pian Ballaur, *Le Grotte d'Italia*, 1982, 4 (XI): 305-318.

2. *pozzo della Neve* .... **—1050 m**
(Campobasso, Molise)

Also called the pozzo di Costa del Carpine, this cave is located at an altitude of 1330 m in the Matese range, in Campochiaro, in the Costa del Carpine zone.

1/25,000 map Campobasso, 162 III NO. 2°02'45.5'; 41°25'22".

The cave is formed in upper Cretaceous limestones (Senonian-Cenomanian).

It was discovered in 1955 by the Circolo Speleologico Romano (C.S.R.) who stopped at –130 m at a small lake. Exploration was not resumed until 1966 when the S. C. Roma reached –300 m. Beginning in 1972 the A. S. Romana undertook the systematic exploration of the pozzo and reached –420 m, then –576 m in 1973. In 1974 they were stopped at a sump (–693 m) which constituted the provisional bottom of the cave. The cave is traversed by a torrential river which forms numerous lakes and ponds as well as sumps. The horizontal trend is often interrupted by wet pits and waterfalls.

In 1981 and 1982 the C.S.R. deepened the cave, reaching –755 m, then – 855 m. In 1984 the C.S.R. discovered a second entrance (1370 m), bringing the depth to –895 m and the length to about 5000 m.

Map: from the work of the A.S. Romana and the C.S. Romano, furnished by L. Ramella.

Bibliography: Bernabai (T.) - L'abisso pozzo della Neve, *Atti Congr. Naz.*, Bologna 1982, pp 346-355.

7. *abisso dei Draghi Volante* .... **–870 m**
(Lucca, Toscana)

Located at an altitude of 1425 m, the cave is found on the north side of Monte Sumbra, at Passo Fiocca, in Vagli di Sotto, in the Apuane Alps.

1/25,000 map Monte Altissimo, 96 ll SO. 2°10'49"; 44°04'52".

This cave is formed in Liassic marbles and Triassic sandstones. The low point in the cave is only 40 m above the resurgence.

It was discovered in 1977 by the G. S. Pipistrelli di Fiesole who explored the entrance area to –30 m. In 1978, they returned to the attack and reached –370 m, which they increased to –490 m in 1979. Two years later the G.S.P. di Fiesole was helped by the G.S.B. and the U.S.B.,

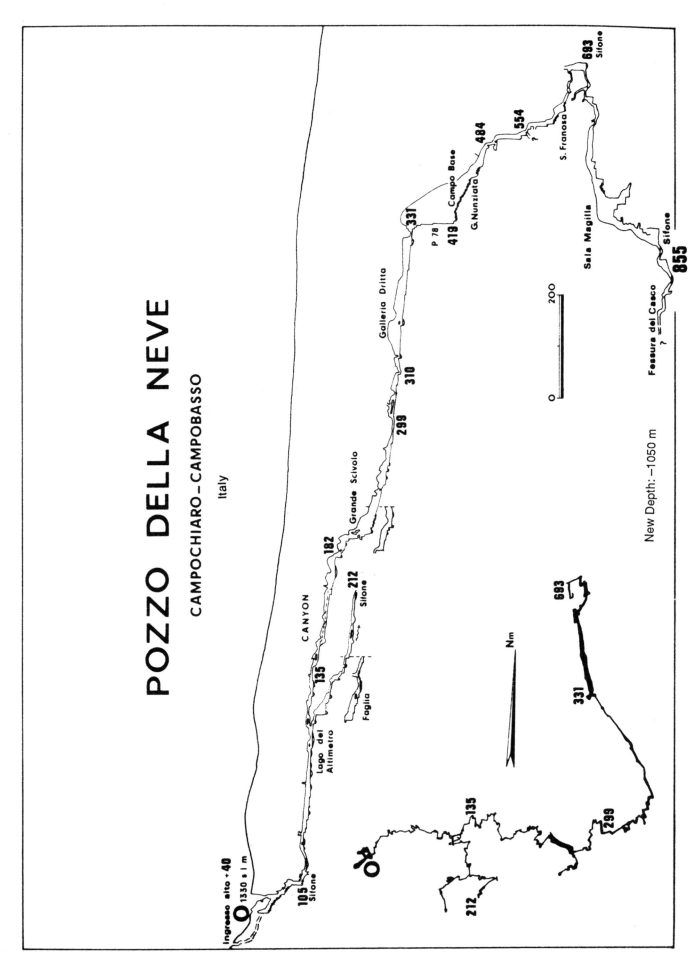

POZZO DELLA NEVE

CAMPOCHIARO – CAMPOBASSO

Italy

New Depth: –1050 m

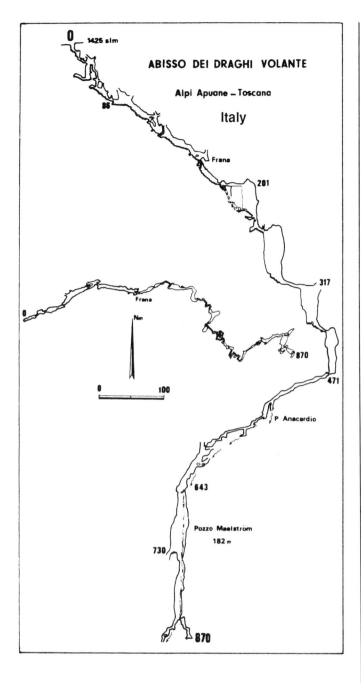

was required before exploration could proceed. Once opened up, the descent to −815 m by the G. S. CAI Lucchese, the G. S. Archeologico Livornese and other clubs proceeded without problems. It was extended to −835 m in 1986.

Map: taken from *Speleologia*, 1986 (14) showing -815 m.

Bibliography: Roncioni (A.) and Campolmi (C.) - L'abisso dello Gnomo, *Speleologia*, 1986 (14): 9-10.

9. *abisso Emilio Comici* (Udine, Friuli) .... **−774 m**
The abisso Emilio Comici or abisso I a Nord del Foran del Muss is located at an altitude of 1993 m, in the Foran del Muss zone, near the Col Sclaf, in the Monte Canin (Julian Alps), Chiusaforte, not far from the Yugoslavian border.

1/25,000 map Monte Canin, 14 ll SE. 0°58'35"; 46°22'40".

It belongs to the same hydrological system as the complesso del Col delle Erbe; the resurgence is the fon-

permitting a considerable increase in depth: –870 m was reached by dropping the terminal 182 m deep pit. Total length 1450 m.

Map: from the G.S.P. Fiesole and the G.S.B., sent by L. Ramella.

Bibliography: *Sottoterra*, 1981, No. 59. Sivelli (M.), Vianelli (M.) - *op. cit.*, 1982.

8. *abisso dello Gnomo* ....(**−835 m, + 35) 870 m**
(Lucca, Toscana)
This new cave of the Valle d'Arnetola opens at an altitude of 1400 m near the mule trail that goes from the bottom of the Valle d'Arnetola to Passo Sella (Alpes Apuanes).

It was discovered on August 13, 1985 by the G. S. Archeologico Versiliese, but digging in the entrance area

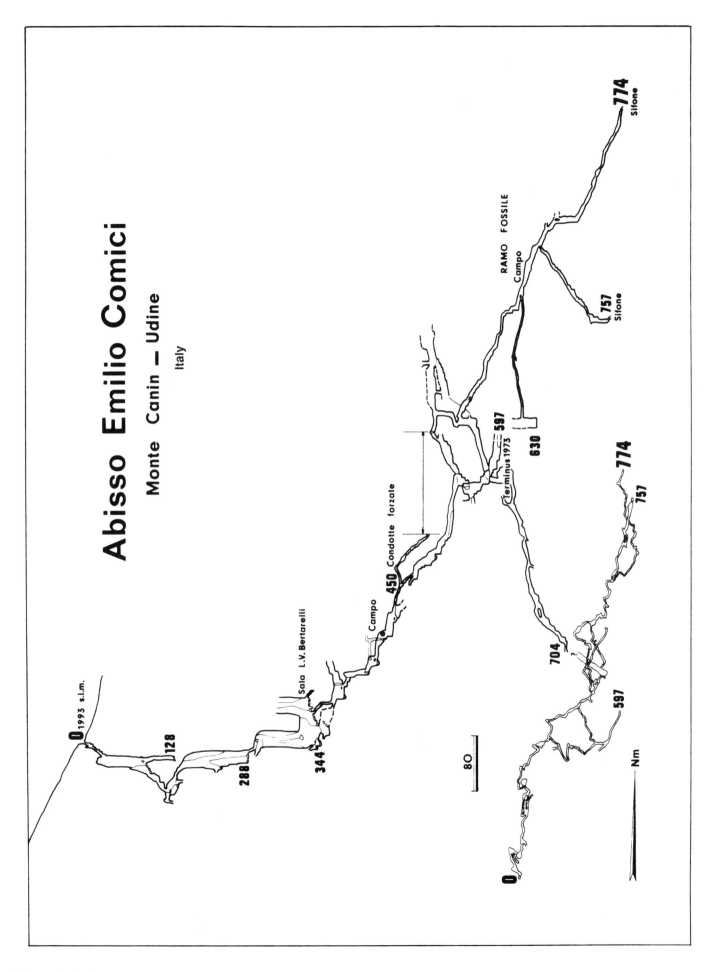

Abisso Emilio Comici

Monte Canin — Udine

Italy

tanon di Goriuda (alt. 868 m). Dachstein limestones (upper Noriano-Rethiano).

The cave was discovered in 1971 by the G. S. CAI Bertarelli di Gorizia who explored it in five seasons. In 1971, they reached –240 m. In 1972 they got to –507 m and reached a first terminus at –597 m in 1973. In 1974 they found ways on at –450 m which split into two branches, each of which ended at sumps, one at –757 m, the other at –774 m.

In 1975 an expedition brought together the G. S. Bertarelli and the Burley Caving Club Survey from Manchester: a new bottom was discovered at –630 m, as well as a way on near the terminal sump at –774 m, but the depth remained unchanged.

Map: from the drawings of the G. S. Bertarelli di Gorizia, sent by L. Ramella

Bibliography: Tavagnutti (M.) - Abisso Emilio Comici. Operazione '78, *Speleologia*, 1979 (1): 44-45.

10. *abisso Presso la Quota 1972* .... **–760 m**
(Udine, Friuli)

This cave has three entrances: S20 (alt. 1915 m), S31 (1928 m), and FDZ 2 (1942 m). It is found in the valle del Boegan of the Monte Canin (Julian Alps, Chiusaforte).

1/25,000 map Monte Canin, 14 ll SE. 0°59'28"; 46°22'24" (upper entrance) and 0°59'28"; 46°22'25".

It is apparently part as the same hydrological system as the complesso del Col delle Erbe and the abisso Emilio Comici (Dachstein limestones).

This cave was discovered in August 1976 by the G.G. XXX Ottobre de Trieste who reached about –180 m. In 1978, the same group increased the depth to –710 m, then came to the terminal sump (still undived) at –760 m the following year.

Map: from the drawings of the G.G. XXX Ottobre de Trieste, sent by L. Ramella.

Bibliography: Besenghi (F.) - Un nuovo abisso sul Monte Canin, *Atti IV Conv. del Friuli-Venezia Giulia*, Pordenone 1979, pp. 301-305.

11. *abisso dei Fulmini* (Lucca, Toscana) .... **–760 m**

The abisso dei Fulmini or Giovanni Leoncavallo is located at an altitude of 1340 m, at Cave Fondone, on the Monte Altissimo (Stazzema) in the Apuane Alps.

1/25,000 map Monte Altissimo, 96 II SO. 2°12'49"; 44°02'59".

The cave is formed in Liassic marbles and Triassic sandstones. A dye trace has established its relation with the sorgente delle Polla, located on the south side of Monte Altissimo.

It was discovered in 1970 by the G. S. Paletn "G. Chierici" who in 1971 reached the depth of –90 m. It had to wait until 1982 for the exploration of the cave to be resumed. Several clubs (G. S. Bolognese, U. S. Bolognese, G.S.A. Versiliese and G.S.P.F.) united their efforts and succeeded in reached –560 m, then –760 m the following year.

This cave, which contains pits (including one of 100 m) alternating with short meanders, ends at a sump, and is considered one of the nicest in this part of the Alps. It is 1050 m long.

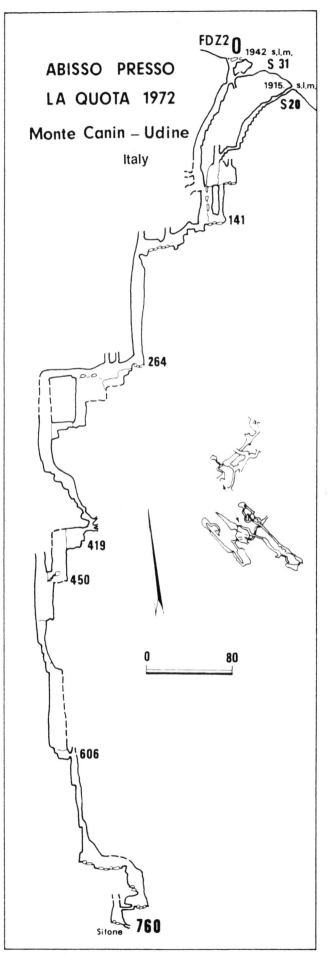

ABISSO PRESSO
LA QUOTA 1972
Monte Canin – Udine
Italy

FDZ2 0 1942 s.l.m.
S 31
1915 s.l.m.
S 20
141
264
419
450
0   80
606
Sifone 760

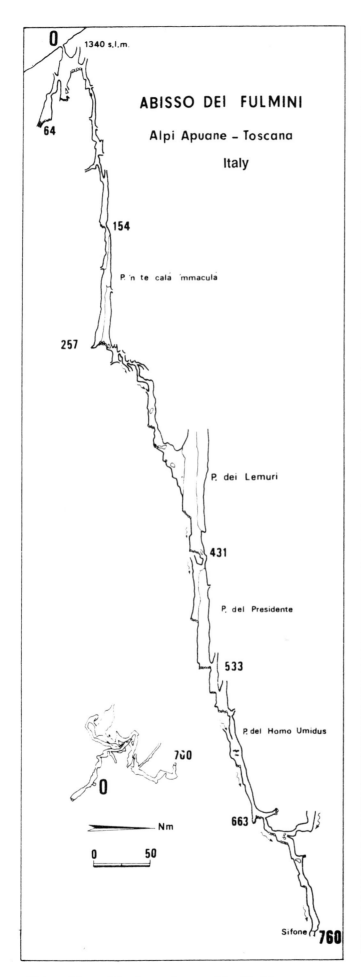

ABISSO DEI FULMINI

Alpi Apuane – Toscana

Italy

0

1340 s.l.m.

64

154

P. in te cala 'mmacula

257

P. dei Lemuri

431

P. del Presidente

533

P. del Homo Umidus

700

0

Nm

0   50

663

Sifone **760**

Map: from the drawings of the U.S.B. and the G.S.B., sent by L. Ramella.

Bibliography: Vianelli (M.) - Apuane: Abisso dei Fulmini, *Speleologia*, 1983 (9): 20-21.

12. *abisso Paolo Roversi* .... **–755 m**
(Lucca, Toscana)

The cave (also called abisso Don Ciccillo) is located in the Carcaraia zone on the north side of Monte Tambura (commune de Minucciano) in the Apuane Alps. It is at an altitude of 1720 m, not far from the Aronde hut. Access is from Giorfigliano to the col of Focolaccia.

1/25,000 map Vagli di Sotto, 96 ll N° 2° 14'47"; 44°06'47". Formed in Lias marbles.

It was discovered on September 3, 1977 by the G. S. Bolognese, who alone explored to –755 m. In the discovery year the G.S.B. reached –165 m. Much progress was made in 1978 when –630 m was reached, then finally –755 m in 1979. In 1983-1984 Polish cavers found new passages at –300 m which descended to about –720 m.

The most remarkable feature of this cave is its 310 m deep terminal pit (see "The Great Pits").

Map: from the drawings of the G.S.B., sent by L. Ramella.

Bibliography: Fabbri (M.) - Abisso Paolo Roversi –755, *Speleologia*, 1979 (2): 48-50; Sivelli (M), Vianelli (M.) - *op. cit.*, 1982.

13. **abisso Modonutti-Savoia** .... **–743 m**
(Udine, Friuli)

This cave is located on Monte Canin, near the Col Lopic, at 1800 m altitude, commune of Chiusaforte.

1/25,000 map Sella Nevea, 14 III SO. 1°02'40"; 46°22'59". Dachstein limestones, upper Trias.

It was discovered on June 30 1985 by the Circolo Speleologico e Idrologico Friulano while ridgewalking who heard water running beneath the rock and snow. Once the snow melted, exploration began in July, with –100 m being reached. Exploration was resumed in September and the depth of –743 m was reached on October 5, stopping at the top of a waterfall when they ran out of rope.

14. *abisso Oriano Coltelli*
(Lucca, Toscana) .... **–730 m**

Another cave in the Apuane Alps. At 1215 m altitude, in the Valle di Arnetola, near Vagli di Sopra (commune of Vagli di Sotta).

1/25,000 map Vagli di Sotto, 96 ll NO. 2°02'26"; 44°05'30". Developed in the same geologic formation as the abisso Paolo Roversi.

Its discovery in 1971 was by the G. S. CAI Lucchese who reached –100 m the same year. The true exploration of the cave was not until 1976. Without particular difficulties, in spite of a stream beginning at –230 m, the G. S. Lucchese and the G. S. Archeologico Versiliese reached the terminal sump at –730 m. The length is 1300 m.

Map: from the drawings of the G.S.L. and the G.S.A.V., sent by Giulio Badini.

Bibliography: Sivelli (M.), Vianelli (M.) - *op. cit.*, 1982.

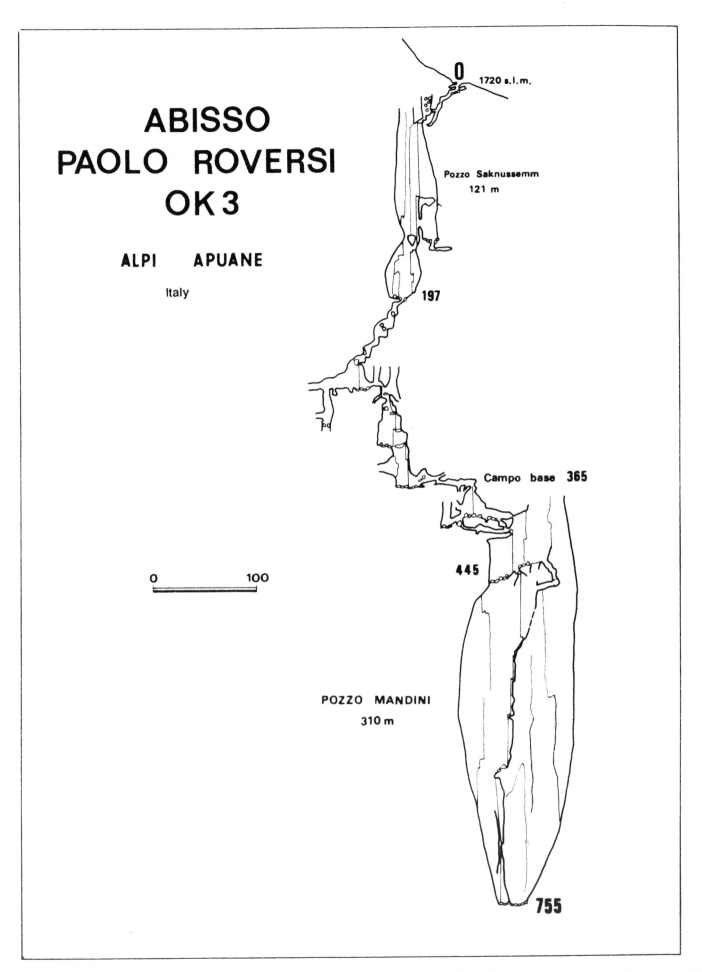

ABISSO
PAOLO ROVERSI
OK 3

ALPI   APUANE

Italy

0   1720 s.l.m.

Pozzo Saknussemm
121 m

197

Campo base   **365**

445

POZZO MANDINI
310 m

755

0   100

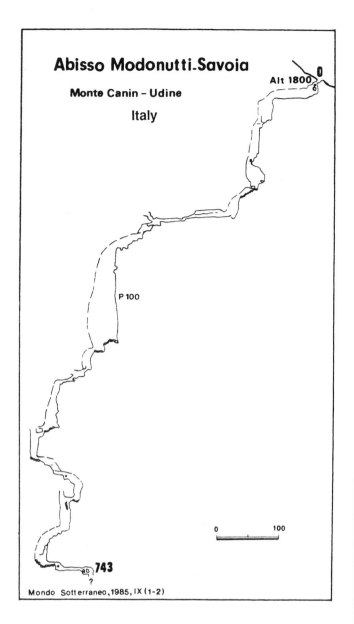

Abisso Modonutti-Savoia

Monte Canin – Udine

Italy

Alt 1800

P 100

0   100

743
?

Mondo Sotterraneo,1985, IX (1-2)

16. *abisso II del Monte Poviz* .... **–720 m**
(Udine, Friuli)

Another cave on Monte Canin, located in the area of the Monte Poviz (Julian Alps, commune of Chiusaforte), at 1873 m altitude. Also known as Gronda Pipote.

1/25,000 Sella Nevea, 14 II SO. 1°02'32"; 46°22'54". Formed in the Dachstein limestone of the upper Trias (Noriano-Rethiano).

The C. G. "Eugenio Boegan" de Trieste discovered the cave in 1979 and explored it in 1980 and 1981 to –720 m, accompanied by the C.S.I. Friulano. It is a cave with small dimension passages and narrow squeezes and awkward pits, the difficulty of which have limited the depth to –720 m.

Map: from the drawings of the C. G. "Boegan" and the C.S.I. Friulano, furnished by L. Ramella.

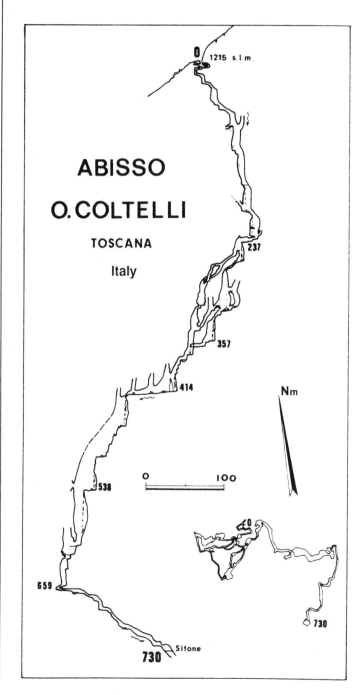

ABISSO

O. COLTELLI

TOSCANA

Italy

0   1215 s.l.m.

237

357

414

Nm

538

0   100

659

730

730   Sifone

15. *abisso ET5* (Udine, Friuli) .... **–726 m**

This un-named cave, located to the "SE della quota 1972", is near the "abisso Presso la quota 1972" (see above), at 1918 m altitude. It is in the valle del Boegan, on Monte Canin (Julian Alpes, commune of Chiusaforte).

1/25,000 map Monte Canin, 14 ll SO. 0°59'31";46°22'18".

Like its neighbor, the abisso ET 5 seems to be hydrologically connected with the resurgences of the complesso del Col delle Erbe, the abisso Emilio Cormici, and the fontanon di Goriuda (alt. 868 m). It is formed in the Dachstein limestones of the upper Trias (Noriano-Rethiano).

The G. G. XXX Ottobre de Trieste discovered the cave in 1980 and explored it to –150 m. In 1983, helped by the C. G. "Eugenio Boegan" de Trieste, they reached –726 m.

Map: from the drawings of R. Besenghi, A. Fedel, T. Ferluga and U. Mikolic, furnished by L. Ramella.

Bibliography: Ferluga (T.) - ET 5 e noi, *Progressione*, 1984 (12): 7-9.

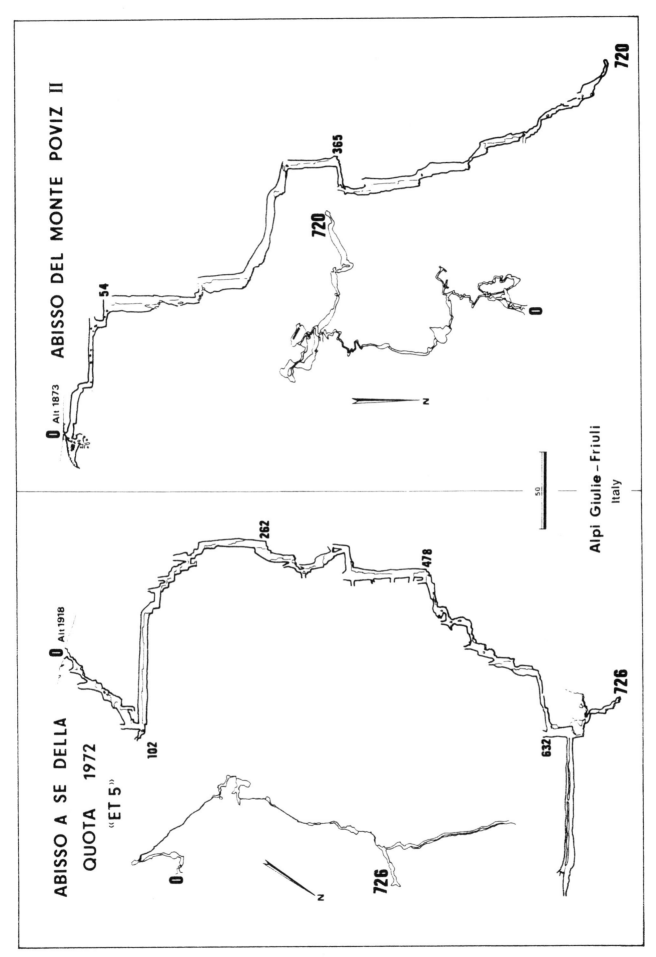

ABISSO DEL MONTE POVIZ II

720

365

720

54

0 Alt 1873

0

N

50

Alpi Giulie – Friuli
Italy

ABISSO A SE DELLA
QUOTA 1972
«ET 5»

0 Alt 1918

262

478

726

102

632

726

726

0

N

Bibliography: Mikolic (U.) - "Gronda Pipote" –720...tanto per non perdere il vizio, *Speleologia*, 1984 (11): 45-46. See also *Progressione*, 1980 (5); 1981 (7) and 1982 (9).

17. abisso Cappa (Cuneo, Piemonte) .... **–710 m**

A cave located near the cliffs of the Bric dell'Omo (2310 m), on the north slopes of the famous Conca delle Carsene, in the Marguareis range, not far from the crest that forms the border with France (commune of Briga Alta). Access is by the road which goes from the col de Tende to the col des Seigneurs. One leaves the road at a place called Collepiano and goes towards the north to cross the border. It is located at an altitude of 2154 m and has a 1 m diameter opening.

1/50,000 map Viève. x 1024.540; y 222.660 (France) or 1/25,000 map Certosa di Pesio, 91 IV SE (Italy).

The cave is hydrologically connected with the resurgence N° 18 du Pis del Pesio, 2300 m to the north (alt. 1340 m, dye trace of July 19, 1968). Formed in the Brianconnais limestones of the Dogger.

Gérard Cappa discovered the cave in 1967 and explored it with the Association Sportive du Bâtiment et des Travaux Publics de Nice. That year the discovery of a 180 m deep pit at –134 m marked the end of exploration. The pit was descended in 1968 but a constriction at –345 m stopped the progress of the A.S.B.T.P.

In 1973 the Centre Méditerranéen de Spéléologie (Nice) and the A.S.B.T.P. resumed exploration and found new passages which added 20 m to the vertical extent. They stopped for lack of rope. In 1974 the A.S.B.T.P. and the C.M.S. continued and reached, by a new series of pits, first –440 m then –662 m (length 2500 m). Additional passages were explored by the C.M.S. in 1975 and 1977 bringing the length to 4500 m but not increasing the depth. In 1979 the C.M.S. got past the old terminus to reach –698 m (length 7116 m). In 1982 the G. S. Piemontese found a new entrance, the abisso 18. In 1983, with the help of the G.S.P. and the A.C. Nice, the C.M.S. reached a sump at –710 m, bringing the length to over 8 km.

Large passages are heading both to the source du Pisio to the north, and to the plan Ambroise to the south (polje at 2109 m alt.), located in France.

Map: from the drawings of the C.M.S., collected by Lucien Béranger, sent by A. Depallens.

Bibliography: *Bulletin des Phénomènes karstiques*, C.M.S., 1976 (1).
Centre Méditerranéen de Spéléologie - Le massif du Marguareis, *Spelunca*, 1976 (3): 113-123.

18. abisso Paolo Fonda (Udine, Friuli) .... **–700 m**

Another deep cave added recently to the many others known in the Monte Canin region. This one is at 1850 altitude, near to the Gilberti hut, in the commune of Chiusaforte (see above, complesso del col delle Erbe).

1/25,000 map Sella Nevea, 14 III SO. 1°00'33"; 46°22'24".

Because of its close proximity to the complesso del Col delle Erbe, the likely resurgence is the fontanon di Goriuda (alt. 868 m). Dachstein limestones of the upper Trias (Noriano-Rothiano).

It was discovered by Maurizio Glavina of the C. G. "Eugenio Boegan" de Trieste in 1984. It was that group (made up of Blanchetti, Fedel, Feresin, Ferluga, Glavina, Lazzarini, Pezzolato, Savio, Sollazzi, Squassino, Sussan and Vascotto) who, in 1984-1985, reached the terminal sump at –700 m.

While the cave could be characterized as a series of narrow pits, its main distinguishing feature is a 286 m deep pit, the deepest in the Monte Canin area.

Map: from the drawings of the C.G.E. Boegan de Trieste, sent by Suzanna Martinuzzi.

Bibliography: P. Squassino, in *Speleologia*, 1985 (13).

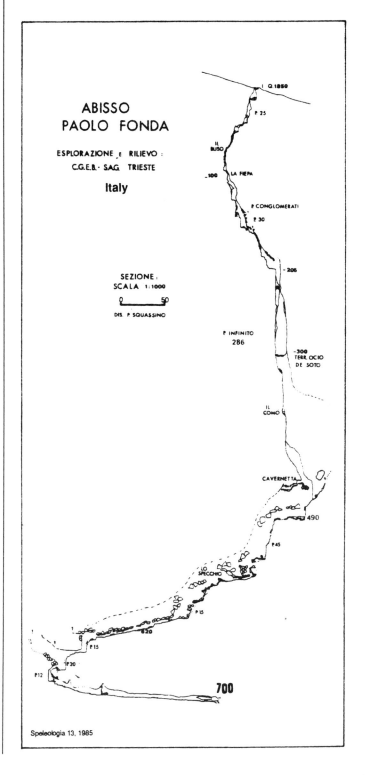

ABISSO
PAOLO FONDA

ESPLORAZIONE E RILIEVO :
C.G.E.B. · S.A.G. TRIESTE

Italy

SEZIONE:
SCALA 1:1000

0_____50

DIS. P. SQUASSINO

Speleologia 13, 1985

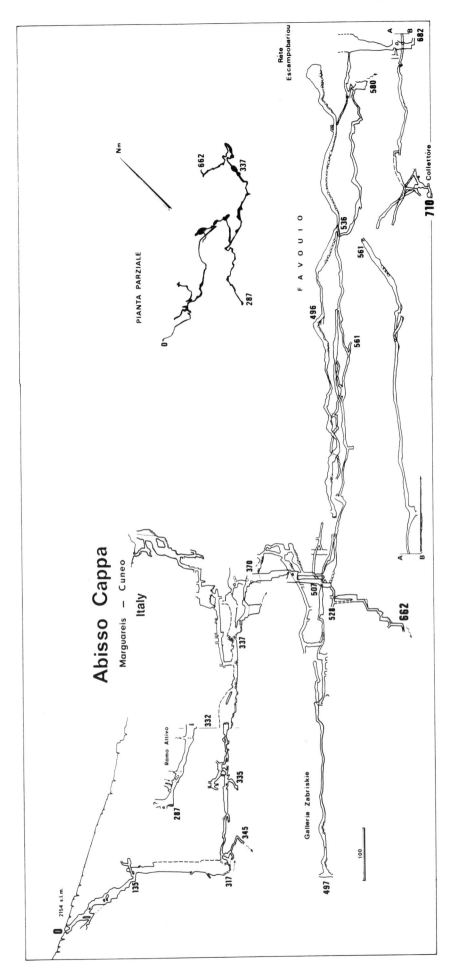

Abisso Cappa

Marguareis — Cuneo

Italy

19. **abisso Carlo Seppenhofer** .... –690 m
(Udine, Friuli)

Cave in the Monte Canin (Alpi Giulie, municipality of Chiusaforte) near the Col Solif, at 1900 m altitude, formed in the Dachstein limestones of the upper Trias (Noriano-Rethiano). The presumed resurgence is the fontanon di Goriuda (alt. 868 m).

1/25,000 map Monte Canin, 14 II SO. 0°58'45"; 46°22'46".

Discovered by the G. S. "L.V. Bertarelli" de Gorizia in 1973 and explored by this group to –196 m in 1974. Exploration was pursued in 1979 by the Gruppo Triestino Speleologico (–355 m), then 1980 (–375 m), 1983 (–480 m), and 1984 (–690 m). There is an ice plug at –120 m. (*Atti II Conv. di Spel. del Friuli-V.G.*, Udine 1975 pp. 152-160; *Speleologia*, 1980 (4) profile to –375 m).

20. **abisso I del Monte Cavallo**
di Pontebba (Udine, Friuli) .... (–677, +13) 690 m

Located in the karst area of Pontebba, on the Austrian border, on the high plateau of the Creta di Rio Secco, at an altitude of 2130 m, about 7 km to the NW of Pontebba. Also called abisso Klondike.

1/25,000 map Pontebba, 14 I SO. 0°47'00"; 46°33'11". Devonian limestone.

The entrance was spotted in April 1985 by the Gruppo Triestino Speleologico who frantically (as they wrote) descended to –677 m (*Speleologia*, 1986 (14) profile). The entrance may be in Austria!

21. **abisso Francesco Simi** (Lucca, Toscana) ... –684 m

Located at an altitude of 1194 m, near to the Colle di Ripanaia, in the Valle di Arnetola (Apuane Alps, commune of Vagli di Sotto), this cave has three entrances. In is formed in Lias marbles.

1/25,000 map Vagli di Sotto, 96 II NO. 2°12'06"; 44°05'26".

The G. S. CAI Lucchese discovered it in 1971 and got to –362 m the same year. In 1976 the Società Adriatica di Scienze de Trieste discovered the way on at –290 m, coming to a sump at –684 m in 1977 (*Speleologia*, 1979 (2) profile; M. Sivelli and M. Vianelli, *op. cit.*, 1982).

22. **abisso di Bifurto** (Cosenza, Calabria) .... –683 m

The abisso di Bifurto is located at an altitude of 920 m, at Fossa del Lupo, in the commune of Cerchiara di Calabria in the Monte del Pollino (limestones of the upper Cretaceous).

1/25,000 map Cerchiara di Calabria, 221 I SE. 3°55'44"; 39°51'44".

The G. S. CAI Piemontese discovered and explored this "inghiottitoio" to –435 m in 1961. The following year, with the G. S. Bolognese and the G. S. Alpi Maritime de Cuneo, they got to a muddy sump at –683 m. In 1978 the G. S. Alpi Marchigiano discovered new passages. When it rains, the cave has flowing water beginning at –343 m (*Grotte, Bull.* G.S.P., 1963 (19) profile: *Actes IVth Congr. Int. Spél.*, 1965, 3 profile; Di Maio in *Rass. Spel. Ital.*, 1966, 18 (1-2) and Courbon, *Atlas des Grands Gouffres du Monde*, 1979, profile).

23. **abisso Sandro Mandini** .... –678 m
(Lucca, Toscana)

This cave with two entrances (alt. 1160 m) is located in the Valle di Ametola, rich in large caves (see above, Oriano Coltelli and Francesco Simi), commune of Vagli di Sotto. Formed in the Lias marbles.

1/25,000 map Vagli di Sotto, 96 II NO. 2°12'27"; 44°05'27".

Discovered in 1978 by the G. S. CAI Lucchese who explored it with the G.S.A. Livornese to –590 m. In 1979 these two clubs pushed their explorations to –678 m. The cave is active beginning at –130 m (*Speleologia*, 1979 (2), profile; M. Sivelli and M. Vianelli, *op. cit.* 1982).

24. **buca di Monte Pelato** (Lucca, Toscana) ..... –676 m

The buca di Monte Pelato, or abisso Gerardo Bagnulo, is located at an altitude of 1260 m, at Pian della Fioba, on the south face of the Monte Pelato, municipality of Stazzema (Apuane Alps, limestones of the lower Lias).

1/25,000 map Monte Altissimo, 96 II SO. 2°13'39"; 44°03'30".

A dye trace on October 4, 1975 showed the resurgence to be the buca di Renara (alt. 310 m).

The cave was discovered in 1973 by the G.S.A. Versiliese, but explored by the G. S. CAI Bolognese who stopped at –188 m, after enlarging several passages. In 1974 the attempts of the G.S.B. did not succeed in getting them past the terminal constriction, which was conquered in January, 1975 by the G. S. Bagni di Lucca who reached –270 m. At –60 m a new descending branch was discovered. It became active beginning at –188 m. From May to September 1975, the explorers pushed down to –656 m. In 1984 the G. S. Piemontese added 20 m to the vertical extent by discovering an upper entrance.

This cave has three branches: – 317 m (constriction), –332 m and –656 m (sump). The pozzo del Centenario, at –457 m, is a 105 m deep pit. The river is reached at the –620 m level (Courbon, *op. cit.*, 1979, profile; M. Sivelli and M. Vianelli, *op. cit.*, 1982).

25. **abisso Cesare Prez** (Udine, Friuli) .... –654 m

At an altitude of 1796 m, between the Foran del Muss and the Colle delle Erbe, on Monte Canin, municipality of Chiusaforte (Giulie Alps, Dachstein limestones and dolomitic limestones of the upper Trias), this cave is hydrologically connected (dye trace) with the fontanon di Goriuda (alt. 868 m. See above, complesso del Col delle Erbe).

1/25,000 map Monte Canin. 14 II SE. 0°59'08"; 46°22'33".

This cave was discovered in August 1969 by the Gruppo Grotte XXX Ottobre de Trieste who, after opening a blowing constriction, reached –478 m, then, in 1970, a sump at –627 m. In 1971 the level of the sump was lower (–631 m) and it was dived, but a narrowing appeared at –654 m (*Boll. G.G. Ott. CAI*, 1970 (1): 20-25 profile; Courbon, *op. cit.*, 1979, profile).

26. **abisso Eunice** (Lucca, Toscana) .... –651 m

Another cave in the Valle di Arnotela, municipality of Vagli di Sotto (Apuane Alps). At an altitude of 1135 m, in the lower Lias marbles.

1/25,000 map Vagli di Soto, 96 II NO. 2°12'30"; 44°05'42".

The cave was discovered in 1981 by the G. S. Savonese who, the same year, with the G. S. Imperiese and the G. S. Bolognese, explored to a sand and clay sump at –651 m (*Speleologia*, 1982 (7) profile; M. Sivelli and M. Vianelli, *op. cit.*, 1982).

27. **abisso Attilo Guaglio** (Lucca, Toscana) ..... –648 m
Very close to the preceding cave, it is located in the Valle di Arnetola, commune of Vagli di Sotto, at an altitude of 1130 m.

1/25,000 map Vagli di Sotto, 96 II NO. 2°12'29"; 44°05'37".

This cave was discovered and explored in 1978 by the G. S. Savonese. Starting at –70 m, a small stream flows which is followed to the terminal sump at –648 m (*Stalattiti e Stalagmiti*, 1978 (16) profile; *Speleologia*, 1979 (7) profile; M. Sivelli and M Vianelli, *op. cit.*, 1982).

28. **abisso Eugenio Boegan** (Udine, Fiuli) .... –624 m
Between the Colle delle Erbe and the Sella Bila Pec, at an altitude of 1875 m, on Monte Canin, commune of Chiusaforte (Alpi Giulie, Dachstein limestones of the upper Trias), the abisso Eugenio Boegan, named after a great leader of Italian caving, who died in 1939, belongs to the same hydrological system as the complesso del Cel delle Erbe (see above): relationship established by fluorescein dye trace.

1/25,000 Monte Canin, 14 II SE. 0°59'30"; 46°22'26".

This is a cave discovered and explored in 1963 by the C. G. "Eugenio Boegan", Società Alpina delle Giulie, CAI, Trieste, to –190 m in 1963, –358 m in 1964, and –487 m in 1966. In 1967 the terminal muddy sump was reached at –624 m. The abisso has two large pits, 150 and 128 m deep (Stabile, in *Alpi Giulie*, 1968 (63) profile; Courbon, *op. cit.*, 1979, profile; *Progressione*, 1984 (12) : 22-23).

29. **abisso Straldi** (Cuneo, Piemonte) .... –614 m
Located close to the abisso Cappa, and part of the same hydrological system (see above), the abisso Straldi is found in the Conca delle Carsene, Marguareis range, municipality of Briga Alta, at an altitude of 2272 m.

1/25,000 map Vièvre 5-6. x 1025.176; y 222.382 (France) or 1/25,000 map Certosa di Pieso, 91 IV SE (Italy).

The Expéditions Spéléologiques Françaises discovered and explored the cave to –117 m in 1953. In 1973 the Centre Méditerranéen de Spéléologie, after a short climb, found the way on and with the G. S. Piemontese reached –545 m the same year (stopping for lack of rope). In 1975 the C.M.S. deepened the cave to –614 m, stopping at some constrictions (*Bull. Phénomènes karstiques*, 1976 (1) profile; *Spelunca*, 1976 (3) profile; Courbon, *op. cit.*, 1979, profile).

30. **abisso Marino Vianello** (Udine, Friuli) ...... –585 m
Also called abisso I a SW del M. Spric, with three entrances (L 18, T 8 and T 11), of which the highest is at an altitude of 1850 m. It is located near the col delle Erbe, on Monte Canin, commune of Chiusaforte (Alpi Giulie, Dachstein limestones, upper Trias) and probably belongs to the same hydrological system as the complesso del Cel delle Erbe (see above).

1/25,000 map Monte Canin, 14 II SE. 0°59'48"; 46°22'51".

Discovered by the Commissione Grotte E. Boegan, S.A.G. de Trieste in 1975. In 1978 the cavers explored to the terminal sump at –585 m, after a series of initial pits

followed by horizontal galleries and meanders, giving a length of about 4500 m (M. Serra, Abisso Marino Vianello, *Progressione*, 1979 (3); 15-18 profile).

31. **buso de la Genziana** (Treviso, Veneto) .... –582 m
Also called abisso Genzianella. The entrance (alt. 1018 m) was opened in 1966 when the "Crosetta-Spert" road was enlarged, and is located on the wooded plateau of Bosco del Cansiglio, municipality of Fregona (Prealpi Carniche, Cretaceous limestones).

1/25,000 map Bosco del Cansiglio, 23 II SE. 0°03'07"; 46°03'27" (geographically, it belongs to the Friuli-Venezia Giulia region).

The G. G. CAI Vittorio Veneto began exploration slowly: in 1967 they reached –92 m, then –190 m in 1968 and –240 m in 1970. In 1972 the G.G.V.V., the G. S. Monfalconese and the G.G. XXX Ottobre de Trieste banded together and successively reached –410 m, –513 m, and –540 m. In November they got to the terminal sump at –582 m (Courbon, *op. cit.*, 1979, profile).

32. **voragine di Monte Bül** .... –557 m
(Como, Lombardia)
The cave (alt. 1375 m) is located in the Palanzone range, on Monte Bül, not far from the grotta Guglielmo, in the municipality of Faggeto Lario. It is formed in the Moltrasio limestones (lower Lias). 1/25,000 map Moltrasio 32 IV SE.

It was discovered in 1981 by the S. C. CAI Erba and explored by this club with the S. C. I Protei de Milano. The same year, they reached –350 m and estimated they had explored to –500 m. In 1982 the same groups reached –520 m. In 1983 they reached a bottom at –557 m. It is a cave with large subhorizontal galleries, followed by deep, generally wet pits, the largest being 83 m deep (*Speleologia*, 1983 (9): 13-16 profile).

33. **buca grande di Monte Pelato** .... –540 m
(Massa, Toscana)
Also called abisso Bologna. This cave is at an altitude of 1270 m, on the west side of Monte Pelato, at the place called Pian della Fioba (Apuane Alps, municipality of Massa). 1/25,000 map Monte Altissimo, 96 II SO. 2°13'36"; 44°03'38".

Marbles of the lower Lias. The opening, discovered in September 1972 by the G. S. Bolognese, was enlarged permitting this group, along with the G. S. Bagni di Lucca, to descend to –527 m, to a constriction, after conquering two deep pits of 103 and 105 m. In June 1973 the constriction was forced by the G.S.B. who were stopped by a narrow fissure at –540 m. It is a vertical type cave that is 720 m long (Courbon, *op. cit.*, 1979, profile: M. Sivelli and M. Vianelli, *op. cit.*, 1982).

34. **abisso des Marrons Glacés** .... –540 m
(Como, Lombardia)
This is a high altitude cave, opening at the bottom of a snowy doline, at an altitude of 2100 m, near the Bogani hut, in the gruppo delle Grigne, municipality of Esino. Its potential depth is about 1800 m.

It was discovered in 1980 by the G. S. Milano and the G. S. CAI Verona. Four clubs explored it, the two discoverers plus the G. S. CAI Trevisiol and the G. S. CAI Malo. In succession the depths of approx –220 m (1980), –305 m

(1981; *Speleologia*, 1981 (6) profile), –420 m (1982) and –540 m (1984) were reached.

35. **voragine di Punta Straldi** .... –539 m
(Cuneo, Piemonte)

Also called abisso del "Perdus". This cave in the Conca delle Carsene, Maguareis range, municipality of Briga Alta, is near the abisso Cappa (see above) and is hydrologically connected according to a dye trace on August 16, 1981 by the Club Martel de Nice. The resurgence is the resurgence of 18 Pis del Pesio (alt. 1340 m). Located at an altitude of 2228 m.

1/25,000 map Viève 5-6. x 1024.515; y 222.395 (France) or 1/25,000 map Certosa di Pesio, 91 IV SE (Italy). Cretaceous then Jurassic limestones.

It was discovered in August, 1959 by the Club Martel CAF de Nice who descended to –230 m the same month. In 1960 the C.M. Nice were stopped at –285 m. The way on was found in 1973 by the Centre Méditerranéen de Spéléologie who stopped at –460 m. The exploration was resumed in 1974 by the G. S. Alpi Maritime de Cuneo who gave up at –539 m at an impenetrable fissure (*Bull. Phénomènes Karstiques*, 1976 (1) profile, *Spelunca*, 1976 (3) profile; Courbon, *op. cit.*, 1979 profile).

36. **abisso del Col Iopic** (Udine, Friuli) .... –525 m

Or abisso "Città di Udine". Located at an altitude of 1900 m, to the north of the Col Lopic, and to the NE of Monte Poviz, on Monte Canin, municipality of Tarvisio (Alpi Giulie, Dachstein limestones of the upper Trias).

1/25,000 map Sella Nevea, 14 III SO. 1°02'49.5"; 46°22'55".

Discovered in 1980, the Circolo Speleologico e Idrologico Friulano (Udine) explored it to –352 m the same year, then –525 m in 1982, stopping at an impenetrable constriction (S. Turco, L'abisso "Città di Udine", *Mondo Sotterraneo*, 1983 VII (1): 15-22, profile).

37. **abisso Gianni Ribaldone** .... –523 m
(Lucca, Toscana)

At an altitude of 1225 m, in a white marble quarry (upper Trias), near Fondone à Arni, on Monte Altissimo, commune of Stazzema (Alpi Apuane).

1/25,000 map Monte Altissimo, 96 II SO. 2°12'39"; 44°02'56".

Opened by a tunnel dug by the quarry workers, the G. S. CAI Lucchese began exploration in 1969, reaching –300 m. In 1970 they reached –434 m and in 1971, with the G. S. Palenologico "G. Chierici", they came to a too-tight constriction at –523 m (Courbon, *op. cit.*, 1979, profile; M. Sivelli and M. Vianelli, *op. cit.*, 1982).

38. **grotta di Chiocchio** (Perugia, Umbria) ...... –514 m

This cave, long known, opens at the side of a stream, the Fosso dell'Andreone, which it drains. Located near the village of Castagnacupa, municipality of Spoleto, at an altitude of 735 m. Neocomian limestones, lower Lias. 1/25,000 map Bilano di Spoleto, 131 III SE. The cave has four entrances close to each other, of which one is man-made.

Exploration began in 1961: the G. S. Spoletino, G. S. CAI Perugia and G. S. Pipistrelli de Terni reached –214 m, after digging open a constriction at –19 m. In 1962 the S.

C. Roma turned back at –250 m. In 1963 the G. S. Perugia and the G. S. Spoletino reached –380 m, then –514 m in 1964 (L. Passeri - La grotta di Chiocchio presso Spoleto, *Universo*, 1968, 68 (2): 1-12; Courbon, *op. cit.*, 1979 profile).

39. **abisso Giovan Battista De Gasperi** .... –512 m
(Udine, Friuli)

Also called abisso I sol Monte Robon, this cave, at an altitude of 1890 m, is located on Monte Robon, in the Massiccio del Canin Oriental, commune of Tarvisio (Dachstein limestones, upper Trias). 1/25,000 map Sella Nevea, 14 III SO.

It was discovered in September, 1975 by the Circolo Speleologico e Idrologico Friulano who descended to –276 m, then to –512 m in 1976 (F. Savoia, Abisso G.B. De Gasperi –512, *Mondo Sotteraneo*, 1977: 9-14, profile).

40. **abisso Eraldo Saracco** .... –510 m
(Cuneo, Piemonte)

This cave in the Marguareis range is situated between the grotta di Piaggia Bella (see above) and the col des Seigneurs, not far from the latter, at an altitude of 2140 m, in the municipality of Briga Alta. It is apparently hydrologically connected to the Piaggia Bella. Jurassic limestones. 1/25,000 map Tenda, 91 III NE.

It was discovered in 1964 by the G. S. Piemontese who reached –80 m. In 1966, with the help of the G. S. Faentino, they got to –457 m in one branch, then in 1967 to –478 m in another branch (ending in a sump) and finally to –507 m in 1968 in a third branch (also ending in a sump). In 1976 the G.S.P. discovered the F 33, abisso del Passi Perduti (alt. 2143 m) which they connected to the Eraldo Saracco on September 26, 1976, near the sump at –507 m. The cave contains a 155 m deep pit (*Atti XIth Congr. Naz. Spel.*, Genova 1972, II: 135-139; Courbon, *op. cit.*, 1979; *Grotte*, 1976, (61) profile).

41. **abisso di Malga Fossetta** .... –492 m
(Schio, Vicenza, Veneto)

Alt. 1777 m. Explored by the G.G. CAI Schio in 1973 (–125 m), 1976 (–399 m), 1977 (–445 m; *Stalattite*, 1977, XI, profile) and 1978 (–492 m).

42. **grava dei Gentili** .... –484 m
(S. Angelo a Fasanella, Salerno, Campania)

Alt. 850 m. By the C. S. Romano in 1963 (–100 m), the C. G. Boegan in 1963 (–124 m), the C.S.R. in 1969 (–130 m), 1971 (–270 m), 1972 (–348 m) and 1975 (–484 m; *Not. C.S.R.*, 1975 (1-2) profile).

43. **abisso dei Campelli** .... –483 m
(Moggio, Como, Lombardia)

Alt. 2020 m. By the G. S. Milano in 1965 (–135 m), C.G.M., G.S.P., G. S. Faentino in 1966 (–353 m), G. S. Lucchese, S. C. Orobico in 1977 (–461 m) and 1978 (–483 m; *Speleologia*, 1979 (1) profile).

44. **abisso di Monte Tre Crocette** .... –480 m
(Varese, Varese, Lombardia)

Alt. 1027 m. By the L.V. Bertarelli in 1909 (–145 m), G. S. CAI Varese in 1977 (–389 m), 1981 (–441 m; *Speleologia*, 1982 (7) profile) and 1983 (–480 m).

45. **buca del Pozzone** .... –475 m
(Vagli di Sotto, Lucca, Toscana)
Or abisso della Cava III. Alt. 1025 m. By the G. S. Lucchese and Bagni di Lucca in 1974 (–430 m) and the G.S.L. *et al.* in 1979 (–475 m, *Speleologia*, 1979 (2) profile).

46. **abisso Paolo Picciola** .... –460 m
(Chiusaforte, Udine, Friuli)
Alt. 1940 m. By the C. G. Boegan in 1965 (–208 m), 1968 (–290 m), 1979 (–384 m) and 1984 (–460 m).

47. **bucca della Mamma** .... –450 m
(Stazzema, Lucca, Toscana)
Or abisso Baader-Meinhof. Alt. 1535 m. By the G.S.P. in 1976 (–160 m) and the G.S.P. and C. G. Boegan in 1978 (–450 m, *Grotte*, 1978 (67) profile).

48. **abisso del Nido** (Asiago, Venetia) .... –446 m
Explored by Gruppo Grotte CAI Schio.

49. **complesso di Monte Vermicano** .... –439 m
(Guarcino, Frosinone, Lazio)
Alt. 1555 and 1609 m. By the S. C. Roma in 1972 (–300 m), 1973 (–385 m) and the C. S. Romano in 1979 (Connection with the upper entrance; *Not. C.S.R.*, 1980 (1-2) profile).

50. **grotta delle Tessare** .... (–418, +20) 438 m
(Piobbico, Pesaro, Marche)
Alt. 1050 and 1125 m. In 1951-1952 by the G. S. Ancona *et al.*, (387 m); 413 m in 1979 and 438 m in 1982.

51. **abisso I di Mogenza Piccola** .... –430 m
(Chiusaforte, Udine, Friuli)
Alt. 1775, 1765, and 1741 m. In 1979-1980 by the C. G. Boegan (*Progressione*, 1980 (5) profile).

52. **abisso Bacardi** .... –430 m
(Frabosa Sottana, Cuneo, Piemonte)
In 1982-1983 by the G. S. Alpi Maritime Cuneo (*Mondo Ipogeo*, 1983, partial profile).

53. **abisso Marcel** .... –430 m
(Briga Alta, Cuneo, Piemonte)
In 1984, explored by French and Italian cavers.

54. **buco del Castello** .... (–422, +7) 428 m
(Roncobello, Bergamo, Lombardia)
Or grotta di Roncobello, alt. 1300 m. By the G. G. Bergamo et al. in 1957-1962 (–385 m), G. G. XXX Ottobre Trieste in 1967-1969 (–422 m; *Rass. Spel. Ital. Mem.*, 1974, XI (1) profile).

55. **abisso delle Pozze** .... –427 m
(Chiusaforte, Udine, Friuli)
Alt. 2078 m. In 1982-1983 by the C. G. Boegan and the AKSIA de Katowice (*Progressione* 1984 (12) profile).

56. **inghiottitoio III dei Piani de S. Maria** ..... –422 m
(Corleto Monforte, Salerno, Campania)
Alt. 1071 m. By the C. G. Boegan in 1968 (–253 m), 1970 (–290 m), 1974 (–364 m), 1975 (–399 m) and 1977 (–422 m; *Atti e Memorie*, 1977, XVII, profile).

57. **"trou souffleur"** .... –420 m
(Briga Alta, Cuneo, Piemonte)
Alt. 2140 m. On the French-Italian border. By the C. M. CAF de Nice (1962-1975; *Bull. Phénomènes Karstiques*, 1976 (1) profile).

58. **abisso degli Increduli** .... (–399, +11) 410 m
(Udine, Udine, Friuli)
Alt. 2005 m. By the C. G. Boegan in 1982 (*Progressione*, 1982 (10) profile).

59. **buca della Pompa** .... –408 m
(Vagli di Sotto, Lucca, Toscana)
Or abisso Pina Boschi, alt. 895 m. By the G. S. Bolognese in 1961 (–50 m) and the G. S. Lucchese in 1972-1973 (*Speleologia*, 1979 (2) profile).

60. **abisso Filologa** .... –405 m
(Briga Alta, Cuneo, Piemonte)
Alt. 2030 m. By the G. S. Piemontese in 1983 (*Speleologia*, 1984 (10) profile).

61. **grotta della Mottera** (Piemonte) .... +403 m

62. **grava dei Gatti** .... –403 m
(Ottati, Salerno, Campania)
Alt. 945 m. By the C. G. Boegan in 1961 (–258 m) and 1963 (–403 m; *Atti e Memorie*, 1963, II, profile).

63. **fontanin del Fratte** .... +396 m
(Claut, Pordenone, Friuli)
Alt. 1000 m. In 1983 by the U. S. Pordenonese and G. S. Sacile.

_____

95. **abisso delle Frane** (Piemonte) .... –300 m

LONG CAVES:

1. **complesso di Corchia** .... 45,000 m
(Lucca, Toscana)
See above and also complesso Fighiera-Farolfi-Antro del Corchia.

2. **complesso di Piaggia Bella** .... approx. 30,000 m
(Cuneo, Piemonte)
See above

3. **grotta di Monte Cucco** .... 26,125 m
(Perugia, Umbria)
See above.

4. **buso de la Rana** .... 22,535 m
(Monte di Malo, Vicenza, Veneto)
This resurgence, at Case Maddalena (alt. 345 m) has been explored since 1887 (350 m by locals), in 1933 by the G. G. Arzignano (1365 m), from 1952 to 1956 by the G. S. "Massalongo" Verona, G. G. "Trevisiol" Vicenza, G. G. Trento and G. G. Schio (3700 m); then 13000 m in 1975, 15,120 m in 1979 by the G. S. CAI Malo, G. S. CAI Trevisiol and C.S. Proteo; in 1981 the length was 21,395 m (*Speleologia*, 1981 (8) schem. plan) and 22 535 m in 1984.

5. **complesso Fiume-Vento** .... 21,000 m
(Genga, Ancona, Marche)

System made up of the grotta del Fiume and the grotte grande del Vento (alt. 205 and 420 m), at Gola di Frasassi. Has two other entrances, of which one is artificial. Show cave explored in 1948 by the G. S. Marchigiano (1000 m), then in 1966 by the G. S. Asci Fabriano, in 1971 by the S. C. Città di Jesi and the G.S.M. In 1972 the connection of the two caves by the G.S.A. Fabriano made the length 12,000 m, which was increased to 17,500 m in 1981 and 21,000 m in 1983.

6. **grotta della Bigonda** ....14,570 m
(Grigno, Trento, Trentino)
Discovered in 1952, the G. G. CAI Trento explored it for 2880 m in 1952, then 3020 m in 1953. In 1979 the G. G. Selva di Grigno resumed exploration and reached 6700 m in 1979, 9020 m in 1980, 9200 m in 1981, 10,070 m in 1982, and 14,570 m in 1984 for a vertical extent of 268 m (−88, +180).

7. **Su Palu** (Urzulei, Nuoro, Sardegna) ....13,120 m
Located in the codula di Luna (alt. 200 m). Discovered in 1980 (S. C. Paris), explored in 1980 (S. C. Paris, G.R.E.S. Paris-VI), 1981 (S. C. Cagliari and G. S. Ragaïe, 1650 m), 1982 (S. C. and G. G. Cagliari, S. C. Oliena, 3595 m), 1983 (10,020 m; *Spelologia*, 1983 (9) plan), 1984 (11,820 m) and 1985 (13,120 m). Vertical extent 292 m (−156, +136).

8. **complesso del Col delle Erbe** ....13,000 m
(Udine, Friuli)
See above.

9. **Is Angurtidorgius** ....9558 m
(Villaputzu/ Ulassari, Cagliari/ Nuoro, Sardegna)
Explored by the S. C. Cagliari: 4950 m (1973), 8800 m (1974) and 9558 m (1984).

10. **grotta di San Giovanni Su Anzu-Ispinigoli** ....8435 m
(Dorgali, Nuoro, Sardegna)
Explored in 1954 (G.G. Nuorese), 1955 (connection by G. G. Boegan and G.G.N.), 1958 (G.G.N., 5200 m), 1963-1966 (G.S.P., 8435 m; *Rass. Spel. Ital.*, 1968 (2) plan).

11. **complesso Tacchi-Zelbio** ....8350 m
(Zelbio, Como, Lombardia)
In 1961 (G. S. Como, 1240 m), 1976-1984 (G. G. Milano, 8350 m).

12. **abisso Cappa** ....approx. 8000 m
(Briga Alta, Cuneo, Piemonte)
See above.

13. **sa grutta 'e Montes Logos** ....7150 m
(Dorgali, Nuoro, Sardegna)
Near to Su Palu, explored in 1980-1982 by the G. S. Milano and various clubs (*Spelologia*, 1981 (6) plan showing 6585 m). Also called Su Spiria.

14. **grotta della Mottera** ....approx. 7000 m
(Ormea, Cuneo, Piemonte)
Explored by the G.S.P. (1961) and the S. C. Tanaro (1982-1984; *Spelologia*, 1984 (11) parital plan).

15. **grotta del Bue Marino** ....approx. 7000 m
(Dorgali, Nuoro)
400 m in 1952-1954 (G. G. Nuorese), 5365 m in 1955-1967 by the G.S.P. *et al.* Extended 1 km beyond the terminal sump in 1982 by the G. S. Ragaïe.

16. **Omber en banda al bül des Zel** ....approx. 7000 m
(Serle, Brescia, Lombardia)
Explored in 1954, 1967, 1977 (3000 m), 1980 (4000 m) and 1984 by the G. G. Brescia.

17. **complesso Spipola-Acquafredda** ..approx. 7000 m
(S. Lazzaro di Savena, Bologna, Emilia-Romagna)
Explored in 1932-1934, 1956, 1958 (connection), 1983. Developed in gypsum.

18. **complesso Mainarda - La Val - Fossa del Noglar** ....6595 m
(Clauzetto, Udine, Friuli)
Explored in 1953, 1954, 1980 (connection), 1983 by the C. G. Boegan and the S.A.S. Trieste (*Atti XIVth Congr. Naz. Spel.*, Bologna, 1982, plan).

19. **grotta di Su Mannau** ....6350 m
(Fluminimaggiore, Cagliari, Sardegna)
Explored from 1954-1982 by the S. C. Cagliari, G. S. Pio XI *et al.* (*Spelologia*, 1982 (7), schem. plan).

20. **grotta nuovo di Villanova** ....6200 m
(Lusevera, Udine, Friuli)
Explored by the Circ. Spel. Idr. Friulano in 1925 (1700 m), 1953-1954, 1983-1985.

21. **complesso Su Bentu-Sa Oche** ....6110 m
(Oliena, Nuoro, Sardegna)
Explored from 1953 by the G. G. Nuorese, with the G. S. Pio XI, G.S.P. (*Spelologia*, 1982 (7) plan).

22. **complesso C1-Regioso** ....6020 m
(Ormea, Cuneo, Piemonte)
See above. Explored by the G. S. Imperiese (1970-1980; *Bol G.S.I.*, 1979 (13) plan showing 5436 m).

23. **buco Cattivo** ....approx. 6000 m
(Genga, Ancona, Marche)
In 1982, by the G. S. Marchigiano and the G. S. Jesino.

24. **grotta di Castel Sotterra** ....approx. 6000 m
(Volpago del Montello, Treviso, Veneto)
1978-1982, G. G. Treviso (*Spelologia*, 1982 (8) plan). Conglomerate.

25. **risorgiva di Eolo** ....5300 m
(Trasaghis, Udine, Friuli)
Or fontanon di Avasinis. In 1980-1984 by the C. G. Boegan.

26. **pozzo della Neve** ....approx. 5000 m
(Campochiaro, Campobasso, Molise)
See above.

27. **grotta del Calgeron** ....4885 m
(Grigno, Trento, Trentino)

28. **grotta di Castelcivita** ....4800 m
(Castelcivita, Salerno, Campania)

29. **grotta di San Giovanni d'Antro** ....4500 m
(Pulfero, Udine, Friuli)
1974-1981, C.S.I.F. (*Speleologia*, 1981 (5) plan).

30. **grotta di Slogli Neri** ....4000 m
(Giustenice, Savona, Liguria)

31. **complesso Mezzogiorno-Frasassi** ....4000 m
(Genga, Ancona, Marche)

32. **grotta Edigio Feruglio** ....4000 m
(Lusevera, Udine, Friuli)

_____

44. **grotta delle Arenarie** ....approx. 3000 m
(Borgosesia, Vercelli, Piemonte)

LUXEMBOURG

Luxembourg has very little limestone. Among the caves that have been explored, the most notable is the **grotte de Moestroff** (Moestroff) which is a maze formed in a grid of fissures. Its length is over 4000 m (*La Vie Souterraine*, 1966 (6) plan).

Elsewhere, sandstone terrains are the most common and it is in these that are found the deepest cave in Luxembourg, the **grotte de Sainte-Barbe** (Schnellert) with a vertical extent of 40 m.

MALTA

A group of independent islands located to the south of Sicily, Malta would have passed unnoticed if not for Leander Tell, who twenty years ago noted the small **ghar Dalam** (Valletta), which is 200 m long.

NORWAY
NORGE

The majority of the limestone caves are located in the county (**fylke**) of Nordland, between latitudes of 65° and 68° North. They are formed in Precambrian marbles.

Their entrances have in general been known to Norwegians and to travellers (geologists, hunters, and fishermen) since the 19th century. There were three major periods of exploration: from about 1870 to 1940 (Norwegian geologists and naturalists principally), 1951-1965 (expeditions by British caving clubs and university groups), and from 1965 to the present when foreign cavers (mostly British, but also Sweedish and French) as well as a growing number of Norwegian cavers have pursued explorations.

The Rana Grotteklubb was created in 1966 and the Mo Speleologisk Selskap in 1968. The first issue of its national review, the Norsk Grotteblad, was published in 1977 and the national society, Norsk Grotteforbund, was founded in 1981. It is in the clubs such as Rana Turistforening Fjellsportgruppe, Båsmo Grotteklubb, and Bodø og Omegn Bre-, Tinde- og Grottegruppe that one finds the most active cavers. Most of the British activity was done by members of the Gritstone, Kendal, SWETC, Westminster, and Wessex caving clubs, in collaboration with their Norwegian colleagues.

The classic reference for Norwegian speleology is the work of Gunar Horn, *Karsthuler i Nordland* (1947). Other useful information can be found in *Norsk Geologisk Tidsskrift, Naturen*, in the thesis of Jean Corbel, *Les Karsts du Nord-Ouest de l'Europe* (1957), in the publications of Shirley and D. St.-Pierre (*Studies in Speleology*, 1965, 1 (5); *Cave Res. Gr. Transactions*, 1966, 8 (1); 1969 (1); *Norsk Grotteblad*, 1977, 1 (3/4); 1982, 3 (9); 1984 (14); *Grottes et Gouffres*, 1984 (93), of Stein-Erik Lauritzen (*Norsk Geografisk Tidsskrift*, 1984, 38 (3/4) etc, or of Ulv Holbye (*Om vern av kalksteingrotter og grotteområder i Norge*, 1974).

David St Pierre.

DEEP CAVES:

1. *Råggejavre-Raige* .... **–620 m**
(Hellemojfjord, Tysfjord, Nordland)
This cave is located at the top of the marble slopes overlooking the deep fjord de Hellemofjord. The upper entrance is at an altitude of about 620 m and only 900 m from the sea. It also has an intermediate entrance, Mistral, at an altitude of 114 m, and a lower entrance located at an altitude of 3 m, allowing a traverse of 617 m vertically.
Coordinates: 16°32'30"; 67°52'55".
Noted by Foslie in 1942, the cave was explored by the Kendal C.C.: in 1968 they stopped at –180 m in the middle of a 138 m deep pit and in 1969 they reached –564 m. In 1979 the connection was made with the lower entrance.
Map: from the drawings of the explorers, sent by John Yeadon (degree 4) and documents sent by D. St. Pierre.
Bibliography: *Kendal C. C. Journal*, 1969 (4) profile to –575 m; *Descent*, 1980 (47) profile to –581 m.

2. **Larshølet** .... **–326 m**
(Reingårdslivatn, Rana, Nordland)
Alt. 394. Discovered around 1870 (Lars Bjørnnes), descended in 1914 (John Oxaal, –82 m), 1934 (Horn, –284 m) and 1951 (Corbel and Railton).

3. **Greftkjelen** .... **–315 m**
(Greftvatn, Gildeskål, Nordland)
Alt. 350 m. Known in 1907 (Helland) and explored from 1971-1972 by William Hulme's Grammer School of Manchester (*Norsk Grotteblad*, 1983 (11); 1984 (14).

4. **Okshola-Kristihola** ....(+161, –139) 300 m
(Vatnam, Fauske, Nordland)

5. **Greftsprekka** .... **–250 m**
(Greftvatn, Gildeskål, Nordland)
Alt. 339 m. Explored by Holbye, Lauritzen, Grønlie in 1976 (–132 m) and 1977.

6. **Lauknesfjellgrotta** .... **–213 m**

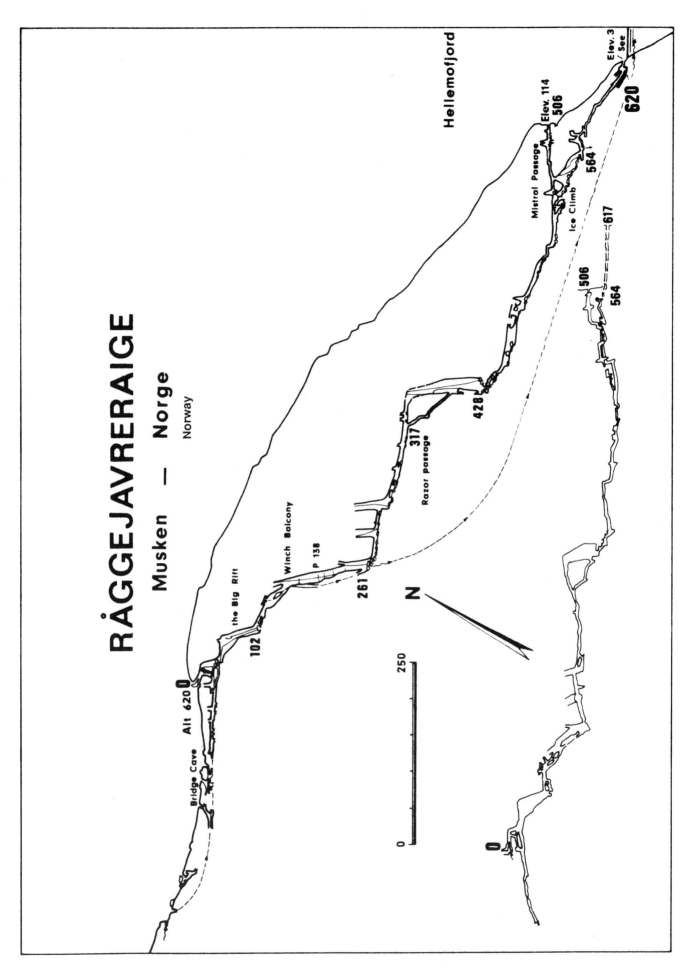

RÅGGEJAVRERAIGE

Musken — Norge

Norway

Hellemofjord

Bridge Cave

Alt 620

the Big Rift

102

Winch Balcony

P 138

261

317

Razor passage

428

506

564

617

506

564

620

Ice Climb

Mistral Passage

Elev. 114

Elev. 3 / See

N

250

0

0

(Hellemofjord, Tysfjord, Nordland)
Alt. 630 m. In 1971 by the Craven Pothole Club (C.P.C.Jl., 1971, 4 (5)).

7. **Østhølet** .... –210 m
(Hellemofjord, Tysfjord, Nordland)
Alt. 650 m. In 1971 by the Craven P. C.

8. **Salthulene** .... –195 m
Sørfjord, Tysfjord, Nordland)
Alt. 620 m. In 1985 by Lauritzen *et al.*

9. **Stordalsgrotta** (Stordal, Bardu, Troms) ..... –184 m
Alt. 960 m. In 1984 by Sveriges Speleolog Förbund *et al.* Depth estimated at –260 m.

10. **Ytterlihølet** .... –180 m
(Brygfjelldal, Hemnes, Nordland)
Alt. about 840 m. In 1975 by the W.H.G.S. of Manchester.

11. **Svarthamarhola** ....(–80, +77) 157 m
(Mefjell, Fauske, Nordland)
Alt. 250 m. In 1970 by Holbye and W.H.G.S.

12. **Nesmølnelvgrotta** .... –133 m
(Saltdalen, Saltdal, Nordland)
Alt. 220 m. In 1971 by the W.H.G.S.

13. **Dunderhølet** .... –125 m
(Dunderlandsdal, Rana, Nordland)
Alt. 600 m. In 1968 by Northern Explor. Gr.

14. **Krystallgrotta** .... –115 m
(Plurdal, Rana, Nordland)
Alt. 460 m. In 1965 by Northern S. C.

15. **Rønåliholet** .... –110 m
(Gråtådal, Beiarn, Nordland)
Alt. about 320 m. Two entrances (connection 1972, SWETC).

---

23. **Jordbekkgrotta** (Rana, Nordland) .... –100 m

LONG CAVES:

1. **Okshola-Kristihola** ....9500 m
(Vatnam, Fauske, Nordland)
Alt. 170 m. Maze system, explored beginning in 1968 by the Kendal C.C., William Hulme's G. S. and U. Holbye. Connection in 1968. Extentions in 1972 and 1974 (*K.C.C. Jl.*, 1969 (4)).

2. **Greftkjelen** ....3725 m
(Greftvatn, Gildeskål, Nordland)
Length estimated at 4890 m. Possible connection with Greftsprekka.

3. **Jordbrugrotta** (Plurdal, Rana, Nordland) ...3000 m
Explored in 1967 by Newill and St Pierre.

4. **Larshølet** ....2900 m

(Reingårdslivatn, Rana, Nordland)
Extended in 1934 by Horn (2300 m) and the Cambridge U.C.C. in 1957 (C.R.G. Trans., 1957, 5 (1), plan).

5. **Greftsprekka** ....2600 m
(Greftvatn, Gildeskål, Nordland)
Mapped by Swedish cavers in December, 1985.

6. **Setergrotta** ....2400 m
(Røvassdal, Rana, Nordland)
In 1939 (Horn, 1500 m) and 1965 (SWETC, *Speleo*, 4 (1) plan). Eight entrances.

7. **Hamarnesgrotta** ....2200 m
(Langvatn, Rana, Nordland)
Seven entrances. In 1874 (Corneliussen, 300 m), around 1914 (Oxaal, 445 m) and 1923 (Natvig, 975 m), then 1934-1939 (Horn, *N.G.U.*, 1947, (165) plan).

8. **Salthulene** (Sørfjord, Tysfjord, Nordland) ...2056 m

9. **Grønligrotta** ....2000 m
(Røvassdal, Rana, Nordland)
Show cave. In 1913 (Oxaal, 1210 m), then Horn (1500 m) and 1969-1970 (Grønlie, E. Haugane, *Norsk Grotteblad*, 1979, 2 (5) plan). Same hydrological system as Setergrotta.

10. **Pikhauggrotta** ....2000 m
(Glomdal, Rana, Nordland)
In 1937 (Horn, 110 m) and 1957-1958 (Cambridge U.C.C., *Cave Science*, 1959, 4 (29) plan).

11. **Store Grublandsgrotta** ....1900 m
(Ivarrud, Hatfjelldal, Nordland)
(*Report of the Ermysted's G. S. Spel. Exped. to Northern Norway, 1967, 1967-1968*, plan). Recent fills (1985) have shortened the cave.

12. **Sverrehola** (Sørfola) ....1900 m
Explored in 1986 by the P. C. Hobye and the Kiieby-Gainstad.

13. **Svarthamarhola** ....1814 m
(Mefjell, Fauske, Nordland)
Ice cave. In 1970 and 1975 (*Grottan*, 1975, 10 (4) plan).

14. **Råggejavre-Raige** ....1810 m

15. **Fiskegrotta** (Plurdal, Rana, Nordland) ....1650 m
In 1968 (*Eldon P. C. Exped to Northern Norway* 1968, 1969, plan).

---

37. **Kvandalhola** (Skjerstad) ....1000 m

POLAND

Polska

The karst terrains in Poland only occupy 2.5% of the surface area of the country. They are localized in the south, partly in the Carpathians (Tatra mountains) where, of the 500 catalogued caves, are found the most significant in Poland, and partly in the Sudety, where the density of caves is less and the ones found are generally smaller. The most extensive region is that of Kraków-Wieluń, but glaciation has inhibited cave development in this area.

It was with archaeology and geology that scientific study of caves began in Poland in the 19th and beginning of the 20th centuries, the oldest dating to 1815 (S. Staszic). The first speleological activities were by the Kraków Anthropological Committee beginning in 1870. J. G. Pawlikowski wrote a cave guide from 1880 to 1890. The first caving club was formed in 1923 in Zakopane. It was after the second world war, under the guidance of S. Zwoliński (Zakopane) and K. Kowalski (Kraków), that Polish caving really developed. In 1950 a club was formed in Kraków, with clubs soon following their example in other cities. Today cavers are united under the Polish Alpine Association (P.Z.A.) and their foreign explorations (notably in Austria) have been numerous in recent years.

DEEP CAVES:

1. **system Jaskini Śnieżnej** .... **−775 m**
(Western Tatra, Nowy Sącz, Zakopane)
The deepest cave in Poland has three entrances (jaskinia Nad Kotlinami, alt. 1876 m; Jasny Awen, 1851 m and jaskinia Śnieżnej, alt. 1703 m) and is located on the NW slopes of Mount Małołączniak (2096 m), in the Western Tatra, not far from the Czeck border. Access is gained by climbing, on the SW side of Zakopane, the Małej Łąka valley which leads to the glacier cirque Niżnia Świstówka. The entrance to the cave is on the tourist map "Tatrański Park Narodowy".

The cave is formed in Triassic limestones and dolomites. The resurgence is 4500 m to the west at Lodowe Źródło (alt. 971 m).

The lower entrance, jaskinia Śnieżnej, was discovered by cavers from Zakopane in May 1959 and explored the same year to about −350 m, then −565 m in 1960. The bottom, at −568 m, was reached on August 31, 1961.

Located 173 m higher, the jaskinia Nad Kotlinami was discovered on July 2, 1966 and connected to the other cave on May 19, 1968, giving a system 741 m deep and 4700 m long.

In 1972, divers from the SW PTTK Warszawa passed the terminal sump; in 1974 they discovered a descending gallery which led to a new sump at −753 m. A third sump was reached in 1981 at −769 m. They dived it in 1983 (−771 m) and in 1985 (−775 m). Explorations were made by the SW PTTK Warszawa, STJ KW Kraków, SG Wrocław and the cavers of Zakopane and of Gliwice.

Map: compilation of available maps furnished by R.M. Kardás, OW PTPNOZ Warszawa.

Bibliography: Koisar (B.) in *Speleologia*, 1969 IV (1): 33-37;
Speleologia, 1967, III (1): 9-49.

Grodzicki (J.) in *Taternik*, 1974 (6) and *Speleologia*, 1975, VIII (2).
Kardás, (R. M.) - Greatest caves of Poland, *Kras i Speleologia*, 1982, 4 (XIII): 103-104.

2. **jaskinia Bańdzioch Kominiarski** ....**(−546, +16) 562 m**
(Western Tatra, Nowy Sącz, Zakopane)
The two entrances to this complicated cave are at altitudes of 1683 and 1456 m, on the NE side of the Kominiarski Wierch (1829 m), 9 km to the SW of Zakopane, in the Western Tatra. Access is made via Kiry then climbing the Kościeliska valley as far as Lodowe Źródło, then going to the SW on the sides of Mount Stoły.

The lower entrance was discovered on July 11, 1968 by cavers of Poznan. In 1968-1969 a vertical extent of 300 m (−250, +50) was explored. In 1973 the explorations were pursued by cavers from Kraków: the vertical extent passed to 350 m (−250, +100), then to 410 m in 1974 (−280, +130). In 1976 the upper entrance, 226 m higher, was discovered, giving a depth of 506 m. This was finally increased to 562 m (−546, +16) in 1977.

Map: from the maps of the teams of OK PTTNOZ Kraków, KKTJ Kraków, AKSIA Katowice. Compilation of St. Kotarba, sent by Jerzy Mikuszewski.

Bibliography: Wiśniewski in *Taternik*, 1977 (2, 3); 1978 (3).
Kardás (R. M.) - *loc. cit.*, p. 104.

3. **system Wysoka-Za Siedmioma Progami** ....**(−251, +209) 460 m**
(Western Tatra, Nowy Sącz, Zakopane)
The altitudes of the seven entrances are 1514, 1503, 1500, 1499, 1492, 1478, and 1465 m. Explored in 1956 (199 m), 1978 (218 m), 1983 (427 m) and 1984. The connection of Wysoka to Za Siedmioma Progami was made in 1981.

4. **jaskinia Kosia** .... **−375 m**
(Western Tatra, Nowy Sącz, Zakopane)
Alt. about 1850 m. Discovered in 1983 and explored in 1985 by the KKTJ Krakow.

5. **jaskinia Wielka Litworowa** ....**(−354, +7) 361 m**
(Western Tatra, Nowy Sącz, Zakopane)
Alt. 1906 m. Explored beginning in 1962: 236 m in 1974, 347 m in 1981 (*Speleofórum* 1985, profile), then 361 m.

6. **jaskinia Ptasia Studnia** .... **−351 m**
(Western Tatra, Nowy Sącz, Zakopane)
Alt. 1627 and 1576 m (Lodowa Litworowa). Explored in 1960-1962. (−295 m, profile in *Stalactite*, 1977, 27 (1)), then in 1981 (-351 m).

7. **jaskinia Czarna**......................**(−158, +141) 299 m**
(Western Tatra, Nowy Sącz, Zakopane)
Alt. 1408, 1326 and 1300 m. Explored in 1961-1964 by A.K.T. Wrocław (223 m), then 1978 (299 m).

8. **jaskinia Miętusia** ....**(−241, +7) 248 m**
(Western Tatra, Nowy Sącz, Zakopane)
Alt. 1273 m. Discovered in 1936, explored in 1937 (−128 m), 1952 (−213 m), 1972 (dive, −241 m) and 1986 (+7 m).

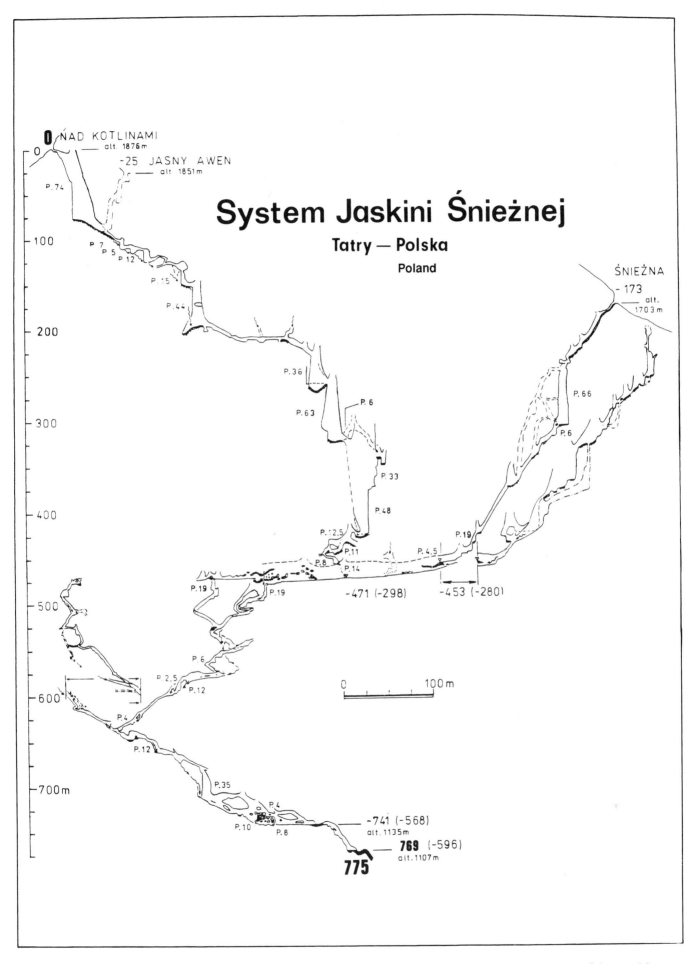

System Jaskini Śnieżnej

Tatry — Polska

Poland

NAD KOTLINAMI
— alt. 1876 m

-25 JASNY AWEN
— alt. 1851 m

ŚNIEŻNA
- 173
— alt. 1703 m

P.74

P.7
P.5 P.12

P.15

P.44

P.36

P.6

P.66

P.63

P.6

P.33

P.48

P.12.5

P.19

P.11

P.4.5

P.8

P.14

P.19

-471 (-298)

-453 (-280)

P.19

P.19

P.6

P.2.5

P.12

0   100 m

P.4

P.12

P.35

P.4

-741 (-568)
alt. 1135 m

P.10   P.8

769 (-596)
alt. 1107 m

775

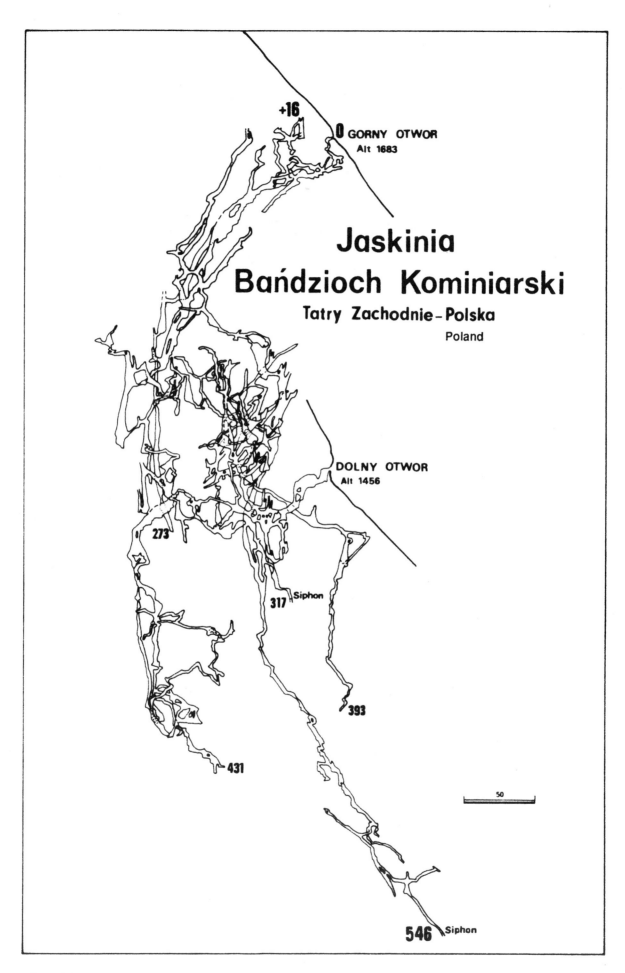

+16

0 GORNY OTWOR
Alt 1683

Jaskinia
Bańdzioch Kominiarski
Tatry Zachodnie - Polska
Poland

DOLNY OTWOR
Alt 1456

273

317 Siphon

393

431

50

546 Siphon

9. **jaskinia Śnieżna Studnia** .... (–155, +25) 180 m
(Western Tatra, Nowy Sącz, Zakopane)
Alt. 1748 m. Discovered in 1969, bottom reached in 1984.

10. **jaskinia Małołącka** .... –166 m
(Western Tatra, Nowy Sącz, Zakopane)
Alt. 1873 m. In 1960 by the STJ KW Krakow (–164 m) and 1982 (–166 m).

11. **jaskinia Miętusia Wyżnia** .... (–105, +50) 155 m
(Western Tatra, Nowy Sącz, Zakopane)
Alt. 1393 m. In 1950 and 1963-1970.

12. **jaskinia Pod Wantą** .... –151 m
(Western Tatra, Nowy Sącz, Zakopane)
Alt. 1793 m. In 1961, 1962 (–144 m, SW PTTK Warszawa) and 1977 (–151 m, ST PTTK, Zakopane).

13. **jaskinia Zimna** .... approx. (–17, +130) 150 m
(Western Tatra, Nowy Sącz, Zakopane)
Alt. 1120 m. In 1957.

14. **jaskinia Naciekowa** .... (–63, +40) 103 m
(Western Tatra, Nowy Sącz, Zakopane)
Alt. 1197, 1186 and 1170 m. In 1959 (–25 m) and 1978-1979.

15. **jaskinia Marmurowa** .... –102 m
(Western Tatra, Nowy Sącz, Zakopane)
Alt. 1770 m. In 1960 by the STJ KW Kraków.

LONG CAVES:

1. **jaskinia Miętusia** .... 9293 m
(Nowy Sącz, Zakopane)
In 1952: 950 m. See above.

2. **jaskinia Bańdzioch Kominiarski** .... 9200 m
(Nowy Sącz, Zakopane)
The length reached about 5 km in 1976, then 8,700 m in 1979, then 9,200 m. See above.

3. **system Wysoka-Za Siedmioma Progami** .. 7074 m
(Nowy Sącz, Zakopane)
From 970 m in 1956, the length of Wysoka reached 2,017 m in 1980, 5,256 m in 1983, 6,200 m in 1984, and 6,702 m in 1985. See above.

4. **jaskinia Wielka Śnieżna** .... 6290 m
(Nowy Sącz, Zakopane)

5. **jaskinia Czarna** .... approx. 6000 m
(Nowy Sącz, Zakopane)
In 1964 (5450 m) and in 1978. See above.

6. **jaskinia Wielka Litworowa** .... 4340 m
(Nowy Sącz, Zakopane)
Explored in 1974 (1150 m), 1979 (2100 m), 1981 (2900 m), 1982 (4210 m). See above.

7. **jaskinia Zimna** (Nowy Sącz, Zakopane) ...... 3450 m
Its exploration began in 1885, then 1913, 1936, 1955, 1967 (2975 m) and 1980-1985. See above.

8. **jaskinia Ptasia Studnia** .... 2990 m
(Nowy Sącz, Zakopane)
Connection with Lodowa Litworowa in 1963. 1540 m in 1979.

9. **jaskinia Kasprowa Niżnia** .... 2350 m
(Nowy Sącz, Zakopane)
Explored in the 1950's (*Speleologia*, 1971, VI (1/2) plan of 2015 m). 2320 m in 1978 and 2350 m in 1981.

10. **jaskinia Szczelina Chochołowska** .... 2320 m
(Nowy Sącz, Zakopane)
1,650 m in 1952, 2,300 m in 1966 and 2320 m in 1980.

11. **jaskinia Naciekowa** .... 1500 m
(Nowy Sącz, Zakopane)

12. **jaskinia Mylna** (Nowy Sącz, Zakopane) ...... 1300 m

13. **jaskinia Bystrej** (Nowy Sącz, Zakopane) ..... 1200 m

14. **jaskinia Magurska** .... 1200 m
(Nowy Sącz, Zakopane)

PORTUGAL

There are four major karst areas in Portugal: 1) the limestone ranges of Estremadura which are the most extensive in the country, being in the form of an arc SW-NE that is 150 km long and 35 km wide when north of Lisbon (culminating at 678 m while the resurgences are at an altitude of 50 m). 2) the Cambrian outcrops of the high Alentejo which are of small dimensions. 3) Algarve, in the south of the country, with modest demensions. 4) the pseudo karsts of Madeira and the Azores which contain lava tubes.

The Swiss E. Fleury, who published *Portugal Subterraneo* (1925), A. and B. de Barros Machado, who published *Inventario des cavernas calcarias de Portugal* (1941) listed over 343 caves, and the biospeleologists J. M. Braga (1892-1972) and K. Lindberg are among the first explorers of Portugese caves. In 1948 the enthusiasm of the Brun de Silveira, Camarate, Franca and Francisco de Abreu permitted the founding of the Sociedade Portuguese de Espeleologia, which was succeeded by the Federação Portuguese de Espeleologia in 1985.

After Ch. Thomas.

DEEP CAVES:

1. **gruta dos Moinhos Velhos** .... –183 m
(sierra de Aire, Leiria, Estremadura)
Discovered in 1948 by F. de Abreu.

2. **algar da Lomba** .... –148 m
(Covao do Coelho, Ribatejo)
Pit of 93 m. Bottom reached in 1977 by S. C. Rodez, Capdenac and Figeac (Soc. Languedocienne Spél., *Portugal 82*, profile).

3. **algar do Ladoeiro** .... –115 m
(Santo Antonio, Estremadura)

4. **algar do Aviaŏ** .... –114 m
(Santarem, Alcanade, Estremadura)
(S.L.S., *Portual 82*, profile).

5. **algar da Mardadinha** .... –110 m
(Santo Antonio, Estremadura)

6. **algar da Manga Larga** (Estremadura) .... –106 m

7. **algar do Maroiço** .... –105 m
(Porto de Mos, Menoiga, Estremadura)

8. **algar das Gralhas VII** .... –100 m
(Santarem, Alacanade, Estremadura)

9. **algar da Agua** .... –100 m

LONG CAVES:

1. **gruta da Almonda** ....5500 m
(Zibreira, Estremadura)

2. **gruta dos Moinhos Velhos**.....................3000 m
(Leiria, Estremadura)

3. **gruta dos Balcoes** (Terceira, Azores) ....2650 m
(*Cascade Caver*, 1980, 19 (11-12) plan).

4. **algar da Lomba** ....2200 m
(Covao do Coelho, Ribatejo)

5. **gruta da Pena da Falsa**...........................2000 m
(Alvados, Estremadura)

6. **Cova da Velha** (Alvados, Estremadura) ......1350 m

7. **gruta del Rei I** (Portunho, Estremadura)....1300 m

8. **gruta del Fabrica del
Tabacco Micaelense** ....approx. 1200 m

9. **algar de Arroteia**.................................1000 m
(Chao das Pias, Estremadura)

ROMANIA

The first known exploration of a Romanian cave was in 1703 when Barbenius, vicar of Risnov, near Brașov, visited peștera Cerna Deal. The most famous Romanian speleologist is Emil Racoviță (1868-1947), who dominated the history of underground research. He studied biology in Paris and met René Jeannel, with whom he founded the review Biospeleologica, in which they published their famous "Enumeration of visited caves". After visiting the caves of Europe, Racoviță returned to Romania to found, in Cluj in 1921, the first biological institute in the world, giving a solid footing to Romanian speleology. Of interest are his *Travaux de l'Institut de Spéologie "E. Racoviță"*,

written originally in French. It was only in the 1960's that speleological exploration began in earnest. Thanks to a system of points that rewards the various clubs, the number of known caves increased from 1600 in 1975 to 4600 in 1979 and 6800 in 1981 and 10,800 in 1986! (C. Goran, *Catalogul sistematic al pesterilor din România*, 1981)

The major karst areas in Romania are located in the southern Carpathians: Banat (Aninei mountains), Olténie (mount Mehedinți), in the Apuseni mountains (mounts Bihor, Pădurea, Craiului), in the eastern Carpathians (mounts Rodnei, Bucegi) and finally to the east, in the Dobrogea.

Romania has a famous ice cave 105 m deep, the **ghețarul de la Scărișoara**, in the Bihor mountains. The deepest shaft is the **Clocoticiul din cîrca Părețeilor** (mount Vîlcan), 121 m deep. The **peștera 6S de la Mînzălești** is a 1257 m long cave formed in salt (Subcarpatii Vrancei). Finally, there is a 102 m long cave formed in loess: **peștera din valea lui Moș Stoian** (southern Dobrogea).

DEEP CAVES:

1. **peștera Tăușoare** ....(–348, +35) 383 m
(m. Rodnei, Văsăud)
Alt. 965 m. Discovered in 1955 (L. Birte) and explored until 1971 by the Focul Viu București "E. Racoviță" București, Cluj-Napoca, Exploratorii Reșița, Liliacul Arad, Avenul Brașov, and Politehnica Cluj-Napoca clubs.

2. **peștera Șura Mare** (m. Sebeș) ....+380 m
Alt. 460 m. In 1984 by Focul Viu București, Vertikum Budapesta and Cristal Oradea.

3. **avenul din Stanul Foncii** .... –339 m
(m. Pădurea Craiului)
Alt. 650 m. In 1981 and 1983 by C.S.E.R. Cluj (*Peștera*, 1984 (1) profile).

4. **ponorul Sîncuta** (m. Pădurea Craiului) ..... –295 m
By C.S.R. Cluj-Napoca.

5. **avenul din Hoanca Urzicarului**
(m. Bihor) ....(–286, +2) 288 m
Alt. 1170 m. By Speodava Dr. Petru Groza, Z. Oradea, C.E.R. București (*Bul. CCSS*, 1982 (6) profile).

6. **avenul Dosul Lăcșorului** (m. Sebeș) .... –268 m
By C.E.R. București (*Trav. Inst. Spéol. "Racovița"*, 1982, XXI, profile).

7. **peștera de la Jghiabul lui Zalion**
(m. Rodnei) ....(–237.5, +4.5) 242 m
Alt. 830 m. In 1980 (*Bul. C.E.R. București*, 1982 (7) profile).

8. **avenul din Poiana Gropii** .... –236 m
(m. Domanului, Văliug, Banat)
Alt. 750 m. In 1970 (*Trav. Inst. Spéol. "Racovița"*, 1971, X, profile).

9. **avenul din dealul Secătura** (m. Bihor) ...... –230 m

10. **peştera Ciur-Ponor-Topliţa de Roşia**
(m. Pădurea Craiului) .... (–203, +25) 228 m
Stream system with eight sumps.

11. **avenul din Şesuri** (m. Bihor) .... –220 m
Alt. 1120 m. In 1978 by Z. Oradea (*Aragonit*, 1978 (1) profile).

12. **peştera Ponoraş** (m. Pădurea Craiului) ...... –212 m

13. **Cetăţile Ponorului** (m. Bihor, Petroasa) ..... –201 m
Alt. 925 m. A great karst site (*Trav. Inst. "Racoviţa"*, 1980, XIX, historical, partial section).

14. **avenul Sohodol II** (m. Bihor) .... –193 m
(*Bul. CCSS*, 1982 (6) profile).

15. **avenul din Cuciulata** (m. Bihor) .... –186 m
Alt 1400 m. (*Nymphaea*, 1978 (6) profile).

16. **avenul Pobraz** (m. Pădurea Craiului) .... –185 m

17. **peştera din pîrîul Hodobanei** ... (–121, +60) 181 m
(m. Bihor)

18. **peştera din valea Stînii** (m. Sebeş) .... –180 m

19. **peştera de la Zăpodie-peştera-Neagră**
(m. Bihor) .... (–162, +16) 178 m
(*Bul. C.E.R. Bucureşti*, 1977-1978 (5) profile).

20. **peştera Ponorici-Cioclovina cu Apă** .... –174 m
(m. Sebeş)
(*Bul. C.E.R. Bucureşti*, 1976 (4) profile).

LONG CAVES:

1. *peştera Vîntului* (m. Pădurea Craiului) ... **32,277 m**
This cave is found in the Bihor district, at an elevation of 325 m. It is near to Şuncuiuş, in the cliff on the left bank of the Crişul Repede, at the place called Poiana Frînturii, above a spring.

The cave was discovered in April, 1957 by Bela Bagameri. About 8900 m were explored between 1957 and 1966. In 1961, to facilitate access, an artificial entrance was made near the natural one. The Cluj-Napoca club was founded in 1966 and devoted itself to exploring the cave. About 15,500 m had been explored by 1969, when mapping of the cave began. The length reached 20,785 m in 1976 and 25,716 m in 1979. Then in 1982 the length reached 30,750 m, followed by 32,277 m in 1984.

The geological setting of the region is crystaline formation on top of conglomerates, then sedimentary Trias formations (massive limestones and white dolomites of the Ladinien-Norien) and finally the Lias. The cave is developed in a perimeter affected by two SW-NE oriented faults. It has an active level and three fossil levels. The cross sections are of the "canyon" type, tubular, or oval. Speleothems are rare but of a great variety (aragonite and gypsum).

The cave is developed linearly, with a range of 3000 m.

Map: from the drawings of the Institut de Spéologie and the C. S. de Cluj-Napoca.

Bibliography: Arpad Szilagyi *et al.*, - Peştera Vîntului, *Bul. Clubului de Speol. "E. Racoviţă" Bucureşti*, 1979 (6): 98-117 and *Trav. Institut Spéol. "E. Racoviţa"*, 1979, XVIII: 259-266.

Ică Giurgiu.

2. **peştera din pîrîul Hodobanei** .... 22,042 m
(m. Bihor)
Cave explored by Z. Oradea. From 5,906 m (1979), the length increased to 17,151 m (1980) then 22,042 m in 1982.

3. **complexul Topolniţa** .... 20,500 m
(podişul Mehedinţi)
Has four entrances and was explored by the Institut de Spéol. and Avenul Braşov (1958-1982) (*Trav. Inst. Spéol. "Racoviţa"*, 1976, XV, plan with 15 km).

4. **peştera Ciur-Ponor-Topliţa de Roşia** ..... 14,291 m
(m. Pădurea Craiului)
Cave with a 4,600 m long stream, explored by three

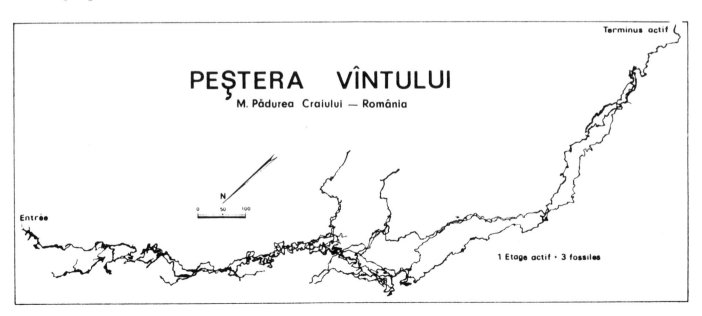

PEŞTERA VÎNTULUI

M. Pădurea Craiului — Romania

Terminus actif

Entrée

N

0   50   100

1 Etage actif · 3 fossiles

clubs from Bucureşti and Cristal Oradea (1981-1984) (*Bul. C.E.R. Bucureşti*, 1983 (8) plan).

5. **peştera Tăusoare** (m. Rodnei) .... 13,261 m
By "E. Racoviţă" Cluj-Napoca.

6. **peştera de la Zăpodie-peştera-Neagrã** ... 12,048 m
(m. Bihor)
By Z. Oradea (1974-1978) (*Bul. C.E.R. Bucureşti*, 1977-1978 (5) plan of 1975 with 10,879 m).

7. **peştera Polovragi** (m. Căpăţînei) .... 10,350 m
(plan with 9,171 m in *Studii şi Cercetări de Speol. Rm. Vîlcea*, 1984 (3)). Explored by Focul Viu *Bucureşti*.

8. **peştera Cornilor** .... 10,200 m
(m. Pădurea Craiului)
In 1981 by C. S. Dr. Petru Groza.

9. **peştera Emil Racoviţă din valea Firii** ...... 8882 m
(m. Bihor)

10. **peştera Ponorici-Cioclovina cu Apă** .... 7890 m
(m. Sebeş)
(*Bul. C.E.R. Bucureşti*, 1976 (4) plan).

11. **Cetăţile Ponorului** (m. Bihor) .... 7500 m

12. **peştera Buhui** (m. Aninei) .... 7429 m

13. **peştera Rătei** (m. Leaota) .... 7200 m

14. **peştera Comarnic** (m. Aninei) .... 6210 m

15. **peştera Şura Mare** (m. Sebeş) .... 6100 m

16. **reţeaua Coiba Mică-Coiba Mare** .... 6085 m
(m. Bihor)
Connection in 1978 (5680 m). In 1983 (*Bul. CCSS*, 1980 (4) plan), by Z. Oradea.

17. **peştera Dîrninii** (m. Bihor) .... 5645 m

18. **huda lui Papară** (m. Trascău) .... 5200 m

19. **peştera Bulba** .... 5160 m
(platoul Mehedinţi, Baia de Aramă)

20. **peştera Meziad** (m. Pădurea Craiului) .... 4750 m
In 1973 by the Inst. de Spél. (*Trav. Inst. Spéol. "Racoviţa"*, 1974, XIII, plan).

21. **peştera din dealul Curecea** .... 4565 m
(podişul Mehedinţi)

22. **grota Zînelor** (m. Rodnei) .... 4349 m
(*Bul. C.E.R. Bucureşti*, 1979 (6), plan with 4004 m).

23. **ponorul Sîncuta** (m. Pădurea Craiului) ...... 4250 m

24. **peştera Limanu** (Dobrogea) .... 4204 m

25. **peştera E. -A. Martel** .... 4130 m
(m. Mehedinţi, Călugăreni)
(*Bul. Speol. CCUTC, BTT*, 1979 (1) plan).

26. **peştera Urşilor de la valea de Brazi** .... 4042 m
(m. Retezat)
(*Bul. C.E.R. Bucureşti*, 1979 (6) plan).

27. **peştera lui Epuran** (platoul Mehedinţi) ...... 4000 m
(*Pestara Epuran*, plan)

28. **peştera de la Fîntîna Roşie** (m. Bihor) ...... 3870 m
(Bul. C.E.R. Bucureşti, 1977-1978 (5) plan).

29. **peştera Ponoraş** .... 3861 m
(m. Pădurea Craiului)

30. **avenul din Şesuri** (m. Bihor) .... 3840 m
(*Aragonit*, 1978 (1) Plan).

31. **peştera Lazului** .... 3801 m
(m. Mehedinţi, Călugăreni)
(*Bul. CCSS*, 1978 (2) plan with 3201 m).

32. **peştera Muierilor** .... 3566 m
(m. Parîng, Baia de Fier)
(*Peşteri din România*, 1976, plan).

33. **gheţarul de la Barsa** (m. Bihor) .... 3474 m

34. **peştera Micula** (m. Bihor) .... 3360 m

35. **reţeaua Lumea Pierdută** (m. Bihor) .... 3322 m
(*Nymphaea*, 1985, XI, plan)

———————

43. **peştera Aurica** (m. Pădurea Craiului) .... 2679 m

SAN MARINO

Since this Atlas has divided the world into political rather than geological areas, the long and deep caves of the State of San Marino (embedded in Italy) are the **abisso Titano** (San Marino) which is 136 m deep and 145 m long and the **genga del Tesoro** (Dirupo di San Mirino) which is 40 m deep. Formed in gneiss, the **risorgente di Rio Marano** (Monte Giardino) is 80 m long.

SPAIN

Espagne • España

Caving activity in Spain has grown dramatically in the last fifteen years. At the end of 1985 there were over 44 caves explored (and generally mapped as well) to depths of over 500 m (with eight passing the 1000 m mark), and ten caves longer than 15 km.

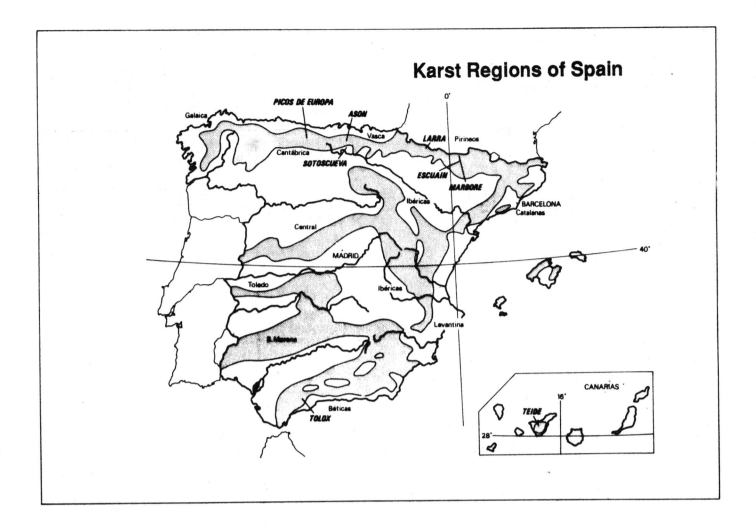

Karst Regions of Spain

Comparing these figures (44 and 10) with those published in 1981 by C. Puch ("Las Grandes Cavidades Españolas", *El Topo Loco*, 1981 (3/5, 226 p., 176 fig.), (25 and 5 respectively), or with those (12 and 2) published by C. Chabert ("Les Grandes Cavités Mondiales", *Spelunca*, 1977, suppl. N° 2, pp 14-17), one sees that there has really been an "explosion" in the pace of caving in Spain in the last nine years.

Foreign participation in this phenomenon (mostly French and British) has been decisive. Cultural expansion increased after WWII and caving was included in this expansion. Some regions, such as País Vasco (Euskadi) and Cataluña were pioneers in this effort.

The area of karst in Spain is vast, as was already noted in 1932 by Eduardo Hernández-Pacheco ("Sintesis fisiológica y geológica de España", *Trab. Museo Nac. Ciencias Nat. - Serv. de Geología*, 1932 (38) 568 p., 27 fig., 7 maps, Madrid) and it is difficult to understand why (apart from the regions mentioned above) the exploration was polarized to the big cities, the main exceptions being small groups formed around karst geologists such as Julivert and Llopis Lladó (Oviedo), Adolfo Eraso, Fernández Rubio, etc.

This began to change in 1970 with the wider dissemination of information, the greater interest in outdoor activities, and the greater availability of automobiles and the building of better roads to gain access to the more remote areas where the deepest caves tend to be located.

Spain has a very complex geology, but a large portion contains more or less soluble karst terrains. This karst potential covers about 100,000 square kilometers distributed as follows: 17% in the Cantabrian, Vascan and Pyrenean cordilleras, 48% in the Ibéricas, 7.5 % in the Catalanas, and the remaining 30% in the Béticas and the Penibéticas of Andalucia ("Tipología de los karsts en España", in N. Llopis Lladó, *Fundamentos de Hidrogeología kárstica*, Barcelona, 1970, pp 253-268).

The very deep caves (and to some extent the long caves) are generally located in the northern mountain ranges. This is because geological and paleogeological conditions have created thick, often steeply dipping carbonate formations in these ranges, while the interior ranges have limestone only at the highest elevations.

Spanish karst is rich and varied, so that often various morphological manifestations can be found very close together. This is in harmony with the wild changes seen in the landscapes throughout this country, which has a surface area equal to that of Texas.

Difficult alpine caves (such as the highest system in Europe, the sima de la Punta Olas, 405 m deep, at an altitude of 3009 m in the Pyrenees de Huesca) are found not far from active caves in the Almeria desert (the evaporite basin of the Río Sorbas), or from long caves at low or medium elevation, such as the cueva del Reguerillo, 8300 m long, near Madrid.

One also finds in this mosaic of Spanish karst significant caves in sandstone (sistema Sabadell –580 m deep, see below), in conglomerate (bofia de Torremàs, –198 m deep, Pirineo leridano; cueva de Fuentemolinos, 4086 m long, Burgos), and in gypsum (sima de Las Yeseras, 4500 m long, near Madrid; túnel del Sumidor, –205 m deep, València).

Carlos Puch.

DEEP CAVES:

1. *Laminako ateak .... –1408 m*
or **puertas de Illamina** (Isaba, Navarra)
This cave is located in the mountain range south of the Piedra de San Martin, in the commune of Isaba. It is found on the north flank of the Sierra de Budoguia, one hour's walk from the Hoya del Portillo de Larra, accessible from Zuriza (via Linza and Petrechema) or from the Belagua valley, via Lapazarra. It is at an altitude of 1980 m.
1/50,000 map Larrau. x 347.55; y 74.35 (France).
Hydrologically, it drains the underwater St. George river, and its resurgence (dye trace of 1970) is the Sorcier or Illamina spring (Laminako lezia), 11 km to the NW, at an altitude of 442 m, with a discharge of 5.64 cubic meters per second, in the gorges of Kakueta (Sainte-Engrâce, France). It traverses the Cretaceous Cenomanian-Turonian-Campanian strata.

Jean-François Pernette and I. Ortilles discovered the entrance in August 1979, during an expedition of the S. C. Frontenac, G. E. Satorrak, and the Institución Principe de Viana de Pamplona. A reconnaisance was made of the first 92 meters. A Franco-Spanish expedition ,"Amalgame 80", formed in 1980, and descended the entrance pits (387 m), then the river passage, discovered beyond a very tight canyon, followed a river passage to a depth of 1192 m. Thus most of the vertical extent was explored in a single season. A group with varying interests, the "Amalgame 81" reached a terminal sump at –1325 m in 1981. Fred Vergier was able to not only dive this sump (50 m), but also the next two (100 m, 50 m) before being stopped by a fourth sump at –1338 m. Dives by Belgian cavers brought the depth to 1353 in 1986. The total length is 11,900 m.
Map: from the drawings of Amalgame, I.P.V., G.E.S., S.E.C.E.M., 1980-1981.
Bibliography: Puch (C.); "Las Grandes cavidades españolas", *El Topo Loco*, Zaragoza, 1981 (3-5): 14-15, 45.
Pernette (J.-F.), R. Maire, "Le Bu 56 ou sima de las Puertas de Illamina, Navarre, *Spelunca*, 1983 (9): 25-34.

2. *sima del Trave* (Cabrales, Asturias) ...... –1380 m
The sima del Trave opens at an altitude of 1920 m, about 500 m west of the Cueto del Trave (2241 m), on the NW side of the Urrieles mountain (central range of the Picos de Europa).
One gains access from Poncebos (alt. 520 m) by the village of Castillo de Bulnes, then the alpine huts of Amuesa (1700 m altitude gain over 6 km). The entrance is the the edge of a vast depression located below the trail that leads to the J. R. Lueje hut.
Coordinates: 1°10'13"; 43°13'04" (U. T. M. ; 349. 28; 787. 22). 1/25,000 map Macizo Central P. E. (Adrados).
The cave is probably part of the drainage system feeding the Farfao de la Viña, a resurgence with an average

flow of 3 cubic meters per second, in the gorges of the Cares (alt. 320 m), 3500 m distant as the crow flies.

A group from the Spéléo-Club de la Seine (S. C. S.) was prospecting in the area of the Cueto del Trave in 1982 at the end of their camp when they found the 309 m entrance shaft to the cave, which they descended to a depth of 160 m (see section on "Great Pits"). The following year, the S. C. S. pursued the exploration and reached –830 m. In 1984 a zone of tight canyons stopped exploration at –1195 m. The following August the bottom of the cave was reached at –1256 m. A length of 2865 m has been mapped.
Map: surveys of the S. C. Seine 1983-1985, using topofil Vulcain, grade 4.
Bibliography: Bigot (J. -Y.), P. Génuite, B. Vidal, "La sima del Trave", *L'Aven*, bull. S. C. S. , 1984 (44): 119, 122; *ibid.* , 1985 (45): 95, 122.
Vidal (B). - Sima del Trave, *Spelunca*, special issue "Les Picos de Europa", 1985, pp. 36-37.

Pat Génuite.

3. *sistema de la Piedra de San Martín .... –1342 m*
(Isaba, Navarra) (–1006, +336)
See France, réseau de la Pierre Saint-Martin. Of the seven entrances, only one, the "historic entrance", is in Spanish territory.

4. *sistema Arañonera* (Torla, Huesca) .... –1185 m
This cave is located in the Peña Otal (or Aroñonera) mountain), which is part of the Sierra Tendeñera. Its two entrances, the grallera del Turbón (alt. 1949 m), and the cueva de Santa Elena (alt. 1391 m) are located on the barranco del Turbón and a bit above the río Ara, respectively. To get to the T.1 (grallera del Turbón), one takes a trail from Llandos de Fénez that crosses the forest to reach a vertical escarpment that overlooks the alpine meadows above the woods. The entrance is then close by.
1/50,000 map U.T.M. 30-8. 0°08'29"; 42°40'51" and 0°07'33"; 42°40'24". Developed in the Montien-Ilerden limestone which is strongly jointed. Santa Elena is the resurgence for the upper drainage of the mountain. A hydrological connection with other nearby caves such as the S.1-S.2 (see above) is evident, although no human connection has been made.

The E.R.E. de Barcelona discovered the T.1 in 1972 and reached –200 m, then –316 m and –470 m. The resurgence was unblocked in 1974 and the connection was made soon thereafter. The exploration of an ascending gallery climbing towards higher caves has occupied the E.R.E. and the E. C. Gràcia since 1979. The length reached 6505 m in 1986.

A new depth –1185 m was achieved in 1987, which is a significant change from the previously published depth of (–600, +101) 701 m.
Map: E.R.E., G.I.E., E.C.G.
Bibliography: *Espeleòleg*, 1981 (32), pp. 63-127, maps. Puch (C.), *op. cit.*, 1981, pp. 20, 53.

5. *sima 56* (Cillórigo-Castro, Cantabria) ..... –1169 m
The cave is at an altitude of 1975 m, on the Cueto de los Senderos in the Andara range, in the western Picos de Europa. From Sotres, one must take the trail which climbs to the Mazarassa mines. Arriving in Casa Blanca (1770 m),

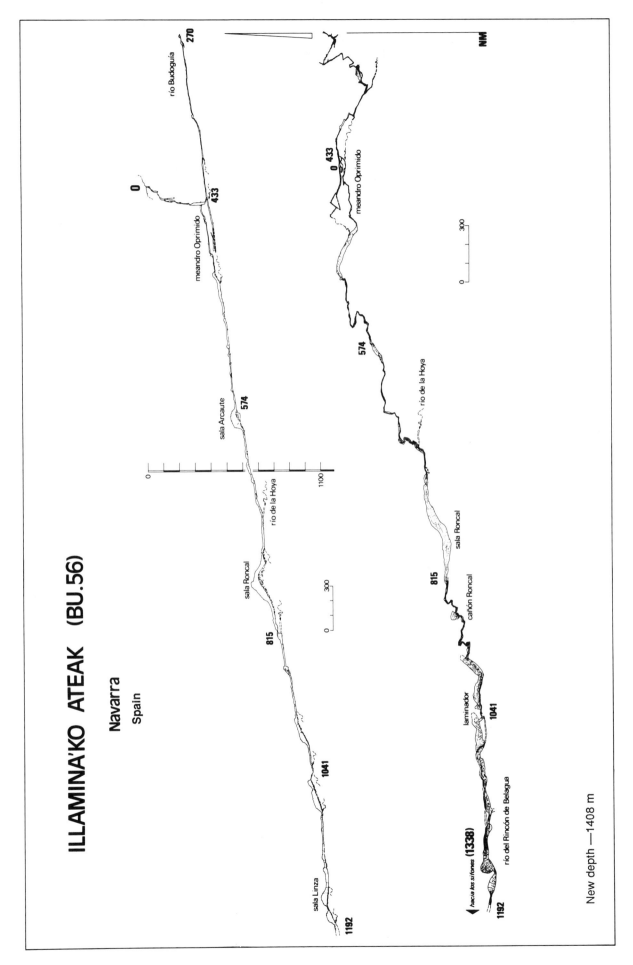

ILLAMINA'KO ATEAK (BU.56)

Navarra
Spain

rio Budoguia

270

meandro Oprimido

0   433

sala Arcaute

574

rio de la Hoya

sala Roncal

815

1041

sala Linza

1192

meandro Oprimido

0   433

rio de la Hoya

574

sala Roncal

815

cañón Roncal

laminador

1041

hacia los sifones (1338)

rio del Rincón de Belagua

1192

NM

300

0

New depth —1408 m

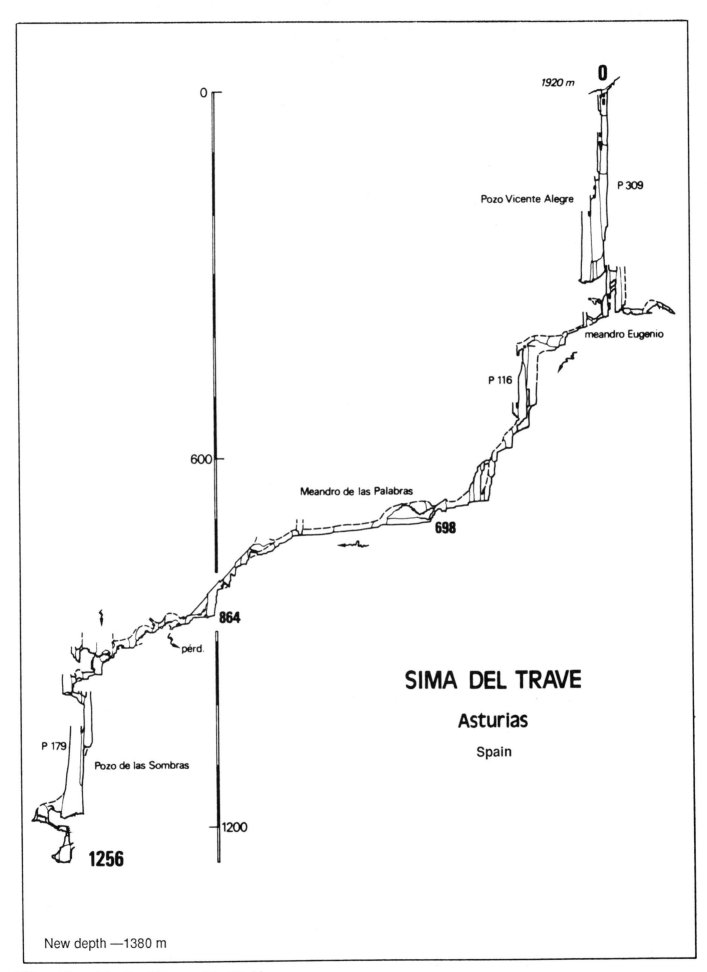

1920 m **0**

Pozo Vicente Alegre P 309

meandro Eugenio

P 116

0

Meandro de las Palabras

600

698

864

pérd.

SIMA DEL TRAVE

Asturias

Spain

P 179

Pozo de las Sombras

1200

1256

New depth —1380 m

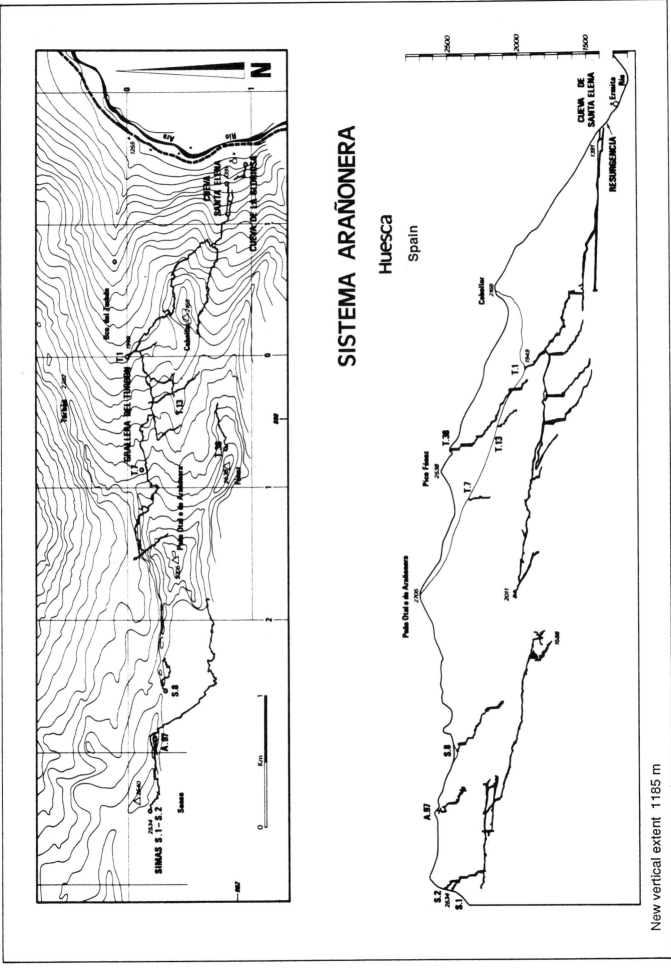

SISTEMA ARAÑONERA

Huesca
Spain

New vertical extent 1185 m

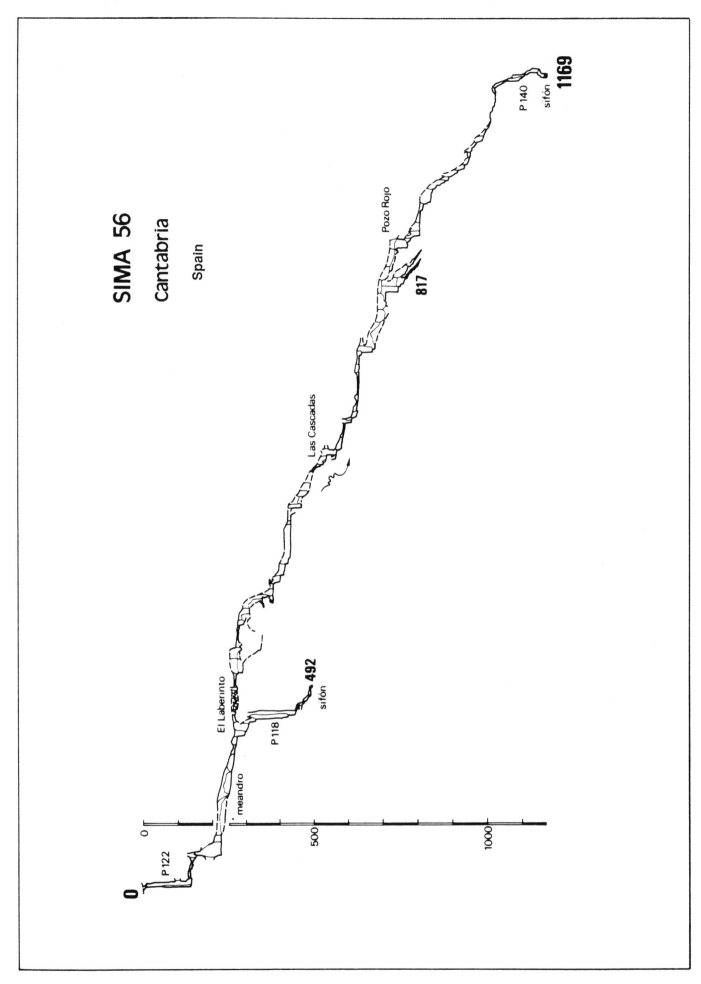

SIMA 56

Cantabria

Spain

having taken a right at the first fork, one passes a lake and some ruined houses, then climbs in the direction of a rock in the form of a bishop's headpiece. After a steep slope, find a cairn: the entrance is 50 m to the west, above a saddle, in the direction of the Cueto de los Senderos.

Coordinates: 4°43'33"; 43°12'30". The cave stream, dye traced in the branch of –492 m, reappears in the cueva del Nacimiento (alt. 760 m) in Tresviso, 5500 meters away. Formed in the Montaña limesone, dated to the late Carboniferous.

The entrance was discovered and explored to the –228 m level by the Lancaster University Speleological Society (LUSS). A tight canyon led the LUSS to the top of a 118 m pit in 1979, but at –492 m they were stopped by a squeeze that was too tight.

A continuation was found in 1980 at the top of the pit. Because of the difficulties involved, exploration was put off until 1982. A camp was made at –300 m and a sump reached at –817 m.

Again, another way on was found in 1983 at –750 m, which this time led the LUSS to a terminal sump at –1169 m, for a mapped length of 4120 m. Additional explorations in 1984 brought the length to 5620 m.

Map: from the surveys of the LUSS and S.E.I.I. (1978-1983).

Bibliography: Puch (C.), *op. cit.* , 1981, pp. 27, 69. Benoit, (P.) *et al.*; "Les Picos de Europa", *Spelunca*, suppl. to N° 19, 1985, pp. 48-49.
L.U.S.S. and S. E. I. I., *Sima 56*, Madrid, 1985, 88 p.

6. *sistema Badalona* (Escalona, Huesca) ... **–1149 m**
The system is comprised of the sima B 15 (alt. 2264 m) and the fuente de Escuaín (alt.. 1115 m) caves in the Sierra de las Sucas. The fuente de Escuaín is the principal resurgence for the Escuaín range, and is located at the foot of the barranco de la Serra, on the right bank, at the confluence with the Gargantas de Escuaín. The sima B 15 is located some 3 km to the NNW, in the lapiaz below the peaks of las Sucas and Puntas Verdes (2618 m).

1/50,000 map UTM 30-9. x 263.84; y 4723.6 (B 15) and x 264.5; y 4720.82 (fuente). Limestones of lower Eocene age.

The G. E. Badalona located the B 25 entrance in 1972 and descended the first pit, having already begun the exploration of fuente de Escuaín in 1968. With the S.I.E. de Barcelona they descended to –244 m in the B 15 in 1976. In 1978, the G.E.B. reached –701 m and was stopped at the top of a 50 m pit. In 1979 an eight day underground camp permitted the Badalonians to explore a river passage down to –961 m and to climb a side stream to –250 m. The same year, they climbed the 75 m tall "cascada Silvia" (25 m) in the fuente, stopping at the foot of another one, 40 m high, the key to the connection.

The connection and first traverse were done by the G.E.B. in 1980. Unexplored pits at higher altitudes give a still greater depth potential to this hydrologic system.

Map: from the surveys of the G.E.B.
Bibliography: Puch (C.), *op. cit.*, 1981, pp. 15, 46.

7. *sistema del Xitu* (Onís, Asturias) .... **–1148 m**
This great deep cave is found in the Cornión range, west of the Picos de Europa. Its main entrance (alt. 1652 m) is in the Jito (or Xitu) de Ario, north of Gustuteru (1812 m). One takes the Los Lagos road from Covadonga. From La Ercina lake (1108 m) a trail leads to the Vega de Ario hut (1600 m). Just before the hut, some 500 m to the west, one encounters the Jito de Ario, and, around 100 meters before an orientation table is the entrance to the cave.

1/25,000 map Macizo del Cornión (Lueje) 04°55'06"; 43°14'04".

Formed in the Montaña limestones of upper Carboniferous age, the cave has three distinct drainages. A dye trace in 1981 revealed that the largest of them drains to the fuente Culiembro, on the right bank of the río Cares. Other probable resurgences are the fuente de la Canal de Trea (830 m), and the fuente de la Canal de la ría de las Párvulas.

The cave was discovered in 1979 by the Oxford University Caving Club (O.U.C.C.), who pursued a first active branch to –354 m. In 1980, the explorers were stopped by a sump at –362 m, at the bottom of a showering pit, but a side passage, noticed the preceeding year, led to a second active branch of ample dimensions. This was descended to –859 m, for a length of 3899 m. In 1981, the O.U.C.C. reached a terminal sump at –1139 m, which was sounded to a depth of 9 m. The length was then 6111 m. Horizontal extensions were found by the O.U.C.C. in 1984 at the –600 m level. Two new entrances were added in 1985, the pozu La Cistra and the pozu Los Caracoles, located at lower altitudes than the orignal entrance.

Map: from the surveys of the O.U.C.C.
Bibliography: Puch (C.), *op. cit.* 1981, pp. 16, 48. Gale (S.), Pozu del Xitu, in "Les Picos de Europa", *Spelunca*, suppl. to N° 19, 1985, pp. 18-20.

8. *sima G.E.S.M.* (Tólox, Málaga) .... **–1098 m**
The sima G.E.S.M. is located in the karst of the Hoyas del Pilar, in the sierras de Tólox and las Nieves, on the Ronda mountain. It opens at an altitude of 1687 m. Access is from the 12 km mark on the Ronda-San Pedro road from Alcántara, by a 12 km trail that passes the Félix Rodriguez de la Fuente hut and continues to the Puerto de las Pilones. Continuing on foot to the Cerro Mateo (1834 m), one arrives at the Hoyas del Pilar.

1/50,000 map N° 1051, Ronda 5°00'23"; 35°41'47". Lias limestones of Jurassic age.

The G.E.S. Málaga discovered the cave entrance on September 23, 1982. It was descended in August, 1973, to a depth of –140 m during a expedition by the G.E.S.M., G.E. STD Madrid, G. E. O. Sevilla, and the S. E. Marbellí. In September, G.E.S.M., S.E.M., and Circ. Gibralfaro O.J.E. pushed down to –315 m. The August 1974 expedition included three Málaga clubs (G.E.S.M., S.E.M., E.C.M.), the G. E.O.).-Sevilla, the G. E. Priego de Córdoba, and the G. G. de la O.J.E. A depth of about –520 m was reached (mapped to –413 m).

In July 1975 the G.E.S.M., with the collaboration of the E.C.M., reached –646 m, a depth which was estimated at the time as –940 m! In September 1977 the G.E.S.M., this time with the E.R.E. de Barcelona, reached –806 m. Reinforced by seven other Spanish clubs, they reached –939 m in July 1978; in September the G.E.S.M. and the E.R.E. were stopped at –1074 m by a sump.

The sump was explored to a depth of 21 m and a length of 195 m in 1979 by the Frenchmen F. Poggia and F. Vergier, aided by P. Courbon, the S.E.S.E. Málaga, G. E.

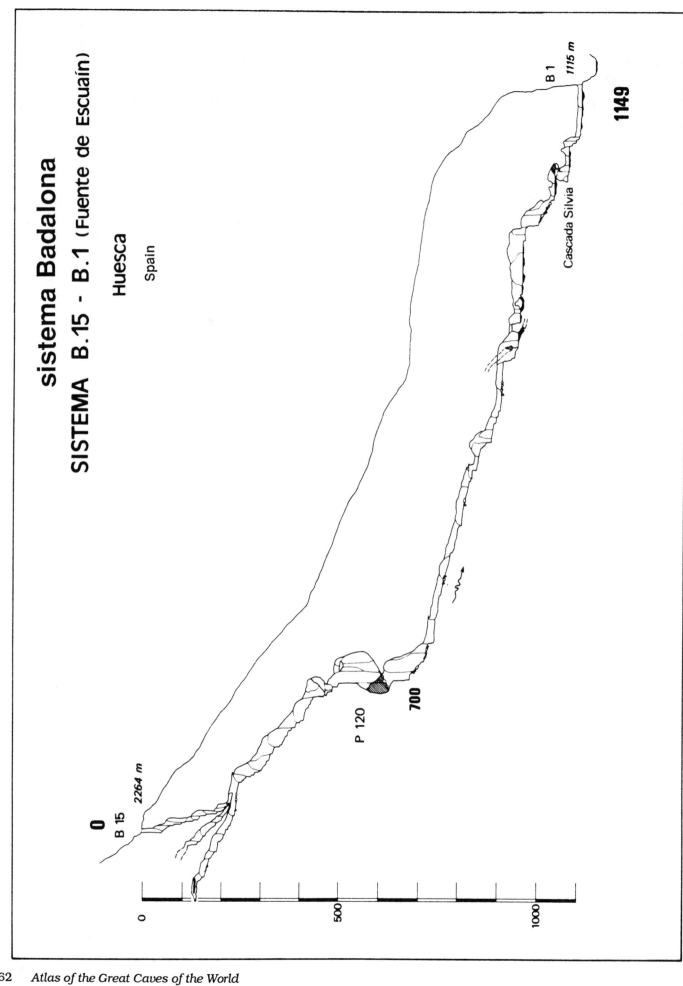

sistema Badalona

SISTEMA B.15 - B.1 (Fuente de Escuaín)

Huesca

Spain

0
B 15
2264 m

P 120

700

1149
B 1
1115 m

Cascada Silvia

0

500

1000

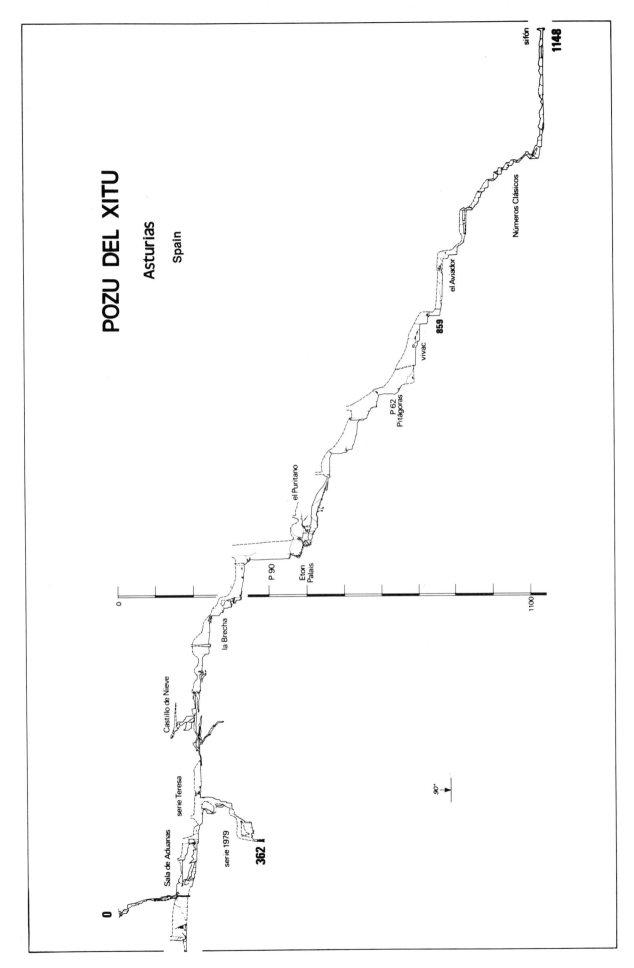

POZU DEL XITU

Asturias

Spain

sifón

1148

Números Clásicos

el Aviador

859

vivac

P 62
Pitágoras

el Puritano

P 90

Eton
Palais

la Brecha

Castillo de Nieve

serie Teresa

Sala de Aduanas

serie 1979

362

0

0

1100

90°

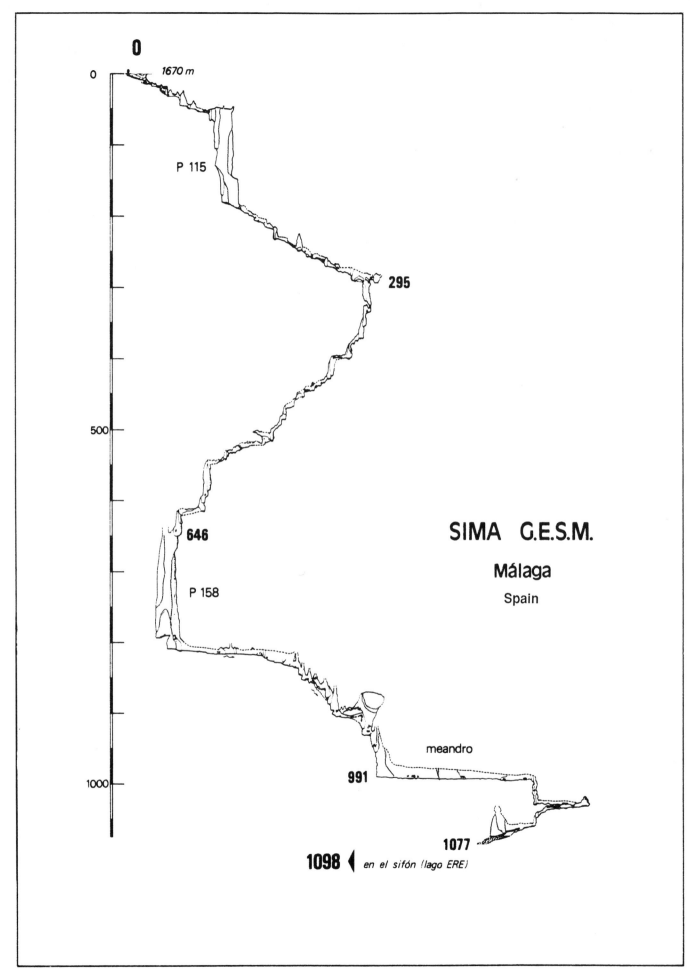

0

0

1670 m

P 115

295

500

646

P 158

SIMA G.E.S.M.

Málaga

Spain

meandro

1000

991

1077

1098 ◄ *en el sifón (lago ERE)*

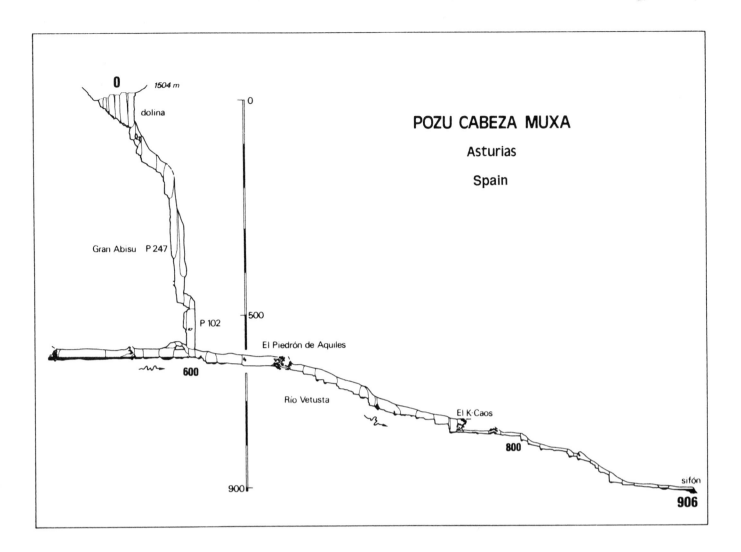

POZU CABEZA MUXA

Asturias

Spain

(Labels on diagram: 0 1504 m; dolina; Gran Abisu P 247; P 102; El Piedrón de Aquiles; Río Vetusta; El K-Caos; sifón; 906; depth scale 0, 500, 600, 800, 900)

Granada, and the G.S.A. Belge. The length was increased to 2718 m.

Map: from the surveys of the G.E.S.M., E.R.E.-C.E.C., G.E.S.-S.E.M. (1973-1979).

Bibliography: Courbon (P.), *Atlas des Grands Gouffres du Monde*, 1979, pp. 65-66, 69.

Puch (C.) - *op. cit.*, 1981, pp. 15-16, 47.

G.E.S. de la S.E.M., "La sima G.E.S.M. - 1098", *Monografías Espeleológicas*, Málaga, 1983 (3): 37 p.

9.　*torca Urriello* ....(–1017, +5) 1022 m
(Cabrales, Asturias)

Torca Urriello is located in the central range of the Picos de Europa. The entrance is located at an altitude of 1860 m, about a hundred meters below the Naranjo du Bulnes (2515 m), at a limestone-dolomite contact.

Access is by a long walk (8 km with 1000 m of elevation gain) beginning at the hamlet of Invernales de Cabao in the Dujé valley, and heading towards the Delgado Ubeda hut.

Coordinates: 04°48'57"; 43°12'06". 1/25,000 map Macizo Central, P.E. (Adrados).

The torca Urriello drains the area of the Jou Lluengu, resurging in a diffuse series of springs between 700 m and 800 m altitude above the village of Bulnes. Another possible resurgence, in spite of its distance 7 km away, is the Farfao de la Viña, at an altitude of 320 m, in the gorges of the Río Cares.

The cave was shown to the G. S. du Camping-Club de France (C.C.D.F.) and to the Centre Loisir Plein Air (C.L.P.A.) de Montpellier in 1979 by the guardian of the hut. They explored it to –189 m. The depth of –373 m was reached in 1980 by the C.L.P.A., the C.C.D.F., and the G. S. Créteil. The following year they were reinforced by the S. C. Seine and exploration and mapping went down to –819 m. The S.C.S. and the C.L.P.A. explored parallel branches in 1982 and were stopped at a sump at –1017 m. The length is 3632 m.

Map: surveys of the S.C.S and the C.L.P.A. in August 1981 and 1982.

Bibliography: Génuite (P.), Torca Urriello, in "Les Picos de Europa", *Spelunca*, suppl. to N° 19, 1985, pp. 38-39.

Pat Génuite

10.　*pozu de Cabeza Muxa* (Onís, Asturias) .... **–906 m**

The entrance to this cave is at an altitude of 1504 m, and is located in the Cornión range (western Picos de Europa). It is located SE of a small hill which extends to the NE from the pico de Cabeza Muxa (1558 m). Access is from the La Ercina lake, taking a trail to Vega Maor, traversing Belbín, Parres, and Arneado, and climbing to the south in the direction of the peak of the same name.

1/25,000 map, Macizo del Cornión (J. R. Lueje). 04°54'57";43°15'09". Montaña limestones of upper Carboniferous age (Namurien).

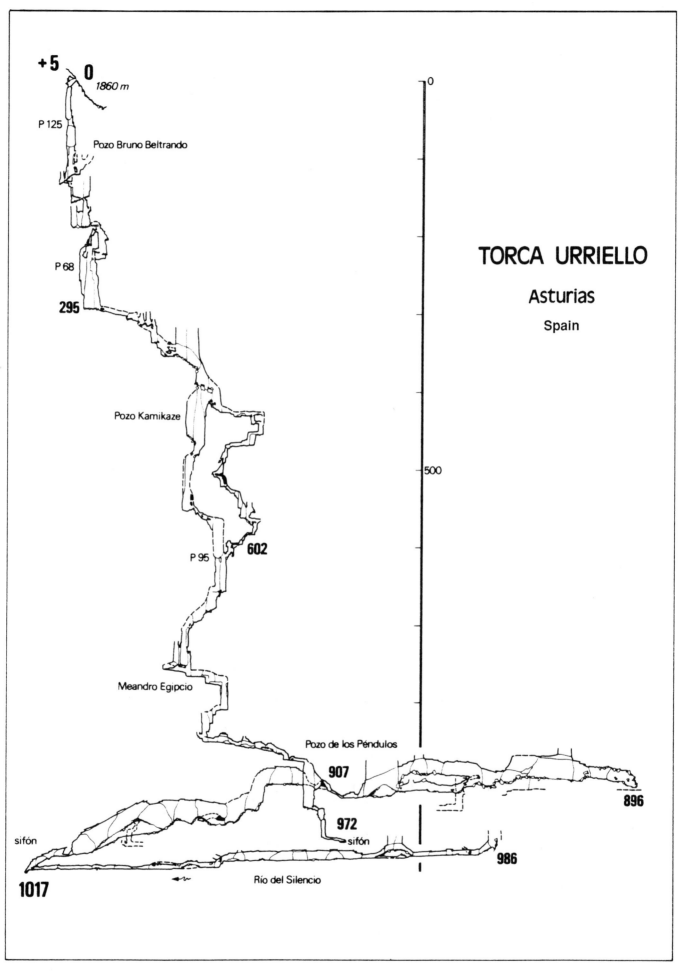

+5 **0** *1860 m*

P 125

Pozo Bruno Beltrando

P 68

295

Pozo Kamikaze

P 95 **602**

Meandro Egipcio

Pozo de los Péndulos

907

972
sifón

sifón

896

986

TORCA URRIELLO

Asturias

Spain

0

500

1017

Río del Silencio

Two clubs, the G. E. Polifemo and the O.J.E. (Oviedo) uncovered the cave in 1973. In 1975, in collaboration with R. Larma of the C. E. de Aragón, they were stopped at –225 m at the top of a huge 247 m pit for lack of rope. The pit was partially explored in 1976 and 1977 by the Asturian cavers.

It was conquered in 1979 by the Sociedad de Investigaciones Espeleológicas del Centro Excursionista "Aguila" de Barcelona (S.I.E.) who had to quit in the middle of a second great shaft at a depth of 536 m, again for lack of rope.

The S.I.E. descended the new 102 m pit in 1980 to discover an active stream passage interrupted by breakdown at the –620 m level.

They returned to the cave in 1982 and found a way through the breakdown. They were not able to gain additional depth until 1983, when they reached –703 m. The exploration the following year was stopped by a 19 m deep terminal sump at -906 m, for a total length of 2630 m.

Map: from the surveys of the S.I.E-C.E.A.

Bibliography: Puch (C.), *op. cit.*, 1981, pp. 22, 57. Benoit *et al.*, "Les Picos de Europa", *Spelunca*, suppl. to N° 19, 1985, pp. 22-23.

11.   *torca del Jou De Cerredo* .... –852 m
(Cabrales, Asturias)

This cave is located to the NE of the Torre Cerredo (2648 m), at an altitude of 2325 m, in the Urrieles range, in central Picos de Europa. Access is from the J. R. Lueje hut (see sima del Trave above), from which the trail at the bottom of the Jou de Cabrones is taken towards the Torre Cerredo. At the end of the Jou climb to the left up slopes that lead to a col, from which one sees the crest on which the entrance is found.

Coordinates; 04°50'18"; 43°12'21". 1/25,000 map Macizo Central P.E. (Adrados).

Hydrologically, it is part of the same system as the sima del Trave.

It was during the last days of their 1983 expedition that the Grup de Exploracions Subterranies del Club Muntanyenc Barcelones discovered the entrance to the cave: they descended to a tight blowing canyon at − 80 m. Shortly thereafter other incursions were made down to about –200 m. The G.E.S.-C.M.B., with the S. C. Frontignan and the S. C. Alpin Languedocien, reached –774 m in 1984, having conquered 109 m and 104 m deep pits. They were stopped at the top of yet another pit.

Exploration was (hopefully) temporarily stopped by breakdown at the bottom of the last pit. Twenty meters has been cleared so far to reach a depth of 852 m.

Map: G.E.S.-C.M.B., S.C.A.L. and S.C.F.

Bibliography: Cano Ventosa (M.), Torca del Jou de Cerredo, in "Les Picos de Europa", *Spelunca*, suppl. to N° 19, 1985, p. 41.
G.E.S., "Picos 84-85", *Sota Terra*, 1985 (6): 49-51.

12.   *sima S1-S2* (Linás de Broto, Huesca) .... **–846 m**
This cave has two entrances (S1, alt. 2473 m, and S2, alt. 2534 m). It is located above the Soaso valley, in the Sierra de Tendeñera. More precisely, it is in the east face of the collado de Año, facing the pic de Tendeñera (2857 m). To get there one must take, 500 m before Linás de Broto, the trail which climbs by the barranco del Sorrosal to the Soaso hut, then continue in a straight line by a steep slope

to the cliffs of the cirque and then to the left in the direction of the Tendeñera.

1/25,000 map Lambert 146-III. x 887.0, y 902.741. Paleocene limestones. Possible connection with the sistema Arañonera (see above).

The entrance was located in 1977 by the G.I.E. and the E.R.E. de Barcelona, who went down to –100 m, then –180 m. In 1979 repeated trips led the explorers to 346 m (thanks to the successful aid climb of "la ventana" at –277 m), then –450 m, and –505 m. Connection with the sistema Arañonera seemed very likely at that time.

In 1980 the G. E. Gràcia explored more than 4 km of passages, with numerous side streams and unexplored pits, leading to a dead end pit at –729 m. A strong air current still hinted at the likelihood of a connection.

An extension was found in 1981 by the G.E. Gràcia giving the cave a vertical extent of –846 m and a length of 6009 m.

Map: from the surveys of the explorers.

Bibliography: Puch (C.), *op. cit.*, 1981, pp. 19, 51.

13.   *torca T-173* .... –831 m
(Cillórigo-Castro, Cantabria) (1)

Located on a rocky escarpment which rises up in the middle of a vast depression, east of the Cueto de los Senderos (alt. 1706 m), in the Andara range, Picos de Europa. Access is directly from the trail from Sotres to the Andara mines, 1.5 km before the Casa Blanca, by climbing up towards the Cueto de los Calabreros. 1/25,000 map Macizo Oriental P.E. (F.E.M.). Coordinates: 04°43'34"; 43°13'19". Carboniferous Montaña limestone.

The entrance was found in 1980 by the Lancaster University Speleological Society. The explorers were stopped by a flowstone blockage at –304 m in 1981. A climb of 20 m permitted them to find a small half-submerged pit with a blowing fissure. The depth of 500 m was reached in 1984. In 1985 a new tight fissure stopped the explorers at –831 m. Side leads at the –600 m and –700 m levels have yet to be explored.

Map: L.U.S.S.

Bibliography: Ibberson (P.), "Tresviso 85. The year of the Dosser", *Caves & Caving*, 1986, 31, pp. 14, 17.
(1) This cave has been baptized "Dosser's Delight" by the L.U.S.S.

Carlos Puch.

14.   *torca de la Laureola* .... approx. –830 m
(Cabrales, Asturias)

This cave is located in the high altitude lapiaz that extends from the Cueto del Trave to the Pico Cabrones. Access is by the same trail that leads to the sima del Trave (see above).

Like its sister, the torca de la Laureola most likely belongs to the watershed of the Farfao de la Viña (alt. 320 m), in the Cares gorges, at the foot of the Urrieles range in the central Picos de Europa.

The entrance was found in 1984 by the Spéléo-Club de la Seine. The first exploration was made by this club in August 1985, to a depth of about 830 m. Mapping stopped at 780 m. The surveyed length was 1255 m.

Map: from surveys of the the the S.C.S., 1985.

Pat Génuite.

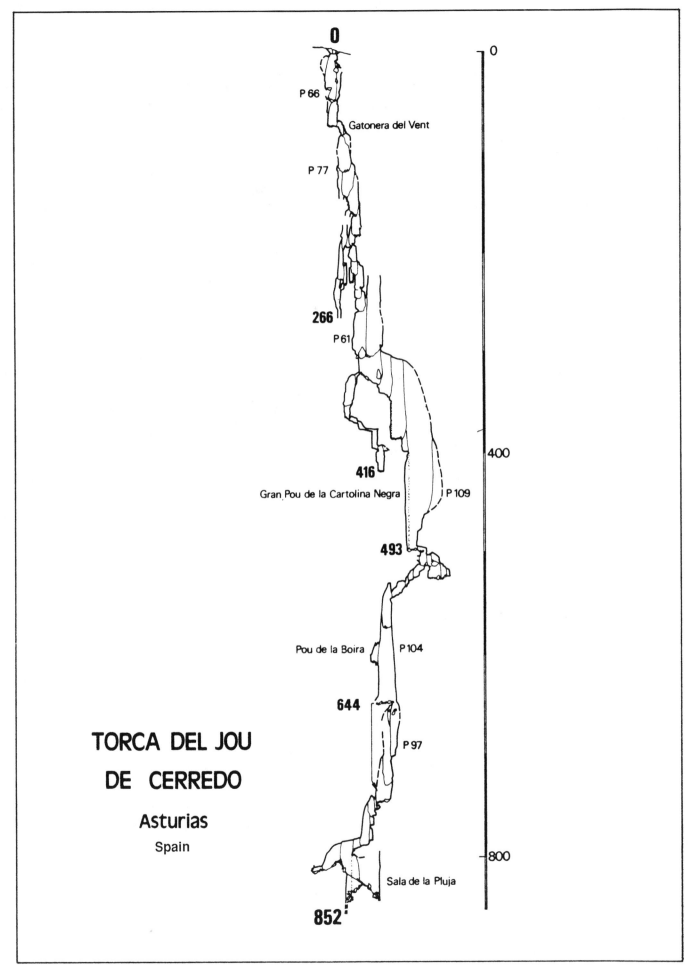

0

P 66

Gatonera del Vent

P 77

266

P 61

416

Gran Pou de la Cartolina Negra

P 109

493

Pou de la Boira

P 104

644

P 97

**TORCA DEL JOU
DE CERREDO**

Asturias

Spain

Sala de la Pluja

852

0

400

800

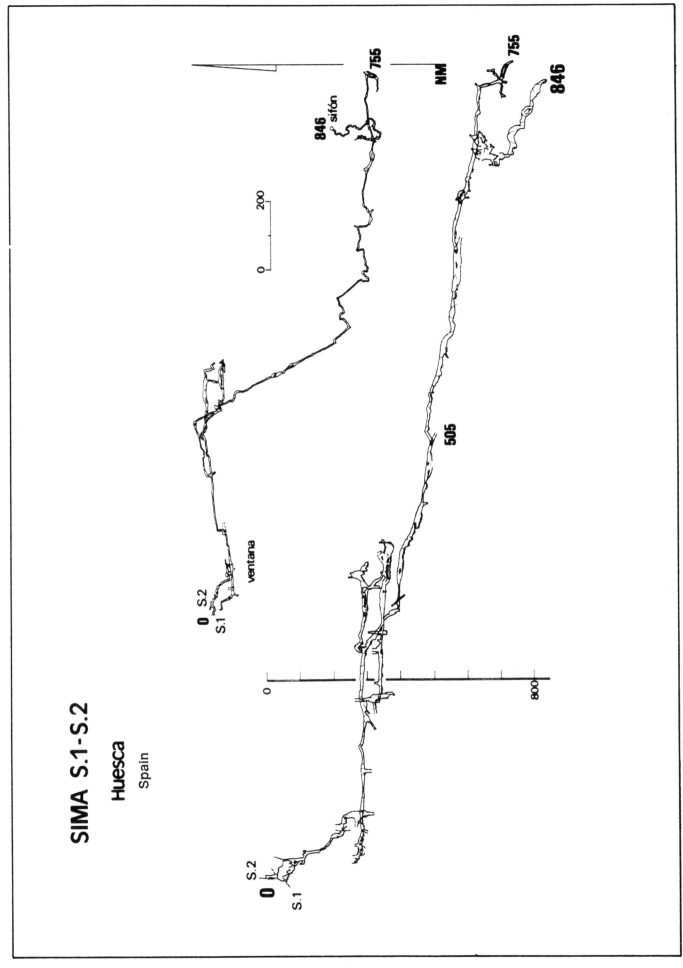

SIMA S.1-S.2

Huesca
Spain

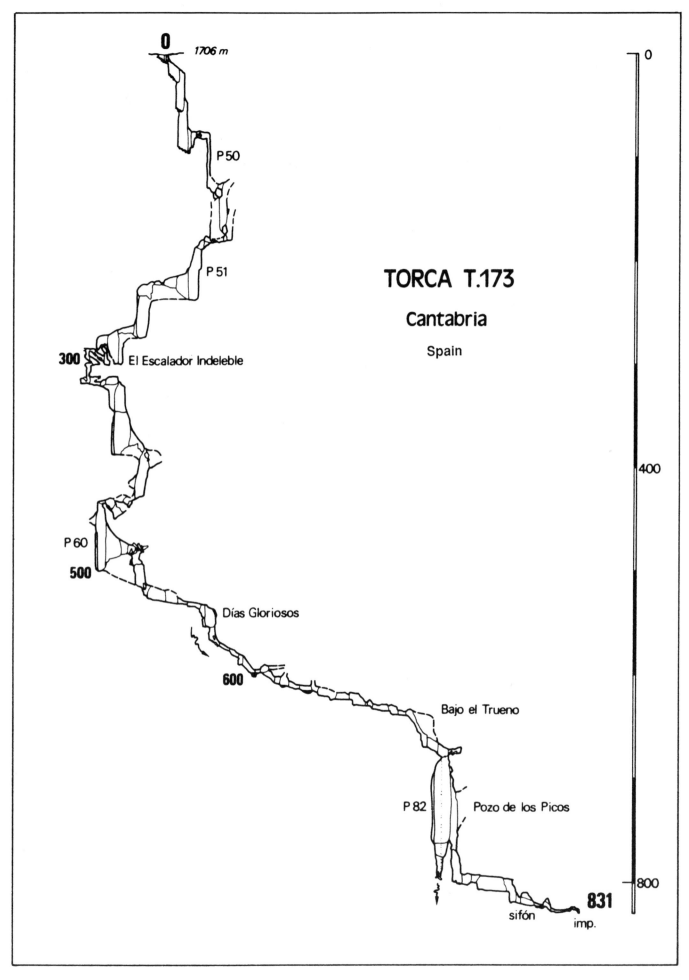

0   *1706 m*

P 50

P 51

300   El Escalador Indeleble

TORCA T.173

Cantabria

Spain

P 60

500

Días Gloriosos

600

Bajo el Trueno

P 82   Pozo de los Picos

831

sifón

imp.

0

400

800

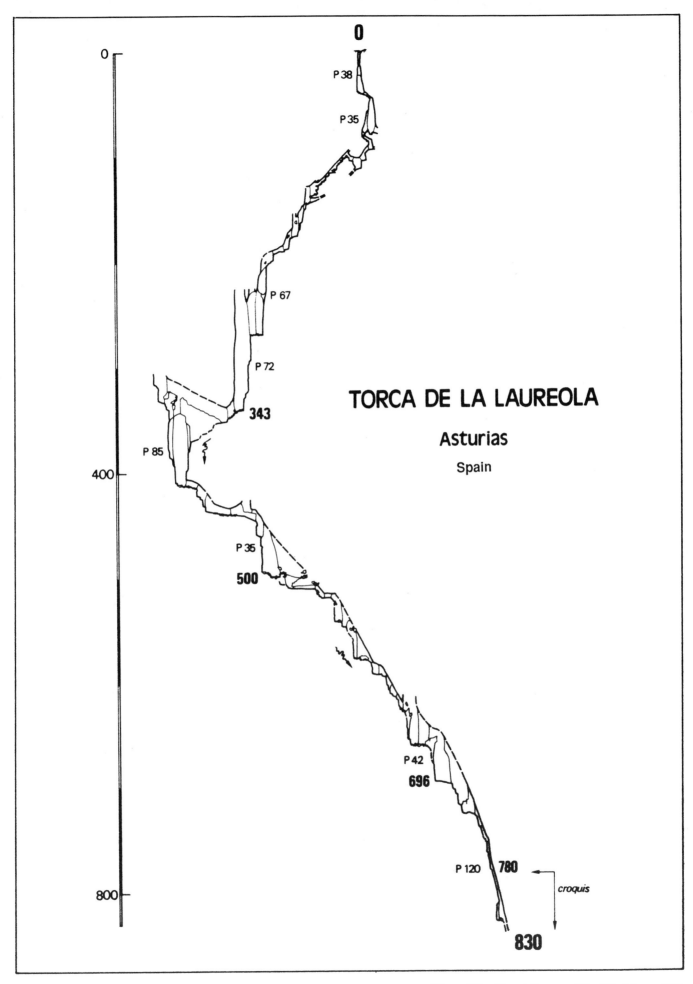

0

P 38

P 35

P 67

P 72

343

P 85

400

TORCA DE LA LAUREOLA

Asturias

Spain

P 35

500

P 42

696

P 120 **780**

croquis

800

830

15. *sistema Garma Ciega-Cellagua* .... **–825 m**
(San Pedro de Soba, Cantabria)

This is a system formed of two caves, the sima de Garma Ciega (alt. 1115 m), and the sumidero de Cellagua (alt. 960 m). The sima is at the foot of the Pico Tejes (1149 m), and the sumidero is to the south, next to the Mazo Chico (1114 m) in the Mortillano range. From Astrana, a trail climbs to the Mazo Grande. At the end of the trail, one climbs towards the collado de la Espina and finds the insurgence in the SW side of the great depression Llana la Cueva. The Garma Ciega is found in the middle of the lapiaz which lies above the large depression to the west. 1/50,000 map N° 59, Villacarriedo. 0°07'15"; 43°14'00" (sima); 0°07'30"; 43°13'45" (sumidero). Urgonian limestones (Aptian-Albian).

The probable resurgence is Las Fuentes (alt. 260 m) in the val d'Asón.

The Société Spéléologique de Bourgogne (S.S.B.) found the sink of Mazo Chico entrance while prospecting in 1965, and descended to –150 m after digging out the entrance. Bad weather and floods made exploration difficult in 1967 and no more than –231 m could be reached. Better conditions in 1968 allowed penetration to about –350 m and the discovery of over 2100 m of new passages.

The Garma Ciega entrance was found in 1969 and captured the attention of the S.S.B., who descended to about –220 m. The connection with the other cave was made on July 22, 1970, with exploration down to about –530 m, stopping at the top of a 7 m waterfall.

A five day bivouac in 1972 permitted the S.S.B. to reach a depth measured at the time as –853 m, stopping at the top of a 15 m drop. They were stopped in 1973 by unstable breakdown at a depth they measured as –916 m, which had to be wrong as it was below the resurgence! The Polish cavers of the P.T.T.K. de Gdynia climbed a 25 m chimney near the collapse in 1974 and reached a descending gallery which lead to a low passage near the terminal sump. They announced –970 m!

A Spanish expedition in 1980 brought together several clubs to resurvey the system, announcing –864 m. Another resurvey in 1984 and 1985 by the A.R.E.S. (Dijon-Paris) and the C.A.S.T.A.R. (Lille-Paris) found a depth of –825 m and a length of 9226 m. The two resurveys lie on top of each other on the plan views.

Map: from the surveys of the S.E.G., G.E.S.-C.E.C., S.E.G.I.M.

Bibliography: Courbon (P.), *op. cit.*, 1979, pp. 66, 70. Puch (C.), *op. cit.* 1981, pp. 16-17, 49.
Balart (D.), J. -L. Montero, "Avance sobre los trabajos topográficos realizados en el sistema Garma Ciega - Sumidero de Cellagua", *Exploracions*, 1984 (8): 65-81.

16. *sistema Cueto-Coventosa* .... **–815 m**
(Arredondo, Cantabria)

This system is composed of three caves: the sima del Cueto (alt.. 980 m), the cueva Coventosa (alt. 185 m), and the cueva Cuvera (alt. 175 m). It is located in the Peña Lavalle range which dominates the west side of the val d'Asón. Coventosa is near the trail which goes from Puente Nuevo to Socueva, 3 km from the Arredondo-Espinosa de los Monteros road. Cueto is on the south flank of the Pico Mosquitero, near the Peña Lavalle.

1/50,000 map N° 59, Villacarriedo. 0°03'47"; 43°15'14" (Cueto); 0°04'42"; 43°15'08" (Coventosa).

The system is formed in the Urgonian limestones of the lower Aptian-Albian, in the Asón series. The morphological similarity of the large passages in Cueto and the cueva Cañuela indicates the existence of a large fossil system.

The cueva Coventosa has been known for many years. It was systematically explored beginning in 1963 by the S. C. Paris (B. Dressler *et al.*), who mapped 2500 m to a sump. In 1965-1966 the S. C. Dijon took up the exploration, extending the length to over 6700 m.

Gérard Juhué and the Dressler brothers located the sima del Cueto entrance on April 4 1966, but the presence of a cairn indicated that it had been previously known to shepherds. A first sounding showed the entrance pit to be 193 m deep. The S.C.D. descended the pit using a winch in August 1966, finding it was actually 302 m deep, and turned around at the –370 m level (see section on "GREAT PITS").

A new expedition by the S.C.D. in 1967 reached –555 m, with the cave still going strong. In 1968 they continued down to –572 m to the ceiling of a huge horizonal passage that they explored for 2 km until a pit blocked the way.

The pit was descended in 1969 to give the cave a depth of –745 m and a length of about 3500 m, with the horizontal gallery continuing past the pit as well. While Coventosa grew to a length of 7020 m in 1971, Cueto gained 10 m in depth and a length of over 4 km. Additional passage found in 1975, in the hopes of a connection, brought the length to 6730 m.

The S.C.D. turned to Coventosa in 1976-1977 and brought the length to about 8500 m. They returned to Cueto in 1978, gaining no depth but an increase in the length to about 9500 m.

On April 13, 1979 the Grenoble cavers from the C.A.F. and the S.C.D. made the connection between Coventosa and Cueto, making a system –805 m deep and 20700 m long. On August 7 of the same year, the S.C.D. connected with cueva Cuvera by diving a sump, bringing the depth to –815 m and the provisional length to 23730 m.

The length was increased to 25,455 m in 1984 and 27,260 m in 1985 as members of the S. C. Paris under the leadership of P. Morverand have searched for a connection with the nearby cueva Cañuela.

Map: from the surveys of the S.C.D., S.G. C.A.F., and the S.C.P.

Bibliography: *Cuadernos de Espeleológia*, Santander, 1973, XI (7); *Scialet*, Grenoble, 1979 (8); *Spelunca*, 1979 (4). Courbon (P.), *op. cit.*, 1979, pp. 66-67, 71. Puch (C.), *op. cit.*, 1981, pp. 17-18, 50.

17. *sima Tere* (Cillórigo-Castro, Cantabria) .... **–792 m**

This cave is located in the Andara range in the eastern Picos de Europa at an altitude of 1820 m.

The entrance is found in the middle of mine tailings, on the east side of the Sara depression, 40 m NE of the conflence of two trails that pass by this basin. To get there, as for the sima 56, one must take the trail that goes from Sotres to the Andara mines, and, after the Casa Blanca, to the La Providencia mines or the Sara depression.

1/25,000 map Macizo Oriental, P.E. (F.E.M.). 4°42'08"; 43°12'27".

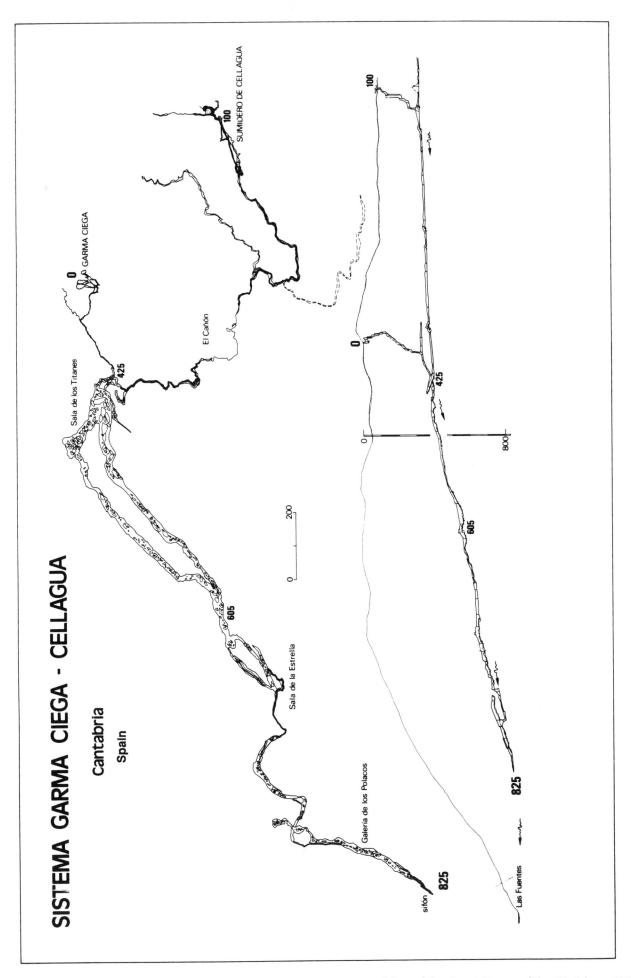

SISTEMA GARMA CIEGA - CELLAGUA

Cantabria

Spain

GARMA CIEGA

0

SUMIDERO DE CELLAGUA

100

El Cañón

Sala de los Titanes

425

605

Sala de la Estrella

Galería de los Polacos

825

sifón

100

0

425

605

825

800

Las Fuentes

200

0

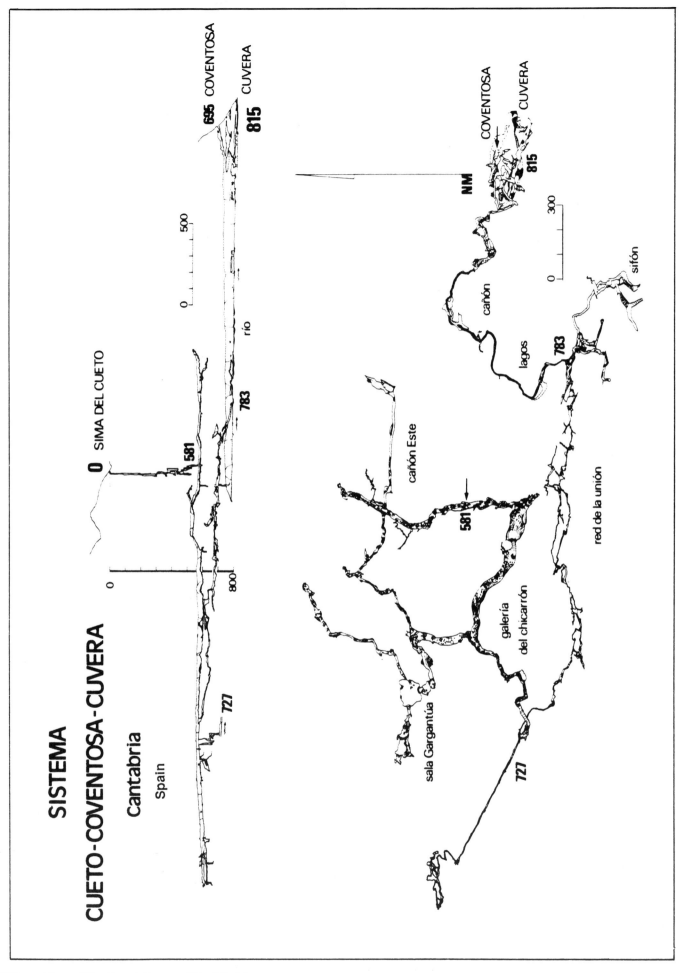

SISTEMA
CUETO-COVENTOSA-CUVERA

Cantabria
Spain

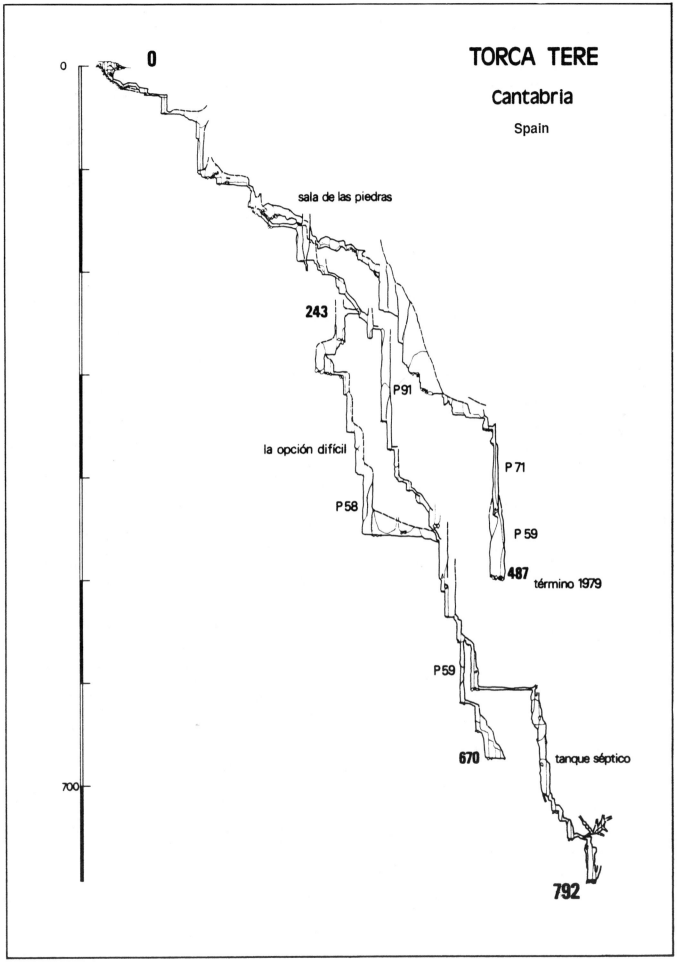

TORCA TERE

Cantabria

Spain

0

sala de las piedras

243

P91

la opción difícil

P 71

P58

P 59

487 término 1979

P 59

670 tanque séptico

792

0

700

Exploration of this cave was temporarily abandoned in 1977 by the Lancaster U.S.S., but taken up again by them in 1979 to reach the bottom of a blocked pit at –487 m. The following year a continuation was found at –130 m that permitted the L.U.S.S. to reach –530 m in 1980. The terminal sump, measured at –792 m, was reached in 1981. The length is about 1000 m.

Map: from the surveys of the L.U.S.S.

Bibliography: Puch (C.), *op. cit.*, 1981, pp. 28, 71. Benoit (P.) *et al*, "Les Picos de Europa", *Spelunca*, suppl. to N° 19, 1985, pp. 50-51.

18. *torca Tejera* (Cabrales, Asturias) .... **–758 m**

The cave is at an altitude of 1340 m and has a double opening. It is located in the Cornión range in the western Picos de Europa, NNE of Lloroso (1792 m). Access is from the Arenas de Cabrales by ascending the Cares river on the south to Poncebos and then climbing from Camarmeña to the hamlet of Ondón. Once on the plateau, one follows a valley on the north to the foot of Lloroso and then reaches the depression were the cave is found.

1/25,000 map Macizo del Cornión (Lueje). 04°52'45"; 43°16'37".

The entrance was found in 1979 by the G. S. Doubs and the S.S.S. de Genève. The G.S.D. made a rapid incursion in 1980 and ended at –184 m for lack of time and equipment. In 1981, an expedition was organized by the G.S.D., the S.S.S. Genève, and the Soc. Amateurs de Cavernes: one descending branch was ended at –368 m at a sump, while another, taking off at the –220 m level, was explored to –635 m, the mapping stopping at –537 m.

A new expedition in 1982 by the same groups reinforced by the S. C. Nyon reached the terminal sump at –758 m. The length is 1900 m.

Map: from the surveys of the G.S.D. and the S.S.S. Genève.

Bibliography: Chorvot (G.), Torca Tejera, in "Les Picos de Europa", *Spelunca*, suppl. to N° 19, 1985, pp. 20-21.

19. *sima del Florero* .... **–723 m**
(Cillórigo-Castro, Cantabria)

This cave is located to the north of sima 56 (see above), in the Cornión range (eastern Picos de Europa). To get there, take a trail 300 m from the Casa Blanca on the left, pass a saddle, and reach the foot of pico Delboro (1883 m) which one must climb, the cave being located under a large overhang bordering the trail, at an altitude of 1810 m.

1/25,000 map Macizo Oriental P.E. (F.E.M.) 04°43'27"; 43°12'58".

This cave could well by hydrologically connected with sima 56.

It was discovered during the 1980 expedition by Lancaster U.S.S. (L.U.S.S.) and Burnley C. C., and explored the same year to –210 m. Exploration was finished the following year by the LUSS with the Sección de Espeleológia de Ingenieros Industriales de Madrid: a "terminal" sump was reached at –723 m.

Map: from the surveys of LUSS, S.E.I.I.

Bibliography: Benoit (P.) *et al*, "Les Picos de Europa", *Spelunca*, suppl. to N° 19, 1985, pp. 52-53.

20. *cueva Buchaquera* ....(–607, +107) **714 m**
(Acumuer, Huesca)

The cave is located on the SE side (called the "Buchaquera") of the Peña Somota, in the Collorada range, above the Aurin river valley. It opens at an altitude of 1960 m in the center of a large lapiaz that is difficult to pinpoint, east and above the village of Villanúa.

1/25,000 map N° 145 (Sallient). 0°26'36"; 42°41'11". Possible resurgences are gruta del Rébeco Candalu (alt. 960 m) and cueva Vieja, both near the village of Villanúa (in the Aragon river valley).

René Jeannel made a first incursion into the cave in 1913. It was rediscovered in 1973 by the G. S. Préhistorique Vosgien d'Epinal (D. Colin, M. Méry, C. Pérignon), who explored it regularly until 1975, succesively reaching –128 m (–92, +36, fossil sections of the cave), –416 m (–380, +36, stopped due to lack of rope), and –626 m (–590, +36). The G.I.E.P.E. de Sabiñánigo reached the terminal sump in 1976, at –607 m. Shortly thereafter, the G.S.P.V. explored an ascending branch to +107 m and mapped the sump area.

Map: from the work of the G.S.P.V.

Bibliography: Courbon (P.), *op. cit.*, 1979, pp. 67, 72. Puch (C.), *op. cit.*, 1981, pp. 19, 52.

21. *sima de Cemba Vieya* .... **–703 m**
(Cangas de Onís, Asturias)

This is another cave in the Picos de Europa, in the western side of the Cornión range, at an altitude of 2200 m. Access is from Covadonga, climbing to the Enol lake and following a trail to the Vega Redonda hut (1540 m), from which another trail leads to the Jou Santu and the Peña Santa de Castilla. Crossing the col la Fragua, one follows another trail to a spring and goes in the direction of the Aguja de Enol, penetrating into the lapiaz. The entrance is at the edge of the glacier, at the foot of the Aguja de Enol.

1/25,000 map Macizo del Cornión. 1°17'20"; 43°12'55". Carboniferous Montaña limestones. The presumed resurgence is the Gueyos de Jungumia (alt. 1440 m approx.).

Members of the Control de Actividades Espeleológicas en los Picos de Europa discovered the entrance in 1974 and descended to 110 m to the top of a great 210 m deep shaft. This was vanquished by the S.E. Ingenieros Industriales de Madrid and the G.E. Polifemo de Oviedo, who reached a depth of about 400 m.

Exploration resumed in 1980 when the S.E.I.I. de Madrid descended the río de los Asturianos to the top of another deep shaft (150 m deep), which was bottomed in 1981.

The S.E.I.I. and collaborators reached the terminal sump at –703 m in 1982.

Map: from the surveys of the S.E.I.I.

Bibliography: Puch (C.), *op. cit.*, 1981, pp. 33-34, 89, 210. Benoit (P.) et al, "Les Picos de Europa", *Spelunca*, suppl. to N° 19, 1985, pp. 24-25.

22. pozo de Cuetalbo .... **–650 m**
(Posada de Valdeón, León)

This is a cave on the lapiaz of the Cuetalbo (2157 m) (Vega Huerta, south face of the Cornión range in the Picos de Europa). (4°58'; 43°12').

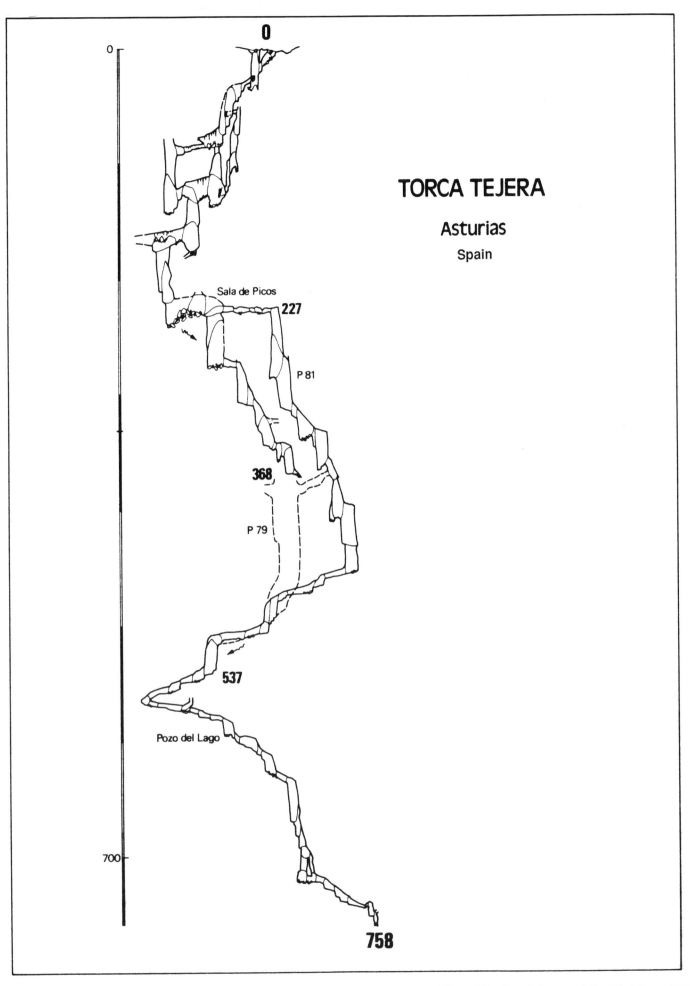

0

TORCA TEJERA

Asturias

Spain

Sala de Picos

227

P 81

368

P 79

537

Pozo del Lago

758

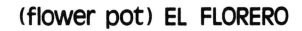

(flower pot) EL FLORERO

Cantabria

Spain

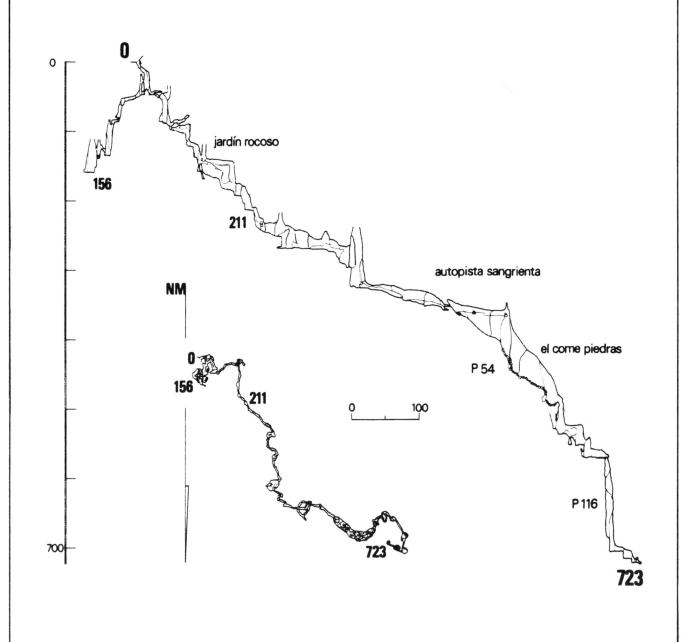

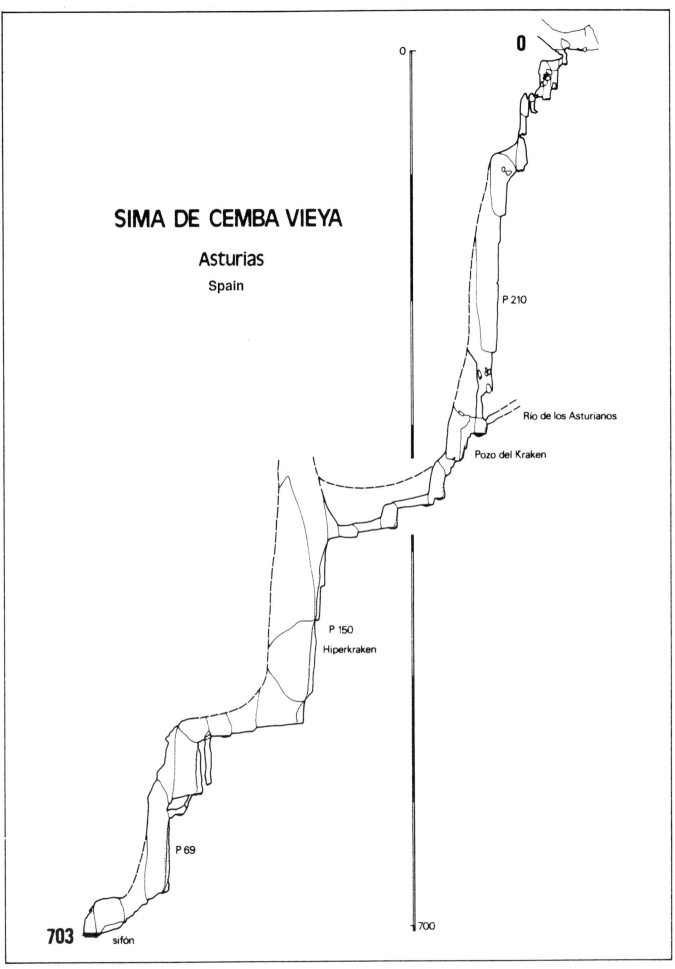

SIMA DE CEMBA VIEYA

Asturias

Spain

0

0

P 210

Río de los Asturianos

Pozo del Kraken

P 150

Hiperkraken

P 69

700

703

sifón

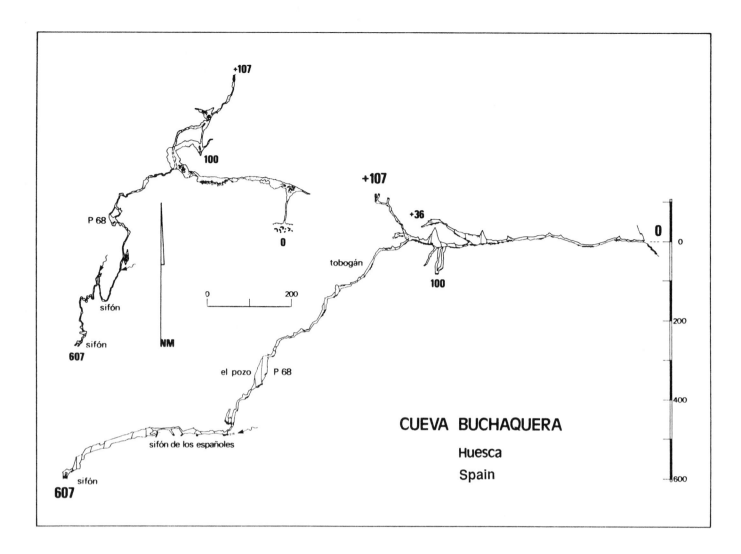

CUEVA BUCHAQUERA

Huesca

Spain

Explored in 1984 by York U.C.P.C. and S.E.I.I. de Madrid to –259 m. In 1985 the explorers reached a sump at –650 m. A side branch at –570 m leads to an undescended pit at –634 m (*Caves and Caving*, 1985 (30), pp 18-20).

23. **sistema Sara** ....(–635, +13) 648 m
(Cillórigo-Castro, Cantabria)

Located in the Andara region (eastern Picos de Europa), this system has two entrances, Sara II (alt. 1820 m) and Sara III (alt. 1833 m), which are connected by a mine gallery that is now blocked. The sima Tere (see above) is very close by. The resurgence is the cueva del Nacimiento (alt. 760 m) de Urdón. 4°42'14"; 43°12'25".

Lancaster U.S.S. descended to –210 m in 1977 and, with the S.E.I.I., reached the terminal sump at –635 m in 1978. The S.E.I.I. verified the connection with an upper entrance (Sara III) at the 281 m deep pit via a mine gallery (Puch, *op. cit.*, 1981, pp. 20-21, 54, 214; *Spelunca*, suppl. to N° 19, 1985, pp 54-55, profile).

24. **sistema Los Gorrinos-Prado La Fuente** ... –640 m
(Amieva, Asturias)

Located in the Cornión range (western Picos de Europa), this cave has two entrances (alt. 1515 m and 1420 m), the first in the Boca de les Abarques, the other on the slope overlooking the Canal de Ozania. The resurgence

(dye trace) is the fuente Prieta (alt. 865 m). 05°00'41"; 43°13'11" and 05°00'38"; 43°13'06".

The sima Prado La Fuente was discovered by the S. C. Orsay Faculté in 1978. They descended to –368 m, then –550 m in 1979 and 560 m in 1980 with the S. C. Aude. On August 28, 1981 the S.C.O.F. and the S.C.A. connected the two entrances, making a system 640 m deep and 3600 m long (P. Benoit, *Spelunca*, suppl. to N° 19, 1985, profile).

25. **sima del Jou Lluengu** .... –619 m
(Amieva, Asturias)

This cave is in the Cornión range, Picos de Europa, at an altitude of 1680 m, above the Canal de Ozania. Resurgence is the fuente Prieta (alt. 940 m). See the sistema Los Gorrinos above. 4°59'44"; 43°12'47".

The S.C.O.F. and the S.C.A. explored the cave in 1980 to –360 m approx. In 1981 they reached the terminal sump at –619 m (P. Benoit, in "Les Picos de Europa", *Spelunca*, suppl. to N° 19, 1985, profile).

26. **pozo Estella** (Isaba, Navarra) .... –614 m

Located in the lapiaz d'Añalarra, in the Pierre Saint-Martin area, at an altitude of 2063 m, not far from the French border (x 348.896; y 76.535). This cave is hydrologically connected to the Illamina system (alt. 442 m, see Laminako ateak above).

The French of the S. C. Frontenac discovered the cave in 1973, and explored it with the I.P.V. Pamplona in 1975 to –420 m. In 1976 the depth was brought to –550 m by the A.R.S.I.P., but a surprise flood in a pit killed F. Zamora. In 1977 the A.R.S.I.P. were stopped at –614 m (Courbon, *op. cit.*, 1979, profile; Puch, *op. cit.*, 1981, profile and plan).

27. sima del Llagu las Moñetas .... –613 m
(Cabrales, Asturias)

Located at an altitude of 1762 m, in the Urrieles (or Central) region of the Picos de Europa, SE of the Peña Castil (2444 m), in the cliffs above the small Las Moñetas lake (1°00'11"; 43°11'28").

It was explored in 1976 by the S. C. Alpin Languedocien to –180 m. In 1977, with other clubs from the Hérault, a first bottom was reached at –595 m, and a second one in 1981 at –613 m, following a 37 m aid climb done in 1979. This cave has a 148 m deep shaft ("Les Picos de Europa", *Spelunca*, supp. to N° 19, 1985, profile and plan).

28. sistema de la Torre de Altaiz................... –608 m
(Camaleño, Cantabria)

This system has three entrances: sima de la Torre de Altaiz (alt. 2090 m), I.25 (alt. 2045 m), and the sima del Jou de Lloroza (alt. 2062 m). It is located in the Urrieles range, SW of the collado de Fuente Escondida (1°08'30"; 43°09'55" and 1°07'50"; 43°09'50"). The Fuente Dé spring (alt. 1100 m) is the probable resurgence, but it could be one of the springs on the right bank of the Caín.

The first entrance (2090 m) was discovered in 1974 by the A.S. Charentaise who descended to –246 m, then –310 m in 1975. In 1978 they reached –345 m. The second entrance (2062 m) was discovered in 1973 and led the explorers to –330 m in 1975 and –443 m the following year. The provisional bottom (–554 m) was reached in 1982. In 1983 the A.S.C. explored a new system in the I.25-I.19 which connected with the Jou de Lloroza on one side and down to –583 m on the other side. The tight fissures at the bottom were pushed to yield another 25 m of depth in 1985 ("Les Picos de Europa", *Spelunca*, suppl. to N° 19, 1985, profile and plan).

29. avenc Badalona .... approx. –600 m
(Escalona, Huesca)

The avenc C.9 opens at an altitude of 2062 m, on the south face of the Sierra de las Sucas, between las Tres Marías (2809 m) and the Puntas Verdes (2621 m). The resurgence (dye trace) is the Fuente de Escuaín (alt. 1115 m). See the sistema Badalona above (x 261.68; y 4724.14).

This cave was discovered in 1970 by the G. E. Badalona. In August, 1974, they reached about –500 m, while about –600 m (sump) was reached in December 1974. In 1980 the C.E.A. de Zaragoza discovered a descending side passage that led to about –650 or –700 m in 1985. (Courbon, *op. cit.*, 1979, profile; Puch, *op. cit.*, 1981, pp. 21, 55).

30. avenc de las Grallas ....approx. –600 m
(Escalona, Huesca)

This cave (C.13), located in the same area as the C.9, was explored in 1978 to 296 m. In 1982 the explorers from the G.E.B. got through the terminal squeeze and descended to approx. –600 m. (*Exploracions*, 1982 (6), p. 136).

31. sima de los Puertos de Ondón ..... approx. –600 m
(Cabrales, Asturias)

This cave is located close to the area where the torca Tejera is found (see above), and is close to the sima del Frailín (–350 m), on the Cabezo de Llorosos (western Picos de Europa).

Discovered by the S.S.S. de Genève in 1983, who descended that year to 310 m. It was pushed in 1984 to about -600 m (*Exploracions*, 1984 (8), p. 122).

32. pozu Cebolleda (Amieva, Asturias) .... –597 m
Located between the sima del Jou Lluengu and the sistema Los Gorrinos-Prado la Fuente, in the SW side of the Cornión mountains in the Picos de Europa (north flank of the Muda de Ozania). This double entrance cave (alt. 1657 and 1640 m) belongs to the fuente Prieta (alt. 940 m) hydrological system (1°19'14"; 43°12'47").

The cave has three descending branches and was explored by the S.C.O.F. to –350 m in 1975 (the year of its discovery), and to a sump at –440 m and to –535 m in the "colector" in 1976. In 1977 a sump was reached at –580 m. In 1978 the discovery of a higher entrance added 17 m to the vertical extent of the cave (Puch, *op. cit.*, 1981, profile; P. Benoit in "Les Picos de Europa", *Spelunca*, suppl to N° 19, 1985, profile and plan).

33. sistema Jorcada Blanca - Las Perdices ... –590 m
(Onís, Asturias)

Also in the Cornión mountains (Picos de Europa), under the col of the same name, to the west of the pic La Verdilluenga (2129 m), it is located at an altitude of 1940 m (1°15'50"; 43°13'20").

The pozu Jorcada Blanca was discovered and explored in 1982 by the Oxford U.C.C. to –520 m (end of the vertical drops). In 1984 the British reached the terminal sump at –590 m. A lower entrance, the pozu las Perdices, was connected in 1984 to the active canyons at the bottom of the cave (*Caves and Caving*, 1983 (22), pp. 12-15, profile and plan; *Exploracions*, 1984 (8), p. 122).

34. sistema Sabadell .... approx. –580 m
(Hoz de Jaca, Huesca)

The cave A.21, or sistema Sabadell is located at an altitude of 2175 m at the base of a cliff, in the Canal de Tresarriú, on the NNW side of the Peña Roya (2589 m), near the Búbal dam, in the central Pyrenees (x 723.25; y 4729.54). Formed in sandstone, its resurgence may be the double spring of El Espumoso (1240 and 1305 m), or the fuentes del Saldo.

It was discovered and explored, after several digs, by the I.E.E. de Sabadell in 1978 and 1979. The announced a depth of approx 600 m, then 626 m (mapped). In 1982, the E.R.E. mapped the cave to –533 m, close to the bottom (*Espeleòleg*, 1982 (34-35), pp. 192-212, profile and plan; *Exploracions*, 1984 (8), pp. 111-119.

35. sistema de Ormazarreta .... –576 m
(Unión Aralar, Navarra)

This system has two separately explored caves that were connected in 1984: Ormazarretako Ieizea II (alt. 1205 m) and Larretxikiko Ieizea II (alt. 1176 m), located in the depressions on the NE sides of the Aparein butte, in the middle of the Navarra region of the Sierra de Aralar (x

578.453; y 4758.685 and x 577.565; y 4758.878). The resurgence is Aiaiturrieta, at Ataun.

Larretxiki II was explored in 1978 by the S. C. Aranzadi (Donostia), who reached a sump at –446 m. In 1981, the S.E.C.V.C. (Donostia) began the exploration of Ormazarreta II. The S.C.A. immediately took the initiative in the work, reaching the underground river and exploring down to –513 m the same year. In 1984 the terminal sump (–576 m) was reached. Shortly thereafter the S.C.A. discovered a short passage (climbs, squeezes) above the terminal sump in Larretxiki and popped out in the Ormazarretako Ibaia river at –500 m (I. Gockoetxea and K. Sansinenea, "El río subterráneo de Ormazarreta", *Príncipe de Viana*, suppl. de ciencias, 1982 (2), pp. 443-456; Puch, *op. cit.*, 1981, pp. 29-77).

36. **torca de las Pasades** .... –567 m
(Ruesga/Arredondo, Cantabria)
This cave is located in the Porracolina range, in the Alto del Tejuelo, near the col de las Pasadas (alt. 925 m) (x 444.835; y 4789.810).

Discovered in 1978 by the S. C. Dijon, who made a reconnaissance to –130 m. In 1979, the S.C.D. reached –430 m, and in 1980 a terminal sump stopped exploration at –589 m.

The mapping was redone shortly thereafter by the E. C. Gràcia de Barcelona, who gave the cave a depth of –567 m. The supposed resurgence is located in the valley of the Río Miera, at an altitude of 200 m (Puch, *op. cit.*, 1981, pp. 23, 61).

37. **sistema Félix Ruiz de Arcaute** .... –563 m
(Fanlo, Huesca)
This system has six entrances; sima Viva El (2778 m), sima Tormenta (2759 m), forca del Gabieto N° 17 (2750 m), N° 18 (2750 m), and N° 28 (2745 m), and the pozo de las Cornejas, all connected together in 1974-1975. Located in the Tallón-Marboré mountains, south of the Picos del Gabieto (3025 m) (x 740.600; y 4730.753 and x 740.900; y 4730.853). The probable resurgence is in the Circo de Cotatuero (alt. 2150 m).

The G. S. Pyrénées (Toulouse) explored the sima Viva El in 1973 to 265 m and the forca del Gabieto to –140 m. In 1974 the two caves were connected and the depth of –502 m reached, after a 165 m deep pit. It was in 1975 that the G.S.P. reached the terminal sump at –563 m and connected the pozo de las Cornejas to the system (*Ouarnède*, 1976 (6), 56 pp., profile and plan; Puch, *op. cit.*, 1981).

38. **Gazteluko Urzuloa I** .... approx. –560 m
(Arechavaleta, Guipúzcoa)
The cave is located above the valley of Gaztelu, in the depression of Degurixa, at an altitude of 980 m, in the Andarto Mendi mountains, not far from the Gaztelu Arroko Ieizea III (–444 m) (1°12'05"; 42°58'16"). The resurgence (dye trace in 1971) is at the manantial de Saratxo (alt. 470 m), near Araoz.

In 1971 the S. C. Aranzadi, led by F. Ruiz de Arcaute, found the cave and descended to –45 m. In 1973 and 1974, the Aloña Mendi E. T. reached –100 m, then –360 m in 1975. A sump was reached by the same group in 1977, helped by the G. E. Esparta de Baracaldo. In 1985 the A.M.E.T. bypassed the sump and descended a pit about 60

m deep. Exploration in progress (Puch, *op. cit.*, 1981, profile and plan).

39. **sistema del Mortero de Astrana**
(Soba, Cantabria) .... (–530, +22) 552 m
The system has five entrances: torca del Mortero (720 m), sumidero del Río Leolorna (743 m), sumidero del Río Cubieja (814 m), sima de Cuesta Cuivo (838 m), and a lapiaz fissure (M.14) (860 m). It is located on the side of the Mortillano mountains (Peña del Cuivo (0°06'59"; 43°12'32" for the Mortero).

The S. C. Dijon in 1961 made a first reconnaissance of the Mortero (where dead animals are tossed) to –141 m. In 1962 the S. S. Bourgogne and the F. J. de Santander stopped at –168 m at the top of the large 178 m deep pit. This was descended for 120 m by the S. C. Paris in 1963. In 1964, the S.S.B. and the F.J. reached –367 m, just before a sump. In 1975 the S.E.C.E.M. (Manresa) descended to –380 m (losing the river) and mapped the bottom section, the tributary coming from the Río Cubieja, and the higher galleries. In 1978 the C.E.A. (Zaragoza) made a complete map (–351 m / 1932 m). From 1976-1979, J. P. Combredet (S. C. Paris) organized several expeditions with clubs from Paris and Lille, deepening the cave and connecting with the upper entrances, making for a system with a vertical extent of –552 m and a length of 7228 m (Puch, *op. cit.*, 1981, pp. 25-26, 65, plan and profile).

40. **sima T.38** (Torla, Huesca) .... –552 m
Located in the Sierra Tendeñera, on the east side of the Pico Fênez, near the summit, higher than the sistema Arañonera (see above), at an altitude of 2490 m (0°08'58"; 42°40'24").

Discovered in 1974 by the E.R.E. and the G.I.E. de Barcelona, these two clubs descended to –60 m (1974), –200 m (1976), and –519 m in 1977, to a tight squeeze. This became impenetrable at –552 m (1978) (*Espeleòleg*, 1981 (32), pp. 117-120; Puch, *op. cit.*, 1981, pp. 26, 66, profile and plan).

41. **sistema del Lago de Alba** .... –530 m
(Benasque, Huesca)
Has two entrances: the cueva de Alba, located above the Baños de Benasque (Río Esera valley), and El Bujerín, located near Alba lake, 530 m higher.

In 1983 the G.E.R.S.A.E.M. (Barcelona) undertook the mapping of the long known cueva Alba and, after a climb of 20 m, discovered a strongly flowing river (0.5 to 0.8 m³/s). The explorations followed one another and 300 m of vertical gain were soon achieved. A small blowing hole was discovered higher up on the mountain in 1984 and the explorers quickly descended 250 m to a large breakdown blockage. At the end of the year the obstacle was passed and a new traverse of –530 m and 2.5 km was opened in the Pyrenees The present length is over 5 km (*Carbonato*, 1984 (4), p. 71, profile; *Exploracions*, 1985, (9), p. 130).

42. **grallera del Puerto de Gistáin** .... –508 m
(Benasque, Huesca)
Also called pozo Luluna. Located between the Gistáin col and the Pico Royo lake, about 50 m below the crest, at an altitude of 2695 m, next to four other caves which together play the role of sub-glacier insurgences (Nevero,

cueva de Plata, sumidero del Puerto de Gistáin, and PR.4) (x 290.050; y 4728.125). The resurgence should be to the east, near the Astos valley, some 700 or 900 m lower. Discovered in 1980 by the S. C Aude, who reached –45 m that year. In 1981 the explorers reached –145 m and –480 m in 1982. A year later, a flooded pit was found to block the way at –508 m. (C. Bès, *Les Posets, Rapp. d'Expéd.* - S.C.A., 1983, 32, pp., *Spéléoc*, 1983 (23)).

43. **morterón II del Hoyo Salzoso** .... –505 m
(Soba, Cantabria)
At an altitude of 880 m in the Mortillano range, in a depression in the valley of Los Trillos (0°07'45", 43°14'47"). The probable resurgence is the manantial de la Punta del Praduco (alt. 180 m), on the right bank of the Río Asón.
The entrance was discovered by F. Chavarria and D. Vergés, who unblocked it. The first three pits were descended to –150 m (date not specified). From May to July 1979 the O. Escolà and the E.R.E. (Barcelona), with the S.E.G.M. Gelera and the S.I.S. de Terrassa, reached –505 m in three trips, stopping at a tight spot. The cave is essentially vertical (Puch, *op. cit.*, 1981, plan and profile).

44. **avenc la Bufona** (Escalona, Huesca) .... –501 m
As with the C.9 (see above), this cave belongs to the hydrologic system of the fuente de Escuaín (alt. 1115 m) (see sistema Badalona above), the avenc opening at 1820 m, downstream from the C.9, in the sierra de Las Sucas (x 261.64 ; y 4723.5).
The G. E. Badalona discovered the entrance in 1971 and reconnoitered the first pits down to –180 m. In 1973 an underground stream was discovered and about –400 m was reached. A sump put a stop to exploration in 1976 (remapped by the S.I.E. de Barcelona in 1979, *EspeleoSie*, 1981 (25), profile and plan).

45. **red del Río Silencio** .... –492 m
(Rasines, Cantabria)
See below.

46. **torca de Jornos II** .... –489 m
(Aramñón range, Carranza, Vizcaya)
alt. 710 m (Puch, *op. cit.*, 1981, profile).

47. **Aitzondoko Ieizea** .... –488 m
(Izarraitz range.Azcoiti, Guipúzcoa)
(*Carbonato*, 1983 (3) profile).

48. **pozo Toneyo** .... –485 m
(s. de Beza, Soto de Sajambre, León)
Alt. 1395 m (*Lo Bramavenc*, 1984 (8)).

49. **sistema del Hoyo Grande** .... –471 m
(Colina range, Soba, Cantabria)
Alt. 1210 m. In 1985 by the S. C. Chablis.

50. **torca L.M. 7** ....–458 m
(Porracolina range, Ruesga/Arredondo, Cantabria).
Alt. 920 m. (*Exploraciones*, 1983 (7)).

51. **sima FREU.3-FR.3** .... –455 m
(Larra, Isaba, Navarra)
Alt. 2085 m and 2050 m (*Bull. ARSIP*, 1974/76, partial profile).

52. **Iñeritzeko Ieizea** .... –454 m
(Illuntzar range, Nabárniz, Vizcaya)
Alt. 643 m (*Kobie*, 1977 (7) profile).

53. **torca de los Morteros-cueva de Imunia** ....(–415, +38) 453 m
(m. Imunia, Espinosa de los Monteros, Burgos)
Alt. 1295 m (Puch, *op. cit.*, 1981, profile).

54. **pozu Conjurtao** .... –452 m
(Cornión range, Onís, Asturias)
(*Caves and Caving*, 1985 (30), profile).

55. **cueva del Vierro** .... –449 m
(Andara range, Cabrales, Asturias)
Alt. 1610 m (*Picos de Eur.*, *20 ans de spéléo.*, S.C.A.L., 1985, profile).

56. **Gaztelu arroko Ieizea III** .... –444 m
(Andarto range, Aretxabaleta, Gulpúzcoa)
Alt. 930 m (Puch, *op. cit.*, 1981, profile).

57. **torca de los Corrales del Trillo** .... –440 m
(Peña Lusa, Soba, Cantabria)
Alt. 1320 m (Puch, *op. cit.*, 1981, partial profile).

58. **sima de la Torre de los Traviesos** .... –435 m
(Cornión range, Cangas de Onís, Asturias)
Alt. 2180 m (Puch, *op. cit.*, 1981, profile).

59. **sima grande de la Torrezuela** .... –430 m
(Cornión range, Amieva, Asturias)
Alt. 2200 m (*Picos de Eur. Amieva 1984*, S.C.O.F., 1985, profile).

60. **sima AN.8** (Larra, Isaba, Navarra) .... –420 m
Alt. 1790 m (*Karstologia*, 1985 (6)).

61. **cueva de la Haza** .... –418 m
(Colina range, Soba, Cantabria)
Alt. 1240 and 1180 m (Puch, *op. cit.*, 1981, plan).

62. **sima de la Porra Altiquera** .... –418 m
(Cornión range, Cangas de Onís, Asturias)
Alt. 1800 m (Puch, *op. cit.*, 1981, profile).

63. **sima de la Punta de las Olas** ......(–415, +2) 417 m
(Perdido range, Fanlo, Huesca)
Alt. 3009 and 2990 (Puch, *op. cit.*, 1981, profile).

64. **sima de la Tartracina-cueva del Fraile** .... –415 m
(Marboré range, Fanlo, Huesca)
Alt. 2570 and 2520 m (*Marboré 1984*, G.S.P., 1985, profile).

65. **Engolidor de las Foyas** .... –409 m
(Las Foyas, Ansó-Fago, Huesca)
Alt. 1720 m (*Espeleolèg*, 1985(36) profile).

66. **pozo de Frontenac** or **AN.6** .... –406 m
(Larra, Isaba, Navarra)
Alt. 2160 and 2145 m (Puch, *op. cit.*, 1981, profile).

67. **torca del Hoyo Medio** .... –405 m
(Alto de las Minas, Rasines, Cantabria)
Alt. 562 m (J. Léon, *Las Grandes Cavid. de Cantab.*, 1986, in press).

68. **sima de la Horcada Verde** .... –402 m
(Urrieles range, Camaleño, Cantabria)
Alt. 2200 m (*Picos de Europa*, A.S.C. 1971-1972, A.S.C., 1983, profile).

69. **sima del Marboré** .... –401 m
(Marboré range Fanlo, Huesca)
Alt. 2920 and 2870 m (from 1975 to 1982 by the G. S. Pyr.).

70. **sima de la Hoya del Portillo de Larra** ...... –400 m
(La Pakiza, Ansó-Fago, Huesca)
Alt. 1767 m (*Bull. ARSIP*, 1977/80 (12/15) profile).

71. **sistema de la Cuvada Grande** .... –400 m
(Valnera range, Espinosa de los Monteros, Burgos)
By the G. E. Edelweiss (Burgos) in 1984-1985.

———————

132. **pozo las Palomares** .... –300 m
(Sierra de Beza, Soto de Sajambre, León)
Note: the **cueva del Viento** (Icod de los Vinos, Tenerife, Canarias) is a lava tube in three segments, the two deepest being 261 and 217 m deep.

LONG CAVES:

1. *ojo Guareña*......................................88,907 m
(Merindad de Sotoscueva, Burgos)
This system unites the following entrances: cueva Palomera, sima Dolencias, sima de los Huesos, cuevo Cornejo, cueva de los Cuatro Pisos, cuevas de San Bernabé, cueva del Moro, cueva la Mina, sumideros del Trema, and sumidero (or "Ojo") del Guareña. It is located north of Burgos, in the Merindad de Sotoscueva territory. It is developed in the limestones and dolomites of the upper Coniacian. The waters of the río Guareña and those of the río Trema partially pass through the system and resurge at La Torcona, near the confluence of the Trema and La Hoz.
1/50,000 map N° 84 (x 602.4, y 936.5, z 740 m (cueva Palomera) and x 604.1, y 936.65, z 750 m (sima de los Huesos).
The cave has essentially always been known (vestiges of the Neolithic age, the Bronze age, and the Celtic periods are evident). Its systematic exploration began only in 1956, led by the G. E. Edelweiss de Burgos. In 1957, with the S. C. Aranzadi (Donostia) and the G. E. Vizcaíno (Bilbao), the second and third levels were discovered. In 1958 the international expedition "O.G.58" brought together five Spanish clubs and groups from five foreign countries: 12,000 m were mapped. In 1963 the "O.G.63" expedition was organized (with seven Spanish clubs), bringing the length to 16,100 m. The 1964 "O.G.64" expedition had the same clubs plus four foreign groups and brought the length to 21,550 m. The explorers from the G.E.E. and groups from Burgos, Córdoba, Sabadell, and others reached 25,430 m in 1967. In 1968, fourteen Spanish groups participated in "O.G.68" to reach 32,056

m. In 1972 the length reached an estimated 50 km. The offical length in 1973 was 46,700 m.
Beginning in 1975 the G. E. Edelweiss undertook a complete systematic survey of the cave, reaching 57 km by 1977 (with about 66 km explored). Their efforts brought the length to 61 km in 1978 and 68 km in 1981. That year also saw the connections with the cueva la Mina and the sumideros del Trema. The sumidero del Guareña was connected in 1985, and the mapped length was increased to 88,907 m.
Map: G. E. Edelweiss (Gr 5D).
Bibliography: G.E.E., *Kaite* 1986 (4-5) (monograph), Burgos, 415 p.
Puch, *op. cit.*, 1981, pp 125-126, 140, 151-181.

2. *Red del Río Silencio* ....53,000 m
(Rasines, Cantabria)
This sytem has five entrances: torca de los Caballos (alt. 550 m), cueva-sima del Escobal (alt. 193 m), Joyu Hondo II (alt. 180 m), torca de la Canal (alt. 160 m), and cueva del Valle (alt. 58 m) (resurgence). The first entrance is in the locality called El Hoyón (a vast doline), near to las Peñas Surbias. The resurgence is located at La Mies, not far from Helguera.
1/50,000 map 20-25. x 470.768; y 4791.610 (Caballos) and x 466.090; y 4794.260 (Valle).
The cave is formed in Urgonian limestones (lower Aptian-Albian).
The entrance to the torca was discovered in 1977 by the S. C. de la M.J.C. de Rodez (France) who descended to –189 m. In 1978, the same group explored the torca to a sump at –404 m (length 3000 m) and the cueva for about 4 km. In 1979, with the G.A.E.S. and the C.A.S. de Bilbao, 13 km were explored in the cueva. In 1980, the same groups connected the two caves, bringing the length to 20,455 m and the vertical extent to 492 m. In 1981, the length reached 30,457 m, then 36,126 m in 1982 and 45,000 m in 1983. In 1985 the length reached 53 km.
Map: S. C.-M.J.C.R., G.A.E.S., C.A.S.
Bibliography: Puch, *op. cit.*, 1981, pp. 28, 73-74.

3. *sistema de la Piedra de San Martín* ....52,000 m
(Isaba, Navarra)
See France.

4. *sistema Cueto-Coventosa-Cuvera* ....28,000 m
(Arredondo, Cantabria)
See above.

5. *sima del Hayal de Ponata* ....27,000 m
(Orduña, Vizcaya-Ayala (Alava))
Also called SI.44. Located at an altitude of 995 m on the NE side of the Sierra Salvada After a short length underneath Vizcaya, it is developed along an EW fault in the province of Alava. 1/50,000 Map N° 86. x 493.144; y 4762.031.
The cave is formed in the limestones of the middle-upper Coniacian.
Discovered in 1983 by the G.E. Alavés (Vitoria-Gasteiz), who discovered 13 km that year. The lengths of 17 km and 24 km were reached in 1984 and 1985 respectively (with the G.E. Edelweiss de Burgos). In early 1986 the G.E.A discovered the continuation in the principal river

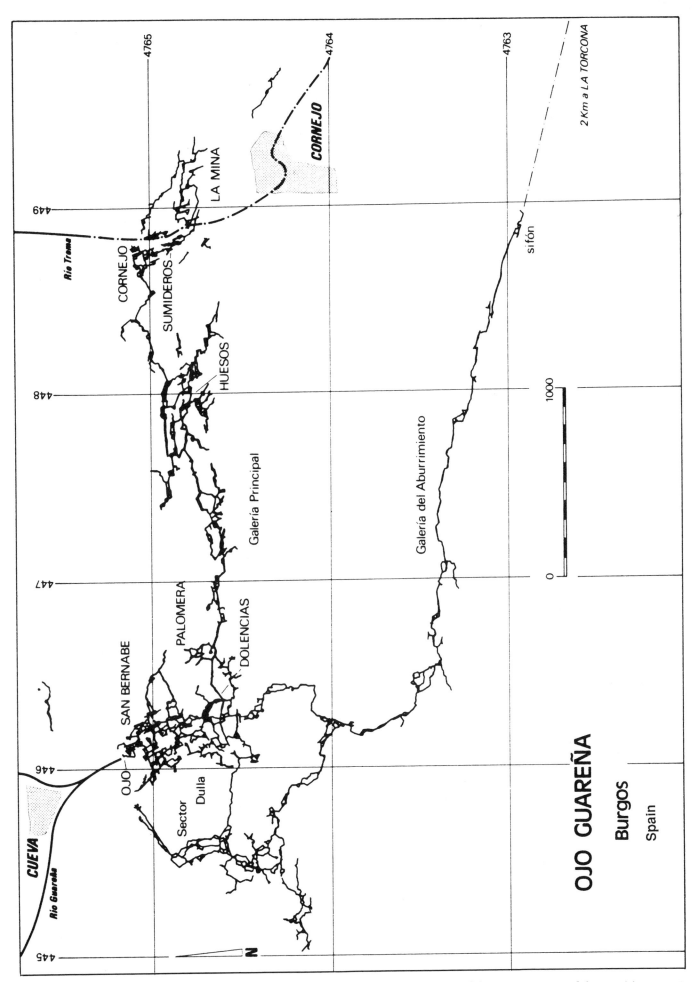

OJO GUAREÑA

Burgos

Spain

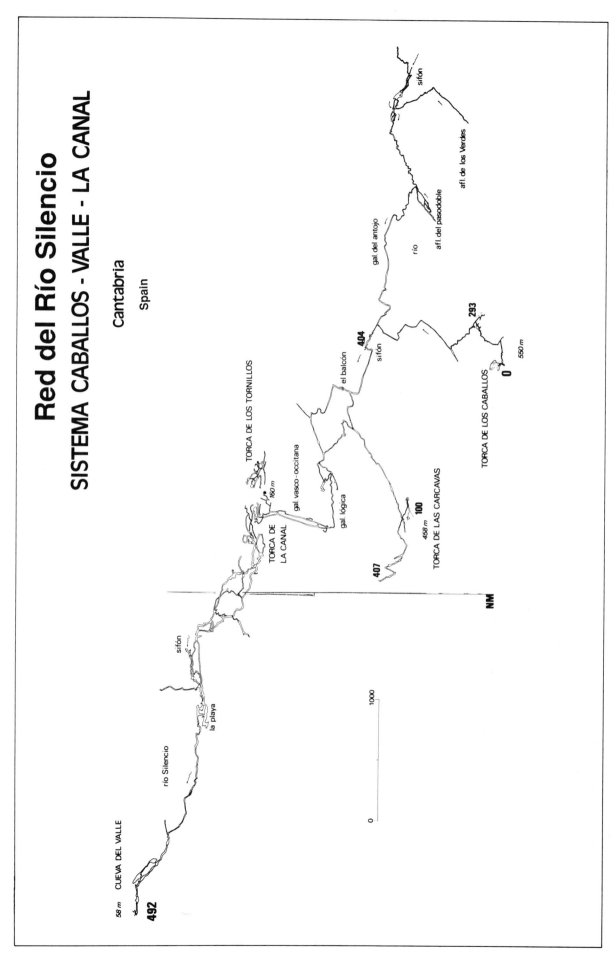

Red del Río Silencio
SISTEMA CABALLOS - VALLE - LA CANAL

Cantabria

Spain

CUEVA DEL VALLE

492

58 m

río Silencio

la playa

sifón

TORCA DE LOS TORNILLOS

TORCA DE LA CANAL

gal. vasco - occitana

160 m

gal. lógica

el balcón

gal. del antojo

sifón

rio

afl. del pasodoble

afl. de los Verdes

sifón

404

293

0

550 m

TORCA DE LOS CABALLOS

100

458 m

TORCA DE LAS CARCAVAS

407

NM

0

1000

passage and reached 27 km, for a depth of 220 m. Most of the cave is developed along a principal axis, following a large fault for over 7 km linear distance, with several ups and downs. Several rivers flow through it.

Bibliography: J. M. López de Ipiña and F. Alangua. "La SI.44, una neuva gran red subterranea en el karst de Sierra Salvada (Alava)", *Kultura*, 1985 (6), 9 pp.

6. ***sistema Azpilicueta-Coterón-Reñada*** ...19,349 m
(Ruesga, Cantabria)

This sytem resulted from the connection of three different caves: torca de Azpilicueta (alt. 475 m), torca de Coterón (alt. 370 m), and cuvío de la Reñada (alt. 180 m), formed within the Beralta and Limón mountains, at La Vega (Matienzo). The río Clarión flows from the Reñada resurgence. It is developed in the lower Aptian-Albian series (limestones, sandstones, marls of Urgonian age).

1/50,000 Map N° 59. x 450.400; y 4794.980 (Azpilicueta) and x 450.450; y 4795.770 (Reñada).

In 1964 and 1965 the S.E.S.S. (Santander) began to explore Reñada and its active resurgence (Comediante). In 1973, Manchester University S.S. discovered the continuation and explored 2 km, then 4 km in 1974 and 5 km the following year. The torca de Coterón was discovered in 1981 (length 3482 m). The connection was made in 1982, making a 13 km long system. The torca de Azpilicueta was descended the same year to –263 m. Finally, in 1985, the British connected the vertical cave to the horizontal galleries of the system, bringing the cave to over 19 km long for a vertical extent of 300 m.

Bibliography: J. Corrin, *The Matienzo 1981 and 1982 Expeditions* 1983, 52 pp., plan.

7. **cueva de los Chorros**
(Riópar, Albacete) .... 16,072 m

This resurgence cave is located halfway up the cliff on the northern side of the Calar del Mundo, at an altitude of 1122 m. It is the beginning of the río Mundo, tributary of the Segura. (1/50,000 map N° 23-24 x 549.2; y 4256.3. Lower Cretaceous). The C. E. Alcoi mapped 2360 m in the cave in 1965 and 7158 m in 1966. In 1977 the C. E. Alicante and Alcoi passed the terminal sump and explored 2 km of passages. In 1978, the length reached 12 km, then 14,477 m in 1979 and 16,072 m in 1983. Unconfirmed reports indicate the current total may be as high as 40 km! (*Lapiaz*, 1981 (7), monog., 48 pp., plan and profile).

8. **cueva de Uzueca** .... 15,845 m
(Solórzano, Cantabria)

Located near Riaño (x 451.950 ; y 4800.020; z 175 m). Explored beginning in 1976 by the M.U.S.S. (J. Corrín, *op. cit.*, 1983, plan).

9. **sistema del Hoyo Grande** .... 15,300 m
(Soba, Cantabria)

See above. Located above the val de Asón, with three entrances: torca del Hoyo Grande (alt. 1210 m) and sumideros de Saco N° 2 and 4. It was explored by the S. C. Dijon (1972 to 1975) and the S. C. Chablis (beginning in 1974), who made the connection in 1981 for a length of 11 km. The length reached 15.3 km in 1985.

10. **cueva de Huertas** (Teverga, Asturias) ...... 14,500 m

Also called cueva de Fresnedo. Explored by the G. E. Polifemo de Oviedo in the 1970's (Puch, *op. cit.*, 1981, plan).

11. **cueva del Soplao** .... 14,500 m
(Rionansa, Cantabria)

Located in the La Florida mines (Valdáliga). Explored from 1976 by the E. C. Cántabro (Santander) (*Bol. Cánt. de Espeleo.*, 1982 (2), plan).

12. **cueva del Rescaño** (Udias, Cantabria) .... 13,500 m

Also called cueva del Río. Located near Cobijón. Explored by the S.E.S.S, the G. E. Montañés (Santander), and the G.A.E.S. (Bilbao). ("El Karst de Udias", *Bol. Cánt. de Espeleo.*, 1983 (3), article, plan).

13. **cueva de Mairuelegorreta** .... 12,340 m
(Cigoitia, Alava)

Explored by the G. E. Manuel Iradier, 1957-58 (9 km) and the G. E. Alavés, 1962, 1964, and 1968 (Puch, *op. cit.*, 1981, plan).

14. **cueva Fresca** (Soba, Cantabria) .... 12,000 m

Explored by the S. C. Dijon (Puch, *op. cit.*, 1981, plan).

15. **Laminako ateak** (Isaba, Navarra) .... 11,900 m
See above.

16. **cueva del Piscárciano** .... 11,000 m
(Hoz de Arreba, Burgos)

By the G. E. Niphargus (Burgos). (Puch, *op. cit.*, 1981, partial plan).

17. **cueva del Tornero** .... 11,000 m
(Checa, Guadalajara)

By the G. E. Minas (Madrid). (Puch, *op. cit.*, 1981, plan).

18. **cueva del Nacimiento** .... 10,000 m
(Cillórigo-Castro, Cantabria)

Or cueva del Agua. Located near Tresviso, in the Eastern Picos de Europa. Explored by the L.U.S.S. (Puch, *op. cit.*, 1981, plan).

19. **sistema Garma Ciega-Cellagua** .... 9226 m
(Soba, Cantabria)
See above.

20. **cueva de la Cañuela** .... 8965 m
(Arredondo, Cantabria)
(Puch, *op. cit.*, 1981, plan).

21. **torca de los Morteros-cueva de Imunía** ...8800 m
(Espinosa de los Monteros, Burgos)
See above.

22. **cueva de la Vieya-los Quesos**
(Cabrales, Asturias) .... 8500 m
(*Espeleología Asturiana*, 1981 (5), article, plan).

23. **cueva del Reguerillo** (Patones, Madrid) .....8268 m
(Puch, *op. cit.*, 1981, plan).

24. **torca del Sedo-la Cuevona-torca de Oñite** (Ruesga, Cantabria) .... 8250 m
("Matienzo", B.R.C.A. *Transactions*, 1981, 8(2) plan).

25. **solencio de Bastarás** (Panzano, Huesca) ... 8243 m
(Puch, *op. cit.*, 1981, plan).

26. **sistema Errekaseku** (Cogoitia, Alava) .... 8000 m
(Puch, *op. cit.*, 1981, p. 133).

27. **cueva de las Breveritas** .... 7922 m
(Icod de los Vinos, Tenerife, Canarias)
(*Cascade Caver*, 1978, 17 (1-2), plan).

28. **complejo Hundidero-Gato** .... 7818 m
(Montejaque/Benaoján, Málaga)
(Puch, *op. cit.*, 1981, partial plan).

29. **sistema del Xitu** (Onís, Asturias) .... 7500 m
See above.

30. **cueva de Don Justo** .... 7323 m
(Frontera, Hierro, Canarias)
(Puch, *op. cit.*, 1981, plan). Lava tube.

31. **sistema del Mortero de Astrana** .... 7228 m
(Astrana de Soba, Cantabria)
See above.

32. **cuevas Sopladoras-cueva del Agua** .... 7010 m
(Soba, Cantabria)
(Puch, *op. cit.*, 1981, plan).

33. **Arrikrutz Koba** (Oñati, Guipúzcoa) .... 7000 m
(Puch, *op. cit.*, 1981, plan).

34. **cueva Cullalvera-torcas Humizas** .... 6750 m
(Ramales, Cantabria)
(Puch, *op. cit.*, 1981, plan, *Bol Cánt. de Espel.*, 1984
(5) partial plan).

35. **sistema Arañonera** (Torla, Huesca) .... 6505 m
See above.

36. **sima S.1-S.2** (Linás de Broto, Huesca) .... 6505 m
See above.

37. **cova do Rei Cintolo** (Mondoñedo, Lugo) .... 6500 m
(Puch, *op. cit.*, 1981, partial plan).

38. **cueva de la Lastrilla** .... 6364 m
(Castro-Urdiales, Cantabria)
(Puch, *op. cit.*, 1981, plan).

39. **Otxabide** (Zubiaur, Vizcaya) .... 6200 m

40. **sistema de la Cuvada Grande** .... 6000 m
(Espinosa de los Monteros, Burgos)
See above.

41. **Ormazarretako Ieizea II - Larretxikiko Ieizea II** (Aralar, Navarra) ..... 6000 m
See above.

42. **cueva de la Vegalonga** .... 5900 m
(Tameza, Asturias)
(*S.U.S.S. Journal*, 1978, 2 (6), partial plan).

43. **sumidero de Monticueva** .... 5850 m
(Voto, Cantabria)
(J. León, *op. cit.*, 1986).

44. **sima 56** (Cillórigo-Castro, Cantabria) .... 5700 m
See above.

45. **pozo Estella** (Isaba, Navarra) .... 5500 m
See above.

46. **cueva del Tobazo** .... 5500 m
(Valderredible, Cantabria)
(*Mesetaria*, 1984 (1), plan).

47. **cueva del Linar** .... 5473 m
(Alfoz de Lloredo, Cantabria)
(Puch, *op. cit.*, 1981, plan).

48. **cova Cuberes** (Pallars Jussà, Lérida) .... 5438 m
(Puch, *op. cit.*, 1981, partial plan). Conglomerate.

49. **Aixako zuloa** (Deba, Guipúzcoa) .... 5000 m

50. **sistema de Alba** (Benasque, Huesca) .... 5000 m
See above.

51. **Pagolusieta** (Ceánuri, Vizcaya) .... 5000 m

52. **cueva del Puerto** (Calasparra, Murcia) ...... 5000 m

53. **sima de la Punta de las Olas** .... 5000 m
(Fanlo, Huesca)
See above.

54. **cueva François** (Soba, Cantabria) .... 4995 m
(Puch, *op. cit.*, 1981, plan).

55. **Gesaltza** (Oñati, Guipúzcoa) .... 4915 m
(Puch, *op. cit.*, 1981, plan).

56. **torca del Mostajo** (Ruesga, Cantabria) ...... 4839 m

57. **complejo Motillas-Ramblazo** .... 4751 m
(Jerez (Cádiz) / Cortes (Málaga)
(Puch, *op. cit.*, 1981, plan).

58. **sima del Marboré** (Fanlo, Huesca) .... 4744 m

59. **red del Carrillo** (Soba, Cantabria) .... 4550 m
(Puch, *op. cit.*, 1981, plan).

60. **cueva de Nerja** (Nerja, Málaga) .... 4537 m
(*Spes*, 1985(4), article, plan).

61. **sistema Fraile-Tartracina** .... 4506 m
(Fanlo, Huesca)
See above.

62. **cueva de San Miguel el Viejo** .... 4400 m
(Ayala, Alava)
(Puch, *op. cit.*, 1981, plan).

63. **cueva del Búho** .... 4392 m
(Puente Viesgo, Cantabria)
(*Exploracions*, 1982 (6), plan).

64. **cueva Honda** (Ampuero, Cantabria) .... 4376 m
(*Bol. Cánt. de Espel.*, 1982 (3), plan).

65. **cueva de las Lamiñas-cueva de Armiña** 4300m
(Ondárroa, Vizcaya)

66. **cueva de la Haza** (Soba, Cantabria) .... 4290 m
(*Cuadernos Espel.*, 1975, XIII (8), plan).

67. **cueva de Riaño** (Solórzano, Cantabria) ...... 4265 m
(*B.R.C.A. Transactions*, 1981, 8(2), plan).

68. **torca del Hoyu las Muñecas** .... 4200 m
(Peñamellera Baja, Asturias)
Spelunca, 1984 (14), plan).

69. **cueva de Fuentemolinos** .... 4086 m
(Puras de Villafranca, Burgos)
(Puch, *op. cit.*, 1981, plan). Conglomerate.

_____

104. **cueva de los Peines** .... 3000 m
(Castro Urdiales, Cantabria)

NOTE: The cueva de los Verdes (Haria, Lanzarote, Canarias) is a lava tube system with a total length of 6100 m, the longest segment being 2565 m long.

SWEDEN
Sverige

Sweden, in spite of unfavorable geographic and climatic conditions, has the characteristic of possessing both karstic caves and non- or pseudo-karstic caves in equal measure. The limestone areas are spread over the totality of the country, but those with the largest caves are located in Northern Lappland and in the Gotland. The most remarkable non-karstic caves are the boulder caves formed in gneiss and granite.

The relatively large speleological knowledge of this country was due to the efforts of the great Swedish speleologist, Leander Tell, who died in 1980. He began his prospecting in 1920 and published the important *Arkiv för Svensk Grottforskning* beginning in 1961. On May 28, 1966 he founded the Sveriges Speleolog-Förbund, which today continues his work. The S.S.F. regularly puts out the bulletin *Grottan*. For French readers, the thesis of Jean Corbel, *Les Karsts du Nord-Ouest de l'Europe* (1957) is a valuable reference.

DEEP CAVES:

1. **Voitasgallo** .... − 140 m
(Vadvetjåkko Nat. Park, N. Lappland)
Alt. 1000 m. Discovered in 1978 and mapped in 1981 by the S.S.F. (*Grottan*, 1980 (1); 1984 (3); *Svenska Grottor*, 1981 (4)).

2. **Lämmelhålet** (Vadvevagge, N. Lappland) .. − 120 m
Alt. 1000 m. Discovered in 1978 and mapped in 1981 by the S.S.F. (same references).

3. **Sotsbäcksgrottan** .... − 110 m
(Vindelfjällen Nat. Res., S. Lappland)
Alt 850 m. In 1969 by the S.S.F. (*Svenska grottor*, 1981 (4)).

4. **Lullehatjårrogrottan** .... − 110 m
(N. Torne Träsk, N. Lappland)
Alt 760 m. In 1954 by G. Rasmusson (*Svenska Grottor*, 1981 (4)).

LONG CAVES:

1. **Lummelundagrottan** .... 3100 m
(Lummelunda, Gotland)
Explored in 1950, 1953-1961 by L. Tell, 1964 by C. -F. Lundevall and G. Olsson, 1972-1985 by the S.S.F. Extensions found since 1980.

2. **Bodagrottorna** (Iggesund, Gävleborg) .... 2606 m
Boulder cave (gneiss), mapped in 1982 by Alf Siden.

3. **Korallgrottan** (Frostviken, Jämtland) .... 2200 m
In 1985 by Torbjörn Doj and the S.S.F. Connected to Skymningsgrottan in 1986(*Grottan*, 1986, 1, plan).

4. **Labyrintgrottan** .... 2100 m
(Vindelfjällen Nat. Res., S. Lappland)
From 1966 to 1976 by the S.S.F. (*Grottan*, 1976 (2) map with 1800 m).

5. **Kåppashåla** (Björkliden, N. Lappland) .... 1800 m
In 1983 and 1984 by H. G. Karlsson and the S.S.F. (*Grottan*, 1983 (4); 1984 (3)).

6. **Sotsbäcksgrottan** .... 1750 m
(Vindelfjällen Nat. Res., S. Lappland)
In 1969 by Y. Freij and E. Johansson (*Svenska Grottor*, 1981 (4)).

7. **Lullehatjårrogrottan** .... 1300 m
(Torne Träsk, N. Lappland)
In 1954 by G. Rasmusson (*Svenska Grottor*, 1981 (4)).

8. **Hoppet** (Vadvevagge, N. Lappland) .... 1000 m
In 1983 and 1984 by A. Lindén and R. Århlin (*Grottan*, 1984 (3)).

9. **Hölickgrottan** (Hornslandet, Gävleborg) ...... 907 m
Boulder cave (gneiss) mapped in 1984 by Alf Sidén.

10. **Voitasgallo** .... 800 m
(Vadvetjåkko Nat. Park, N. Lappland)

11. **Marmorlabyrinten** .... 570 m
(Stipok, Padjelanta Nat. Park, N. Lappland)

12. **Lämmelhålet** (Vadvevagge, N. Lappland) ..... 550 m

13. **Strångbergsgrottan** (Gällö, Jämtland) .... 510 m
Fissure cave (granite).

14. **Örnnästet** (Iggesund, Gävleborg) .... 503 m
Boulder cave (gneiss).

15. **Nedre Kåppasjokkgrottan** .... 455 m
(Björkliden, N. Lappland)

16. **Östra Jordbäcksgrottan** .... 450 m
(Kåtaviken, S. Lappland)

SWITZERLAND

Suisse/ Schweiz/ Svizzera

Switzerland, with a surface are of only 41,295 km², is nonetheless rich in limestone ranges. It is a country of high mountain karsts, found principally in the extension of the French Prealps, the Western Oberland, Diablerets (Vaud), Glarner Alpen (with the Sivellen), and Muotathal (Schwyz). These have been explored relatively recently (beginning in about 1970). The first karst regions explored were those in the NW of the country, covering a large portion of the French speaking cantons (Vaud, Valais, Jura, Neuchâtel).

Among the few known documents relating to early exploration, one must cite the description of the grotte de Môtiers by Jean-Jacques Rousseau in 1763 (read Raymond Gigon, *Inventaire spéléologique du canton de Neuchâtel*, 1976, 224 p.) and the works of the Genevan scientist Horace-Benedict of Saussure.

In the 19th century, the caves of the Jura mountains piqued the curiosity of several isolated explorers (L. de Buch, Ebel, Pictet, Thury, Hugi, Desor...). As with most European countries, most of the speleological work to date has taken place in the 20th century. The club of the "Boueux" was founded in 1930 in Geneva and had G. Amoudruz and E. Buri for members (read Jean-Jacques Pittard, *Explorateurs de l'ombre* (1965), 176 p.) and was principally interested in the Jura caves. The "Boueux" grew into the Swiss Speleological Society (S.S.S. or S.G.H.) when, at the beginning of the second world war, some of the members were asked by the Swiss army to be part of a mountain patrol group. Georges Amoudruz was the first president of the S.S.S.

H. Stauber put out a list of 478 Swiss caves in 1936, while J. Schnoerr began an inventory for the canton of Neuchâtel. This work, pursued by Maurice Audétat, was extended to all of French-speaking Switzerland. Still in the 1930's, the group Amis de la Nature explored the caves in the canton of Vaud under the guidance of Chenuz.

The German-speaking cavers began earlier, but confined themselves to a more limited range: the Nidlenloch (Kormann in 1895) which in 1909 became the deepest known cave in the world (–376 m), or the Hölloch in which Aloïs Ulrich began exploration in 1875, to be followed later by M. Widmer and P. Egli.

After the second world war the pace of exploration began to accelerate. In 1949 H. Nünlist and Alfred Bögli resumed work in the Hölloch. Large systems with deep pits began to be discovered throughout the high mountains (Faustloch, Bärenschacht, Sieben Hengste, ...).

Today the S.S.S. is still the national organization representing the various local clubs internationally. It organizes a high quality congress every four years, and publishes *Stalactite* twice a year.

DEEP CAVES:

1. ***Siebenhengste-Hohgant-Höhlensystem***
(Eriz/Beatenberg/Habkern, Bern) .... **–1020 m**
This large system extends under three Swiss municipalities: Eriz, Habkern and Beatenberg, in the canton of Bern. It presently has more than 15 entrances located in the Sieben Hengste (1955 m) and the Hohgant (2163) mountains which rise to the north of Interlaken, 10 km to the west of Thunerse. Two of the entrances are in the mountain next to Hohgant. Access is via Habkern, by climbing up the Lombach valley.

The upper entrance, Z49 (alt 1810 m) is located on the west side of the valley de Wagenmoos, 200 m to the west of the huts at the elevation of 1771 m (1/25,000 map Beatenberg. x 628.975; y 117.625).

One of the resurgences (dye trace of 1972), Gelber Brunnen (alt 558 m) is located on the bank of the Thunersee; the other, Bätterrich, is in the lake itself. Also draining to these resurgences are K2, Bärenschacht, and H6/JCB Höhle (see below).

The first explorations in the system date to 1966 when the S. C. Jurassien discovered the gouffre de la Pentecôte (P23, alt. 1707 m) and the gouffre de la Glacière (P27, alt. 1710 m). In 1967 the G. S. Lausanne descended to –213 m in the later. In 1968 the gouffre Johnny (P26, 1700 m) was discovered by J. Wunderli and in 1970 the trou Victor (P51, alt 1732 m) was discovered by V. Courtois. Also in this year Belgian cavers (S.S. Wallonie, Centre Routier Spéléo de Bruxelles) began to take an interest in the area. They were joined by the S.S.S. Lausanne and the S. C. des Montagnes de Neuchâtel.

In April, 1972 the P26 and P51 were connected by the S.C.M.N. The same year, the Belgians discovered the gouffre Dakoté (P53, alt 1762 m) which was linked to the others in July. The depth of the system reached –450 m, stopping at a sump.

On July 6, 1974 the S.S.S.L. connected the P29 to the system, which gave a length of 17,000 m, and in 1975 (year in which the terminal sump was dived for 10 m in depth by the G.S.L.) the length was increased to 22 km, then 27 km in 1976 and 30 km in 1977, with no gain in depth. More Swiss clubs (S.G.H. Bern, S.G.H. Interlaken, S.G.H. Basel) joined in the exploration, as well as a new Belgian club, the C.I.P.S.

The Z49 (alt 1910 m) was discovered in 1977 by the G.I.P.S. and connected in on August 19, 1978, while new passages discovered by the C.R.S. were found to bypass the terminal sump. The depth reached –816 m (length 32,448 m), then –828 m in 1979 (length 42,000 m). Also in 1979 the G.S.L. added three new entrances: H 1a, H 1b, and CCC 2.

The F1 (Hohgant) was discovered in 1981 and descended to –583 m in 1982 by the H.R.H. before being connected to the system on December 22, 1982, adding 17

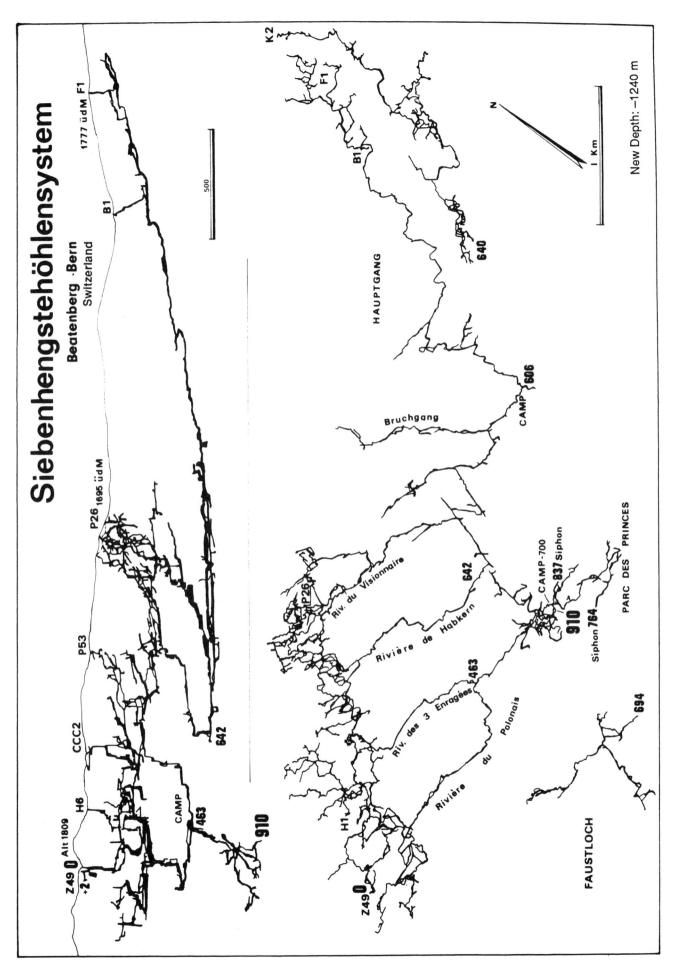

Siebenhengstehöhlensystem

Beatenberg · Bern
Switzerland

1777 üdM F1

B1

500

P26 1695 üdM

P53

CCC2

H6

Z49 Alt 1809

+2

CAMP

463

910

642

K2

F1

B1

640

HAUPTGANG

Bruchgang

CAMP 606

N

1 Km

New Depth: −1240 m

P26

Riv. du Visionnaire

Rivière de Habkern

642

463

CAMP -700

837 Siphon

910

Siphon 764

PARC DES PRINCES

Riv. des 3 Enragées

Rivière du Polonais

H1

Z49 0

694

FAUSTLOCH

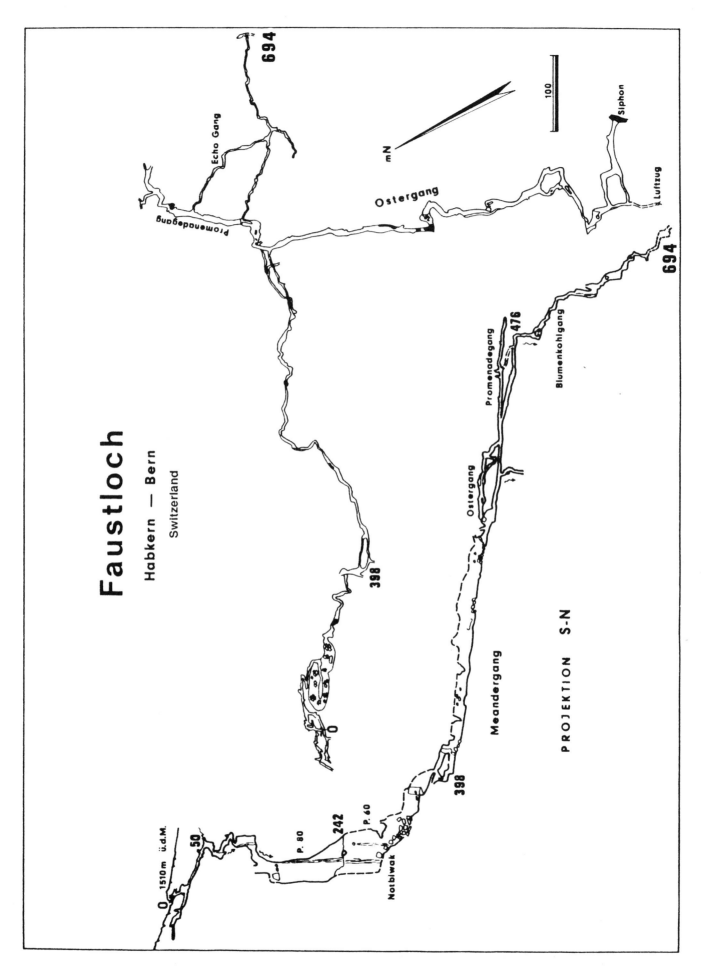

Faustloch

Habkern — Bern

Switzerland

PROJEKTION S-N

km to the length of the system and bringing the total to 65 km. The B1 (Hohgant) entrance was discovered in 1983, while the Swiss remapped the deep parts of the system. The length reached 70,500 m in 1984 and over 80 km in 1985, with a vertical extent of 912 m.

In 1986 the Belgians extended the deep part of the cave, coming within 50 m of Faustloch. Meanwhile the Swiss raised the high point and connected with the B6/5. Further explorations brought the vertical extent to 946 m and the length to over 100 km.

Hugo Maler discovered the Faustloch entrance in June, 1970. This blowing hole (alt. 1510 m) was enlarged by the S.G.H. Bern who reached –60 m in 1971. The S.G.H. Bern and S.G.H. Basel returned in 1974 to fight strong waterfalls and bad rock, reaching –400 m. Exploration in 1975 and 1976 brought the depth to –694 m and the length to 4709 m. The terminal sump was dived in 1987 by Ph. Rouiller and the connection to Sieben Hengste completed, creating a system over 110 km long with a vertical extent of 1020 m. This is the first cave in the world that is both over 1000 m deep and 100 km long.

Maps: sent by Ph. Rouiller. Faustloch shown separately.

Bibliography: Hof (A.), Jeannin (P. -Y.), Rouiller (P.) - Sieben Hengste-Hohgant höhle, *Le Trou*, 1984 (34), n spécial. Bibliographie.
Janz (W.), P. Rouiller, U. Wiamer in *Cavernes*, 1974 (3): 74-79.

Ph. Rouiller

2. **Bärenschacht** (Beatenberg, Bern) .... –956 m

This cave (alt. 1505 m) belongs to the same hydrological system as the Sieben Hengste (x 628.590; y 173.980). It was discovered in 1963 by P. and K. Grossniklaus who reached –64 m in 1964. The S.S.S. Interlaken and S.S.S. Bern pursued the exploration, reaching –100 m in 1964 and –218 m in 1965. A sump at –565 m was reached in the winter of 1972-1973. It was dived for 5 m in December 1973 and found to be impenetrable. (F. Knuchel in *Stalactite*, 1973 (2): 17-25; Courbon, *Atlas des grands gouffres du monde*, 1979, pp. 186-187, profile; Klingenfuss. in *Höhlenpost*, 1985 (69): 8-22.).

The terminal sump was dug open and passed in 1986. This 56 m long, very narrow sump made exploration beyond very difficult, but over 4 km of large passages have been mapped beyond it by Ph. Rouiller and two others. The water table was reached at –956 m, but many leads remain.

3. **Hölloch** ....(–120, +747) **867 m**
(Muotathal, Schwyz)

Hölloch is located in the limestones of the Schwyz Alps, to the east of the Quatre-Cantons lake, between the Uri lake and the Linth valley. The lower entrance (alt. 734 m) is found in the Muota valley which one reaches easily from Schwyz, a town which is 15 km or so from Muotathal. From the gasthaus "Höllgrotte" to the entrance is a 10 minute walk.

1/50,000 map 1172 Muotathal. x 702.720; y 203.710.

The Hölloch is the overflow route for the waters which resurge at Schleichender Brunnen (alt. 638 m, flow 300-1200 l/s in winter, 2000 l/s in summer). An upper entrance to the cave was dug open on August 12, 1980 (alt. 1260 m). In 1985, drills and dynamite gave the system a third entrance. It is formed almost entirely in the Schrattenkalk limestones (lower Cretaceous, Urgonian), and slightly in the Seewerkalk (upper Cretaceous). The size of the catchment basin for the Schleichender Brunnen is 22 km². The potential vertical extent is over 1200 m. At high altitude there are several known caves (among them Schwyzerschacht and Disco-schacht, the later only 10 m from the Hölloch!) that have not yet been connected to the system below them. If connected they would add 286 and 599 m respectively to the vertical extent.

For quite some time the Hölloch was neck and neck with the Flint Ridge Cave System (today part of the Mammoth Cave System) for the title of longest cave in the world. The cave has been explored from the bottom upward, the high point being 747 m above the lower entrance (alt. 1481 m).

In 1875 Aloïs Ulrich penetrated for the first time into the Hölloch. From 1899 to 1907 the explorations led by Widmer gave a surveyed length of 4280 m (total explored length estimated at 7 to 9 km) and a vertical extent of 265 m (–95, +170). Martel visited the cave and it was the subject of the doctoral thesis of Paul Egli (1904).

In 1946 Alfred Bögli undertook to study the tourist section of the cave. The systematic exploration of the Hölloch by the S.S.S. began in 1948, they organized a first underground camp in 1949. In 1950 and 1951 the members of the Club Alpin Suisse devoted 19 expeditions to the cave, reaching a length of 25 km in 1952 and a vertical extent of 280 m.

In 1953 the Arbeitsgemeinschaft-S.A.C.-Höllochforschung (A.G.H.) was founded under the scientific direction of Alfred Bögli and the technical direction of N. Nünlist. Because of the danger created by the summer floods, explorations and underground camps take place only in winter (from August 15 to 24, 1952 A. Bögli and three companions were trapped by a sudden flood. After 224 hours, they finally saved themselves by their own efforts!).

The length and vertical extent progressed regularly: 37000 m in 1953, 48000 m in 1954, 55000 m in 1955 (vertical extent 390 m: –95, +295), 61 km in 1956, 65 km in 1957, 68 km in 1958, 70 km in 1959 (A. Bögli assumes technical and scientific direction of the A.G.H.), 71 km in 1960, and 74 km in 1961. For a short time, Hölloch became the longest cave in the world, but the pace of exploration began to slow, only 1 km more may have been explored in 1963 (vertical extent 425 m) and 1234 m in 1964.

1965 was an important year: a dig, by O. Wüest, using explosives, gave access to a passage leading to the upper levels; the vertical extent increased to 531 m (–95, +436) and the length to 80,927 m.

In 1966 the A.G.H. reached 85,333 m, then 93,397 m in 1967. In 1968 the Hölloch may have become the first cave in the world to pass 100 km in length; the vertical extent reached 678 m.

In 1969, 103,705 m were explored, the high point being 631 m above the entrance (vertical extent 740 m), then 109,182 m in 1970, 112,128 m in 1972 and 120,561 m in 1973, with a vertical extent of 808 m (+699).

From 1974, the pace of new discoveries slowed a bit: 123,823 m length, for a vertical extent of 828 m (–116, +712), then 124,007 m in 1975 and 129,525 m in 1976.

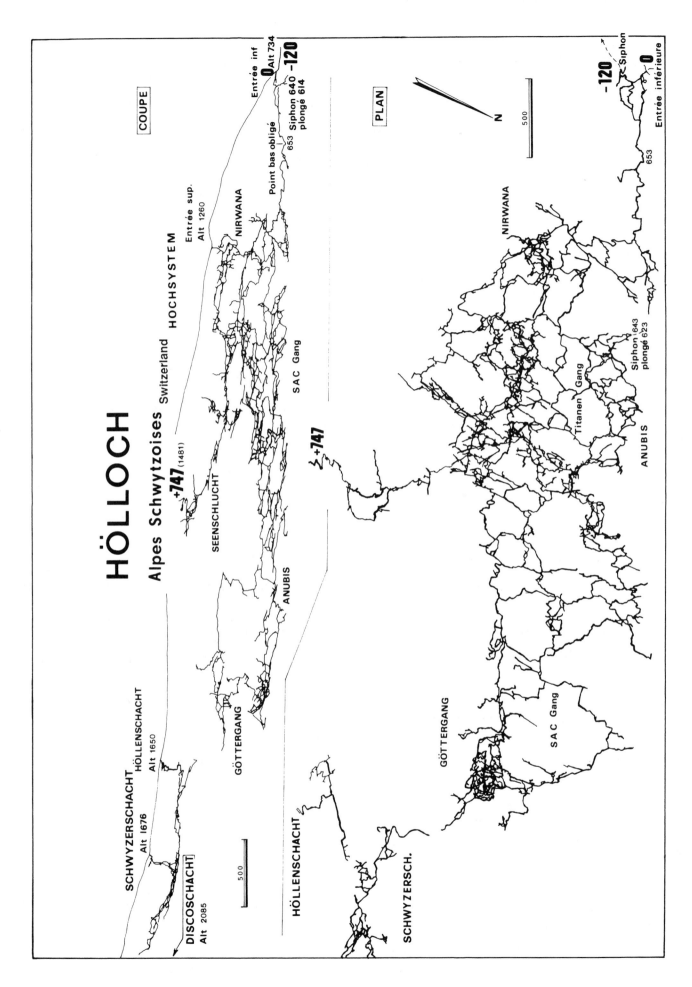

HÖLLOCH

Alpes Schwytzoises Switzerland

COUPE

HOCHSYSTEM

+747 (1481)

↓+747

SCHWYZERSCHACHT
Alt 1676

HÖLLENSCHACHT
Alt 1650

DISCOSCHACHT
Alt 2085

500

HÖLLENSCHACHT

SCHWYZERSCH.

GÖTTERGANG

SEENSCHLUCHT

ANUBIS

NIRWANA

Entrée sup.
Alt 1260

Entrée inf
0 Alt 734 −120

Siphon 640
plongé 614

Point bas obligé

653

SAC Gang

PLAN

N

500

−120   0

Siphon

Entrée inférieure

653

NIRWANA

Titanen Gang

Siphon 643
plongé 623

ANUBIS

GÖTTERGANG

SAC Gang

The length progressed to 133,169 m in 1977, 135,625 m in 1978, 139,434 m in 1979, 142,100 m in 1980 and 147,110 m in 1982. The length figures were revised at this point to give a new value of 133,050 m.

The vertical extent (–120, +747 m) represents for the caver a climb of +828 m, since one must pass by a low point at the altitude of 653 m.

Map: from the drawings of the A.G.H., sent by Ph. Rouiller.

Bibliography: Bögli (A.) - Le Hölloch et son karst / Das Hölloch und sein karst, *Stalactite*, suppl. N 14, 1970, 110 p. plan. h.t.; *Stalactite*, special issue 1186.

Ph. Rouiller

4.   K 2 (Habkern, Bern) .... **–640 m**

The K 2 is located at an altitude of 1893 m on the west side of the Hohgant mountain, above of Innerbergli lapiaz. It is in the municipality of Habkern.

1/25,000 sheet 1208, Beatenberg. x 633.590; y 181.693.

This cave belongs to the same hydrological system as the Sieben Hengste-Hohgant. If a connection were made, it would be the upper entrance of a 1335 m deep system. It drains the east side of the Innerbergli lapiaz, parallel to the F1, which drains the west side.

Discovered before the first snow of 1980, thanks to a systematic investigation of the "K" zone in the Innerbergli lapiaz, it was descended to a constriction at about –155 m (Höhlenforschung Region Hohgant and S.S.S. Basel). In 1981, the depth of –630 m was reached in six tries. The cave, previously linear, then became mazy. A last expedi-

tion reached the terminal sump at –640 m from an underground bivouac; the length reached 5 km.

In 1982, the K 2 was left in favor of the F1. Until the end of 1985 only three expeditions entered the cave. They explored the mazy section, bringing the length to 6800 m in 1984 and 7540 m in 1985. A lower entrance, the 203, was discovered in 1983.

Two connection possibilities exist: on the one side the Sieben Hengste-Hohgant is only 170 m away, on the other the Haglätschhöhle comes close to the K 2 in several places. The terminal sump has not yet been dived.

Map: from the drawings of the H.G.H., sent by Ph. Rouiller.

Bibliography: Rouiller (Ph) - F1-K2, *Stalactite*, 1983 (1): 28-39.

Ph. Rouiller

5.   *réseau de la Combe du Bryon*
(Leysin, Vaud) ....(–499, +123) 622 m

The system is made up three caves, the gouffre du Chevrier (alt 1711 m), the grotte Froide (alt. 1710 m) and the boulevard Appenzellois (alt. 1717 m). It is located in the municipality of Leysin. The entrance of the gouffre du Chevrier, longest known of the three, is found in the rocky cirque of Bryon, being the southernmost of a dozen or so openings which are found at the base of the rocky cliff. The system is formed in upper Jurassic limestones. The stream (flow 300 to 500 l/s) which is lost at –443 m in the gouffre du Chevrier reappears some 600 m lower at the captage de Fontannet (municipality of Aigle, 6500 m away).

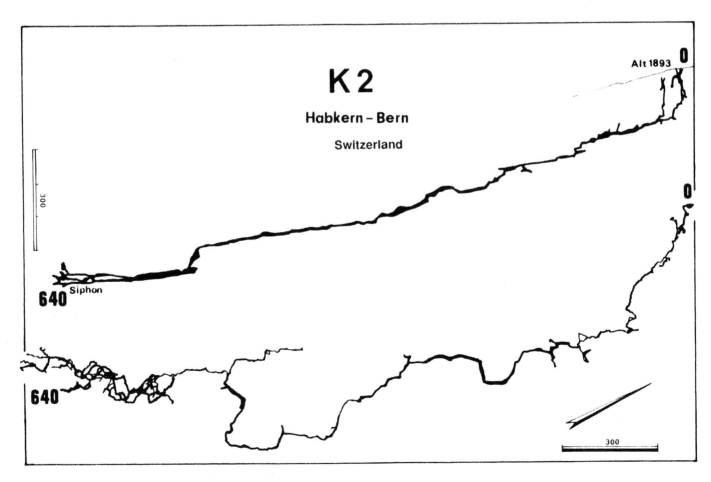

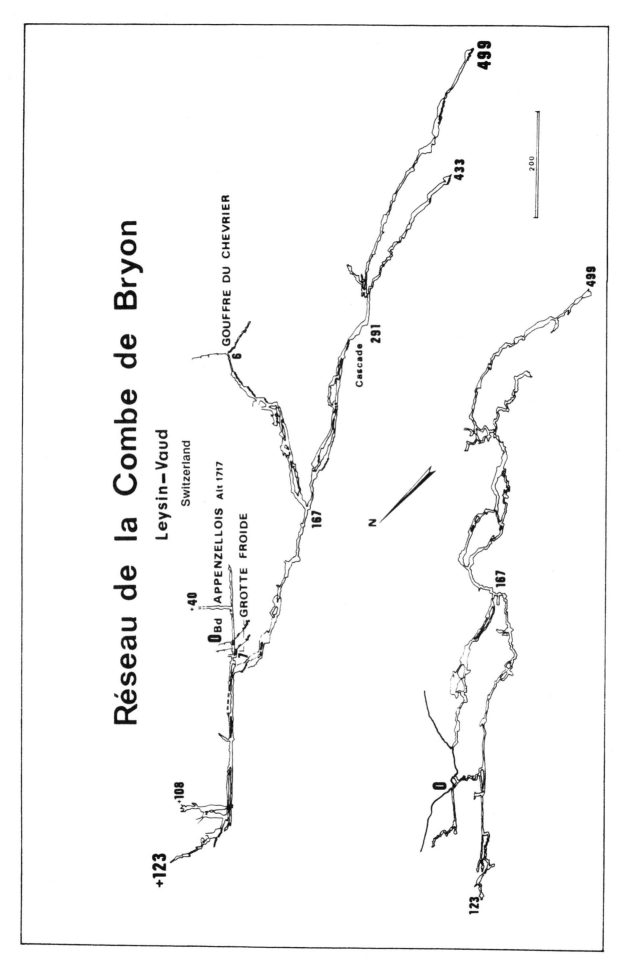

Réseau de la Combe de Bryon

Leysin–Vaud

Switzerland

GOUFFRE DU CHEVRIER

APPENZELLOIS Alt 1717

GROTTE FROIDE

1/50,000 map no. 1265, Les Mosses. x 567.985; y 135.665 (Chevalier).

The Chevrier entrance has been known for many years (it is mentioned in the 1932 review by the C.A.S.). In 1949 a team of local cavers (P. Schüler, J.-P. Graf, Greiner, Riedler, Toyanoff) became interested in the cave and descended to –45 m. In 1950 they dropped the 60 m pit at the and of the entrance passage and reached –165 m. In 1951, helped by Vaud cavers from the C.A.S., they got to –309 m. In January, 1952 they were stopped by a sump at –367 m.

In January, 1955 Jean-Paul Graf and a team from Nyon discovered an upper passage that permitted them to reach the bottom of the cave at –514 m on February 5 and 6 of the same year.

In September and October 1976 C. Brandt of the G. S. Lausanne dived the sump at –367 m and descended to –474 m in the passages beyond.

The grotte Froide was discovered in 1960 by the S.S.S. Lausanne and connected to Chevrier on November 25, 1979 by the G.S.L. (connection by diving). Upstream, the explorers were able to ascend to +123 m in 1981.

The Chevrier was entirely remapped and the low point remeasured as –499 m. The length was 4,388 m at the end of 1985.

Map: from the drawings of the G. S. Lausanne, sent by Ph. Rouiller.

Bibliography: Martin (R.), Graf (J. -P.) - Grotte-gouffre du Chevrier, *Actes du Ist Congr. Suisse de Spél.*, 1962, pp. 23-30.
Baron (JJ. -P.), *Spéléologie du canton de Vaud*, éd. Attinger, 1969.
Dutruit (J.) - Le réseau de la Combe du Bryon, *Le Trou*, 1982 (28): 10-15, bibliographie.

Ph. Rouiller

6.   **Kobelishöhle** (Alt St-Johann, St-Gallen) .... –546 m
Kobelishöhle is found in the high pastures of Hinter Selun in the Churfisten range (2279 m), at an altitude of 1700 m.

1/25,000 map 1134, Walensee. x 736.355; y 224.525. The probable resurgence is Rinquelle (alt. 603 m), 5,500 m away.

The entrance was found in 1963 by the Ostschweizerische Gesellschaft für Höhlenforschung (O.G.H.) who descended the first pit (154 m) in 1965. In 1967 they stopped around –280 m. In 1969, helped by the S.C.M.N and the S.V.T., a sump was reached at –343 m. In 1976 several clubs banded together and found a way on (stopping at –344 m) which, in 1977 was found to end at –367 m in a closed room. In 1983 the Bale cavers did a climb in this room to find a way on (stopping around –380 m). In 1985 the explorers were stopped by an inpenetrable sump at –533 m. The length reached 2,130 m (P. Rouiller, in *Höhlenpost*, 1977 (45)).

7.   **Schneehühnersystem** .... –477 m
or **réseau des Lagopèdes** (Flühli, Luzern)
Formed in 1977 by the connection of the P 55 Schrattenfluh (alt. 1660 m) with the P 68 (alt. 1885 m) (*Cavernes, S.-C. Montagnes de Neuchâtel*, 1976 (1) profile to –252 m).

8.   **réseau des Mortheys** .... approx. –470 m
(Charmey, Fribourg)
Alt. 2000 m (*Actes 6th Congr. Suisse*, 1978, profile to –350 m).

9.   **A 2** (Beatenberg, Bern) .... approx. –470 m

10.   **gouffre de Cornette de Bise** .... –460 m (?)
(Vouvry, Valais)

11.   **Schwyzerschacht** (Muotathal, Schwyz) ..... –448 m

12.   **Seichbergloch** (Grabs, St.-Gallen) .... –446 m
(*Stalactite*, 1978 (2) profile).

13.   **Nidlenloch** (Oberdort, Solothurn) .... –418 m
Explored in 1895, 1909 (– 376 m), 1937-1938. Resumed in 1978.

14.   **Charetalpschacht 514** .... –401 m
(Muotathal, Schwyz)

15.   **gouffre du Petit-Pré** .... –389 m
(Mont Tendre, Bière, Vaud)
Explored from 1959 to 1964 (SSS Lausanne, Genève) (*Stalactite*, 1980 (1) profile).

16.   **gouffre glacé de Chézette** .... –380 m
(Vouvry, Valais)

17.   **Selun-Höhlensystem** .... approx. –378 m
(Alt St. Johann, St-Gallen)

18.   **Schrattenhöhle** (Kerns, Obwalden) .... 358 m

19.   **Windloch** (Klöntal, Glarus) .... 349 m
Alt. 1500 m. +250 m by the Zürcher S.P.C. in 1975.

20.   **Charetalpschacht G 126** .... –334 m
(Muotathal, Schwyz)
(*Ouarnède*, Toulouse, 1977 (9) profile to –300 m).

21.   **Beatushöhle** (Beatenberg, Bern) .... +312 m

22.   **Blockschacht** .... approx. –300 m
(Alt St-Johann, St-Gallen)

23.   **Rauchloch** (Alt St-Johann, St-Gallen) ...... –280 m

24.   **gouffre des Corneilles** .... –276 m
(Albeuve, Fribourg)
(*Stalactite*, 1984 (1) profile).

25.   **Hälliloch** (Beatenberg, Bern) .... –267 m
(*Stalactite*, 1983 (1) profile).

26.   **Charetalpschacht 302** .... –266 m
(Muotathal, Schwyz)
(*Ouarnède*, 1977 (9) profile to – 234 m).

27.   **Disco-schacht** (Muotathal, Schwyz) .... –261 m

28.   **Schwarzberghöhle** (Silenen, Uri) .... –261 m
(*Sous Terre*, 1982-83 (22) plan).

29. **Hakenschacht B 7** (Erlenbach, Bern) ...... − 254 m

30. **JCB/H6 Höhle** (Beatenberg, Bern) .... − 253 m

31. **Blätzschacht** (Muotathal, Schwyz) .... − 252 m

LONG CAVES:

1. **Hölloch** (Muotathal, Schwyz) .... 133,050 m
See above.

2. **Siebenhengste-Hohgant-
Höhlensystem** .... approx. 110,000 m
(Beatenberg/Eriz/Habkern/Bern)
See above.

3. **Schwyzerschacht** (Muotathal, Schwyz) ... 13,211 m
See above.

4. **grotte de Milandre** (Boncourt, Jura) .... 10,520 m
Explored by the S.S.S. Jura.

5. **Beatushöhle** (Beatenberg, Bern) .... 9086 m
Explored by the S.S.S. Bern/Interlaken (*Actes 7ᵗʰ
Congr. Suisse*, 1982, plan with 8,414 m). See above.

6. **K 2** (Habkern, Bern) .... 7540 m
See above.

7. **Nidlenloch** (Oberdorf, Solothurn) .... 7540 m
(*Stalactite*, 1978 (1) plan with 5 km). See above.

8. **réseau du Poteux** .... approx. 7500 m
(Saillon, Valais)
Explored by the S.S.S. Valais (*Actes 6ᵗʰ Congr. Suisse*,
1978, plan).

9. **Windloch** (Klöntal, Glarus) .... 7100 m
Explored by the Zurcher SPC (*Stalactite*, 1976 (1) plan
with 5386 m). See above.

10. **réseau des Mortheys** .... approx. 5500 m
(Charmey, Fribourg)
See above.

11. **Schrattenhöhle** (Kerns, Obwalden) .... 5120 m
See above.

12. **Bärenschacht** (Beatenberg, Bern) .. approx. 5000 m
(See above).

13. **Haglätschhöhle** (Habkern, Bern) ... approx. 5000 m
See above. Sieben Hengste hydrological system
(*Stalctite*, 1980 (2) plan of 3,580 m).

14. **Neuenburgerhöhle** (Flühli, Luzern) .... 4720 m

15. **réseau de la Combe du Bryon** .... 4388 m
(Leysin, Vaud)
(*Stalactite*, 1983 (2) plan). See above.

16. **Lauiloch** (Muotathal, Schwyz) .... 4317 m

17. **Schneehühnersystem** (Flühli, Luzern) ...... 4162 m
(*Stalactite*, 1978 (2) plan). See above.

18. **Apollohöhle** .... approx. 4000 m
(St-Antönien, Graubünden)

19. **réseau de Covatannaz** .... approx. 4000 m
(Sainte-Croix, Vaud)

20. **Mäanderhöhle** (Eriz, Bern) .... 3200 m

21. **Lachenstockhöhle** (Innerthal, Schwyz) ...... 3200 m

United Kingdom

A country with a strong speleological tradition, the United Kingdom has been, along with France, Italy, Austria, and Yugoslavia, one of the birthplaces of speleology. It is difficult to pinpoint a date for the beginning of caving, for as far back as the 16th century some caves in Somerset and Derbyshire were being explored by anonymous miners. One can nonetheless mention some important names and dates which have been important in the development of British caving.

Although Lloyd dropped the Eldon Hole (− 67 m) in Derbyshire in 1770, it is really in the 19th century that speleology took off. In 1837, Birbeck, Farrel, and Metcalfe began exploring caves in Yorkshire. At the same time prehistorians, notably Boyd Dawkins who wrote a major book *Cave Hunting* (1874), hunted throughout caves looking for human and animal remains. Beginning in 1885 H.E. Balch, pioneer of the Mendips, started investigating the famous Wookey Hole resurgence, a cave which had essentially always been known and in which exploration continues today.

The Yorkshire Ramblers' Club was founded in 1892. It was very active in the beginning of the 20th century. In 1895 E.-A. Martel succeeded in bottoming the famous Gaping Gill. This event gave a great boost to vertical caving in Britain. In 1897 Martel published his important work: *Irlande et Cavernes Anglaises* (*Ireland and English caves*).

The first cave dives were in 1935 in the Mendips (Wookey Hole, Swildon's Hole). Balcombe and Sheppard founded the Cave Diving Club in 1946, a group which has become very active in subsequent years.

As in other European countries, British caving made a great leap forward after the end of the second world war; fellwalking (caves in England are found on fells, not ridges!), digging, and pushing tight leads led to important discoveries. Today progress is being made by increasingly audacious dives (Keld Head, Ingleborough Cave), extensive digs, and by connections between caves.

The British Cave Research Association, born of the union of the Cave Research Group of United Kingdom and the British Speleological Association, with its publications *Caves and Caving* and *Cave Science*, is today in the dynamic forefront of British caving. Numerous reference works are available today; notably *British Caving* (1962), *The Science of Speleology* (1976), the regional references *Limestone and Caves of Northwest England,...of Mendip*

Hills,...of the Peak District and *...of Wales*, as well as local guides, notably the five volumes of *Northern Caves* by D. Brook *et al.*, covering the Yorkshire Dales, including Lancashire and Cumbria.

Tony Waltham.

DEEP CAVES:

1. **Ogof Ffynnon Ddu** .... –308 m
(Penwyllt, Powys, South Wales)

2. **Giant's-Oxlow System** .... –214 m
(Castleton, Derbyshire)

3. **Lost John's-Gavel System** .... –211 m
(Gragareth, Lancashire)
Dived to – 65 m in the terminal sump.

4. **Gaping Gill System** .... –203 m
(Ingleborough, Yorkshire Dales)

5. **ogof Daren Cilau** .... –186 m
(Llangattock, Powys, South Wales)

6. **Penyghent Pot**.................................... –184 m
(Penyghent, Yorkshire Dales)
Alt. 424 m (D. Brook *et al.*, *Northern Caves*, 1972, 2, profile to – 178 m).

7. **Peak-Speedwell System** .... –182 m
(Castleton, Derbyshire)

8. **ogof Agen Allwedd** .... –180 m
(Llangattock, Powys, South Wales)

9. **Reyfad Pot** .... –179 m
(Belmore, Fermanagh, Ulster)

10. **Longwood Swallet** (Mendip, Somerset) ...... –175 m

11. **Meregill Hole**...................................... –173 m
(Ingleborough, Yorkshire Dales)
Four entrances, alt. 396 to 383 m (D. Brook, *op. cit.*, 1975, 3, profile).

12. **Gingling Hole** .... –169 m
(Fountains Fell, Yorkshire Dales)
Alt. 473 m (D. Brook, *op. cit.*, 1976, 2, profile). Explored in 1966.

13. **Dale Head Pot** .... –168 m
(Penyghent, Yorkshire Dales)
(D. Brook *et al.*, *op. cit.*, 1976, 2, profile to –165 m).

14. **Swildon's Hole** (Mendip, Somerset) .... –167 m

15. **Tatham Wife Hole**................................. –162 m
(Ingleborough, Yorkshire Dales)
Alt. 435 m (D. Brook *et al.*, *op. cit.*, 1975, 3, profile to –156 m).

16. **Black Shiver Pot** .... –159 m
(Ingleborough, Yorkshire Dales)

Alt. 409 m. (*U. Leeds S. A. Explorations Journal*, 1969, profile to –157 m). Connection possible with Meregill Hole.

17. **Long Kin West Hole** .... –159 m
(Ingleborough, Yorkshire Dales)
Alt. 451 m. (D. Brook *et al.*, *op. cit.*, 1975, 3, profile to –161 m).

18. **Nettle Pot** (Castleton, Derbyshire) .... –159 m
Alt. 462 m (T. Ford, *Caves of Derbyshire*, 1974, profile).

19. **Notts-Ireby System** .... –152 m
(Gragareth, Lancashire)
Alt. 396 m. Connected with Notts Pot in 1976.

20. **King Pot** (Whernside, Yorkshire Dales) .... –151 m

21. **West Kingsdale System** .... –151 m
(Gragareth, Yorkshire Dales)
Sump of Keld Head dove to –21 m.

22. **Manor Farm Swallet** .... –151 m
(Mendip, Somerset)
Alt. 223 m (*Proc. Univ. Bristol S. S.*, 1974, 13 (3), profile).

LONG CAVES:

1. *Ease Gill System* ....52,500 m
(Gragareth, Cumbria-Lancashire)
Ease Gill System extends both under Casterton Fell and Leck Fell, which are separated by the Ease Gill Beck, to the north of the Ingleton-Kirkby Lonsdale road, about 4 km to the east of Casterton. There are 14 entrances, of which 8 are in the Ease Gill valley: Top Sink (348 m), Boundary Pot (336 m), Pool Sink (332 m), Bore Hole (328 m), Slit Sinks (326 m), County Pot (319 m), Oxford Pot (320 m) and Link Pot. On Casterton Fell lie Cow Pot (305 m), Lancaster Pot, Bull Pot of the Witches (293 m), and Hidden Pot (296 m). Pippikin Pot (320 m) lies on Leck Fell.

Exploration of the system essentially began in 1946 (except Cow Pot which was explored in 1899 by the Yorkshire Ramblers' Club) with the discovery of Lancaster Pot by G. Cornes and it has been continuous since. Exploration has been the work of numerous clubs of which the British S.A., Cave Diving Group, Happy Wanderers C.P.C., Northern P. C., Northern C. C., Red Rose C.P.C. and U. Leeds S. A. have been prominent. An important date was December 16, 1978 when Pippikin Pot and Lancaster Pot were connected by B. Hrindij, increasing the length from 33 km to 45,400 m (52,500 m in 1984).

To the west, the connection with Aygill Cavern has not yet been accomplished, but connections to Lost John's-Gavel System (to the east) and Notts-Ireby System are thought to be imminent.

Map: from the map of A. Waltham and D. Brook "Three Counties Cave System", B.R.C.A., 1980, updated in 1985 and sent by A. Waltham.

Bibliography: Eyre (J.), Ashmead (P.) - Lancaster Pot and the Ease Gill Caverns, *Cave Res. Gr. Trans.*, 1967, 9 (2); D. Brook *et al.*, *Northern Caves*, Dalesman Books, 1975, Vol. 4.

Ease Gill Cave System

Lancashire — England

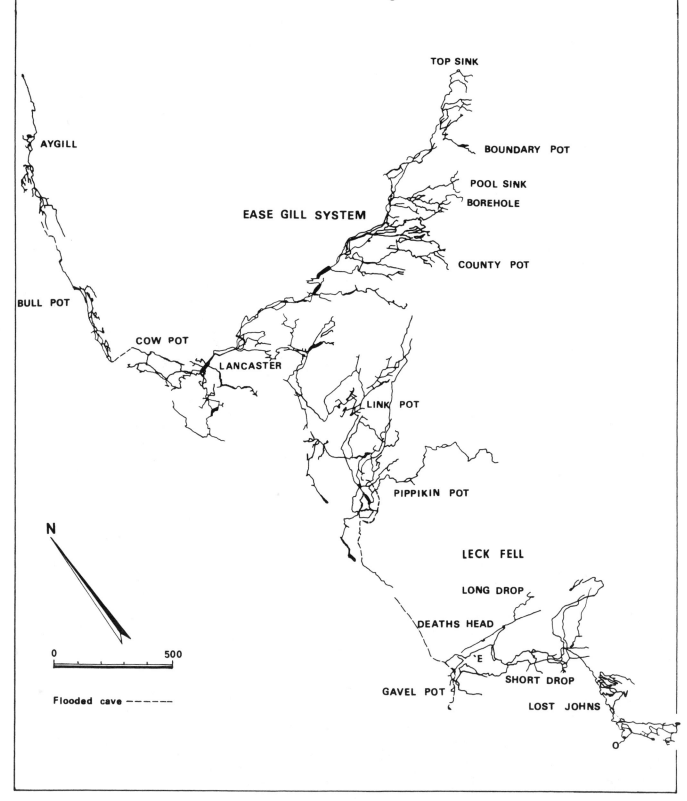

TOP SINK

AYGILL

BOUNDARY POT

POOL SINK

BOREHOLE

EASE GILL SYSTEM

COUNTY POT

BULL POT

COW POT

LANCASTER

LINK POT

PIPPIKIN POT

LECK FELL

N

LONG DROP

DEATHS HEAD

0   500

'E

SHORT DROP

GAVEL POT

Flooded cave — — — — —

LOST JOHNS

O

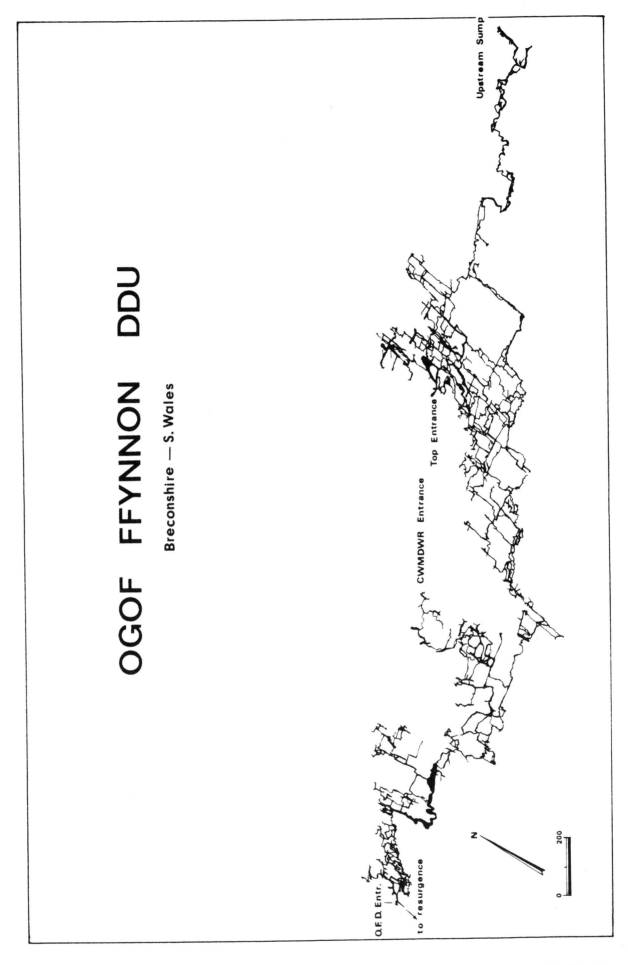

OGOF FFYNNON DDU

Breconshire — S. Wales

Upstream Sump

CWMDWR Entrance

Top Entrance

O.F.D. Entr.

to resurgence

N

0   200

2. *Ogof Ffynnon Ddu* ....43,000 m
   (Penwyllt, Powys, South Wales)

   This system has four entrances and is located above Craig y Nos, to the east of the Tawe river, which empties into the sea at Swansea.

   Efforts to enter by the resurgence (Cave Diving Group 1946) failed and it was the digging open of nearby ogof Ffynnon Ddu by Peter Harvey and Ian Nixon that opened the system for exploration, which has been continuous to the present day and has principally been the work of the South Wales Caving Club. In 1952 Bill Little and Lewis Railton were caught in a flood and had to wait 60 hours before being rescued.

   Cwmdwr Quarry Cave, opened by the quarry workers, then reopened by cavers, was connected to the system in 1966. One year later, the Top entrance was also dug open and connected into the system.

   Cave diving and digs have played a major role in the exploration of ogof Ffynnon Ddu. Beside Railton, Alan Coase, Paddy O'Reilly and Colin Fairbairn are among those who have devoted much effort. In 1953, about 6,400 m were known, then 15,000 m in 1967, 31,120 m in 1969, 37,000 m in 1972, 38,500 m in 1974, and finally 43,000 m in 1984.

   Hydrology: ogof Ffynnon Ddu drains the Byfre River which sinks at Pwll Byfre, some 500 m from the upstream terminus in the cave. The resurgence is close to the lower entrance, a few meters from the banks of the Tawe river.

   Map: by L. Railton from 1951, then P. O'Reilly from 1967 to 1969, sent by A.C. Waltham.

   Bibliography: O'Reilly (P. and S.), Fairbairn (C.) - *Ogof Ffynnon Ddu*, S. Wales C. C., 1969, XII-52 p.,map.

3. **Ogof Agen Allwedd** ....29,000 m
   (Llangattock, Powys, South Wales)

   Discovered in 1949 and explored regularly since. Map showing 24,800 m in *BRCA Bull.*, 1977 (15). Gently dipping system with no pits and active passages ending in sumps.

4. **Ogof Daren Cilau** ....24,000 m
   (Llangattock, South Wales)

   Explored in 1957 (250 m) and 1963 (750 m). Main explorations commenced in 1984. Possible connections with Agen Allwedd and Ogof Craig-a-Ffynnon.

5. **Dan yr Ogof** ....15,000 m
   (Breconshire, South Wales)

   Resurgence known since 1912, run as a show cave and principally explored by the South Wales C.C.

6. **Gaping Gill System** ....14,700 m
   (Ingleborough, Yorkshire Dales)

   Stream system entirely traversable by cavers and containing ten entrances (Gaping Gill, 396 m; Flood Entrance Pot, 396 m; Rat Hole, 402 m; Disappointment Pot, 405 m; Stream Passage Entrance, 415 m; Bar Pot, 395 m; Wades Entrance, 398 m; Car Pot, 396 m; Ingleborough Cave, 259 m and Beck Head Cave, 259 m). While only 954 m were known in 1896 and 4 km known in 1937, the connection with Ingleborough Cave in 1983 increased the length from 11,300 m to 14,300 m.

7. **West Kingsdale System** ....11,700 m
   (Gragareth, Yorkshire Dales)

   Stream cave with eleven entrances, including Swinsto Hole (383 m) and Rowten Pot (362 m), and 3200 m of underwater passage. The connection of the West Kingsdale system and Keld Head (253 m) was made on July 6, 1978 by C. Yeadon and O. Statham after diving a 1830 m long sump (*Univ. Leeds S. A. Explorations Journal*, 1969, map with 6,500 m).

8. **Lost John's-Gavel System** ....11,100 m
   (Gragareth, Lancashire)

   Connections between Lost Pot, Rumbling Hole, Gavel Pot (1975) and Coal Cellar Hole. Possible connection with Ease Gill System and Notts-Ireby System by diving (*Trans. BCRA*, 1983, 10 (4) plan).

9. **Peak-Speedwell System** ....10,300 m
   (Castleton, Derbyshire)

   Connection of Peak Cavern (191 m) and Speedwell cavern (248 m) in 1970, by passing a sump (T. Ford, *Caves of Derbyshire*, 1974, map with 8046 m).

10. **Mossdale Caverns** ....10,000 m
    (Grassington, Wharfedale, Yorkshire Dales)

11. **Langcliffe pot** ....9600 m
    (Wharfedale, Yorkshire Dales)

12. **Little Neath River Cave** ....8200 m
    (Ystradfellte, Powys, South Dales)
    (*Proc. Univ. Bristol S. S.*, 1971, 12 (3) plan.).

13. **Swildon's Hole** (Mendip, Somerset) ....8000 m

14. **Ogof Craig-y-Ffynnon** ....8000 m
    (Gwent, Llangattock, South Wales)
    Explored beginning in 1976.

15. **Notts-Ireby System** ....7500 m
    (Gragareth, Yorkshire Dales)

16. **Saint Cuthbert's Swallet** ....7000 m
    (Mendip, Somerset)

17. **Marble Arch System** ....6500 m
    (Cuilcagh, Fermanagh, Ulster)

18. **Reyfad Pot** (Belmore, Fermanagh, Ulster) ..6500 m

19. **Goyden Pot System** ....6200 m
    (Nidderdale, Yorkshire Dales)

20. **White Scar Cave** ....6100 m
    (Ingleborough, Yorkshire Dales)
    (*B.R.C.A. Trans.*, 1977, 4 (3), plan).

21. **King Pot** (Whernside, Yorkshire Dales) ......5700 m

22. **Stump Cross Caverns** ....5600 m
    (Wharfedale, Yorkshire Dales)

23. **Prod's-Cascade System** ....4800 m
    (Cuilcagh, Fermanagh, Ulster)

24. **Giant's-Oxlow System** ....4730 m
    (Castleton, Derbyshire)
    (T. Ford, *Caves of Derbyshire*, 1974, map).

25. **Noon's-Arch System** ....4500 m
    (Belmore, Fermanagh, Ulster)

26. **Magnetometer Pot** ....4400 m
    (Fountains Fell, Yorkshire Dales)

27. **Birkwith System** ....4200 m
    (Penyghent, Yorkshire Dales)

28. **Dub Cote Cave** ....approx. 4000 m
    (Penyghent, Yorkshire Dales)

29. **Fairy Hole** (Weardale, Durham) ....3900 m

30. **Birks Fell Cave** ....3600 m
    (Wharfedale, Yorkshire Dales)

31. **Meregill Hole** ....3600 m
    (Ingleborough, Yorkshire Dales)
    Possible connection with Black Shiver Pot (Brook *et al.*, *op. cit.*, 1975, 3, schem. map).

32. **Dow-Providence System** ....3600 m
    (Wharfedale, Yorkshire Dales)

33. **Rift-Large Pot System** ....3500 m
    (Gragareth, Yorkshire Dales)

34. **Baker's pit-Reeds system** ....3400 m
    (Buckfastleigh, Devon)

35. **Wookey Hole** (Mendip, Somerset) ....3350 m

36. **Otter Hole** ....3200 m
    (Chepstow, Wye valley, Gwent, South Wales)

37. **Blea Gill Cave** ....3200 m
    (Whernside, Yorkshire Dales)

38. **Penyghent Pot** ....3080 m
    (Penyghent, Yorkshire Dales)

39. **Bagshaw Cavern** (Bradwell, Derbyshire).....3050 m

40. **Ogof Llyn Parc** (Clwyd, North Wales) ....3000 m
    Explored in 1984-1985.

U.S.S.R.
Sojuz Soveckich
Socialisticeskich Respublik

Because of its huge surface area (22,274,900 km²), there are many karst areas in Russia. In his *Pescery na Territorii SSR* (Moscow, 1973), Cikisev divides them into 26 regions and 39 districts, of which the most famous are, from west to east, the Carpathians (No. 7 and 14) with the gypsum caves of Podolia, the Crimea (15), the Caucasus (16 and 17), the Alaj (24) and the Gissarsk (25) in the Turkestan, the Pamir (27) in the Tadzikistan, the Atlaj (28-30), and the Sajan (31,33). The districts No. 26 (western Tadzikistan) to 39 (South Sikhote-Alin) are spread along the northern borders with Afghanistan, China, and Mongolia. The other districts, not cited here, are spread along the Ural mountain chain.

In contrast to other countries, speleology is a young enterprise in the Soviet Union, which began in the 1950's. Nonetheless, the caves of the Urals and the Ukraine were the subject of descriptions in 1703 (Remezov on Kurgurskaja) and 1721 (Racinski on Kristal'naja). Also in that century, written descriptions were made by Tatiscev, Genin, Lemonosov, Rickov and Lepekhin and Pallas. The number of documents produced by geographers, geologists and travellers increased in the 19th century. We might cite those of N. G. Popov (Perm), P. Kepen, Golovinskij, Dimitriev and Listoce (Crimea), Batalin (Caucasus, 1857-1861), Guebler (Altaj), etc.

In the start of the 20th century, the Crimea still elicited studies (Kruber, 1915) and the caves of far-away places (central Asia or western Siberia) attracted explorers (N. Veber, V. Arsenev, ...). Between 1917 and 1940 the caves of central Asia, the Urals, and the Caucasus were studied in detail. In 1927 a hydrological expedition under the direction of P. Vasilievskij and P. Geltov had as its goal the study of the caves of the Crimean karst (they attempted to descent a 145 m deep pit in Catyr-dag).

Several karstology conferences (Perm, 1947 and Moscow, 1956) helped invigorate speleology. Institutes were formed in the Urals and Caucasus. In 1958, the work of amateurs began to interest the specialists. Speleological clubs were created in Crimea, Moscow, Perm, Sverdlovsk, Krasnojarsk, Lvov, Ternopol, Novo-Sibirsk, and Tbilissi. They are coordinated by the Central Section of Speleo-Tourism of the Central Council for Tourism. These organizations work in collaboration with the Institutes of the Academy of Sciences and the Ministry of Geology (Institute of Mineral Resources).

Today, explorations, projects, and discoveries are too plentiful to count. Since 1973 (Olomouc congress), thanks to Viktor Dubljanskij, Vladimir Iljukhin (tragically killed in 1982), Alexander Klimchouk and V.E. Kisselyov, Soviet cavers communicate the results of their explorations regularly with their western colleagues. Among the large caving literature, in addition to the work of Cikisev already cited, a few of the most important are: Dubljanskij (V.N.) *Karstovye pescery Ukraniy*, Kiev, 1980, 180 p., Dubljanskij (V.N.) and Iljukhin (V.V.), *Putesestvija pod zemley*, Moscow, 1981, 192 p., Dubljanskij (V.N.), *Pescery Kryma*, Simferopol, 1977, 128 p.

DEEP CAVES:

1. **V. Pantjukhina** .... –1508 m
   (Bzybskij, Bol'soj Kavkaz, Abkhazian SSR)
   Cave explored jointly by Soviet and Czech cavers (Orcus Group from Bohumin). The depth of –360 m was reached in 1980, then –550 m in 1981 and –650 m in 1982. A boulder choke at –1025 m was reached in 1986, then a depth of –1465 m in 1987. Exploration continues actively.

2. Snežnaja-Mežonnogo .... –1370 m
(Bzybskij, Bol'šoj Kavkaz, Abkhazian SSR)

This system, the fifth deepest in the world (1988), is located in the speleological region of the Great Caucasus, in the Bzybskij mountain range, Khipstinskij massif, not far from the town of Gudauta. The altitude of the upper entrance (the Mežonnogo shaft) is about 2000 m, while the entrance of the Snežnaja shaft is 35 m lower.

The upper part of the cave is developed in massive and thick-bedded dolomitized limestones. The lower half is in brecciated limestones of the Lower Cretaceous. The rocks dip monoclinally towards the south at an angle of 20 to 40 degrees.

The system is formed by the connection of two shafts, Snežnaja and Mežonnogo, and is mainly vertical down to –550 m (the deepest pit is 160 m deep). Beyond, it is mainly inclined, consisting of a major gallery with an underground river and numerous tributaries. The inclined galleries are complicated by shallow pits and big collapse halls. The end of the system is boulder choked. The underground river flows at a rate of 70-300 l/s at low water. During floods its discharge is increased many fold.

The Snežnaja was discovered in August, 1971 (T. Gužhva, V. Glebov) and descended to –450 m (Speleoclub of the Moscow University, M. Zverev, leader). They reached – 720 m in the summer of 1972. In 1977-1978 the exploration was pushed to –960 m by the Moscow cavers, led by D. Ussikov and A. Morozov.

In 1979 D. Ussikov and T. Nemčenko reached – 1230 m (for a length of about 2200 m), while the Mežonnogo shaft was discovered and descended to –180 m by the Speleoclub of the University of Moscow (led by Yu. Šakir). In 1980, the exploration of Snežnaja reached –1320 m (length about 8000 m), that of Mežonnogo –490 m (Moscow cavers). In 1981 the depths of the caves were –1335 and –570 m respectively (cavers from Moscow and Leningrad).

It was in 1983 that the connection was made, starting from Mežonnogo (cavers from Moscow and Leningrad), establishing a vertical extent of –1370 m and a length of 19,000 m. The volume is 1.6 million cubic meters.

In 1984 the first complete trip from the upper entrance to the boulder choke was made (led by A. Morozov).

Map: Speleoclub of Moscow (1972-1984), sent by A. Klimchouk.

Bibliography: Ljudkovskij (G.V.) et al., "About Investigating the Snežnaja shaft, the deepest karst cave of the USSR", Doklady AN SSR, 1981, 259: 2.
Dubljanskij (V.N.), Iljukhin (V.V.) "Largest karst caves and shafts of the USSR", Moscow, ed. Nauka, 1982.
Mavljudov (B. R.), Morozov (A. I.) "The Snežnaja precipice", Peščery, 1984.

from text by A. I. Morozov (Moscow).

3. system of V. V. Iljukhina .... –1240 m
(Arabika, Gagrinskij, Abkhazian SSR)

The elevation of the upper entrance (Perovskaja shaft) is 2310 m; the lower entrance, Volč'ja, is at 2290 m. They are located on the Arabika plateau, in the Gagrinskij mountain range, in the Abkhazian Republik, Great Caucasus speleological region.

The cave is developed in greyish bedded limestones of the Upper Jurassic, which dip to the SW at 30 to 40 degrees. In 1984 a dubious dye trace indicated a possible relation to the resurgences of Kholodhaja Rečka (alt 50 m) and Reproa (alt 2 m!). In 1985 a new dye trace confirmed the earlier results. This gives a vertical extent of 2300 m to the deepest hydrological system known in the world! (see Kujbyševskaja).

Inclined galleries developed along the bedding predominate in the upper part of the system. A series of parallel pits between the levels of –100 m and –180 m is replaced by a system of inclined narrow meanders between the levels of –180 and –230 m. The cave then divides into four branches, each with series of wet pitches. The main branch drops into an active inclined gallery at –720 m which has some small drops and leads to a sump at –950 m. The low water flow rate is 15 l/s.

The upper entrance, Perovskaja, was discovered in 1980 and explored to –235 m by the Perovskij Speleoclub (P. S) led by V. Iljukhin. In 1981 the depth of –287 m was reached in Perovskaja (P. S. led by A. Efremov) while the Volč'ja entrance was discovered and explored to –160 m (cavers from Kiev led by A. Klimchouk).

The two caves were connected in 1982, and shortly thereafter named after V. V. Iljukhina. Numerous side passages were discovered but the depth remained unchanged (P. S. and cavers from Rostov led by M. Djakin).

In 1983 cavers from the P. S. and from Rostov and Leningrad led by O. Padalko explored three separate branches to –265, –400 and –600 m. A first sump was reached in 1984 (depth –950 m, length 3890 m) which four divers passed in 1985 (sump 40 m long, 10 m deep) to find 200 m of passage leading to a new 15 m deep sump which marked the beginning of an underwater passage. This sump (55 m long) was passed to lead to a series of wet pitches (deepest 59 m) and a third sump of –1220 m. The length was 5000 m in 1985.

Map: Perovskij Speleoclub of Moscow (1983-1984) sent by A. Klimchouk, showing exploration through 1985 only.

Bibliography: Efremov (A. P.) et al., "The new karst cave of the Arabika massif", Peščery Gružli, Tbilissi, 1981, 9.

From notes by M. M. Djakin, V. E. Kiesseljov and S. B. Iljukhin (Moscow, Leningrad).

4. Kujbyševskaja .... –1110 m
(Arabika, Gragrinskij, Abkhazian SSR)

This great cave is on the same plateau as the V. V. Iljukhina (see above) at an altitude of 2180 m. The two caves belong to the same hydrological system. A dye trace in 1984 using fluorescein showed the stream at –800 m to resurge at Kholodnaja Recka (alt. 50 m, flow 2 m^3/s) and at Reproa (alt. 2 m, flow 2.5 m^3/s), as well as in a submarine resurgence.

The cave is formed in Jurassic limestone in an anticlinal arch. Vertical pits 60 to 170 m deep predominate in the upper part of the system (down to –550 m). From the level of –570 m the inclined meanders begin with some small drops, boulder chokes and collapse halls above them. The largest hall, the Kiev room, has a volume of about 600,000 cubic meters. The flow of the stream is 30 to 50 l/s at low water. At –740 m there is a 90 m deep boulder choke, followed by a new series of pits.

In 1979, the year of its discovery, the cave was explored to –150 m by Kiev cavers. In 1980 the boulder

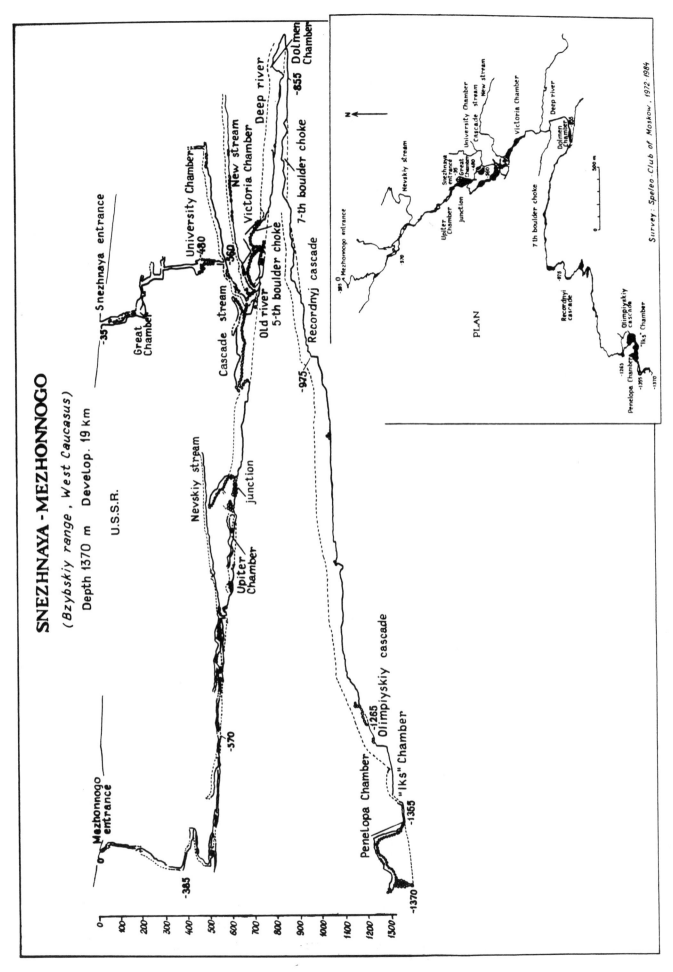

SNEZHNAYA - MEZHONNOGO

(Bzybskiy range, West Caucasus)

Depth 1370 m   Develop. 19 km

U.S.S.R.

Snezhnaya entrance −35

Great Chamber

University Chamber −480

Cascade stream −560

New stream

Victoria Chamber

Old river

5-th boulder choke

Deep river

Dolmen Chamber −855

7-th boulder choke

Recordnyj cascade −975

Mezhonnogo entrance 0

Nevskiy stream

−385

Upiter Chamber

junction −570

Penelopa Chamber −1265

Olimpiyskiy cascade

"Iks" Chamber −1355

−1370

0
100
200
300
400
500
600
700
800
900
1000
1100
1200
1300

PLAN

N

Mezhonnogo entrance −385 0

Nevskiy stream −370

Upiter Chamber

junction

Snezhnaya entrance −35

Great Chamber −480

Cascade stream −560

University Chamber

New stream

Victoria Chamber

Deep river

Dolmen Chamber −855

7th boulder choke

Recordnyi cascade −975

Olimpiyskiy cascade

"Iks" Chamber −1355

Penelopa Chamber −1265

−1370

0   500 m

Survey: Speleo-Club of Moscow, 1972-1984

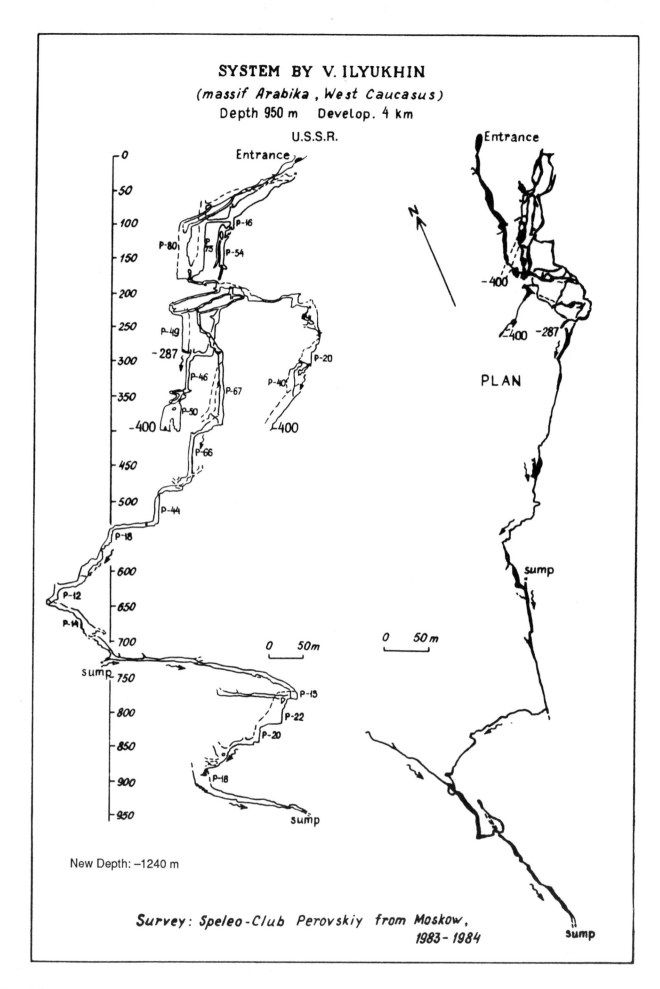

SYSTEM BY V. ILYUKHIN
(massif Arabika, West Caucasus)
Depth 950 m   Develop. 4 km

U.S.S.R.

Entrance

0
-50
-100

P-16

P-80   P-75   P-54

-150

-200

-250

P-49

-287

-300

P-46

P-20

P-67

P-40

P-50

-350

-400

-400

P-66

-450

-500

P-44

P-18

-550

-600

P-12

-650

P-14

-700

sump

-750

P-15

-800

P-22

P-20

-850

-900

P-18

-950

sump

New Depth: −1240 m

Entrance

N

−400

−400   −287

PLAN

sump

0   50m

0   50m

sump

Survey: Speleo-Club Perovskiy from Moskow,
1983-1984

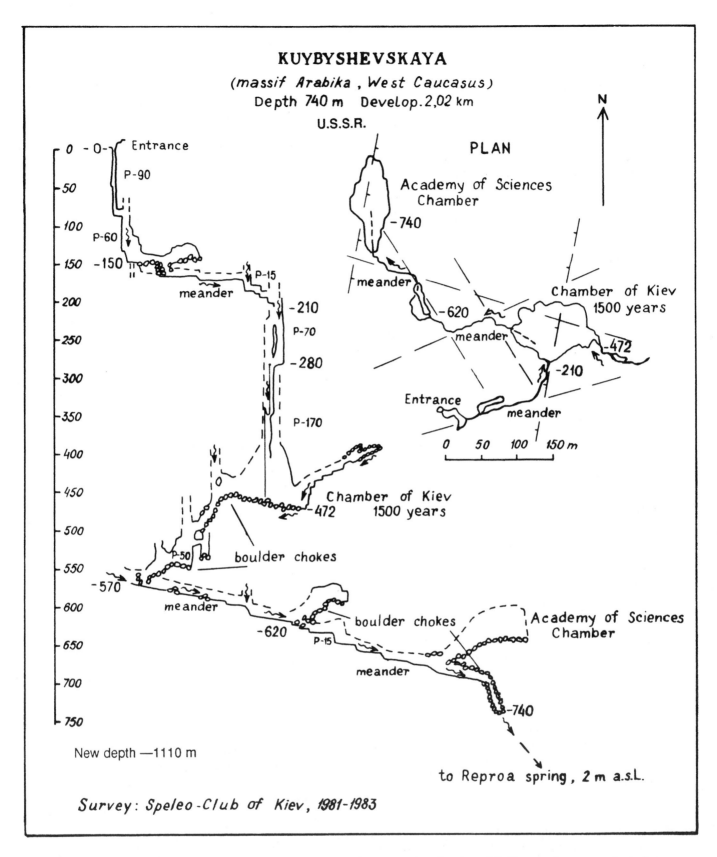

KUYBYSHEVSKAYA

(massif Arabika , West Caucasus)

Depth 740 m   Develop. 2,02 km

U.S.S.R.

New depth —1110 m

Survey: Speleo-Club of Kiev, 1981-1983

PLAN

N

Academy of Sciences Chamber

−740

Chamber of Kiev 1500 years

−472

Entrance

meander

−620

meander

−210

meander

0   50   100   150 m

Entrance

P-90

P-60

−150

meander

P-15

−210

P-70

−280

P-170

Chamber of Kiev 1500 years −472

boulder chokes

P-50

−570

meander

−620

P-15

boulder chokes

meander

Academy of Sciences Chamber

−740

to Reproa spring, 2 m a.s.l.

choke at −150 m was pushed by Kiev cavers led by V. Rogožnikov to reach −210 m.

In 1981 A. Klimchouk led them to −480 m, and in 1982 the boulder choke at −740 m was reached (length 2020 m, volume 805,000 m³).

A series of expeditions in 1983 and 1984 (including three weeks of blasting with explosives) by Kiev cavers led by A. Klimchouk and S. Kuzmenko concentrated on passing the terminal boulder choke, which was achieved in 1985. New passages, pits, and rooms were discovered by the Kiev cavers. They stopped at −970 m at the top of a pit for lack of material. In 1986 they discovered a series of short pits and meanders and three large rooms. They were stopped at −1110 m by a boulder choke.

Map: Speleoclub of Kiev (1981-1983) sent by A. Klimchouk, showing vertical extent of 970 m.

Bibliography: Klimchouk (A.B.), Rogožnikov (V.) "On the impact of the Late Quaternary glaciation on the karst development in the Arabika massif", *Izvestija*, V.G.O., 1984, 116:2.

from text by A. B. Klimchouk (Kiev).

5. *Kievskaja* .... –990 m
(Gissarsko-Alajskaja, Uzbekskaja SSR)

Kievskaja is located at an altitude of 2400 m, on the Kyrktau plateau located in the Zeravsanksij mountain range (Tian-Shan), in the Gissarsko-Alajskaja speleological region, in Uzbekistan.

The shaft is developed in massive and bedded dolomitized Silurian limestones which dip towards the south at an angle of 15-20°. It consists of alternating waterfall pits 20 to 90 m deep and inclined narrow meanders 30 to 140 m long. The stream flows from –100 m and has a flow of 15 l/s at low water. The cave ends at a sump.

Kievskaja was discovered in August, 1972 by cavers from Kiev led by A. Klimchouk, who descended to –270 m. In August, 1973 they reached –500 m and in August, 1975 –700 m (cavers from Kiev led by T. Krapivnicova and A. Klimchouk).

In July-August, 1976 cavers from Kiev and Tomsk (led by A. Klimchouk and V. Cujkov) came to a sump and estimated the depth at –1030 m (adjusted to –950 m in 1977 by the national expedition led by V. Iljuhin) and cavers from Crimea and Perm (chief G. Pantjukhin) reached an estimated –1080 m (adjusted to –990 m). The sump was dived in 1977.

The cave was mapped four times from 1978 to 1983, showing depths of –980 to –1030 m. The final figure is –990 m and the volume is 36,000 cubic meters. The length is 1820 m.

Map: Speleoclubs of Kiev, Moscow, Tomsk, and Perm (1975-1983) sent by A. Klimchouk.

Bibliography: Klimchouk (A.) *et al.*, "The abyssal karst of the Kirktau plateau and the deepest karst shaft of the USSR - Kievskaya", *Gidrogeologia i karstovedenie*, Perm, 1978, 8; "The deepest karst cavity in the USSR", *Izvestija*, V.G.O., 1978, 8; "The karst of the Kirktau massif", Publ. Inst. Geol. Nauk AN USSR, Kiev, 1981. Dubljanskij and Iljukhin, *op. cit.*, 1982.

6. **Moskovskaja** .... –970 m
(Arabika, Kavkaz, Abkhazian SSR)

Explored to –380 m in 1985, then to –630 m in 1986 where the stream (40 to 200 l/sec) disappears into boulders. A parallel route was followed to a tight fissure of –970 m. Explored by cavers from Moscow.

7. *Napra* .... –956 m
(Bzybskij, Bol'šoj Kavkaz, Abkhazian SSR)

Napra is located in the same speleological region (the Great Caucasus) as Snežnaja, in the Bzybsky massif, in the mountains of the same name, in the Abkhazian Republic. The altitude is approximately 2350 m.

The cave is formed in massive limestones of the Upper Jurassic. It is a cave of waterfalls and wet pits, with some parallel pits between –450 and –620 m. The inclined part of the system begins at –870 m, where one encounters a

stream with a low water flow of 50 l/s, ending at two sumps. The end of the cave is at a boulder choke.

In August, 1980 the Speleoclub of Krasnojarsk (S. K.), Z. Zaliev and Yu. Kromm, discovered and explored the cave to –500 m. In August, 1981 the main branch of the cave was explored to –956 m and several side branches were found to dead end (S. K., led by V. Melnikov and S. Musijačenko).

In July and August, 1983 the S.K. and cavers from Čeljablinsk (leader Z. Zaliev, P. Minenkov, diver) explored the side streams and dove the sumps at –900 and –940 m.

Map: Speleoclub of Krasnojarsk (1982-1983) sent by A. Klimchouk.

from text by S. T. Musijačenko and V. P. Melnikov (Krasnojarsk).

8. **Pionerskaja** .... –815 m
(Bzybskij, Abkhazian SSR)

Cave explored by the Speleoclub of Krasnojarsk who reached – 700 m in 1983, then – 800 m in 1984. The terminal sump was dove for 15 m by P. Minenkov.

9. **Forel'naja** .... –740 m
(Bzybskij, Bol'šoj Kavkaz, Abkhazian SSR)

Depth reached in 1984 by cavers from Krasnojarsk, Moscow and Celjabinsk. In 1985 attempts were made to clear the terminal boulder choke.

10. **Grafskij Provel** .... –710 m
(Bzybskij, Bol'šoj Kavkaz, Abkhazian SSR)

Explored in 1986 by cavers from Tomuk, who stopped at the top of a pit.

11. *Ural'skaja* .... –565 m
(Bajsuntau, Gissarsko-Alajskaja, Uzbekskaja SSR)

This is a cave of Central Asia opening at an altitude of about 3200 m. It is located in the Gissarsk mountain range, in the Bajsuntau massif, in the Gissarsko-Aajskaja speleological region (Uzbekistan Republic).

The cave is developed in Upper Jurassic limestones and intusions of the underlying alevrolites. The limestone dips to the west at 10 to 15 degrees. A dye trace has shown the resurgence to be the Macaj spring (alt. 1400 m).

The upper part of the cave is a series of pits to the –305 m level. An inclined meander then follows the bedding to a sump at –565 m. The low water flow rate is 5 to 6 l/s.

Cavers from Sverdlovsk led by A. Ryzkov discovered the cave in 1981. In 1982-1983 the cave was completely explored and mapped and the terminal sump was dove by cavers from Sverdlovsk with the participation of A. Merzljakov, I. Novikov and A. Višnevskij.

Map: Speleoclub of Sverdlovsk (1981-1982) sent by A. Klimchouk.

from text by Yu. E. Lobanov (Sverdlovsk).

12. **Parjaščaja Ptica** .... –535 m
(Fičt plateau, Kavkaz, RSFSR)
(Dubljanskij, Iljukhin, *op. cit.*, 1982, profile).

13. **Ručejnaja-Zabludšikh** .... –510 m
(Alek, Bol'šoj Kavkaz, RSFSR)

System explored in the 1970's. Profiles in Dubljanskij, Iljukhin, *op. cit.*, 1982.

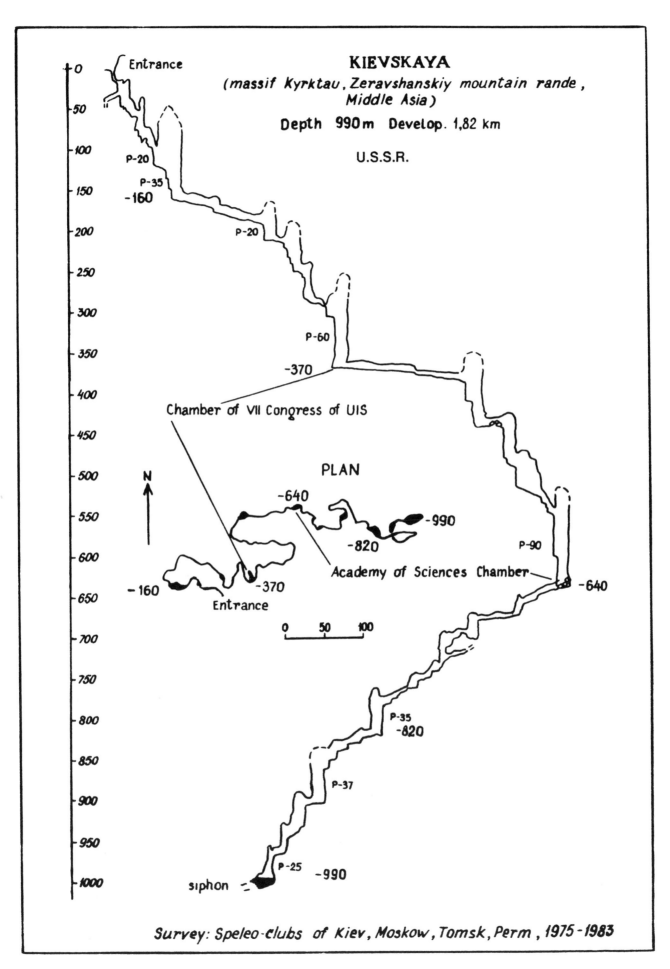

KIEVSKAYA
(massif Kyrktau, Zeravshanskiy mountain rande, Middle Asia)

Depth 990 m Develop. 1,82 km

U.S.S.R.

Entrance

P-20

P-35

-160

P-20

P-60

-370

Chamber of VII Congress of UIS

PLAN

N

-640

-990

-820

P-90

-160

-370

Entrance

Academy of Sciences Chamber

-640

0   50   100

P-35
-820

P-37

siphon

P-25   -990

Survey: Speleo-clubs of Kiev, Moskow, Tomsk, Perm, 1975-1983

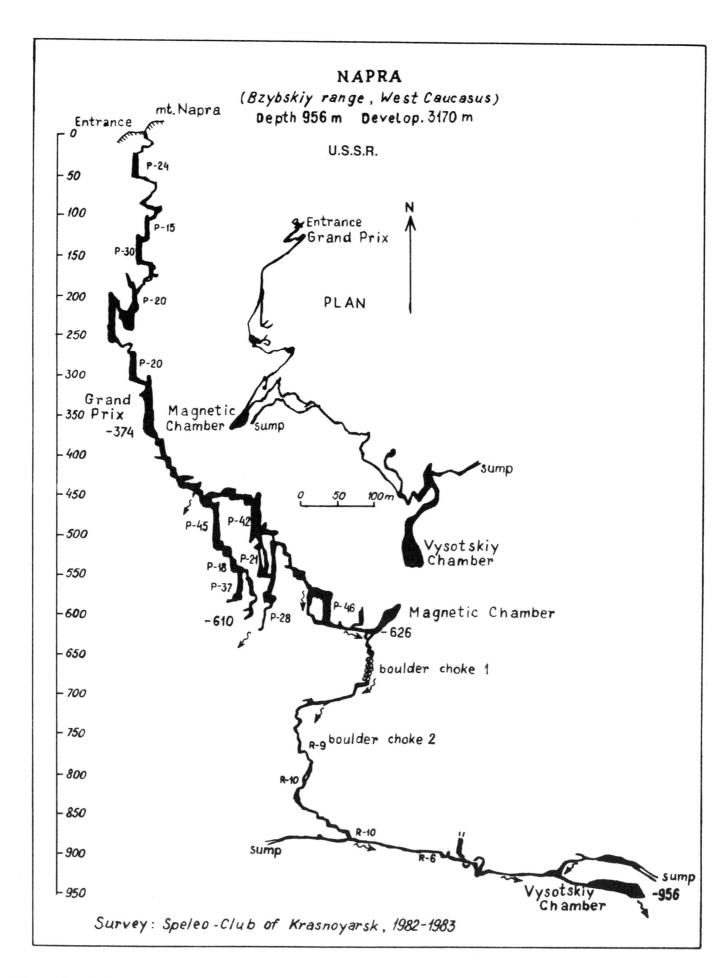

NAPRA
(Bzybskiy range, West Caucasus)
Depth 956 m   Develop. 3170 m

U.S.S.R.

PLAN

N

Entrance   mt. Napra

0

50

P-24

100

P-15

150

P-30

Entrance
Grand Prix

P-20

200

250

Magnetic
Chamber

sump

P-20

300

Grand
Prix
-374

350

sump

400

450

P-45   P-42

500

Vysotskiy
Chamber

P-18   P-21

550

P-37

Magnetic Chamber

-610   P-28

600

P-46

- 626

650

boulder choke 1

700

750

boulder choke 2

R-9

800

R-10

850

R-10

900

Sump

R-6

sump

950

Vysotskiy
Chamber

-956

0   50   100m

Survey: Speleo-Club of Krasnoyarsk, 1982-1983

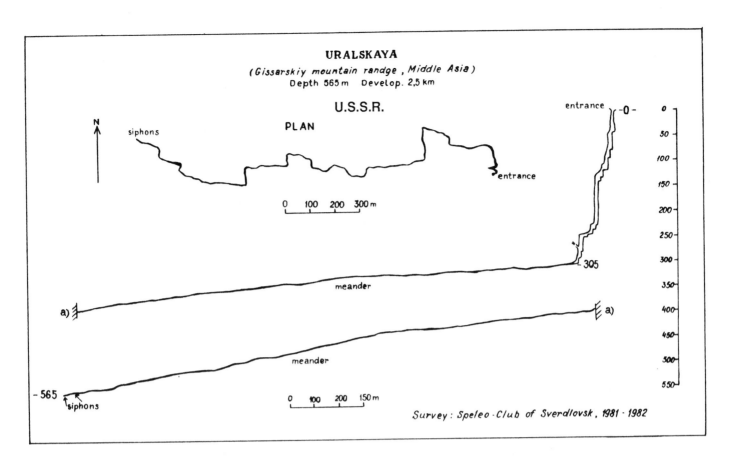

URALSKAYA

(Gissarskiy mountain randge , Middle Asia)
Depth 565 m   Develop. 2,5 km

U.S.S.R.

PLAN

siphons

entrance
-0-

entrance

0   100   200   300 m

meander

a)   a)

meander

-565
siphons

0   100   200   150 m

Survey: Speleo-Club of Sverdlovsk, 1981-1982

14. **Soldatskaja** .... –500 m
(Karabi, Krymskij, Ukrainskaja SSR)
(Courbon, *Atlas des Grands Gouffres du Monde*, 1979, pp. 178, 181).

15. **Osenaja-Nazarovskaja** .... –500 m
(Alek, Bol'šoj, Kavkaz, RSFSR)
(Courbon, *op. cit.*, 1979, pp. 178, 182).

16. **Majskaja** (Džentu, Kavkaz) .... –500 m
Depth of –450 m reached in 1980 (Dubljanskij/Iljukhin, *op. cit.*, 1982, profile to –450 m).

17. **Noktjurn** .... –452 m
(Bzybskij, Bol'šoj Kavkaz, Abkhazian SSR)
Depth reached in 1981.

18. **Oktjabr'skaja** .... –450 m
(Alek, Bol'šoj, Kavkaz, RSFSR)
Dubljanskij/Iljukhin, *op. cit.*, 1982, profile).

19. **Aleksinskogo** .... –450 m
(Bzybskij, Bol'šoj Kavkaz, Abkhazian SSR).

20. **Suvenir** .... –430 m
(Bzybskij, Bol'šoj Kavkaz, Abkhazian SSR)
(Dubljanskij/Iljukhin, *op. cit.*, 1982, profile).

21. **Neždannaja** (Kavkaz, RSFSR) .... –420 m
(Dubljanskij/Iljukhin, *op. cit.*, 1982 profile).

22. **Akhtiarskaja** .... (–390, +20) 410 m
(Kavkaz, RSFSR)
(Dubljanskij/Iljukhin, *op. cit.*, 1982, profile).

23. **Kaskadnaja** .... –406 m
(Čatyr dag, Krymskij, Ukrainskaja SSR)
Depth reached in 1979 (Czech cavers from Bohumín)
(Dubljanskij/Iljukhin, *op. cit.*, 1982, coupe).

24. **Vesenniaja** .... –403 m
(Bzybskij, Bol'šoj Kavkaz, Abkhazian SSR)
Depth reached in 1982 (cavers from Tomsk).

25. **Nakhimovskaja** .... –372 m
(Karabi, Krymskij, Ukrainskaja SSR)

26. **Genrikhova Bezdna** .... –360 m
(Arabika, Kavkaz, Abkhazian SSR)

27. **Studenčeskaja** .... –350 m
(Bzybskij, Bol'šoj Kavkaz, Abkhazian SSR)

28. **Ekologičeskaja** .... –330 m
(Altaj, Sibirskaja)

29. **P-1/7** .... –330 m
(Bzybskij, Bol'šoj Kavkaz, Abkhazian SSR)

30. **Skol'naja** .... –320 m
(Bol'soj Kavkaz, RSFSR)

31. **Absoljutnaja** .... –320 m
(Lagonaki, Kavkaz)

32. **Kan'jon** .... –320 m
(Bzybskij, Bol'šoj Kavkaz, Abkhazian SSR)

33. **Rostovskaja** .... –318 m
(Zagedan, Bol'šoj Kavkaz, Abkhazian SSR)

34. **Geografifieskaja** .... –310 m
(Alek, Bol'šoj Kavkaz, RSFSR)

LONG CAVES:

1. *Optimističeskaja* .... **165,000 m**
(Dnestrovsko-Pričernomorskij, Ternopol, Ukrainskaja SSR)

Optimističeskaja is the second longest cave in the world and is the longest cave in the world formed in gypsum. It is found in the Dnestrovsko-Pričernomorskij speleological region (also known as Podoliya). It is at an altitude of 260 m, to the south of Ternopol, not far from the town of Borščov. It is north of the Dnestr Black Sea, in the Ukrainian Republic.

The cave covers a surface area of 215,000 m² and is formed in Neogenic age horizontal gypsum beds, the thickness of which does not exceed 20 m. The gypsum is overlain by limestones, clays, and loams with a total thickness of 50 m.

The cave is a labyrinth of horizontal passages on three levels. Seven relatively isolated regions can be distinguished in the cave. They show differences in the azimuthal structure of the networks, morphology, and passage directions. There are lakes and intermittent streams in some regions.

The cave was discovered in 1966 by excavation of the sinkhole by the Speleoclub of Lvov, led by M. Savčin and O. Solyar.

From 1966 to 1984 there have been 55 expeditions by the Lvov Speleoclub to explore and survey the cave. The length of the cave increased from year to year: 8976 m in 1966, 20,342 m in 1967, 30,444 m in 1968, 51,608 m in 1969, 55,242 m in 1970, 62,000 m in 1971, 92,132 m in 1972, 105,200 m in 1973, 109,000 m in 1974, 110,840 m in 1976, 125,829 in 1977, 131,467 m in 1978, 142,400 m in 1979, and 157,000 m in 1985. The volume is about 480,000 cubic meters.

The explorations have been under the direction of M. Savčin. Participants have included cavers from Kiev, Minsk, Ternopol, Odessa, the Institute of Mineral Resources (Simferopol), the Institute of Geological Sciences of Kiev, as well as cavers from Bulgaria (1967, 1969), Poland (1970, 1972, 1973), Yugoslavia (1971, 1972), East Germany (1974), Hungary (1975) and Czechoslovakia (1982).

In 1977, Optimističeskaja was connected with Vetrovaja by the Speleoclubs of Lvov and Odessa, led by L. Sučovej and M. Savčin.

Map: Speleoclub of Lvov (1965-1984) sent by A. Klimchouk.

Bibliography: Čikišev (A.G.) "The Caves of the Territory of the USSR", Moscow, ed. Nauka, 1973.
Dubljanskij (V.N.), Smolnikov (B.M.) "Karst-geophysical Investigations of Karst Cavities of the Pridnesrovskaja Podoliya and Pokut'je", Kiev, ed Naukova Dumka, 1969.
Savčin (M.P.), Gunevskij (I.M.) "Optimističeskaja Cave on the Podoliya" in "Karst of the Ukraine", Kiev, 1970, 4.
Dubljanskij (V.N.), Lomaev (A.A.) "Karst Caves of the Ukraine", Kiev, ed Naukova Dumka, 1981.
Dubljanskij (V.N.), Iljukhin (V.V.) *op. cit.*, 1982.
from text by M. P. Savčin (Lvov).

2. *Ozernaja* ....107,000 m
(Dnestrovsko-Pričernomorskij, Ternopol, Ukrainskaja SSR)

Located in the same region as Optimističeskaja, it is the world's second longest gypsum cave. It was discovered in 1940 and mainly explored beginning in 1963 by cavers from Ternopol. The cave passed the 100 km length mark in 1975. The layout of its labyrinth of passages is very similar to that of its neighbor (Dubljanskij/Iljukhin, *op. cit.*, 1982, map).

3. *Zoluška* ....82,000 m
(Dnestrovsko-Pričernomorskij, Moldavskaja SSR)

The Dnestrovsko-Pričernomorskij speleological region has a third very long cave, also in gypsum: Zoluška. The entrance (alt. 120 m) is located in the territory of the Moldavia, but the main part of the cave in in the territory of the Ukraine SSR. It covers an area of 170,000 m².

The cave is laid in the horizontal Neogenic gypsum strata about 30 m thick. The gypsum is overlain by limestones, marls and loams with a thickness of 50 m.

The entrance was opened 35 years ago by quarrying and the cave was changed from being phreatic to being vadose by pumping, which lowered the water table.

The cave is a labyrinth formed by horizontal galleries and passages in the upper part of the strata. Some pits 15 to 17 m deep lead to the lower part of the cave which is flooded and not yet studied.

Large galleries (widths 3 to 6 m, heights 3 to 8 m) are predominant in the central part of the labyrinth. Smaller passages are found on the periphery. Oval form cross sections are typical. The main passage directions are 30 to 60°, 90°, and 290-310°. There are many lakes in this cave, and it has an anomalously high CO_2 concentration in some areas (up to 4% by volume).

Cavers from Černovcij led by V. Koržik and V. Andrejčuk began exploration in 1976. By 1981 they had mapped 40,000 m, and 70,000 m by 1983. In 1984 they began mapping a new part of the cave, reaching a length of 80,000 m by January 1, 1985. The volume is about 590,000 cubic meters.

Map: Speleoclub of Černovcij (1976-1984) sent by A. Klimchouk.

Bibliography: Koržik (V.P.) "Zoluška, the new large gypsum cave", *Doklady An USSR*, Kiev, 1979, series B, 11.
Dubljanskij and Lomaev, *op. cit.*, 1981.
Dubljanskij and Iljukhin, *op. cit.*, 1982.
from text by V. P. Koržik and V. N. Andrejčuk (Černovcij).

4. **Kap-kutan/Promežutočnaja** ....50,300 m
(Kugitang, Gissarsko-Alajskaja, Uzbekskaja SSR)

Longest limestone cave in the USSR. Exploration is actively in progress.

5. *Orešnaja* ....36,000 m
(Sajanskaja, Krasnojarsk, RSFSR)

This is the world's longest conglomerate cave. It is located in the Sajanskaja speleological region, in the Badžejskij massif, in the Krasnojarsk district (R.S.F.S.R.), at an altitude of about 600 m.

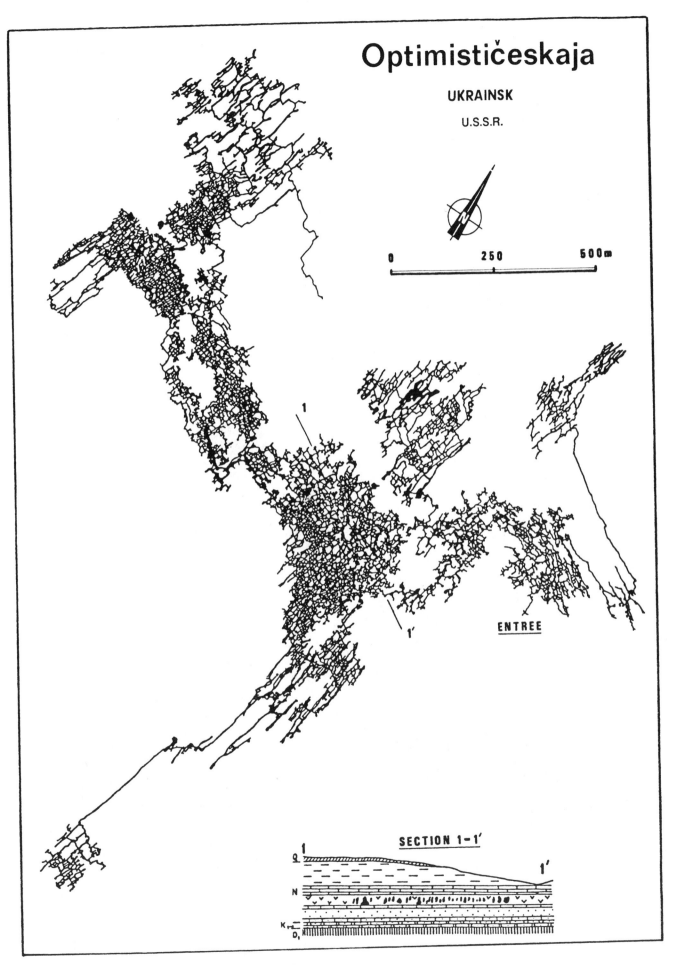

Optimističeskaja

UKRAINSK

U.S.S.R.

0   250   500m

1

1'

ENTREE

SECTION 1-1'

Q

N

K

O₁

1

1'

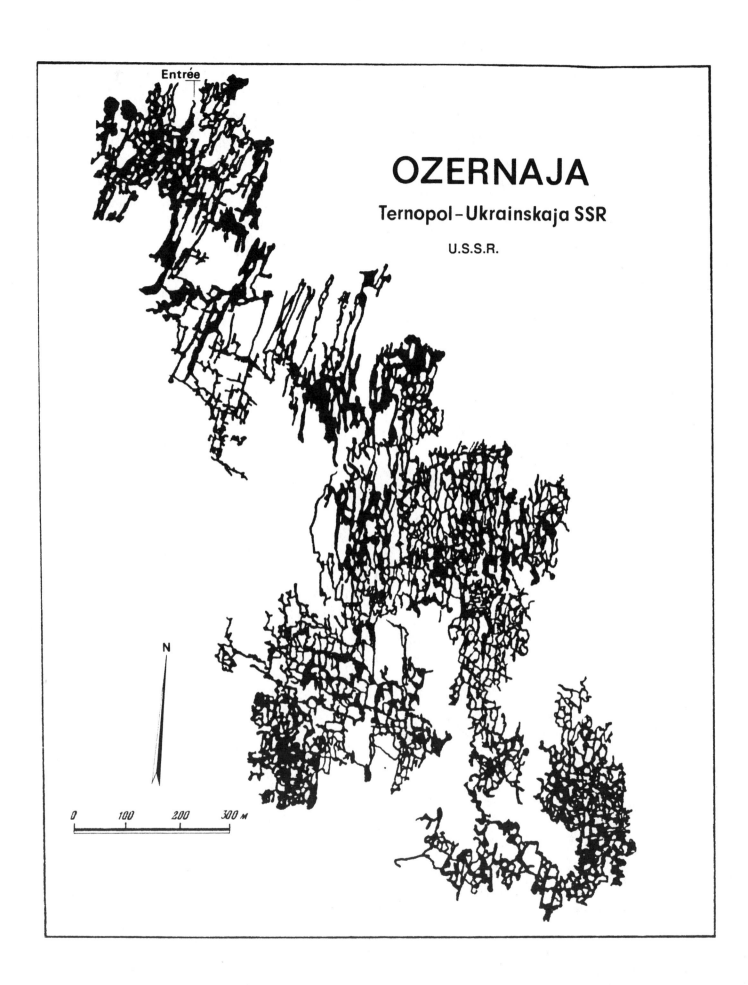

Entrée

OZERNAJA

Ternopol – Ukrainskaja SSR

U.S.S.R.

N

0   100   200   300 м

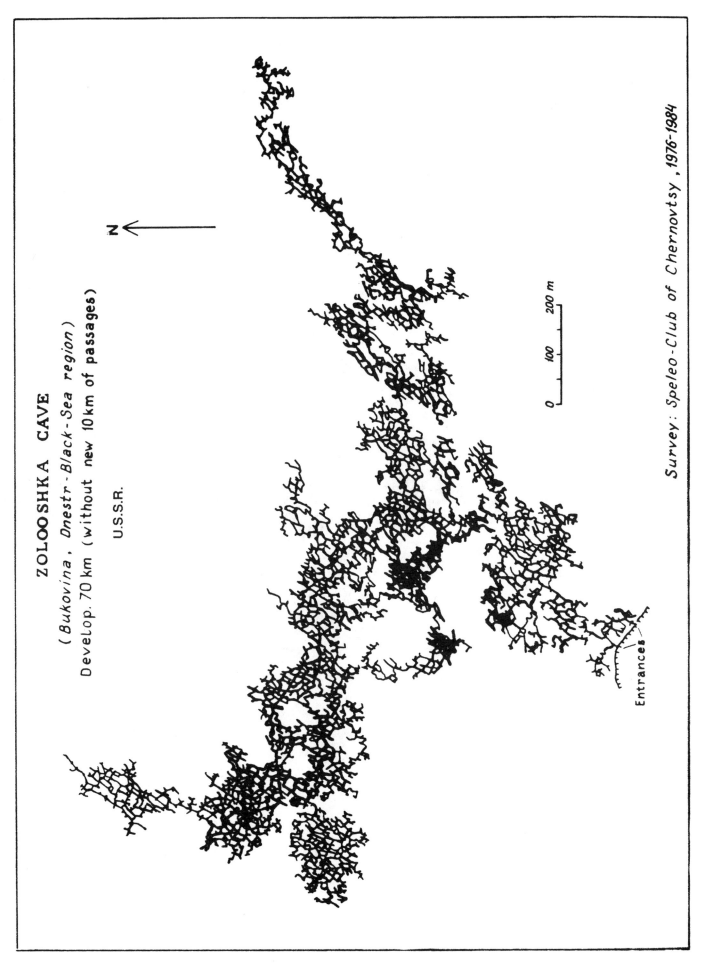

ZOLOOSHKA CAVE

(Bukovina, Dnestr-Black-Sea region)
Develop. 70 km (without new 10km of passages)

U.S.S.R.

N

200 m
100
0

Entrances

Survey: Speleo-Club of Chernovtsy, 1976-1984

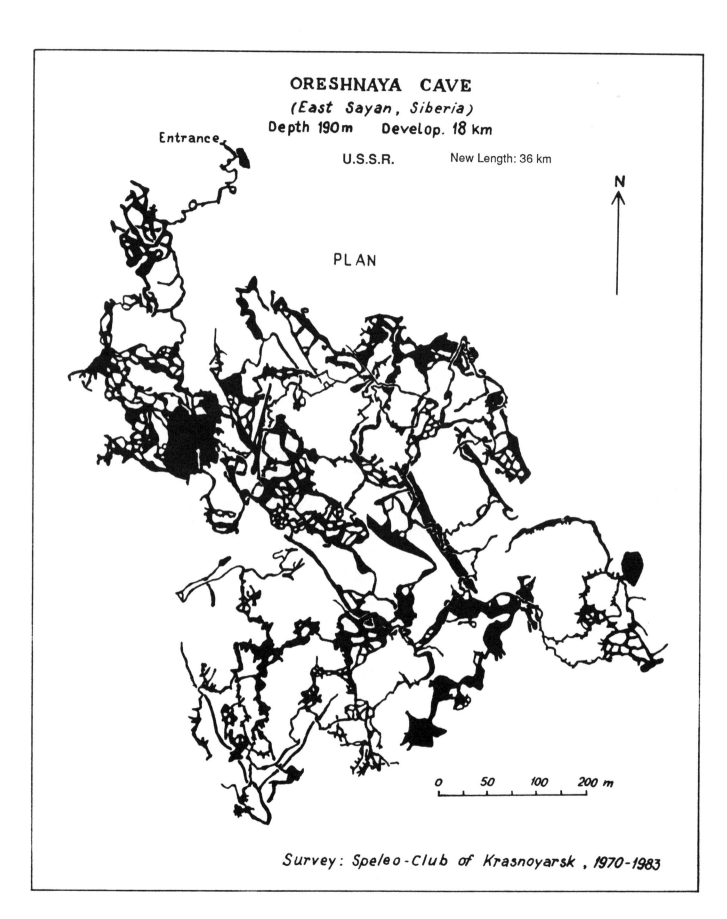

ORESHNAYA CAVE
(East Sayan, Siberia)
Depth 190m   Develop. 18 km

U.S.S.R.   New Length: 36 km

Entrance

N

PLAN

0   50   100   200 m

Survey: Speleo-Club of Krasnoyarsk , 1970-1983

The cave is developed in Ordovician conglomerates with a thickness of about 1200 m. They dip about 50 to 60 degrees to the WSW.

This is a labyrinth in several dimensions, dominated by large horizontal and inclined galleries. One also encounters several rooms and pits. Between –100 and –150 m, there are suspended lakes and sumps (maximum depth 35 m, maximum length 100 m.).

The entrance was discovered in 1964 when the Speleoclub of Krasnojarsk (S.K.) led by M. Dobrovolski and V. Bobrin reconnoitred the entrance galleries. From 1969 to 1971 exploration and mapping (by the S. K., the Krasnojarsk Geologic Office, V. Konosov, N. Larionov, S. Bijulbas, V. Vaščevič, R. Cykin, Z. Cykina, ...) showed a cave 11 km long and 190 m deep. From 1972 to 1983, the discovery of new sections brought the length to 18 km (volume: 186,000 cubic meters). By 1988 the length reached 36 km.

Map: Speleoclub of Krasnojarsk (1970-1983) sent by A. Klimchouk.

Bibliography: Cykin (R.A.) *et al.*, "The Caves of the Krasnojarsk", Krasnojarsk, 1974.
Cykin (R.A.), Cykina (Z.L.) "The Karst of the Eastern Part of the Altay-Sajan Folded Region", Novosibirsk, ed Nauka, 1978.
Dubljanskij/Iljukhin, *op. cit.*, 1982.

from text by R. A. Cykin (Krasnojarsk).

6. **Kristal'naja** .... 22,000 m
(Dnestro-Pričernomorskij, Ternopol, Ukrainskaja SSR)
Gypsum cave, mainly explored in 1962, 1963, and 1979 (Dubljanskij/Iljukhin, *op. cit.*, 1982, map).

7. **Ml'inki** .... 19,100 m
(Dnestrovskio-Pričernomorskij, Čortkov, Ukrainskaja SSR)
Gypsum cave explored from 1960 to the present (Dubljanskij/Iljukhin, *op. cit.*, 1982, map with 15,000 m).

8. **Snežnaja** .... 19,000 m
(Bzybskij, Bol'šoj, Kavkaz, Abkhazian SSR)
See above.

9. **Kulogorskaja-1-2** (Pinego) .... 14,100 m

10. **Krasnaja** .... 14,000 m
(Simferopol, Krymsky, Ukrainskaja SSR)
Known in the 19th century under the name of Kizil-Koba. (Dubljanskij/Iljukhin, *op. cit.*, 1982 map).

11. *Gaurdakskaja* .... **11,010 m**
(Kugitang, Gissarsko-Alajskaja, Turkmenskaja SSR)
This cave (alt. 540 m) is located in the Gissarsko-Alajskaja speleological region, in the Kugitang mountain chain, in SE Turkmenistan.

The entrance area is formed in gypsum while the rest of the horizontal part of the cave is developed in the Kimmeridgian-Oxford limestones.

Except for the entrance hall, Gaurdakskaja is mostly composed of horizontal galleries developed along faults and fractures with directions of 300° (for the big passages), 330°, 0°, and 20°. There are many dripstone and crystal formations of calcite, aragonite, and gypsum. The air temperature is between 18 and 24° C.

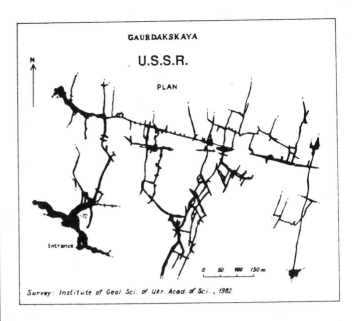

GAURDAKSKAYA

U.S.S.R.

PLAN

N

–72

Entrance

0 50 100 150 m

Survey: Institute of Geol. Sci. of Ukr. Acad. of Sci., 1982

The entrance to the cave was exposed by quarrying of sulphur deposits in 1973. Exploration by cavers from Kiev led by N. Jablokova and V. Rogožnikov brought the length to 7080 m in 1974. A new survey showing 11,010 m using more precise methods for engineering purposes was made in 1982 by the Institute of Geological Sciences, Kiev, led by A. Klimchouk. The depth is –92 m and the volume 83,000 cubic meters.

Map: Institute of Geology of Kiev, Academy of Sciences of the Ukraine (1982) sent by A. Klimchouk.

Bibliography: Lazarev (I.I.) "Geological and Mineralogical Peculiarities of the Gaurdakskaja Karst Cave", *Peščery*, Perm, 1976.
Dubljanskij, Iljukhin, *op. cit.*, 1982.

from text by A. Klimchouk (Kiev).

12. **Voroncovskaja** (Karkaz, RSFSR) .... 10,640 m

13. *Jaščik Pandory* .... **10,100 m**
(Salairo-Kuzneckij, Krasnojarsk, RSFSR)
The cave is at an altitude of about 670 m in the limestones of the Upper Proterozoic. Labyrinth in many dimensions, with predominantly inclined galleries. The horizontal section is around the –170 to –180 m level and is partially flooded.

In 1981 the Speleoclub of Tomsk (V. Vlasenko, M. Kopytov, A. Karpukhin) began exploration and mapped 4500 m, then, with the help of the Speleoclubs of Novokuzneck and Krasnojarsk, 10,100 m in 1983. Several sumps were dived (P. Minenkov).

Map: Speleoclubs of Tomsk, Novokuzneck and Krasnojarsk (1981-1983) sent by A. Klimchouk.
from text by M.V. Kopytov and V.V. Vlasenko (Tomsk).

13. **Sumgan-Kutuk** (Južno-Ural, RSFSR) .... 9860 m
(Dubljanskij/Iljukhin, *op. cit.*, 1982, map with 8200 m).

14. **Div'ja** .... 9720 m
(Kamsko-Srednevolzskaja, Ural, RSFSR)
(Dubljanskij/Iljukhin, *op. cit.*, 1982, map).

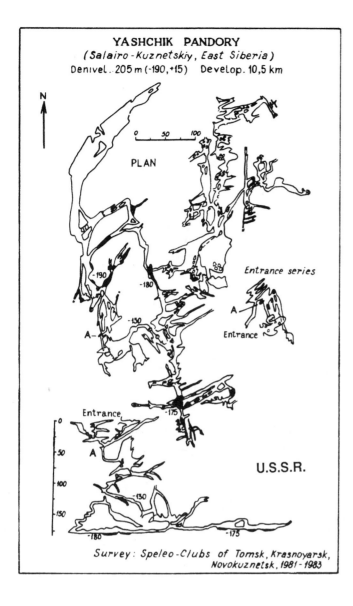

YASHCHIK PANDORY
(Salairo-Kuznetskiy, East Siberia)
Denivel. 205 m (-190, +15)   Develop. 10,5 km

N

PLAN

0   50   100

-190

-180

-130

Entrance series

A

A

Entrance

A

Entrance   -175

0
-50
-100
-150

-130

-180   -175

U.S.S.R.

Survey: Speleo-Clubs of Tomsk, Krasnoyarsk, Novokuznetsk, 1981-1983

15. **Verteba** ....7820 m
(Dnestrovsko-Pričernomorskij, Ukrainskaja SSR)
Gypsum (Dubljanskij/Iljukhin, *op. cit.*, 1982 map).

16. **Kizelovskaja-Viašerskaja** ....7600 m
(Kamsko-Srednevolžskaja, Ural, RSFSR)

17. **Kulogorskaja 1-2** ....7195 m
(Baaldajsko-Kulojskaja, RSFSR)
Gypsum. In 1984.

18. **Kinderlinskaja** ....6700 m
(Ural, RSFSR)
(Dubljanskij/Iljukhin, *op. cit.*, 1982, map with 5700 m).

19. **Nazarovskaja-Osennaja**..........................6500 m
(Alek, Bol'šoj Kavkaz, RSFSR)

20. **Khašim-Ojik** ....6100 m
(Kugitang, Gissarsko-Alajskaja, Uzbekskaja SSR)
In 1985.

21. **Badžejskaja** ....6000 m
(Sajanskaja, Krasnojarsk, RSFSR)
(Dubljanskij/Iljukhin, *op. cit.*, 1982, map).

22. **Konstitucionnaja** ....5880 m
(Pinega, Baldajsko-Kulojskaja, RSFSR)
(Dubljanskij/Iljukhin, *op. cit.*, 1982, map).

23. **Kungurskaja** ....5600 m
(Kamsko-Srednevolžskaja, Ural, RSFSR)
1965 (Dubljanskij/Iljukhin, *op. cit.*, 1982, map).

24. **Olimpijskaja** ....5500 m
(Baldajsko-Kulojskaja, Dvinsko-Mezenskaja, RSFSR)
(Dubljanskij/Iljukhin, *op. cit.*, 1982, map).

25. **V. V. Iljukhina** ....4816 m
(Arabika, Gagrinski, Abkhazian SSR)
See above.

26. **Kumičevskaja** (Valdaj-Kuloj) ....4170 m

27. **Abrskila** (Kavkaz, Abkhazian SSR) ....4000 m

28. **Absoljutnaja**...4000 m
(Lagonaki, Kavkaz)

WEST GERMANY
Bundersrepublik Deutschland

The principal German karst areas are located south of Stuttgart (Schwäbische Alb), east of Nürnberg (Fränkische Alb), and in the south, on the Austrian-German border (Bavarian Alps). West Germany has several gypsum caves, notably in the southern Harz, the longest being Höllern (Markt Nordheim) at 1040 m. There are also 41 show caves. The history of German speleology is distinguished by an emphasis on archeological and paleontological research; in 1608 human skeletons were discovered in Steinbachhöhle. In the 18th century, when the first topographic surveys were done (Einhornhöhle, 1734, Sontheimer Höhle, 1791), caves with bones were studied by Nagel, Leibniz, and especially J.-F. Esper (Muzzendorf cave). The first show cave opened in 1803 (Nebelhöhle). The first caving club, the Schwäbische Höhlenverein, was founded in 1889 by Hendriss and Gusmann. The Verband Deutscher Höhlen und Karstforscher was formed between the two world wars.

DEEP CAVES:

1. **Geburtstagsschacht** .... –698 m

2. **Ze-Schlinger** .... –585 m

3. **Kargrabenhöhle** .... –447 m

4 **Mickerloch** .... –348 m

5. **Salzgrabenhöhle** ....(–90, + 180) 270 m

6. **Canyonhöhle**	–264 m
7. **Zwei-fuß-schacht**	–238 m
8. **Lindwurm-Loch Ness system**	–230 m
9. **Sonntagshornhöhle**	–230 m
10. **Polypenhöhle**	–215 m
11. **Polterschacht**	–213 m
12. **Cannstatter schacht**	–200 m
13. **Jägerkreuzschacht**	–200 m

Note: All of the above caves are found in the Bavarian Alps.

LONG CAVES:

1. **Salzgrabenhöhle** .... 7797 m
   (Simetsberg, Steinernes Meer, Bavarian Alps)

2. **Fuchslabyrinth** (Hohenlohe) .... approx. 6000 m

3. **Kluterhöhle** .... 5700 m
   (Westfalen, Ennepetal-Sauerland)

4. **Falkensteiner höhle** .... approx. 5000 m
   (Württemberg, Schwäbische Alb)

5. **Mordloch** .... 4200 m
   (Württemberg, Schwäbische Alb)

6. **Heilenbecker höhle** (Sauerland) .... 3000 m

7. **Frickenhöhle** .... 2814 m
   (Farchant, Oberbayern, Bavarian Alps))

8. **Ze-Schlinger** (Bavarian Alps)) .... approx. 2500 m

9. **Erdmannshöhle** .... 2148 m
   (Hasel, Weher, Dinkelberg)

10. **Attendorner Tropfsteinhöhle** .... 2000 m
    (Sauerland)

11. **Windloch bei Kauerheim-Alfeld** .... 2000 m
    (Fränkische Alb)

Yugoslavia

Jugoslavija

The origin of the word karst (kras in Yugoslavian) comes from the country around the Trieste bay called Carsus (Carso). In more recent times scientists became interested in karstic phenomena as they traveled the major route linking Vienna to the Mediterranean that passed through the barren and rocky areas, full of dolines and poljes, between Postojna and Trieste.

The first to write extensively about our caves was N. Gučetić in 1584. Some caves in the Ottoman part of the Balkans were described by the traveller E. Čelebija in 1660. In the later half of the 17th century, J.V. Valvasor researched and wrote about karst phenomena. In 1689 he published the first cave map and wrote about karst drainage and the formation of speleothems. He published more than sixty cave descriptions in his work on the Carniole. Among those who have followed him, we should mention the mathematician Nagel, from Vienna, the doctor B. Hacquet, from Idrija, and the Jesuit Gruber. In 1774, when it was still more accessible to Westerners, A. Fortis studied the Dalmate karst and published several cave maps. The same area was also described by J. Lovrić, from Kinj, in 1776.

Postojnska jama, opened to tourists in 1818, contributed to the popularization of karst. In the second half of the 19th century, the formation of the Vienna Speleological Society gave an impetus to scientific research on the poljes of the Carniole which were regularly flooded. At the end of the century kartology began to develop: in 1869 at Postojna, the first Yugoslavian speleological society was formed, Antrom. In 1910 a commission to explore caves was formed with, which J. Poljak did important work. In Serbia, J. Cvijić began his work at the end of the 19th century which led, in 1924, to the creation in Belgrade of a speleological section in the geological society. A. Boué was the first to describe the karsts of Bosnia and Herzégovinia, while the formation, in 1888, of a section of natural sciences in the Regional Museum of Sarajevo permitted the continuity of exploration in this area. The first reports of the Vjetrenica cave date back to 1889.

It was after the second world war that modern speleology developed throughout Yugoslavia: in 1954 the first national congress was organized at Postojna. It brought together representatives from all the republics, and as a result the Association of Yugoslavian Speleologists was formed. Many cavers are organized in connection with alpine or tourist clubs. Today, there are several thousand cavers whose work is described in periodicals such as *Naše Jame, Naš Krš, Speleolog* or the scientific journals *Acta Carsologica, Krš Jugoslavije.*

Some of the major karst areas are Slovenia, Hrvatska, Serbia, and Macedonia. Karst forms 45% (80,000 km²) of the surface are of Hrvatska, 35% of Bosnia and Hercegovina (18,000 km²), 72% of Crna Gora (9,500 km²), 45% of Slovenia (9,000 km²), while in the two other republics less than 10% of the surface area is karsitified. One finds traces of karstification at Triglav (2864 m) and up to 2960 m below sea level in the Dugi otok.

The Dinenc karst extends from Italy to Albania and covers a region 650 km long and 200 km wide. It can be divided into three parts:

1. The littoral zone, an orogenetic karst which includes the classic karst. Submerged springs are its main characteristic, those of Ljuta (300 m³/s) having the largest flow rates;

2. The high karst, parallel to the first, also orogenetic (Mesozoic series) with a thickness reaching 5 km in places. Mountains and plateaus alternate, with poljes making up the low areas (the largest is the Lićko polje, 464 km²).

3. The fluvio-karst zone (merokarst) which skirts the high karst, characterized by water flowing on the surface.

The alpine karst includes the NW of Slovenia (Trias limestones and dolomites, with layers over 2000 m thick).

The Croato-sloven karst includes a part of Gorski Kotar, Kordun and Bela Krajina (Cretaceous and Trias limestones). The high altitude Montenegran karst is made up of mountains and vast plateaus cut by deep canyons (over 1000 m deep). The largest lapiaz areas are found in the Durmitor and Orjen areas.

The are also isolated karsts in the NE of Slovenia, NW Croatia, east of the Bosnia, to the SW and east of Serbia, and in central Macedonia (notably the marble karst of the Jakupica massif where one finds the highest Yugoslavian polje, at an altitude of 2170 m).

There are presently over 16,670 caves catalogued (5,600 in Slovenija, 5500 in Hrvatska, 2000 in Crna Gora, 1820 in Serbia, 1450 in Bosnia and Hercegovina). An unusual speleological feature is the 500 m deep collapse doline Crveno jezero, near Imotski.

A. Kranjc, M. Kranjc and F. Malečkar.

DEEP CAVES:

1. jama na Vjetrenom Brdu .... **–897 m**
(Durmitor, Crna Gora)

Also called jama u Vjetrenim Brdima. The "cave of the wind mountain" is located in the Durmitor massif, near Zabljak, at an altitude of 2180 m. It was discovered in 1984 by a group of Polish cavers who descended to –350 m. In August 1985, an international expedition "Durmitor 85" gathered together French, British, Plish, and Yugoslavian cavers, who pushed exploration to a sump at –897 m.

Map: from the expedition notes in the report "Medunarodna Speleološka Ekspedicija "Durmitor 85", 1985.

2. brezno Pri Gamsovi Glavici .... **–773 m**
(Bohinj, Radovljica, Slovenija)

A cave located in the Julian Alps, not far from Radovljica, at an altitude of 1610 m. It is located in the Pršivec massif (1761 m), to the NW of Bohinsko jezero (Bohinj lake), on the bank of which can be found the resurgence. The cave is formed in Trias limestones. It was discovered in 1969 by the J.S. P.D. Železničar (caving section of the Alpine Club). This club descended to –178 m in 1970 and to –360 m in 1971. With the D.Z.R.J. Ljubljana (Speleological Society of Ljubljana), they pursued exploration in 1972 and in 1973, with the help of the J.D. Dimnice (Speleoclub of Dimnice), reaching a depth of − 440 m.

In 1978 the D.Z.R.J. Ljubljana, after enlarging constrictions and moving breakdown at the –570 m level, reached the depth of –621 m. The following year, they reached –768 m, for a length of 3112 m. In 1980 an international expedition, with cavers from Wroclaw and Yugoslavians from the J. D. Dimnice and D.Z.R.J. Ljubljana, undertook a project to clear the breakdown at the end of the cave, but only made a small gain in depth, to –773 m.

In 1981 the D.Z.R.J. Ljubljana found a split at –590 m and in 1984 another at − 120 m, while beyond the old terminus at –440 m passages were found descending to –474 m. The length increased to 3912 m, but the depth remained unchanged.

Map: from the drawing of the explorers, sent by F. Malečkar.

Bibliography: Pirnat (J.), Planina (T.) - Brezno pri Gamsovi Glavici v Julijskih Alpah, *Naše Jame*, 1974 (15): 47-55.

Andjelic (J.), Malečkar (F.) - Brezno pri Gamsovi Glavici v Julijskih Alpah, *Naše Jame*, 1979 (20): 40-58.
Malečkar (F.) - Brezno pri Gamsovi Glavici..., *Novice*, 1980 (15): 18 and *Pionir*, 1980 (7).

from text by A. Kranjc and S. Morel.

3. Pološka jama .... **–704 m**
(Tolmin, Tolmin, Slovenija)

The three entrances to the cave are located on the slopes of mount Osojnica (1380 m) at altitudes of 1281 m (Odkopani vhod), 738 m and 720 m (Pološka jama). They overlook the valley of Tolminka which runs to the south towards the nearby town of Tolmin, in the Julian Alps.

The cave is formed in Trias limestones. A dye trace has shown the resurgence to be the springs of Tolminka (alt 680 and 540 m).

The lower entrances were discovered in 1924 by the P.K. Krpelj (Alpine Club) of Tolmin who ascended 24 m. In 1964 I. Kenda of the J. K. Idrija succeeded in forcing the terminal constriction and finding a way on. In 1965 the J.K. (Speleoclubs) Idrija and Luka Čeč of Postojna reached +95 m where a new constriction slowed exploration; this was enlarged the same year by the J.K. Idrija and Ljubljana-Matica. In 1966, all the clubs joined together and pushed to +225 m.

In 1970 J. Russum (of Happy Wanderers Club, England) and A. Lajovic (of the J. S. P. D. Železničar) cleaned a boulder choke at +129 m and increased the vertical extent to 465 m (+310, –156); the length then reached 8020 m. In 1971 the J.K.L.-M. and the J.S. P. D. Ž., after some difficult climbs, progressed in altitude to +519 m, for a vertical extent of 674 m.

In July, 1974 the J.K.L.-M. opened up a small pit and added the upper entrance, Odkopani vhod, which increased the vertical extent to 704 m and the length to 10,800 m.

Map: from the drawings of the explorers, sent by Primož Krivić.

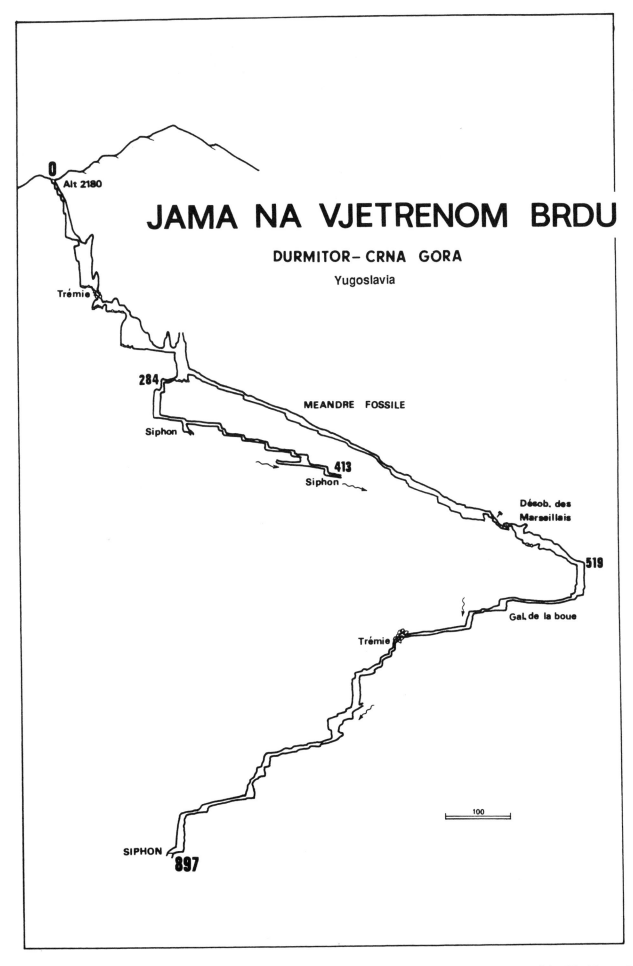

JAMA NA VJETRENOM BRDU

DURMITOR– CRNA GORA

Yugoslavia

0
Alt 2180

Trémie

284

Siphon

MEANDRE FOSSILE

413

Siphon

Désob. des
Marseillais

519

Gal. de la boue

Trémie

100

SIPHON
897

POLOŠKA JAMA

JULIJSKE ALPE – JUGOSLAVIJA

Yugoslavia

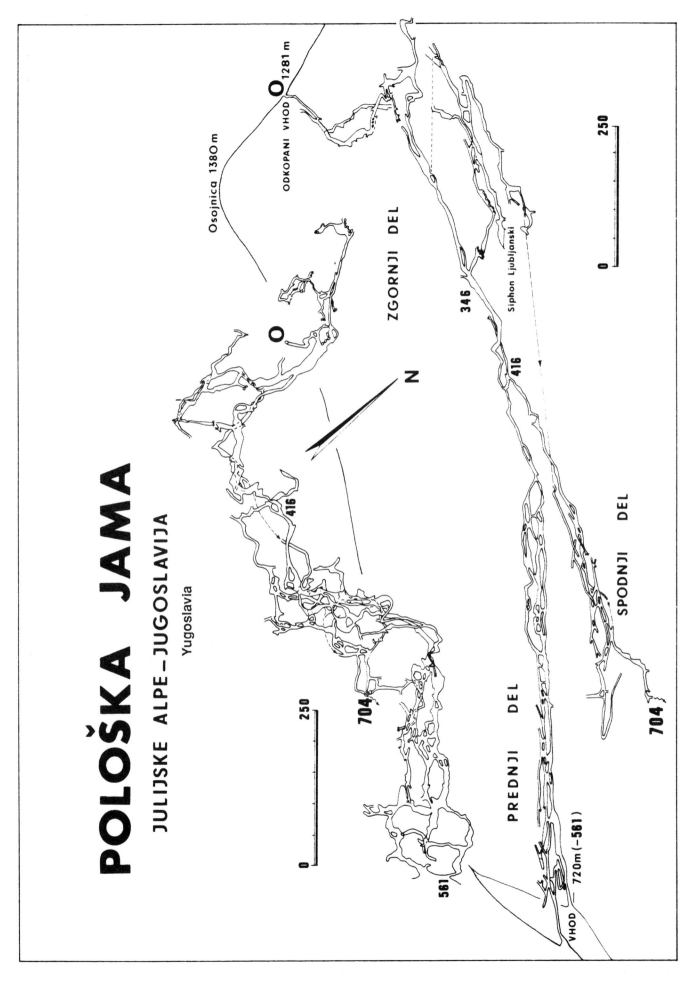

Osojnica 1380 m

ODKOPANI VHOD ○ 1281 m

ZGORNJI DEL

346

416

Siphon Ljubljanski

N

250

0

250

0

416

704

561

PREDNJI DEL

SPODNJI DEL

704

VHOD — 720m (–561)

Bibliography: Habič (P.) - Pološka jama..., *Naše Jame*, 1971 (12): 23-24.

Habič (P.) - Krivić (P.), "New discoveries in the Polog cave", *Naše Jame*, 1972 (13): 98-108.

Lorek (Z.) - Polskie partie w Polosce, *Alpiniste Complet*, Katowice, 1981 (3) : 32-35.

from text by A. Kranjc and S. Morel.

4. *jama U Malom Lomnom dolu* .... —650 m
(Durmitor, Crna Gora)

This difficult cave with many constrictions was explored in 1984 by a Polish team to — 605 m. In 1985 the depth of — 650 m was reached during the international "Durmitor 85" expedition. The explorers stopped at a constriction.

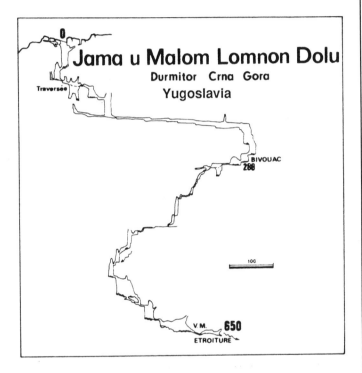

5. Majska jama............................(–581,+11) 592 m
(Bohinj, Radovljica, Slovenija)

This cave is formed in upper Triassic limestones. The resurgence is in the Bohinj lake (alt. 526 m) at an altitude of 1705 m, in the Julian Alpes (see brezno pri Gamsovi Glavici). Discovered in 1958 by the J.S. P.D. Ž who explored the entrance zone in 1975. In 1979, the D.Z.R.J. Ljubljana reached –230 m, after clearing obstructions, then –420 m in 1980, in spite of a tight meander. In 1982, with the J. D. Radek and the S.W.C.C., they descended to –469 m and finally to –581 m in 1983, having followed a stream from the –380 m level to the terminal sump (Paternu (M.), Pintar (G.), Majska jama, *Naše Jame*, 1983 (25) : 50-54.).

6. Stara Skola .... –576 m
(Dalmacija, Biokovo, Hratska)

7. Vilimova jama .... –565 m
(Biokovo, Makarska, Hrvatska)

Altitude 1630 m. Explored in 1984 to –396 m (*Spelunca*, 1985 (17) profile) by a Czech team.

8. ponor na Bunjevcu .... –554 m
(Lika, Velebit, Hrvatska)

This swallow cave (Jurassic limestone) is located at 1200 m altitude in the Velebit massif (1758 m) to the north of the town of Zadar. It was found in 1973 by the S.D. P.D. Velebit Zagreb, then explored by this club in 1976 to –445 m and in 1977 to –534 m (*Speleolog*, Zagreb, 1978-1979, XXVI-XXVII, profile). Later, the discovery of an upper entrance added 20 m to the depth.

9. M-16 .... –547 m
(Tolmin, Tolmin, Slovenija)

The M-16 shaft (Julian Alps, north of Tolmin) is at an altitude of 1840 m and formed in Upper Trias limestones. The resurgences are the Pod Boko springs (alt. 620 to 660 m), located on the banks of the Tolminka river. It was discovered in 1983 by the J.S. P.D. Tolmin and explored by this club to –420 m in 1983 and to –547 m in 1984 (F. Šušteršič, M-16 v družbi najglobljin, *Delo*, 19.11.1984, p. 13).

10. brezno pri Leški Planini .... –536 m
(Radovljica, Slovenija)

Located on the Jelovica plateau, at 1120 m altitude, in Trias limestones. The resurgence is the Lipnica River spring (alt 500 m). Discovered in 1976 by the I.Z.R.K. (Institute of Karst Researches) of Postojna. In 1978 the D.Z.R.J. Kranj reached –536 m (D. Preisinger, Brezno pri Leški planini, *Naše Jame*, 1979 (20) : 76-78, profile; A. Krancj and F. Malečkar, Brezno pri Leški Planini-dopolnilne raziskave, *Naše Jame*, 1982 (23-24) : 73-78).

11. jama pod Kamenitim Vratima .... –520 m
(Biokovo, Makarska, Hrvatska)

This cave, explored by various clubs in 1980, is located in the Blokovo mountains, above Makarska, to the SW of Split. It is a succession of pits, the deepest of which is 220 m.

12. brezno Presenečenj .... –472 m
(Mozirje, Mozirje, Slovenija)

Alt. 1030 m. Discovered in 1974, explored to –177 m in 1974, –324 m in 1976, and –472 m in 1977 by the J.K. Črni Galeb of Prebold (*Naše Jame*, 1978 (19) profile).

13. jamski sistem u Obručinama .... –471 m
(Durmitor, Zabljak, Crna Gora)

14. Jojkinovac .... –460 m
(Grmeč, Bosna i Hercegovina)

15. Solunska Glava .... –450 m
(Jakupica, Makedonija)

16. Velika ledenica v Paradani .... –385 m
(Ajdovščina, Ajdovščina, Slovenija)

Alt. 1090 m. Cave with ice exploited in 1873 and explored in 1917 (–92 m, Michler, Kunaver), 1949-1951 (–120 m, D.Z.R.J. Ljubljana), 1977 (–242 m, J. D. Logatec) and 1978 (–385 m, J. D. Logatec) (*Naše Jame*, 1979 (20) profile).

17. brezno Martina krpana .... –370 m
(Bohinj, Slovenija)

18. **Biokovk** .... –359 m
(Dalmacija, Biokovo, Hrvatska)

19. **ponor Pepelarica** .... –358 m
(Lika, Velebit, Hrvatska)

20. **jama kod Rašpora** .... –355 m
(Rašpor, Istra, Cićarija, Hrvatska)
Alt. 695 m. Explored in 1924-1925 by the Commissione Grotte Boegan (Soc. Alpina delle Giulie) de Trieste (Abisso Bertarelli). Depth fixed at –355 m by the J.K.L.-M. (*Naše Jame*, 1969 (11) profile).

21. **Habečkov brezen** .... –353 m
(Črni vrh, Idrija, Slovenija)
Alt. 668 m. Explored in 1926 by the Assoc. XXX Ottobre de Trieste (abisso del Montenero). Depth revised in 1954 (–366 m) by the Slovenians and the Triestans. Again revised to –353 m in 1977 by the J.D. Logatec (*Naše Jame*, 1978 (19)).

22. **Kavkina jama** .... –350 m
(Tolmin, Tolmin, Slovenija)
Alt. 1830 m. Explored in 1976 (–81 m), 1977 (–267 m), 1978 (–306 m) (*Spelunca*, 1979 (4) profile) and 1982 (–350 m) by the J. D. Tolmin.

23. **Mraznik** (Jakupica, Makedonija) .... –350 m

24. **Jesenska jama** .... –338 m
(Kamnik, Slovenija)

25. **Jazben** or **na Banjščicah** .... –334 m
(Nova Gorica, Slovenija)
Alt. 595 m. Explored in 1925 (–74 m) and 1928 (–334 m) by the S.A.G. de Trieste (abisso di Verco). Depth revised in 1958 and 1968 by the D.Z.R.J. Ljubljana (*Naše Jame*, 1959 (1-2); 1978 (19)).

26. **jama pod Gračišče II** .... –329 m
(Brač, Pražnice, Hrvatska)
Alt. 485 m. Explored in 1958-1959 (S.D. Hrvatska). Vertical pitch of 233 m according to the measurement of 1978 (*Spelunca*, 1979 (4) profile).

27. **Klanski ponor** .... –320 m
(Primorje, Rijeka, Hrvatska)

28. **jama u Dubašniči** .... –320 m
(Mt Kučaj, Srbija)

29. **Puhaljka jama** .... –318 m
(Lika, Velebit, Hrvatska)

30. **Golerjev pekel** .... –317 m
(Solčava, Mozirje, Slovenija)
(*Naše Jame*, 1978 (19); 1979 (20), profile).

31. **Todorova jama** .... –316 m
(Pivska planina, Crna Gora)

32. **Duboki Do** .... –310 m
(Njeguši, Crna Gora)

Alt. 880 m. Descended in 1916 by G. Lahner (Martel, *Nouveau traité des eaux sout.*, 1921, profile p. 184). According to 0 point: –332 m, –340 m.

33. **Klemenškov pekel** .... –310 m
(Solčava, Mozirje, Slovenija)

34. **Ljubljanska jama** .... –310 m
(Kamnik, Kamnik, Slovenija)
(*Naše Jame*, 1974 (15) map).

LONG CAVES:

1. *Postojnski Sistem* .... 19,495 m
(Postojna, Slovenija)
Postojnska jama is a show cave with a worldwide reputation. It is located a few kilometers north of the town of Postojna (which is 52 km south of Ljubljana). It is part of a system that includes four other caves: Otoška jama (length 632 m), Magdalena jama (length 1359 m), Črna jama (length 3029 m) and Pivka jama (length 1844 m).

Postojnki sistem contains the underground flow of the upstream section of the Pivka river, which is swallowed at an altitude of 511 m into Postojnska jama (the entrance to the tourist cave is at 520 m) and resurges from Planinski jama (alt. 447 m). It is formed in Cretaceous age limestones from the dinaric chain. This area (from which the word "karst" originates, being "kras" in Yugoslavian) is that of classic karst, with many examples of well known features such as poljes, dolines, and sinkholes.

The history of exploration is very rich and is tied to the beginnings of serious speleology in the 19th century. The first visitors date to at least 1213, based on inscriptions on the walls. The first mention in literature dates to possibly 1624, or, more certainly, to 1680-1681.

Systematic exploration began in 1818, with the first trips to the lower parts of the system by the worker Luka Čeč. From 1851 to 1856, A. Schmidl and J. Rudolf followed the underground course of the Pivka. They mapped 5850 m.

From 1885 to 1893 the caving club Antron de Postojna pursued exploration. In 1890, Postojnska jama was connected to Otoška jama, discovered in 1889. In 1893 E.-A. Martel visited Postojna and, with the Antron cavers, made the connection between Otoška jama and Magdalena jama on September 20.

In parallel, Pivka jama was explored by Schmidl and Rudolf in 1852 (for 950 m) and 1853, and Črna jama was explored in 1852 (for 493 m) and 1893 (600 m by Kralgher and Ruzicka). These two caves were only connected in 1966 by divers from the D.Z.R.J. Slovenija, bringing the length of the ensemble to 4088 m.

From 1905 to 1911, F. Muehihofer and A. Perko explored Magdalena jama. New passages were discovered in 1922 to 1923 by M. Vilhar (Kristalni rov, Ozki rov) and from 1947 to 1956 by the D.Z.R.J.S. and the I.Z.R.K. (Vilharjev rov, Matevžev rov).

Between 1975 and 1983, the D.Z.R.J.S. and the D.J.P. Proteus explored beyond the sumps in Crna, Pivka, and Magdalena jama.

Although the five caves were connected by an artificial tunnel 495 m long (dug between the two world wars), it was not until September 6, 1980 that divers from the D.J.P. Proteus made the long-hoped-for connection between

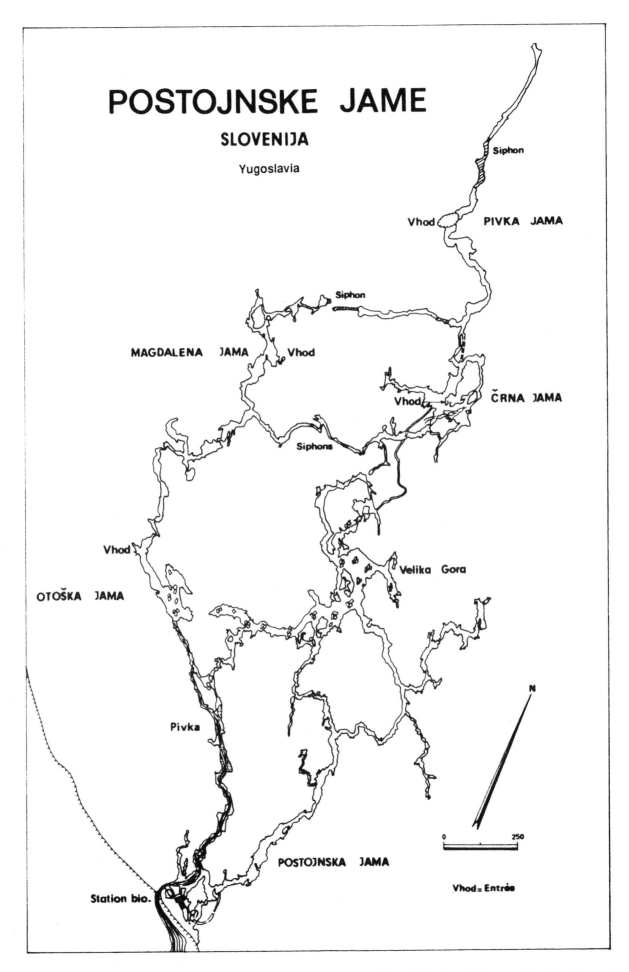

POSTOJNSKE JAME

SLOVENIJA

Yugoslavia

Siphon

Vhod · PIVKA JAMA

Siphon

MAGDALENA JAMA · Vhod

Vhod → ČRNA JAMA

Siphons

Vhod

Velika Gora

OTOŠKA JAMA

Pivka

N

0   250

POSTOJNSKA JAMA

Station bio.

Vhod = Entrée

Magdalena jama and Črna jama, by passing the Zvezni rov sumps.

Exploration today is in the downstream section of the system, in the hope of connecting with the resurgence, Planinska jama, some 5 km away.

There was been quite a variation in the reported length for the system, but a recent calculation of Samo Morel, using the most up-to-date maps, has fixed the length at 19,495 m. In addition, P. Jakopin has begun a volume calculation for Postojnska jama stricto sensu (Velika gora and Koncertna dvorana); over a length of 452 m, he obtained 303,924 cubic meters!

Map: from the drawings of Gallino, Petrini, Sartori (1933-1934), Hribar, Michler, Gosdoparič, Krašenec, sent by A. Kranjc.

Bibliography: Clüver (Ph.) - *Italia antiqua*, Leyden, 1624: 206-208.
Schloenleben (J. -L.) - *Carniola antiqua et nova*, Laibach, 1680-1681: 122-123.
150 let Postojnske jame, Postojna, 1968.
Habe (F.) - The guide books of the Postojna Cave, *Naše Jame*, 1969 (10): 15-32.
Shaw (T. R.) - *History of cave science*, Crymych, 1979, 2: 435-490.
Kranjc (A.) and M. Kranjc - The cave of Postojna in the speleological literature, *Naše Jame*, 1981 (227: 107-113).
Kranjc (A.) - Postonjska jama, un seul système, *Grottes et Gouffres*, 1981 (79): 28-29.
Martel (E. -A.) - Sous terre (sixième campagne). 1893, *Annuaire du C.A.F.*, 1893, pp. 305-325.

from text by Andrej Kranjc.

2. **Sustav-Dula-Medvedica** .... 15,701 m
(Ogulin, Hrvatska)

3. **Pološka jama** .... 10,800 m
(Tolmin, Tolmin, Slovenija)

4. **Sustav-Panjkov ponor-Kršlje** .... 9352 m
(Kršlje, Hrvatska)

5. **Kačna jama** .... 8470 m
(Divača, Sežana, Slovenija)
Explored from 1888 to 1895 (Hanke, Marinitch...) then 1972-1973 by the J. K Logatec. Contains a 278 m deep pit.

6. **Križna jama** .... 8163 m
(Lož, Cerknica, Slovenija)
Explored in 1824-1825, 1853 (Schmidl), 1878-1879, 1926, 1927-1934.

7. **Vjetrenica pećina** .... 7503 m
(Popovo polje, Zavala, Bosna i Hercegovina)
Explored in 1858, 1904, 1924-1928.

8. **jama pod Gradom (Predjama)** .... 7446 m
(Postojna, Postojna, Slovenija)
Explored in 1689 (Valvasor), 1853 (Schmidl), 1912-1922, 1952-1957, 1964 (M. Boon), 1983-1985 (D.Z.R.J. Luka Čeč).

9. **Velika i Mala Karlovica** .... 7307 m
(Cerknica, Cerknica, Slovenija)
In 1847, 1888 (V.Putick), 1921 (L. Bertarelli), 1927, 1931, 1962, 1964-1969.

10. **Veternica pećina** .... 6654 m
(Zagrebačka gora, Medvednica, Hrvatska)

11. **Jopicá pećina** .... 6564 m
(Krnjak, Karlovac, Hrvatska)

12. **Ušački pećinski** .... 6200 m
(Pešter, Srbija)

13. **Planinska jama** .... 6156 m
(Planina, Postojna, Slovenija)
Explored in 1848, 1850 (Schmidl), 1887 (Putick), 1932 (F. Anelli), 1949, 1966, 1970-1972.

14. **Bogovinska pećina** .... 5390 m
(Kučaj, Srbija)

15. **Škocjanske jama** .... 5088 m
(Divača, Sežana, Slovenija)
Explored in 1815, 1823, 1839, 1851 (Schmidl). Explored by Hanke and Marinitch (1885, 1887, 1890, 1893) who reached the terminal sump. System with eight entrances, including Mahorčičeva jama (swallow) and Velika dolina (alt. 428 m). Resurgence at Duino (Trieste, Italy, 34 km distant).

16. **Najdena jama** .... 4986 m
(Planina, Logatec, Slovenija)
Explored in 1937-1939, 1963-1971, 1978-1980.

17. **Dimnice** .... 4640 m
(Markovščina, Sežana, Slovenija)
Explored in 1905, 1968-1978, 1981-1982.

18. **Cerjanska pećina** (Niš, Srbija) .... 4240 m

19. **brezno pri Gamsovi Glavici** .... 3962 m
(Bohinj, Radovljica, Slovenija)

20. **Pešterica-Vetrena dupka** .... 3020 m
(Jerma, Srbija)

VI
AUSTRALASIA

AUSTRALIA

The first limestone discovered in New South Wales was located at Walli by the explorer Evans on May 21, 1815. While probably not the first limestone discovered in Australia, it does mark a beginning of karst documentation in Australia.

Early Australian explorers featured prominently in the early history of Australian caving areas, include Evans, Oxley, Mitchell, Grey, Eyre and Forrest to name a few.

With the discovery of the major tourist caves of Australia at areas such as Jenolan, Noracoorte, Wombeyan, Yarrangobilly, Buchan, Yallingup, Yanchep, Abercrombie, Mole Creek, Hastings and Chillagoe, the documentation of Australian caves began in earnest.

The Government Surveyor Oliver Trickett, although preceded by other cave mappers (such as Wilkinson and Leigh) and various geologists, is perhaps the closest counterpart to Martel in France: in the beginning of the 20th century he mapped and described most of the known caves of New South Wales. He also wrote the first guide to the show caves. Many of the great caves in other areas are due to the efforts of such guides as J. Wilson and J. Wibard (Jenolan), A. Bradley (Yarrangobilly), L. Guymer (Bungonia), W. Redden (Noracoorte) and F. Moon (Buchan), all of whom took an exceptional interest in caves and their exploration.

Speleology fell somewhat dormant in the 1920's, but was reawakened in the 1930's by an increasing number of newspaper reports and diaries kept by keen bushwalkers such as O. Glanfield.

In the 1940's Captain J.-M. Thomson led numerous expeditions to the Nullarbor plain which were popularized in magazines such as *Walkabout*.

The oldest caving club is the Tasmanian Caverneering Club, formed on September 13, 1946 and still active today. The Sydney University Speleological Society was formed on September 22, 1948, while the Sydney Speleological Society was formed on March 23, 1954. The Australian Speleological Federation was born on December 28, 1956, and today has over thirty member clubs, including at least one from each state.

The creation of these groups marked the true beginnings of Australian speleology. The Newsletters and Journals that they publish (*Speleo Spiel, Journal of the S.S.S, The Western Caver, Spar, Nargun, Tower Karst, Oolite, S.U.S.S. Bull., Down Under, C.E.G.S.A. Newsletter*) form a first class set of caving literature.

Cave diving began in the late 1950's in areas such as Jenolan and in Tasmania.

The cave research bulletin *Helictite* first appeared in October, 1962 and provided a regular forum for the writings of Joe Jennings, who arrived in Australia in 1952 and for over 32 years had a strong influence on Australian speleology, inspiring both professionals and amateurs.

In the 1960's, numerous groups directed their efforts towards saving menaced caves at Bungonia, Mount Etna, Colong, Precipitous Bluff, and Texas, bringing out numerous publications in the following decade (*Bungonia Caves, Mount Etna Caves*, etc.).

The 1960's also saw the publication of the A.S.F.'s *Speleo Handbook* listing many of Australia's caves with brief descriptions. This was also the beginning of the deep cave penetrations in Tasmania.

The number of caving publications increased dramatically in the 1970's. This period also saw the foundation of the Jenolan Caves Historical and Preservation Society on July 8, 1972. Expeditions to other countries, such as New Zealand and Papua New Guinea began around this time.

Cave discoveries on the Franklin and Gordon Rivers in Tasmania eventually resulted in the cancellation of plans to build a dam which would have flooded them. These caves contain ancient archeological deposits, and recently cave paintings about 20,000 years old have been discovered.

In the 1980's interest has been focused on Tasmania (where the 20 deepest caves in Australia are found) and on the Nullarbor, a karst plain with an area of 120,000 km², where large underwater systems are being found.

from text by Ross Ellis.

DEEP CAVES:

1. ***Anne-a-Kananda*** .... **-373 m**
(Mount Anne, Tasmania)

Australia's deepest cave is situated in the Northeast ridge of Mount Anne. This is Australia's main alpine karst area and one of the most remote caving areas, requiring expedition style preparations to explore. Mount Anne is 1425 m high and subject to foul weather conditions.

The karst area of Mount Anne is reached by following the Gordon River Road from Maydena (86 km west of Hobert) then the Scotts Peak Road to Gelignite Creek. A walking trail leads from here via Sandfly Creek to the summit of Mt. Anne (about 16 km). The track to the karst area branches off to the left before the summit and enters the bush as the dolomite is reached. The enormous doline entrance of Anne-a-Kananda can easily be seen at this point. The best part of a day's walking is required to reach the area from the road.

Anne-a-Kananda was first entered in 1967 and explored to a depth of about 32 m (the bottom of the doline, which was used as a camp site). In September, 1982 a passage behind the campsite was pushed and resulted in numerous discoveries over several weekends. The Dessicator section was explored to a depth of 300 m by the Northern Caverneers, and the Tasmania Caverneering Club bottomed the cave on a trip at Easter 1983 (for a depth of –373 m). Explorations since then have located more passages and shafts, the Dessicator section having been pushed to –342 m with potential for more. The length is about 3000 m.

The Mount Anne Massif consists mainly of a vast dolerite sill, underlain by quartzite schists with massive limestone deposits beneath the NE ridge. The Pre-Cambrian dolomite dips at a steep angle forming a prominent ridge about 10 km long and with a relief of approx. 600 m, probably the greatest depth potential in Australia. The cave is formed partly in dolomite and partly in conglomerate. Lake Tink is about 700 m below the summit of Mt. Anne and could be the resurgence for the waters from the caves, but no connection has been proven. There is a swallet which drains the lake and is believed to connect with a resurgence about 3.5 km SE of Lake Tink.

Map: from the drawings of the Tasmania Caverneering Club, sent by Ross Ellis.

Bibliography: Bunton (S.), Eberhard (R.) - *Vertical Caves of Tasmania, a Caver's Guidebook*, Adventure presentations, Miranda, 1984, 81 p.

Ross Ellis.

2. *Ice Tube - Growling Swallet System ......* –354 m
(Junee-Florentine, Tasmania)

This system has four entrances: Growling Swallet, Ice Tube, Pendant Pot, and Slaughter-house Pot. It is located in the Junee-Florentine area near Maydena, about 75 km west of Hobart via Norfolk and Westaway. The caves are generally within Mt. Field National Park, but access is via privately-owned logging roads and permits must be obtained from Australian Newsprint Mills at Maydena.

Florentine Valley Road is followed for about 17 km to the Eight Road turnoff, which is then followed to its end (about 2 km). Growling Swallet is a 20 minute walk from the road, while Ice Tube is another 30 minutes further on.

The entrance to Ice Tube was discovered in 1980 and over the next couple of years was extended until the bottom was reached in June, 1982, giving it the title of Australia's deepest cave for a while. In 1983 the Mainline Extension of Growling Swallet was connected to Ice Tube and the through trip is the deepest and most difficult in Australia, involving a descent of 354 m and the negotiation of over 3 km of wet passage.

Pendant Pot had been known for many years, but the connection to Growling Swallet was not made until a sump was dived from Growling Swallet in 1984.

Slaughter-house Pot was first explored in 1980 to a depth of about 80 m. A squeezeway was pushed in 1983 and the connection to the Windy Rift section of Growling Swallet made. Growling Swallet held the Australian deepest cave record for ten years from 1957 to 1967. It was first explored to a sump at –176 m in 1957 by T.C.C. members. The short section of cave down to the sump (about 750 m length) was all that was known of Growling Swallet until

1981 when a rift was followed into Trapdoor Streamway and eventually into Mainline. Exploration is still continuing, and the cave now has over 8 km of passageway mapped and the potential to become Australia's longest cave.

The Junee-Florentine karst area receives a tremendous amount of rainfall. Almost every cave is a swallet taking large amounts of water. Several caves have been linked with the main resurgence (Junee Cave) by dye tracing methods. Junee cave is very short, ending at a deep sump that was dived in 1982 to reveal a second sump 200 m further on. Growling Swallet is the largest stream sink in the area and is about 9 km from this second sump.

The limestone plateau has an area of about 80 km² and is covered by a dense tropical forest. The caves occur in the Gordon limestone of Ordovician age which is massive, hard, dense, and well jointed. The limestone has a maximum relief of 200 m in the Florentine Valley and 400 m above the Junee Rising.

Map: from the drawings of the Tasmanian Caverneering Club, sent by Ross Ellis.

Bibliography: Bunton (S.), Eberhard (R.) - *op. cit.*, 1984.
Davies (D.) - Growling Swallet, *Speleo Spiel*, 1981, (168): 6-8.

from text by Ross Ellis

3. Khazad-Dum-Dwarrowdelf .... –323 m
(Junee-Florentine, Tasmania)
Three entrances (JF 5, JF 4, JF 14). Bottom reached in 1971, starting from the JF 4, by the T.C.C. (–314 m). Connection with JF 5 in 1973 (–323 m; *Caving International Magazine*, 1980 (6-7) profile).

4. Serendipity .... –282 m
(Junee-Florentine, Tasmania)
Explored in 1982-1983 by the T.C.C.

5. Cauldron Pot .... –263 m
(Junee-Florentine, Tasmania)
In 1973 by the T.C.C. (*Speleo Spiel*, 1973 (78) profile).

6. Owl Pot .... –244 m
(Junee-Florentine, Tasmania)
1980. Western Australia S.G.

7. Tassy Pot .... – 238 m
(Junee-Florentine, Tasmania)
Bottom reached in 1981 by the Sydney U.S.S. (*Speleo Spiel*, 1973 (77), profile to – 231 m).

8. Arrakis .... –235 m
(Tasmania)

9. Mini-Martin Cave - Exit Cave .... –220 m
(Ida Bay, Tasmania)
Discovered and explored in 1966 and 1967. Connection with Exit Cave on August 19, 1967.

10. Milk Run (Tasmania) .... –208 m

11. Sesame I - Sesame II Caves .... –207 m
(Junee-Florentine, Tasmania)

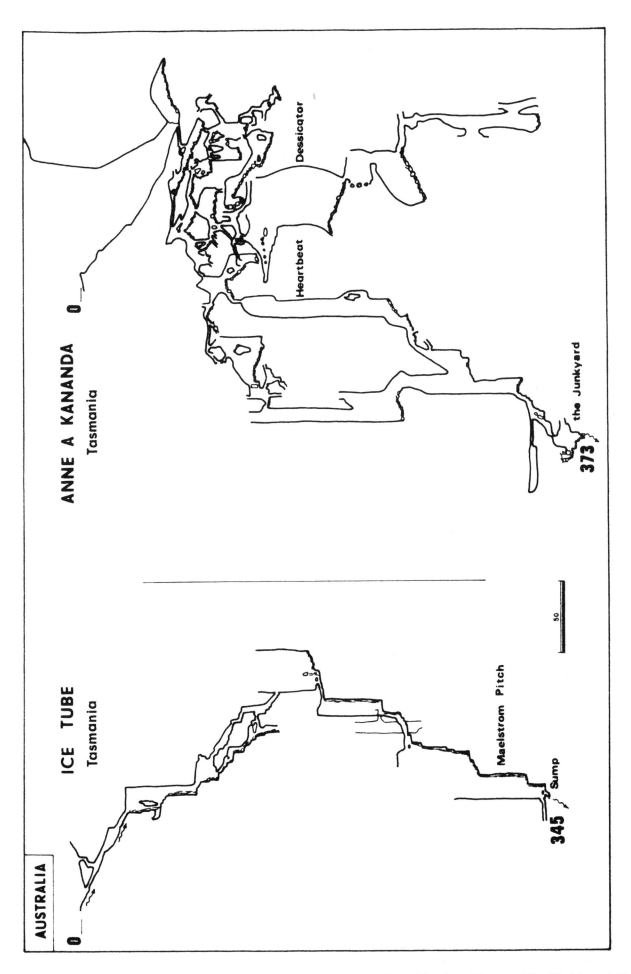

ANNE A KANANDA
Tasmania

0

Dessicator

Heartbeat

the Junkyard

373

ICE TUBE
Tasmania

0

AUSTRALIA

Maelstrom Pitch

Sump

345

50

12. **Flick Mints Hole** .... −204 m
(Junee-Florentine, Tasmania)
Explored in 1985.

13. **Midnight Hole - Mystery Creek Cave** ...... −203 m
(Ida Bay, Tasmania)
Explored and connection in 1968.

14. **Porcupine Pot** .... −202 m
(Tasmania)

15. **The Chairman** .... −197 m
(Junee-Florentine, Tasmania)
(*Bull. Sydney U.S.S.*, 1976, 16 (7) profile to −165 m).

16. **Cyclops** (Tasmania) .... −192 m

17. **Big Tree Pot** .... −189 m
(Ida Bay, Tasmania)
(*Jl. Sydney S. S.*, 1981, 25 (2) profile to −120 m).
Bottom reached in 1982 by the T.C.C.

18. **Peanut Brittle Pot** .... −186 m
(Tasmania)

19. **Udensala** .... −181 m
(Tasmania)

20. **Lost Pot** (Tasmania) .... −175 m

21. **Eagles Nest Caves System** .... −174 m
(Yarrangobilly, New South Wales)
(*Spai*, 1974 (40)).

22. **Top Hole** (Tasmania) .... −174 m

23. **Three Falls Cave** .... −158 m
(Junee-Florentine, Tasmania)

24. **Kellar Cellar** .... −155 m
(Mount Anne, Tasmania)
Deepest Australian pit, dropped in 1971 (*Spar*, 1971 (5) profile).

LONG CAVES:

1. **Mini-Martin Cave - Exit Cave** .... 16,000 m
(Ida Bay, Tasmania)
Cave known a long time and explored from 1966 (length 6,400 m) to 1970 (16,000 m; *Australian Spel. Federation Newsletter*, 1971 (51) map).

2. **Cora-Lynn Cave** .... 13,300 m
(York Peninsula, South Australia)
Discovered around 1910 (*Cave Explor. Gr. of South Australia Nl*, 1973, 18 (2) map with 5375 m.).

3. **Mullamullang Cave** .... 10,800 m
(Nullarbor Plain, Madura, West Australia)
Explored from 1964 (length 3300 m) to 1966 (length 9000 m; *Cave Exploration Group of South Australia Occasional Papers*, 1966, N° 4, map) and 1976.

4. **Ice Tube - Growling Swallet System**
(Junee, Florentine, Tasmania) .... approx. 8000 m
See above.

5. **Jenolan Show Caves** .... 7245 m
(Jenolan, New South Wales)
Cave with a northern section (4755 m) and a southern one (2490 m).

6. **Easter Cave** .... 7004 m
(Augusta, West Australia)
In 1980 by the Western Australian S. G.

7. **Mimbi Cave** .... 7000 m
(West Australia)
(*Illawarra Spel. Soc. Nl*, 1980, 2 (3)).

8. *Cocklebiddy Cave* .... **6500 m**
(Nullarbor Plain, Western Australia)
This cave is located in the center of the famous Nullarbor Plain, in the area known as Naretha. Access is via the Eyre Highway from Adelaide to the Cocklebiddy Motel. From there a track to the NNW leads after 30 km to the 120 m long entrance doline.

This cave is submerged for almost all of its length, and is formed is Eocene limestones. The cave water moves very slowly (a few centimeters per year) through galleries that average 30 m in diameter.

The cave was used as a spring by aborigines 20,000 to 40,000 years ago. It was noted in 1930 by Captain Thomson who, in 1932, followed the terminal lake to a sump.

Diving began in 1972, when the sump was explored for 100 m, then for 500 m in 1974 and 840 m in 1975. In May, 1976, Hugh Morrisson and Beilby got through the 1000 m long sump to make the cave 1370 m long. In 1977, Morrisson and Jones brought the length to 2134 m by diving in the second sump. In May 1979 the length reached 3139 m, of which nearly 3000 m were sumped. Diving in the second sump took 9 divers (including Morrisson, Dekkers and Jones) and 58 bottles. In September, 1982 a new expedition with 16 divers including Morrisson, R. Allum and P. Rogers (with 80 bottles) succeeded in passing the second sump, bringing the length to about 4200 m (with 3500 m sumped).

In September, 1983, five members of the S. C. Paris, led by F. Le Guen (E. Le Guen, J. Krowicki, V. Borel, S. Goutière) attacked the third sump and were stopped at a narrow passage, having used 20 bottles to dive 5540 m. In October of the same year H. Morrisson, helped by P. Rogers, went 240 m beyond the place where the French had stopped (using 30 divers and 120 bottles). The length is now 6500 m, of which 5780 m are sumped.

Map: from F. Le Guen, *Spelunca*, 1984 (15).
Bibliography: Le Guen (F.) - Expédition Nullarbor 83, *Spelunca*, 1984 (15): 17-20.
N.S.S. News, June 1984.

9. **Lannigans-Onslow Cave** or **Colong Caves**. 6000 m
(Kanangra, New South Wales)
(*Cooranbong Spel. Assoc. Reports*, 1959 (3) plan; *Anthodite*, 1984 (2) map).

10. **Queenslander Cave - Cathedral Cave** ...... 6000 m
(Mungana, Chillagoe, Queensland)
In 1974 (*Tower Karst*, 1976 (1) map).

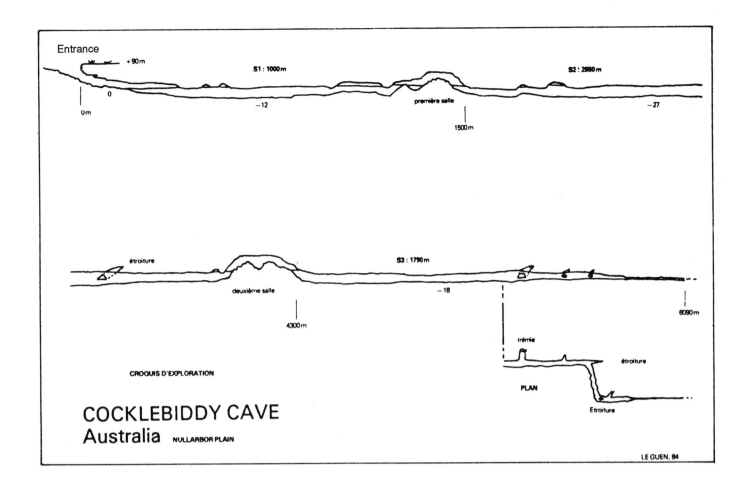

Entrance

+90 m

S1 : 1000 m

S2 : 2880 m

première salle

0

−12

1500 m

−27

0 m

étroiture

S3 : 1790 m

deuxième salle

−18

6090 m

4300 m

trémie

CROQUIS D'EXPLORATION

étroiture

PLAN

Etroiture

COCKLEBIDDY CAVE

Australia NULLARBOR PLAIN

LE GUEN, 84

11. **Herberts Pot** ....5730 m
(Mole Creek, Tasmania)
In 1976 (*Southern Caver*, 1977, 8 (3) map; *Caving Internat. Magazine*, 1980 (6-7) plan).

12. **Johannsens Cave** ....5225 m
(Rockhampton, Queensland)
(Sprent, *Mount Etna Caves*, Univ. Qld Spel. Soc., 1970).

13. **Kalkadoon Cave** ....5038 m
(Camooweal, Queensland)

14. **Old Homestead Cave** ....4500 m
(Western Australia)

15. **Eagles Nest Caves System** ....3600 m
(Yarrangobilly, Kosciusko Nat. Park, New South Wales)
(*Spar*, 1974 (40)).

16. **Mammoth Cave** ....3510 m
(Jenolan, New South Wales)
(Dunkley, Anderson - *The exploration and speleo-geography of Mammoth Cave*, Jenolan, Sydney U.S.S., 1971; *The Caves of Jenolan*, 1976 (2) map).

17. **Victoria Cave - Fossil Cave** ....3060 m
(Noracoorte, South Australia)
(*C.E.G. of S.A. Occ. Papers*, 1976 N° 5).

18. **Royal Arch Cave** ....3050 m
(Chillagoe, Queensland)
(*Sydney Spel. Soc. Occ. Papers*, 1969, N° 3, map).

19. **Anne-a-Kananda** ....3000 m
(Mount Anne, Tasmania)
See above.

20. **Buchan Show Caves** ....2885 m
(Buchan, Victoria)
(Mill *et al.*, - *Victorian Caves and Karst*, Melbourne, 1980).

21. **Taplow Maze Cave** ....2632 m
(New South Wales)
(*Oolite*, 1984, 14 (1) map).

22. **Weelawadfi Cave** ....2460 m
(Eneabba, West Australia)

23. **Wet Cave - Georgies Hall** ....2290 m
(Tasmania)
(Bunton, Eberhard, *op. cit.*, 1984).

24. **Niggle Cave** ....2250 m
(Camooweal, Queensland)
(*Spar*, 1972 (20)).

25. **Sand Cave** ....2210 m
(Noracoorte, South Australia)

26. **Kubla Khan Cave** ....2081 m
(Mole Creek, Tasmania)
(Bunton, Eberhard, *op. cit.*, 1984).

27. **Croesus Cave** ....2030 m
(Mole Creek, Tasmania)
(Bunton, Eberhard, *op. cit.*, 1984).

FIJI

There are caves known in several of the islands that make up the Fiji archipelago, some formed in limestone, others in lava. Several of them have been mapped by the Geological Survey. The longest known is **Wailotua Cave** (Wailotua, Viti Levu, length 1670 m) mapped by T. and J. Gilbert (*Mém. S.C. Paris*, 1983 (9) plan).

Following this are **Salialevu Cave** (Taveuni, length 920 m) and **Udit Cave** (Wainibuka, Viti Levu, length 790 m; T. Gilbert, *Cave Science*, 1984, 11 (2)).

from text by P. Strinati and R.M. Bourke

NEW CALEDONIA

In addition to some biospeleological investigations by people such as V. Aellen and P. Strinati in 1977, this French colonial island was the object of an Australian expedition in 1975 that found two major caves (see *Caves of New Caledonia*, report, 1976): the **grotte d'Adio** (Poya) or grotte de Kaluirer (–120 m deep and 3900 m long) and the **grotte du Koumac** (3700 m long).

NEW ZEALAND

There are two major cave areas in New Zealand:

1. **Waitomo**, made famous by the glow worms, where one finds mostly vadose caves formed in Oligocene limestones less than 100 m thick. The caves are generally heavily decorated in the upper levels. The area is much visited due to its proximity to urban centers.

2. **North-West Nelson**, in South Island, where one finds both little-explored Oligocene limestones which have extensive horizontal systems, and Ordovician marbles (Mt. Owen, Mt. Arthur, and Takaka Hill). It is in the latter that the deep caves are found, not surprising as the depth potential is over 1000 m. Exploration in these mountains only goes back to the 1960's and has been the work of a relatively small number of cavers.

The era of digging has already begun, but cave diving is still rare. The caving population is about 300 cavers and has been relatively stable since around 1960. The clubs are grouped together by the New Zealand Speleological Society (P. O. Box 18, Waitomo Caves), which puts out a bulletin three times a year.

Cathy Worthy.

DEEP CAVES:

1. *Nettlebed Cave* .... **–889 m**
(Mount Arthur, Nelson, South Island)

Nettlebed Cave (alt 330 m) is located on the east side of Mount Arthur, about 500 m to the SW of Pearse Rising (resurgence), above Eyle Creek, which is a tributary of the resurgence.

The entrance was discovered in 1969 by members of the New Zealand Speleological Society (N.Z.S.S.) and the first explorers were Fred Kahl, P. Cromley, T. Rodgers and D. Wheeler, but no major discoveries were made: in 1973 the map showed 1300 m of passage and a vertical extent of 29 m. A dye trace by Paul Williams in 1978 showed the potential of the cave: intensive explorations in 1978-1979 increased the vertical range to 289 m (length 5900 m), then to 341 m in 1979-1980 with the climbing of a 53 m waterfall (Jacob's Well) using a fifteen day long camp. The 1980-1981 expedition included eight Australians and began in October, 1980. An underground camp was made at the known end of the system (Salvation Hall) and the height of +430 m was reached, the length increasing from 9100 m to 12,800 m.

From December, 1981 to January, 1982, the discovery of the Hammer Heights allowed the explorers to reach the height of +597 m (626 m total vertical extent) and a length of 16,457 m. In January, 1983 a film was made ("Two Days to Soft Rock Cafe") and the height and length progressed to +628 and about 18,500 m respectively. "Orwells revenge" was discovered in April 1984, leading to about 20 km of mapped passage and a vertical extent of 667 m. In May 1985 the length increased to 21,016 m. In February, 1986, Funk Hole Rock Fall was climbed, Goodbye Yellow Brick Road (a 130 by 100 m room) was discovered, and a climb was begun at the end of the room. Blizzard Pot was found in the mid-1980's and connected in 1986 to Nettlebed by cavers from the Wellington Caving Group, making a system 889 m deep and over 24,400 m long.

Nettlebed is part of a system formed in Ordovician marbles. The dye trace of 1978, into the swallow of Grange Slocker, shows a vertical potential of 976 m. The resurgence, Pearse Rising, has a low flow rate of 2 m³/sec, while in periods of flood the Nettlebed entrance acts as an overflow route.

Map: N.Z.S.S., sent by Cathy Worthy, showing vertical extent of 693 m.

Bibliography: Pugsley (C.) -Caves of the Mount Arthur region, New Zealand, *Caving Int. Magazine*, 1979 (4): 3-10.
Caving Int. Magazine, 1981 (11): 52-53.
Wright (A.) - Nettlebed: New Zealand's deepest and longest cave, *Caving Int. Magazine*, 1981 (13): 40-43.
New Zealand Speleological Bull., starting in 1978.

2. **Bulmer Cavern** ....approx. –728 m
(Mt. Owen, Nelson, South Island)

Explored in 1985 and 1986 by the N.Z.S.S. The connection with Castlekeep, 50 m above the Bulmer entrance, made a system approximately 728 m deep. Contains a 105 m shaft and many large rooms. Exploration in progress, with leads remaining.

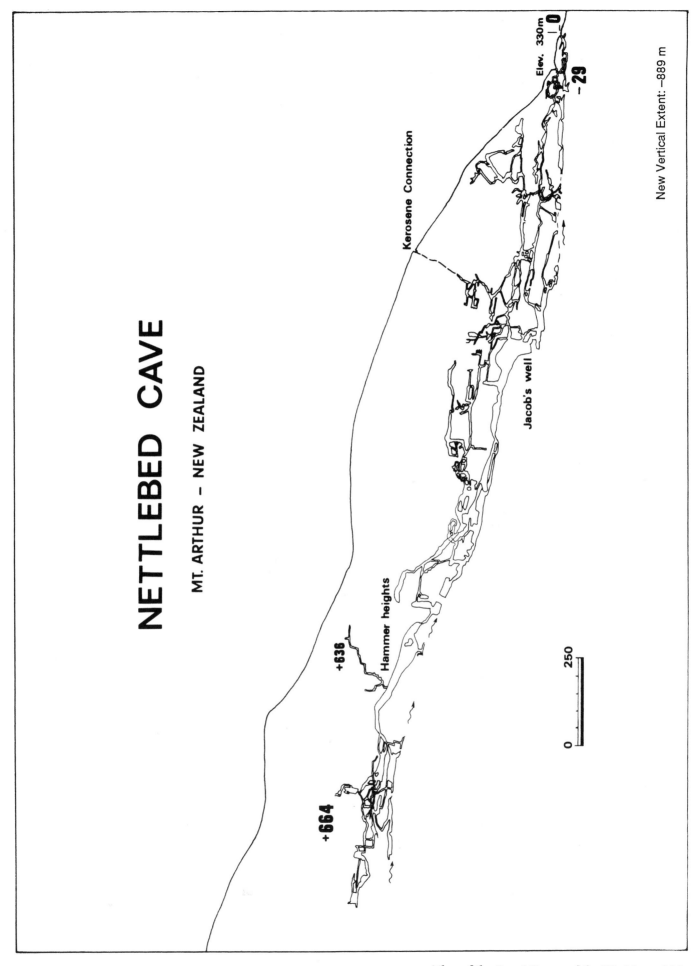

NETTLEBED CAVE

MT. ARTHUR – NEW ZEALAND

Elev. 330 m

0

−29

Kerosene Connection

Jacob's well

Hammer heights

+636

+664

250

0

New Vertical Extent: −889 m

3. **H. H. Hole** .... –623 m
(Mt Arthur, South Island)
Explored to this depth in 1985 by members of the N.Z.S.S.

4. **Greenlink Cave** .... –402 m
(Takaka Hill, Nelson, South Island)
Explored in 1974 by the Nelson S.G. (–287 m), then in 1978-1979 (*C.I.M.*, 1980 (9) schem. profile).

5. **Windrift Cave** .... –362 m
(Mt. Arthur, Nelson, South Island)
Explored in 1985 and in January 1986. Possible connection with Nettlebed.

6. **Harwood Hole - Starlight Cave** .... –357 m
(Takaka Hill, Nelson, South Island)
Through cave explored in 1959 by the N.Z.S.S. (*Journal Sydney Spel. Soc.*, 1976, 20 (1) profile).

7. **Gorgoroth Cave** .... –346 m
(Mt. Arthur, Nelson)
In 1971-1972 by the N.Z.S.S. (*C.I.M.*, 1979 (4) profile). Belongs to the same hydrological system as Nettlebed Cave.

8. **Gormenghast** ....approx. –325 m
(Mt. Owen, Nelson)

9. **Blackbird Hole** .... –315 m
(Mt. Arthur, Nelson)
1971, N.Z.S.S. (*C.I.M.*, 1979 (4) profile).

10. **Laghu Hole** .... –307 m
(Mt. Arthur, Nelson)

11. **Curtis Ghyll Cave** .... –291 m
(Mt. Owen, Nelson)

12. **Ed's Cellar Cave** .... –259 m
(Takaka Hill, Nelson)

13. **Giant's Staircase Cave** .... –259 m
(Mt. Owen, Nelson)

14. **Middle Earth Cave** .... –249 m
(Takaka Hill, Nelson)

15. **Helix Hole** .... –237 m
(Mt. Arthur, Nelson)

16. **Tralfamadore** .... –224 m
(Mt. Owen, Nelson)
1977, Sydney S. S. (*Jl. Sydney S. S.*, 1978, 22 (2) profile).

17. **Aurora Cave** .... –215 m
(Te Anau, South Island)

18. **Turk's Torrent** (Mt. Owen, Nelson)........... –214 m

19. **Summit Tomo Cave** .... –213 m
(Takaka Hill, Nelson)

20. **Coriolis Chasm** .... –210 m
(Mt. Arthur, Nelson)
1971, N.Z.S.S. and Australians (*C.I.M.*, 1979 (4) profile).

21. **Corkscrew Cave** .... –210 m
(Takaka Hill, Nelson)

22. **Olympia Cave** .... –207 m
(Takaka Hill, Nelson)

LONG CAVES:

1. **Nettlebed Cave** ....24,400 m
(Mt. Arthur, Nelson)
See above.

2. **Honeycomb Cave** ....13,150 m
(Northwest Nelson, South Island)

3. **Gardners Gut Cave**...........................11,890 m
(Waitomo, North Island)

4. **Bulmer Cavern** ....approx. 11,000 m
(Mt. Owen, Nelson)

5. **Metro Cave** ....8000 m
(Punakaiki, South Island)

6. **Aurora Cave** ....6400 m
(Te Anau, South Island)

7. **Mangaone Cave** ....6193 m
(Gisborne, North Island)

8. **Mangawhitikan System** ....6160 m
(Waitomo, North Island)

9. **Xanadu Cave** ....5010 m
(Punakaiki, South Island)

10. **Moonsilver Cave** (Cobb Valley)......approx. 5000 m

11. **Fred Cave** ....4440 m
(Waitomo, North Island)

12. **Windrift Cave** ....4410 m
(Mt. Arthur, Nelson)

13. **Thunderer Cave** ....4200 m
(Puketiti, North Island)

14. **Kuratani** (Waitomo, North Island) ....3960 m

15. **Ruakiri Cave** ....3764 m
(Waitomo, North Island)

16. **Eech Cave** ....3700 m
(Piopio, North Island)

17. **Waipura Cave** ....3560 m
(Waitomo, North Island)

18. **Karamu Cave** ....3535 m
    (Hamilton, North Island)

19. **Kairimu Cave** ....3440 m
    (Waitomo, North Island)

20. **Cloaca Maxima** ....3400 m
    (Waitomo, North Island)

NIUE

Because of their many formations, the caves of the island of Niue offer a rich tourist potential which has been noted by many visitors. The longest, **Anatola** (Lakepa) is about 300 m long (*New Zealand Spel. Bull.*, 1979 (111)), while **Vaikona** (Liku) is about 180 m long. A shaft, **Anapala** (Hakupu) was explored by the Swiss caver O. Knab to –50 m, of which 24 m are submerged.

PALAU
Belau

As for many Pacific islands, the Palau islands are either volcanic or limestone. Archaeologists were first attracted to the prehistoric paintings of the caves of Aulong Island. The biospeleologist Tom Iliffe (from Bermuda) led an expedition sponsored by the National Science Foundation to investigate the fauna of the archipelago in 1985.

Long Caves:
1. **Lake Eleven Cave** (Ngermeuangel) ....168 m
   Submerged cave, 37 m deep (*Underwater Speleology*, 1985, 12 (6)).

2. **Chie Malk** (Ngergheu) ....approx. 150 m

3. **Cave Dive Cave** (Ngeteklou) ....131 m
   Submerged cave (*Underwater Speleology*, 1985, 12 (6)).

PAPUA NEW GUINEA

The largest karst region in P.N.G. extends from the Gulf of Papua on the southern side of the island of New Guinea Northwest into Irian Jaya (Indonesian controlled New Guinea). It includes the Muller Range, Hindenburg Range and the Star Mountains, and contains three systems longer then 20 km, and one over 500 m deep. A second major karst occurs on the Huon Peninsula. It is characterized by its high altitude (over 4000 m) and massive potential for deep caves, but reconnaissance trips by P.N.G. based and French cavers have found no significant systems there.

The karsts of New Britain have rugged relief, very high rainfall and contain massive dolines. They are the site for numerous important cave systems, including Muruk (637 m), the deepest cave in P.N.G. There are also important karst regions on New Ireland and Bougainville. The Keriaka Plateau of Bougainville contains Benua Cave which is one of the largest cave chambers in the world (5 million cubic meters). Numerous other karst regions occur in P.N.G., some of which contain important caves.

Caves have been used by P.N.G. villagers for tens of thousands of years for shelter, hunting, burial, and art sites. Some of this usage continues. Systematic exploration by speleologists commenced in 1961 when P.N.G.-based Australian (and some British) cavers started explorations. This activity increased during the 1970's, but has declined in the 1980's with the departure of many expatriates from P.N.G. The first overseas expedition was an Australian one to the Star Mountains in 1965. Between 1972 and 1976 five Australasian (Australia/New Zealand/P.N.G.), one British, and one Japanese expedition came to P.N.G. and made a number of significant finds. Between 1978 and 1985, twelve expedition-sized parties have come from Australia/New Zealand, Belgium, Britain, France, Japan, Spain, Switzerland, and the U.S.A. The Australasian expeditions to the Muller Range and the French ones to New Britain have made the most important discoveries.

At present, exploration by P.N.G.-based cavers continues only on Bougainville. Interest from Australia and New Zealand has declined because of increasing bureaucratic difficulties and the great expense and effort needed to organize a major expedition to P.N.G. The focus of major discoveries has shifted from the New Guinea mainland to New Britain where French, British and Swiss explorers are active.

In 1971 there was only one documented major cave in P.N.G. - the giant chamber of Benua on Bougainville. After the last fifteen years of systematic exploration, there are now 35 caves documented as longer than 1000 m, including one 51.8 km long. There are 63 caves deeper than 100 m, including one 637 m deep. There are a number of very large chambers and long cave pitches, the largest dolines in the world, the greatest underground rivers in the world and some caves located at up to 3800 m above sea level. The potential for future discoveries continues to be very great.

R. Michael Bourke.

DEEP CAVES:

1. ***Muruk*** .... **–637 m**
   (Nakanaï Mts, East New Britain)
   The doline-swallow of Muruk is at an altitude of 1320 m in the wooded karst between the Galowe River and the Wanung River. Access is from the village of Galowe by the Solomon Sea (Jacquinot Bay) and taking the trail which goes to the old village of Mopuna (750 m), then to that of Malpe (810 m), both villages having been abandoned in 1945 at the end of WWII.

   From Malpe head to the west and follow the crest to about 1350 m altitude, then take the dry valley for about 100 m down to the cave entrance. It is a 7 km hike that takes 4 hours from Malpe (18 km and 12 hours from Galowe).

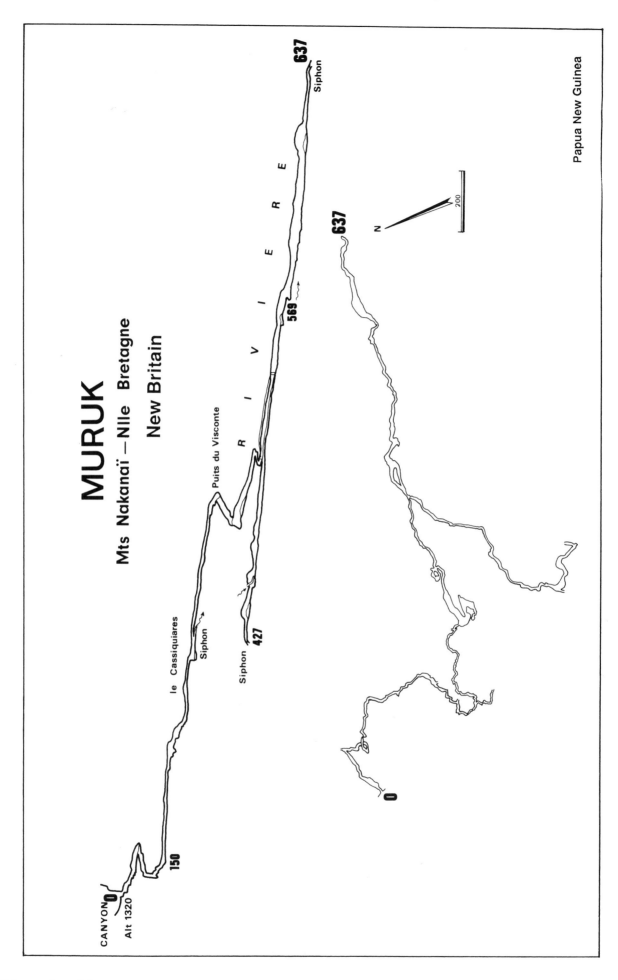

MURUK

Mts Nakanaï — NIle Bretagne
New Britain

Puits du Visconte

le Cassiquiares

Siphon

Siphon

427

569

637

637

Siphon

0

150

CANYON 0

Alt 1320

R I V I E R E

N

200

Papua New Guinea

Muruk was discovered and explored in 1985 by a French team from the F.F.S., "Papou 85". Located by studying maps and aerial photographs, it was reached after five days of trail cutting from the upper camp. The cave was explored and mapped in six days and required the use of about 600 m of rope.

Muruk is apparently part of the watershed for Mayang, a resurgence which cascades into the upstream gorges of the Galowe at an altitude of 400 m. The low water flow is estimated at 50 m³/sec. The resurgence is impenetrable.

Map: from the drawings of the "Papou 85" team (topofil Vulcain, degree 4b), sent by P. Génuite.

Pat Génuite.

2. **Mamo kananda** .... –528 m
(Muller Range, Southern Highlands)
See below.

3. **Bibima** .... –494 m
(Porol escarpment, Chimbu)
Swallow cave explored in 1972 by the P.N.G. C.E.G. (*Niugini Caver*, 1973 (1) profile).

4. **Gambo** .... –478 m
(Nakanaï Mts, Nutuve, East New Britain)
Explored in 1984 by a British expedition. Upstream of Nare cave.

5. **Minye** .... –468 m
(Tuke, Nakanaï mts, East New Britain)
The Minye doline-swallow is located at an altitude of 935 m in the covered karst found halfway between the villages of Tuke and Kapguena. Access is from the village of Pomio on the Salomon Sea (Jacquinot Bay) by climbing towards the northern part of the island. One passes through the villages of Nutuve and Piove to arrive at Kapguena. The doline is then a two hour walk to the north. The trip from Pomio, about 50 km away, takes one or two days.

Two other caves have been connected to the system: **Oro** (alt 895 m) is 800 m east of Minye, while **Tevi** (alt 915 m) is 600 m to the NE.

Long known to the locals, Minye was first explored in 1968 by two cavers from Port Moresby who descended to –60 m. In 1978 a reconnaissance team from the F.F.S. (France) descended the 366 m deep entrance drop (see "The Great Pits") and explored 2500 m in a stream passage. In March, 1984, a French/Swiss expedition "Mégadolines 84" filmed this impressive drop.

In 1985 the French F.F.S. "Papou 85" expedition returned to tackle the main river passage, which proved technically very difficult and ended in a sump. Oro and Tevi were connected to the Tuke room and the stream passage after descents of 230 and 220 m, bringing the length to 5421 m.

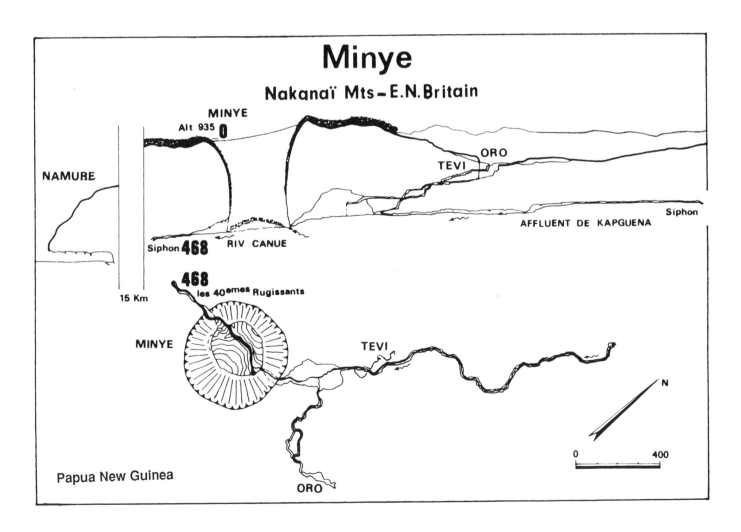

The huge Minye doline (volume 26 million m³) opens onto an underground river that has a low water flow estimated at 15 to 30 m³/sec and resurges at Namure cave, 2 km distant, near the village of Tuke. A distance of 1400 m and a vertical difference of 87 m separate the sumps in the two caves.

Map: from the maps of 1978 and 1985, sent by P. Génuite.

Bibliography: Borough (C.J.) - A large cave and doline near Tuke village. Pomio sub-district, New Britain, *Niugini Caver*, 1973, 1 (2): 25, 26.
Maire (R.), Martinez (D.) - Minye, *Spelunca*, "spécial Nouvelle-Guinée", 1981, suppl to N° 3.

Pat Génuite.

6. **Kavakuna II** .... –457 m
(Nakanaï Mts, East New Britain)
Explored in 1979 by the Swiss (–320 m) and 1980 by the French (*Spelunca*, suppl. to N° 3, 1981, profile).

7. **Malemuli** .... –420 m
(Mamo, Muller Range, Southern Highlands)
Explored in 1982 by the "Muller Range 82" expedition.

8. **Bikpela Vuvu** .... –414 m
(Nakanaï Mts, East New Britain)
Explored in 1980 by the French (*Spelunca*, suppl to N° 3, 1981, profile).

9. *Nare* .... **–400 m**
(Ire, Nakanaï Mts, East New Britain)
Nare is located in the NE part of the Nakanaï Mountains in New Britain. It is at an altitude of 600 m, 6 km to the NNW of Nutuve (1 or 2 hours walk). From the last village, Ire, a trail leads to the cave in 30 minutes of walking.

The cave consists of a huge collapse pit giving access to a large underground river (with a flow over 10 m³/sec). It has long been known to the locals. It was located by aerial reconnaissance in the 1970's and was visited in 1978 by a French expedition which explored both upstream and downstream for some distance. The pit was measured to be from 120 to 150 m across and 250 m to 315 m tall, giving a volume of 4.7 million m³ (see "The Great Pits").

In March-April, 1980, a second French expedition worked on exploring the river, one of the most difficult in the world (to be compared with sumidero Yochib, Chiapas, Mexico). Downstream, they stopped more than 2000 m from the entrance, bringing the length to 4500 m. In 1984, a British team tried to go further, but were stopped after only 100 m at a constriction with only a few centimeters of air space above the raging torrent.

Map: after *Spelunca*, suppl. to N° 3, 1981.

Bibliography: Maire (R.), Sounier (J.-P.) Naré (– 400 m) in "Papouasie-Nouvelle-Guinée"; *Spelunca*, suppl. to N° 3, 1981, pp 12-15.

10. **Kavakuna I** .... –392 m
(Nakanaï Mts, East New Britain)
F.F.S., 1980 (*Spelunca*, suppl. to N° 3, 1981, profile).

11. **Kukumbu** .... –388 m
(Whiteman Range, West New Britain)
Part of the same hydrologic system as Arrakis (–468 m) explored in 1985 by the F.F.S. (*Spelunca*, 1985 (20) profile).

12. **Terbil tem** .... –354 m
(Telefomin, Western Province)
Explored in 1975 by the British NG 75 expedition (*B.C.R.A. Trans.*, 1976, (3/4) profile).

13. **Atea Kananda** .... approx. (–150, +200) 350 m
(Muller Range, Southern Highlands)
See below.

14. **Arem tem** .... –334 m
(Olsobip, Central Highlands, Western Prov.)
Explored in 1978 by a British expedition.

15. **Camp III Hole** .... –330 m
(Telefomin, Western Prov.)
Explored in1975 by the NG 75 British expedition (*B.R.C.A. Trans.*, 1976 (3/4) profile).

16. **Leiwaro Kundu** .... –330 m
(Mt Kaijende, Enga)
Stream cave with two entrances. Explored in 1982 by the international "Mt Kaijende" expedition.

17. **Guimbe** .... –320 m
(Nakanaï Mts, East New Britain)
Explored in 1985 by the F.F.S.

18. **kananda Heiowa Heia** .... –314 m
(Muller Range, Southern Highlands)
Explored in 1973 by the N.S.R.E. (*Papua New Guinea Spel. Exped.*, N.S.R.E., 1973, 1974, profile).

19. **uli Guria** .... –314 m
(Muller Range, Southern Highlands)
Explored in 1973, N.S.R.E. 1973 (*N.S.R.E. 1973 op. cit.*, 1974, profile).

20. **Liklik Vuvu** .... –288 m
(Nakanaï Mts, East New Britain)
Explored in 1980 by the F.F.S.

21. **Ipaku** .... – 284 m
(Whiteman Range, West New Britain)
See Kukumbu.

22. **Pavie** .... –265 m
(Ire, Nakanaï Mts, East New Britain)
Explored in 1984, by a British team. Same hydrological system as Nare.

23. **Kille** .... –260 m
(Ire, Nakanaï Mts, East New Britain)
Explored in 1984 by a British team. Belongs to the same hydrological system as Nare.

24. **Kururu** .... –256 m
(Nakanaï Mts, East New Britain)
Explored in 1985 by the F.F.S.

Nare

Nakanaï Mts-E.N.Britain

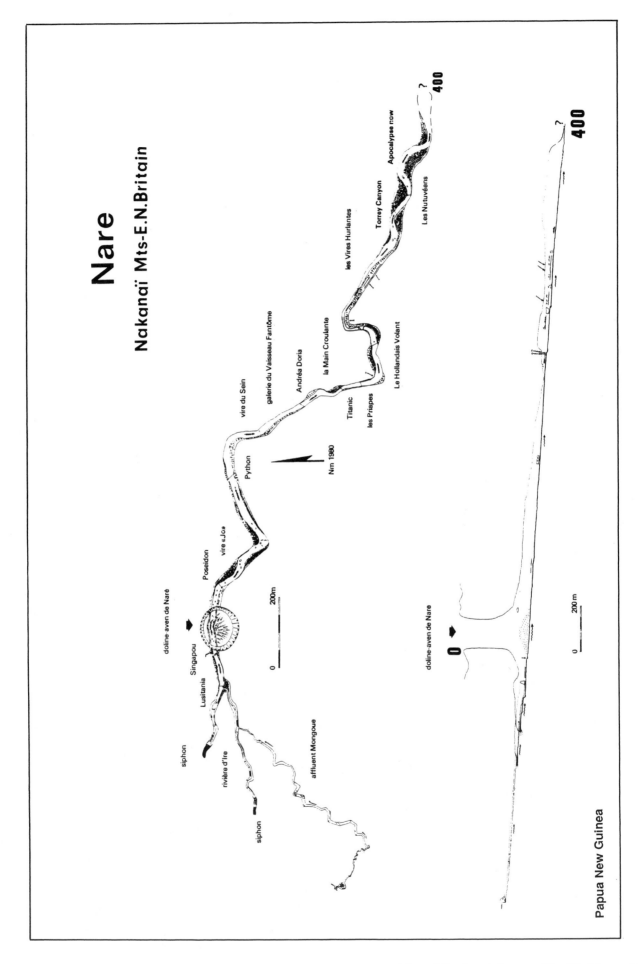

Papua New Guinea

25. **Luse** .... –224 m
(Nakanaï Mts, East New Britain)
Explored in 1980 by the F.F.S. (*Spelunca*, suppl. to N°
3, 1981, profile).

26. **Darua Muru** .... –214 m
(Porol escarpment, Chimbu)
Explored to –187 m in 1973 (P.N.G. C.E.G.) and to
–214 m in 1978 (Spanish cavers) (*Espeleóleg*, 1980 (30)
map).

27. **Kavakuna VI** .... –204 m
(Nakanaï Mts, East New Britain)
Explored in 1980 by the F.F.S. (*Spelunca*, suppl. to N°
3, 1981, profile).

28. **Lemerigamas** .... –203 m
(Lelet Plateau, New Ireland)
Explored in 1979 by a Swiss team, and in 1982 by
"Muller 82".

29. **Ngoma kananda** .... –203 m
(Muller Range, Southern Highlands)
Explored in 1972 by "Muller 82".

30. **Langlang tem** .... –200 m
(Telefomin, Western Prov.)
Explored in 1975 "NG 75" (*B.C.R.A. Trans*, 1975 (3/
4) profile).

31. **uli Eta Riya** .... –200 m
(Muller Range, Southern Highlands)
Explored in 1978, "Atea 78" (*Caves of the Muller
Range*, 1980, profile).

LONG CAVES:

1. *Mamo kananda* ....54,800 m
(Muller Range, Southern Highlands)
The Mamo plateau in which Mamo kananda is formed
is only 4 km to the west of the Atea (see Atea Kananda,
below, for the location). It is separated from the Atea
plateau by a dry valley in which the Yu Dina river used to
flow.

Difficult access (numerous dolines, dense vegetation)
has slowed the rate of exploration of the long caves of this
area, first checked out by the Niugini Speleological Research Expedition (N.S.R.E.) in 1973 and later by the 1976
Muller Range Expedition.

In 1978 some members of the Atea 78 expedition (see
below) set up a camp (Mamo IV) on August 12, in one of the
many promising dolines in the area, Hadia Ndunongairi.
From inside this cave, a surface pit, Hadia Yaneabogairi,
was discovered, making for easier access to the system.
When the expedition left the area on August 25, over 8400
m had been mapped for a depth of 176 m. Numerous side
leads were left unchecked.

A large expedition, "Muller 82" returned in 1982 to
continue work in the system. A huge entrance, kananda
Pugwa, was discovered to the NW of Hadia Yaneabogairi,
and connected to the system on August 16. By the end of
the expedition, the length had increased to 54,800 m for a
depth of 528 m. The number of entrances was increased
to 24. Several close calls involving flooding occurred

during this expedition.

Mamo kananda is formed in Miocene limestones and
collects much of the water from the Mamo plateau. The
drainage is to the south, giving possibilities for the resurgence at the Nali gorge.

Map: from *Spelunca*, 1983 (12).

Bibliography: James (J.) and Dyson (H. J.) (dir.) -
Caves and Karst of the Muller Range, Atea 78, Australia,
1980, 150 p.

Worthington (S.) (pers. comm.).

2. *Atea Kananda* ....34,500 m
(Muller Range, Southern Highlands)
From the landing strip of Kelabo (30 km to the NW of
Koroba), a road leads to Harerage. From there, it takes a
day of walking to cross the Muller Range and reach the Atea
to the SE of Harerage, having passed successively through
Gewane, Hegaroroge, Hulitanda, Ponganepo, and Geloro.

The entrance doline (200 m by 100 m) swallows the Yu
Atea river which resurges 650 m lower, about 2-3 km to the
south at Nali. Nali is also the resurgence of the Yu Dina.
The limestones are of Miocene age.

The first caving trip to the area was by the Australians, the N.S.R.E. 73. They reconnoitred a vertical extent
of 120 m in the Atea Kananda (–70, +50).

In 1976 a small expedition, "Muller Range Expedition
76", returned to the Atea and explored 4500 m, discovering
six other entrances in the doline. Given the importance of
the cave, a large expedition, "Atea 78", returned in 1978,
after two years of preparation. It gathered together 50
participants from five countries (Australia, New Zealand,
United States, Great Britain, and Papua New Guinea).
They were helped by 14 members of the Duna tribe. Under
the direction of Kevin Wilde, 6 tons of material were
brought to the base camp of Atea Gana Anda. During July
and August, 26 km were added to the length of the system,
bringing it to 30,500 m. The vertical extent reached 308 m
(–153, +165).

Another large expedition in 1982, led by Julia James,
assembled 59 cavers from the same countries as before
and employed 12 Duna. In spite of the dry conditions,
fewer discoveries were made: the vertical extent progressed to about 350 m (–150, +200) while the length
increased to 34,500 m.

Map: from the plan of Atea 78.

Bibliography: James (J.), Dyson (H.J.) (dir.) - *Caves
and Karst of the Muller Range*, Atea 78, Australia, 1980,
150 p.

3. **Selminum tem** ....20,500 m
(Hindenburg Range, Western Prov.)
Entirely explored in 1975 by the British expedition
"NG 75" (*B.R.C.A. Trans.*, 1976 (3/4) map).

4. **Ipaku-Kukumbu** ....11,026 m
(Whiteman Range, West New Britain)
Belongs to the Arrakis hydrological system (length
16,318 m). Separated by a 1000 m long doline (Kukumbu)
(*Spelunca*, 1985 (20) profile).

5. **Liklik Vuvu** ....6800 m
(Nakanaï Mts, East New Britain)
Explored in 1980 by the F.F.S. (*Spelunca*, suppl. to N°

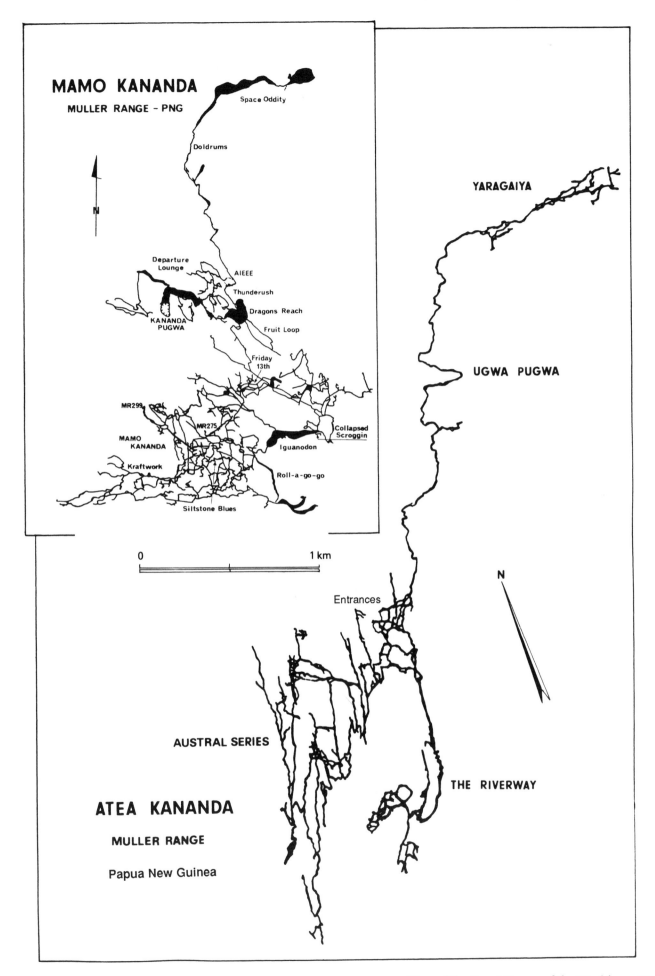

MAMO KANANDA

MULLER RANGE - PNG

Space Oddity

Doldrums

N

Departure
Lounge

AIEEE

Thunderush

Dragons Reach

KANANDA
PUGWA

Fruit Loop

Friday
13th

MR299

MR275

MAMO
KANANDA

Collapsed
Scroggin

Iguanodon

Kraftwork

Roll-a-go-go

Siltstone Blues

0   1 km

YARAGAIYA

UGWA PUGWA

Entrances

N

AUSTRAL SERIES

THE RIVERWAY

ATEA KANANDA

MULLER RANGE

Papua New Guinea

3, 1981, map).

6. **Gambo** ....6000 m
(Nakanaï Mts, East New Britain)
Explored in 1984 by the British.

7. **Minye** ....5420 m
(Nakanaï Mts, East New Britain)
See above.

8. **Nare** ....4600 m
(Nakanaï Mts, East New Britain)
See above.

9. **Muruk** (Nakanaï Mts, East New Britain) .....4574 m
See above.

10. **Kavakuna II** ....3500 m
(Nakanaï Mts, East New Britain)
Explored in 1979 (Swiss) and 1980 (French) (*Spelunca*, suppl. to N° 3, 1981, plan).

11. **Leiwaro Kundu** ....3500 m
(Mt. Kaijende, Enga)
See above.

12. **Guimbe** ....2770 m
(Nakanaï Mts, East New Britain)
Explored in 1985 by the F.F.S.

13. **Arrakis** (swallow) ....2764 m
(Whiteman Range, West New Britain)
This length includes an 800 m long canyon!

14. **Kururu** ....2630 m
(Nakanaï Mts, East New Britain)
Explored in 1985 by the F.F.S.

15. **Bikpela Vuvu** ....2600 m
(Nakanaï Mts, East New Britain)
Explored in 1980 by the F.F.S. (*Spelunca*, suppl. to N° 3, 1981, plan).

16. **Pimbiriga kananda** ....2500 m
(Muller Range, Southern Highlands)
Explored in 1982 by the "Muller 82" expedition. Belongs to the same hydrological system as Atea kananda.

17. **Pavie** ....2250 m
(Nakanaï Mts, East New Britain)
Explored in 1984 by a British team.

18. **Irukunguai** (Irapui) ....2120 m
(Poroi escarpment, Chimbu)
Explored in 1964, 1972 and 1978.

19. **kananda Heiowa Heia** ....2000 m
(Muller Range, Southern Highlands)
1500 m explored in 1973 by the N.S.R.E.

SAMOA and SISIFO
(Western Samoa)

In the heart of the Polynesian archipelago, these small and distant islands are politically separated from the American Samoas. The island of Upolu contains several lava tubes, including **Pe'apa'a** , 1110 m long (*Trans. B.C.R.A.*, 1979, 6 (3) plan) and **Falemaunga**, 558 m long (*New Zealand Spel. Bull.*, 1974 (90) plan).

SOLOMON ISLANDS

The British Solomon Islands are south of the island of Bougainville in Papua New Guinea, but politically separate. They have been sporadically explored by Australian cavers. The limestone is upper Miocene and lower Pliocene age.

The deepest cave known is **Kolokofa** (Santa Isabel island) which is 80 m deep.

LONG CAVES:

1. **Bishop Selwyn's cave** ....658 m

2. **Mbao Hol** ....364 m
(Honiara, Guadalcanal island)
Explored, as was Kolokofa, in 1976 by the N.S.W. Institute of Technol. S.S.

TONGA

A list of 22 caves explored or located in 1952 and 1953 by Ken Pawson from Calgary can be found in volume 6, N° 1, of the *Canadian Caver* (1974). Among those, the deepest is **Ana Ahu** (about –60 m), a vertical pit with a waterfall which was descended on September 16, 1952, by Ken Pawson and Prebend Kauffman. In 1986 a six-person British expedition "Tonga 86", led by D. J. Lowe, surveyed 19 caves. The deepest was **Rift Cave** (approx. –120 m) and the longest was **Fish Cave** (approx. 510 m long and 45 m deep).

VANUATU
(New Hebrides)

The volcanic rocks on this archipelago contain poorly prospected lava tubes (Maewo island, for example). Karst phenomena have been observed in the reef limestones of several islands (Vanua, Mota, Maewo, Erromango, Espiritu Santo). On the island of Vaté, **Siviri** is about 125 m long (B. Gèza, in *Spelunca Mémoires*, 1963 (3)).

Index